Rock and Mineral Analysis

CHEMICAL ANALYSIS

A SERIES OF MONOGRAPHS ON
ANALYTICAL CHEMISTRY AND ITS APPLICATIONS

VOLUME 27

A WILEY-INTERSCIENCE PUBLICATION

JOHN WILEY & SONS
New York / Chichester / Brisbane / Toronto

Rock and Mineral Analysis

SECOND EDITION

WESLEY M. JOHNSON
British Columbia Ministry of Energy,
Mines, and Petroleum Resources

JOHN A. MAXWELL
Geological Survey of Canada

A WILEY-INTERSCIENCE PUBLICATION

JOHN WILEY & SONS
New York / Chichester / Brisbane / Toronto

Library of Congress Cataloging in Publication Data:

Johnson, Wesley M. (Wesley Moore), 1941–
  Rock and mineral analysis.

  (Chemical analysis, ISSN 0069-2883 ; v. 27)
  Previous ed. (1968) by John A. Maxwell.
  "A Wiley-Interscience publication."
  Includes bibliographies and index.
  1. Rocks—Analysis. 2. Mineralogy, Determina-
tive. I. Maxwell, John Alfred, 1921–
II. Title. III. Series.

QE438.J64 1981   552'.06   81-1659
ISBN 0-471-02743-X   AACR2

Printed in the United States of America

10 9 8 7 6 5 4 3 2 1

PREFACE

This second edition of *Rock and Mineral Analysis* is, like the first, intended to be a laboratory reference book which will provide the practicing analyst with an up-to-date treatment of the problems associated with the analysis of geological materials. The prediction made in the preface to the first edition, that there would be rapid evolution of methods incorporating instrumental techniques, has been realized with, if anything, an acceleration in the anticipated rate of change, and this revision reflects the dramatic developments of the last decade.

The organization of the book has been altered, but not its attention to details of specific analytical methodology. This revised edition is intended to complement the first edition in that it describes in some detail the instrumental analysis of rocks and minerals [chiefly by atomic absorption spectroscopy (AAS) and X-ray fluorescence spectroscopy (XRF)], whereas the previous emphasis was very largely on the classical "wet" methods. At the same time, however, this edition is intended to be complete in itself, and material from the early chapters of the first edition has been included, albeit in a revised and updated form. Special attention has been given to the consideration of new advances in sampling theory, and the discussions of standard reference materials and of sample decomposition techniques have been significantly expanded.

The preliminary considerations associated with the analysis of rocks and minerals, such as equipment, reagents, standard reference materials sampling theory, sample preparation, and sample decomposition, are covered in Part I, including, as Appendix 1, a compilation and discussion of data on international standard reference materials by Mr. Sydney Abbey of the Geological Survey of Canada.

Part II describes analytical methods for the determination of those commonly requested analytes that require individual methods of preparation or detection, such as carbon, sulfur, ferrous iron, moisture, and fluorine. Several of these methods are reproduced from the first edition, but where, in our view, they should be superseded by new techniques and better methods, the latter have been included also. Parts III and IV are devoted to the determination of major, minor, and trace elements in rocks and minerals by AAS and XRF methods, respectively, areas only touched upon in the first edition. Each is divided into a section on general considerations, followed by a statement, on an element by element basis, of the analytical guidelines. A final section, Part V, presents a less detailed description of the application of emission spectrographic, neutron activation, fire assay, and electron microprobe analysis (the latter contributed by Dr. A. G. Plant of the Geological Survey of Canada) to rock and mineral analysis.

The selection of elements that are covered in the chapters dealing with the AAS and XRF techniques is based upon our appreciation of the variety of requests that the rock and mineral analyst might well expect to encounter most frequently. It is also based upon the supposition that a virtually complete rock or mineral analysis can be made using the classical "wet" methods supplemented by either AAS or XRF techniques. Thus if a laboratory has AAS and basic laboratory equipment but no XRF facility, the information given is sufficient to enable the analyst to make the complete analysis that is commonly requested.

Because of the many detailed textbooks already available, we have not attempted to discuss the theoretical aspects of any of the instrumental techniques included here; the references (updated to the beginning of 1980) have been selected so that the analyst can go directly to them for this kind of information. The discussion has been restricted to the practical considerations involved in the application of these instruments (including calibration, likely sources of error, the selection of instrumental conditions, and sample preparation) and to the procedures themselves. In the chapters on AAS and XRF the elements are considered individually in terms of the particular method(s) and/or problems associated with each one. The determination of major, minor, and trace level constituents is presented in sufficient detail for someone familiar with the instrument in question to be able to carry out the analysis of rocks or minerals for most of the elements usually encountered at almost any concentration level.

We thank our colleagues Mr. Sydney Abbey and Dr. A. G. Plant for contributing the sections on standard reference materials and microbeam analytical methods, respectively, and, with Mr. G. R. Lachance and Dr. R. S. Young, for their critical comments on portions of the manuscript. We gratefully acknowledge, in the appropriate places, the kind permission of those who have allowed us to reproduce previously published material.

We are also grateful to the British Columbia Ministry of Energy, Mines and Petroleum Resources and to the Geological Survey of Canada for their encouragement and support in the preparation of this revised edition. The contents are, however, the sole responsibility of the authors.

Victoria, British Columbia
Ottawa, Ontario   WESLEY M. JOHNSON
July 1981   JOHN A. MAXWELL

CONTENTS

LIST OF ABBREVIATIONS USED

AAS	Atomic absorption spectroscopy
BCM	British Columbia Ministry of Energy, Mines, and Petroleum Resources
ES	Optical emission spectroscopy
ED-XRF	Energy dispersive X-ray fluorescence spectroscopy
GSC	Geological Survey of Canada
HCL	Hollow cathode lamp
ICP-ES	Inductively coupled plasma optical emission spectroscopy
NAA	Neutron activation analysis
SRM	Standard reference material
WD-XRF	Wavelength dispersive X-ray fluorescence spectroscopy

Rock and
Mineral Analysis

PART I

Preliminary Considerations

CHAPTER 1

THE NATURE OF THE ANALYSIS OF ROCKS AND MINERALS

1.1 GENERAL

The importance of chemical analyses of rocks and minerals to geological and related studies is now so well established that there is no need to begin here by justifying the effort, time, and expense required to provide them; the authors are well content to let the arguments presented by earlier advocates, such as the late Professor Arthur Holmes (Ref. 1, pp. xi–xviii), speak for them.[2, 3] The quantity of such analyses has increased at an exponential rate since instrumental means of analysis such as atomic absorption spectroscopy (AAS)* and X-ray fluorescence spectroscopy (XRF) have begun supplanting the older chemical techniques. The demand, of course, increases with (or just ahead of) the analyst's ability to produce, and the widespread use of computers for the statistical treatment of analytical information has only enhanced this demand.

The chemical analysis of such naturally occurring substances as minerals and ores was very much in vogue as far back as the eighteenth century[4] and, according to Hillebrand et al. (Ref. 2, p. 793), "the composition of the ultimate ingredients of the earth's crust—the different mineral species that are found there—was the favourite theme of the great workers in chemistry of the earlier half of the nineteenth century. . . . As an outgrowth of the analysis of minerals and closely associated with it came the analysis of the more or less complex mixtures of them—the rocks." This resulted in the early development of a systematic analytical scheme which underwent continuous modification in the light of the increasing knowledge of the principles of analytical chemistry and the development of new and improved laboratory facilities. Such famous names as Berzelius, Wöhler, Rose, Klaproth, and Scheele led in direct succession to Washington, Clarke, Hillebrand, Lundell, and Groves, to whom we owe the classical methods of rock and mineral analysis that are still widely used. But these methods were found to be too time-consuming and, starting about 1947, various "rapid" analytical schemes were proposed as alternatives.[5–11] These, in turn, are being replaced by the increasingly widespread application of instrumental methods.

* See "List of Abbreviations Used."

3

These instrumental methods include XRF, AAS, and optical emission spectroscopy [both dc arc excited (ES) and inductively coupled plasma excited (ICP-ES)] as well as electron microbeam techniques. There is a wide range of quality reported in the literature for the results obtained by each of these methods in rock and mineral analysis. In the hands of skilled and knowledgeable analysts, however, ICP-ES, XRF, and AAS techniques can produce results which are just as accurate and precise as those of the classical analysts for most constituents, and the productivity is markedly improved at the same time. The difficulties once impeding the attainment of accuracy and precision with AAS and XRF methods have been alleviated by the use of computers for data manipulation which permits the incorporation of drift correction and polynomial calibration curves into programs designed to produce concentration results from raw machine data.[12]

There have been numerous publications in the last 10 years dealing with empirically obtained[13-18] and with theoretically predicted[19] interelement coefficients for use in making matrix corrections to XRF data (see Sect. 10.4.6). The use of computers is essential for this purpose, and it enables the XRF technique to be used to obtain large numbers of reliable determinations on major, minor, and trace constituents in rock and mineral samples. Both wavelength dispersive and energy dispersive spectrometers are in use; greater success has been achieved to date with the former, but recent advances in energy dispersive X-ray spectrometer technology have increased its capability to produce reliable data, although the determination in rocks or minerals of trace amounts of the elements below atomic number 21 (Sc) is still not satisfactory.

Plasma and dc arc ES methods are both rapid and broad in scope and are capable of the level of quality attained by the colorimetric, XRF, or AAS methods. The advent of the plasma excitation source holds significant promise of an improvement in sensitivity which will be welcomed. This new source has recently been adapted by Govindaraju and co-workers to their previously described automated system.[20] Ingamells and Suhr[21, 22] have described a rapid chemical and spectrochemical scheme of analysis, designed primarily for silicates, but adaptable to other classes of materials, that is simple in operation but capable of producing results having a high degree of accuracy. The proposed combination of techniques is particularly useful with small samples, and it is possible to make a complete analysis with as little as 50 mg of material. Preconcentration of trace elements to improve the sensitivity of determination is a useful technique, and cation exchange chromatography has been used for many years, as have coprecipitation and solvent extraction. A recent application of the first of these to the determination of the major and minor constituents of rocks has been reported by Strelow et al.[23]

The development of electron microbeam techniques[24, 25] has added a new dimension to what the geologists can demand of analysts. For example, it is possible to obtain elemental analyses of minute crystals or mineral inclusions,[26] to study zoning in minerals,[27] and to investigate chemical bonding in minerals.[28] The application of these techniques has assisted in the identification of numerous new minerals and in the quantitative determination of their major elemental constituents.[29]

Mass spectrometry[30] has, like most other types of instrumentation, increased in complexity and consequently in the scope of its application. The isotopic age dating of minerals is benefiting from increased reliability as a result of this, and a growing number of isotope pairs is available to establish increasingly remote dates of geological events.

Each of the previously mentioned techniques has its strengths and weaknesses, and none is a panacea capable of being used to determine all of the elements over all ranges of concentration likely to be encountered. It is necessary that an analyst be sufficiently familiar with each to be able to choose the best technique for a particular task, one that can vary from a request to determine fluorine at the parts per million level or to doing a complete analysis of a sample of hornblende. The constituents to be determined will, in most instances, strongly influence the choice of method, and thus an understanding of the type and number of constituents likely to be encountered is essential.

1.2 THE CONSTITUENTS TO BE DETERMINED

The question, "What constituents should be determined?" is one that defies a simple answer because, earth scientists being what they are, the rock and mineral analyst will eventually be called upon to determine about three quarters of the elements in the periodic table, in concentrations ranging from those of major constituents to less (often much less!) than 1 ppm. This is, to some extent, a recent tendency, due to an increased awareness of the possibilities of the optical emission spectrograph, the growing application of the electron microprobe to the study of individual mineral grains, and to the emphasis being placed upon the distribution of trace elements among the various physical components of the lithosphere. Table 1.1, which is reprinted here in slightly modified form from *Outlines of Methods of Chemical Analysis* (Ref. 31, Table 8, p. 15), shows the elements most often encountered in rock analysis. Among the additional minor constituents frequently determined now are Rb, Cs, Be, Ar, Y, La and the rare earths, Th, Nb, U, Ag, B, Sn, Pb, As, Sb, Se, Te, Hg, Mo, and Bi; less frequent requests are received for the Pt group and Au. Many of these additional elements can be determined routinely by

TABLE 1.1   Elements Most Often Encountered in the Chemical Analysis of Rocks.

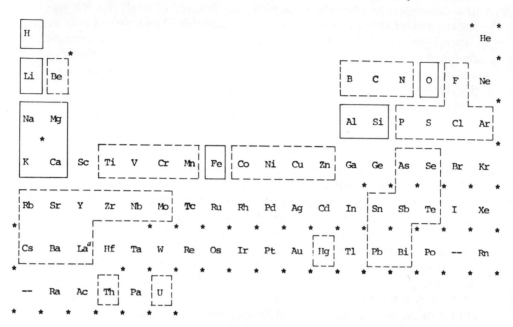

a Also elements 57–71.
Heavy blocks enclose major constituents.
Heavy broken blocks enclose common minor constituents.

ES, and many laboratories such as the Geological Survey of Canada (GSC) and the British Columbia Ministry of Energy, Mines and Petroleum Resources (BCM) have found it of value to make at least a qualitative examination of nonroutine samples before proceeding with either a chemical or an instrumental analysis. It alerts the analyst to possible sources of difficulty or interference in subsequent work, and the photographic plate or film serves as a useful (and permanent) reference for possible future questions about the trace element composition of the sample.

There is no problem about the choice of constituents in a mineral analysis. The use that is to be made of the data, such as the calculation of the formula from the analytical results, or perhaps a demonstration of the degree of solid solution of one mineral in another, together with the wide variations in composition that exist among minerals, will dictate the elements to be determined. This also applies to such specialized geological materials as meteorites. With regard to rocks, however, it has been the long-standing practice to determine between 13 and 18 constituents for a "complete" rock analysis, of which SiO_2, Al_2O_3, Fe_2O_3, FeO, CaO, MgO, Na_2O, K_2O, H_2O^+, H_2O^-, TiO_2, MnO, P_2O_5, S, and CO_2 (or C) are the most commonly determined and BaO, SrO, Cl, F, and the loss on ignition less frequently so. The chief reason for this is that most of these constituents are required for normative calculations, when such are made, and for convenience in comparing one analysis with another. There is still a tendency to request a "complete" analysis in order to provide oneself with what is considered to be a measure of the value of a set of results, that is, the closeness of the total to 100%. The fallaciousness of this reasoning is discussed at some length later in this chapter, but it can be stated here that the choice of constituents to be determined should be governed by the particular petrographical, mineralogical, or geochemical goal, rather than by long-established but disproven and outdated custom.

Modern geochemical studies, such as the sampling of unmineralized rocks in the vicinity of a known ore deposit in an attempt to identify a dispersion halo, or the sampling of outcrops representative of a batholith to establish background values for selected elements and dates of mineral deposition or crystallization, can require the analysis of large suites of samples for a wide variety of elements. The most common elements selected for these types of studies include F, S, Cl, Ni, Cu, Zn, As, Mo, Ag, Au, Hg, and Pb, but Ar, K, Rb, Sr, Zr, Nb, Cs, Ba, and U are frequently chosen as well. Current geochemical theories concerning which elements act as "indicators" for potential ore deposits of various sorts are in a state of flux, and the preceding list can be expected to have additions and deletions for a given study or set of geological parameters.

Very few laboratories can expect to have the good fortune to be equipped or staffed to meet all of these possible analytical demands. A judicial selection of equipment, however, and a thorough knowledge of the techniques available can go a long way toward making the analyst's job a much easier one with fewer frustrations and delays. AAS and XRF, for example, can each be utilized for the determination of the majority of the elements listed previously. With AAS, however, one can determine Cl only indirectly, but it is used routinely for the determination of Hg in concentrations of a few parts per billion. XRF will, on the other hand, enable the direct measurement of Cl, but it cannot detect Hg below a lower limit of several parts per million. Neither technique, of course, can be used for the determination of Ar for the K–Ar geological dating scheme, but both can yield results for K which are more than adequate for this purpose.

Constituents such as loss on ignition, FeO, H_2O+ or H_2O-, and CO_2 must be measured independently and thus are not readily amenable to automated techniques. Although these extra determinations are needed in most "complete analyses," they are expensive in terms of an analyst's time and should be requested only when absolutely necessary. It is possible to supply various combinations of constituents by methods that will best serve the particular need in terms of completeness, accuracy, and speed. It would be to the obvious advantage of both the analyst and the submitter to discuss the requirements and to select only those determinations that are required before any work is started on a suite of samples.

1.3   THE STATEMENT OF THE ANALYSIS

One of the critical aspects of the interaction between the analyst and the sample submitter is the report of the analysis. The analyst should ensure that the report contains sufficient information to preclude the misinterpretation of the data by the submitter. The *order* in which the constituents are reported, their *form*, the *number of significant figures* used in the reporting, and whether they are reported on a "wet" or a "dry" basis are all part of this information.

The most common *order* used is that suggested by Washington (Ref. 32, p. 61). He based his conclusions upon a premise that is no longer tenable, that is, that "a rock (or mineral) analysis . . . is primarily intended for, and almost exclusively used by, petrographers," the latter having been elbowed out of the way by the geochemists with their insatiable demand for new and more abundant data. This does not change his dictum, however, that "what (one) needs especially is an arrangement which shall bring the essential chemical features—both the percentage figures and the molecular ratios—prominently and compactly before the eye, so that the general chemical character and the relations of the various constituents may be seen at a glance. It is also of importance that the arrangement be such as to facilitate comparison of one analysis with another." His order for listing the constituents of the analysis is divided into two parts—the main portion containing the principal constituents and a subordinate part containing the remainder. The main portion is as follows:

$$SiO_2 \qquad MgO \qquad H_2O^+$$
$$Al_2O_3 \qquad CaO \qquad H_2O^-$$
$$Fe_2O_3 \qquad Na_2O \qquad CO_2$$
$$FeO \qquad K_2O$$

The list begins with the chief acid radical and the constituent usually present in the largest amount. The trivalent elements follow, then the divalent ones, with ferrous iron and magnesium together because they are closely associated in ferromagnesian minerals and the calcium last because it is common to these latter and to the feldspars, particularly the plagioclase feldspars. These are logically followed by Na and K in that order. H_2O+ and H_2O-, both given because the latter is not an essential constituent and will vary with the sample environment, together with CO_2 serve as a measure of the freshness of the material and complete a list of constituents that will, for most rocks, constitute about 95% of the sample. This order is usually followed, the major discrepancy being that CaO is placed before MgO in keeping with the analytical order in which it is determined in the classical scheme of analysis.

The order of the subordinate part follows the principle expressed in the main portion:

TiO_2	Cl	MnO
ZrO_2	F	BaO
P_2O_5	S, FeS_2	SrO
V_2O_5	Cr_2O_3	Li_2O
SO_3	NiO, CoO	C

The acid radicals begin the list, followed by the subordinate metallic oxides, again in the order of R_2O_3, RO, and R_2O. This order is by no means universally used: MnO is frequently placed after P_2O_5 (and the two are often placed between H_2O- and CO_2), C is usually found with Cl, F, and S, and SrO is often placed after CaO. Many of the subordinate metallic oxides listed here, plus many others not listed, are often determined by emission spectroscopy, and by convention these are usually reported in the elemental form. In order to avoid unnecessary conversion of elements to the oxide form, it is customary to include in the oxide statement only those spectrographically determined constituents needed to correct the values for major constituents, and to list the others separately as the elements, usually in parts per million rather than in weight percent.

The correction for the oxygen equivalent of S, Cl, and F, when one or all of these are present, is given after the summation, from which it is subtracted to give the final total.

There are now some exceptions.[33, 34] The inclusion of Cr, Ni, and Co in the alphabetical (by symbol, not by name) listing of trace elements is now common practice,[34-36] and the trace elements are reported in parts per million or, if warranted, parts per billion of the element.

Mineral analyses may be expressed in the format given above, but it is often more convenient and illustrative to list the constituents in the order of their importance in the mineral structure. Such specialized geological materials as meteorites, which may contain silicate, sulfide, and metallic phases, require a different format, although the silicate portion is usually expressed in the form outlined previously.

With regard to the *form* in which the constituents should be presented, Washington (Ref. 32, p. 62) urged the use of oxides, on the grounds of convenience and practicability sanctioned by (then) long and universal usage. Although written at the turn of the present century, these ideas have continued to dominate the field. The reporting of the constituents as *oxides* with appropriate application of oxygen equivalents, although subject to questioning from the viewpoint of the physical chemistry of rocks and minerals, remains the most useful system of reporting yet suggested. Because oxygen is the major constituent of most rocks and many minerals, this method of expression enables one qualitatively to judge the completeness of the analysis at a glance. The total, although it may be misleading and must be evaluated in the light of such variables as the analyst and the methods used, does provide at least an indication of the probable accuracy of the analysis. The resulting convenience in the comparison of analyses cannot be ignored, particularly when one has had occasion to work with older analyses in which the constituents are expressed as a variety of compounds. This, however, is not so important at the present time because of the ready availability of computers prepared to convert the statement of the analysis into any form and order for which a program can be written.

The *number of significant figures* that should be given for an analytical result has been a topic of discussion for a number of years. The usual custom of reporting experimental data is that, unless a statement giving a measure of uncertainty is included, the last figure given is correct to within 2 or 3 (Ref. 37, p. 404). That is, if the value 65.2% is given for SiO_2, with no information as to the measure of the uncertainty, it is understood that the most probable result lies between 64.9% and 65.5%. If 65.20% SiO_2 is given, then the most probable result will lie between 65.17% and 65.23%. There is, however, a list of suggestions to be used as a guide in reporting numerical data which states, in part, that the precision and the accuracy of any measurement should be given.[38]

Only rarely are the constituents of a rock or mineral analysis now reported to more than four significant figures. Indeed, as Chalmers and Page[39] have pointed out, reporting more than three significant figures for classical silicate techniques is difficult to justify, and this is also true for the results from instrumental techniques.

The act of rounding off,* according to Chayes,[40] is a destruction of

---

* In rounding off, a number ending in 5 must be rounded off to the nearest even number. For example, 42.55 is the mean of 42.6 and 42.5, and should be rounded to 42.6, but 42.45, the mean of 42.4 and 42.5, should be rounded to 42.4. This practice will preclude introducing a bias due to rounding up only (Ref. 41, p. 60).

observational information which cannot be retrieved by the user of the data. It is the analyst's responsibility, however, to ensure that only information which is meaningful is reported and that the report includes enough information to allow the user of the data to determine the degree of uncertainty in the results. The means by which one can establish the precision and accuracy of a set of results by a statistical analysis of the data obtained is the next topic to be discussed.

1.4 PRECISION AND ACCURACY

The most important thing to remember about a quantitative analytical result is that it can only be an *estimate* of the concentration of the analyte present. A measure of the reliability of that estimate is given by statistically derived statements of the precision and accuracy of the measurement which should accompany the estimate when it is reported as an analytical result.

1.4.1 Precision

The precision of an estimate is a measure of the agreement between replicate values. The smaller the difference between these values and the arithmetic mean of all the values, the greater is the precision of the measurement or analytical method. A quantitative estimate of the precision can be stated in terms of the variance and the standard deviation of a set of results. Given a set of N results $x_1, x_2, \ldots x_N$, the mean of these results is defined as:

$$\bar{x} = \frac{x_1 + x_2 + x_3 + \cdots + x_N}{N}$$

The standard deviation of a set of measurements is the square root of the sum of the squares of the deviation of each measurement from the mean divided by the number of degrees of freedom (i.e., the number of measurements minus 1)

$$s = \sqrt{\frac{(x_1 - \bar{x})^2 + \cdots + (x_N - \bar{x})^2}{N - 1}}$$

For a more detailed discussion of precision the reader should consult standard statistical texts.[41-44]

The question that remains is, however, what kind of precision can be expected in routine rock and mineral analysis? The answer is a complex one, involving a consideration of the analyte and its concentration, the overall sample composition, as well as the technique that is being used to make the measurement. In the determination of major elements using classical techniques, a good analyst was expected to perform as shown in Table 1.2 (part I); an actual example of this quality of performance is given in Table 1.2 (part II). Table 1.2 (parts III and IV) gives an indication of possible variances to be expected for between-laboratory and within-laboratory replicate analyses. Table 1.2 (part V) and Table 1.3 give an indication of the quality of results that can be obtained from the use of AAS. XRF is capable of producing equally precise results (Table 1.3), and Govindaraju et al. use ES techniques very effectively.[20] This last technique is used most often for trace element determinations, although the introduction of plasma sources is changing this generalization. It should be born in mind, when comparing results in this way, that intra-method procedural variability, such as different digestion techniques for AAS or different matrix corrections in XRF, is also present.

It should be pointed out that the AAS and XRF techniques tend to give slightly higher results for SiO_2 than does the classical gravimetric scheme. The results for SiO_2 in Table 1.3 show this to be the case in the recent SY-2*, SY-3, and MRG-1 collaborative study summarized by Abbey et al.[47] In addition, the recommended values for older standards such as W-1 and G-1 have changed over the years, and slightly higher values for SiO_2 have been recommended as more instrumentally derived results are reported. The relative standard deviations of the results are also different (Table 1.3), with the results obtained by the classical gravimetric method showing slightly less interlaboratory consistency than the results from the two instrumental procedures.

Although these techniques would seem to provide results of similar quality for major and minor elements, the situation changes markedly when concentrations in the parts per million, or even parts per billion, range are of interest. As the detection limit for any particular technique is approached, the precision approaches ± 100%. The detection limit will vary from element to element for any given technique and for particular samples, and this will be dealt with in some detail later. In general the detection limits attainable by AAS and ES techniques are below 10 ppm and are lower than those obtained by XRF. A good indication of the precision obtained by these three methods is found in Table 1.4. It can be seen that the interlaboratory agreement is better for AAS in most cases, but all three methods show examples of both poor precision and poor agreement with the recommended value. Wet chemical colorimetric methods for trace element determinations are capable of low detection limits, but they are often slow and involved and open to numerous ma-

---

* See Appendix 1 for explanation of sample symbols.

TABLE 1.2 Precision of Replicate Determinations.

	I	II Basalt		III G-1			IV G-1			V MRG-1			VI Tonalite (T-1)		
		1	2	$\bar{x}$	s	C	$\bar{x}$	s	C	$\bar{x}$	s	R	$\bar{x}$	s	C
SiO_2	±.20	48.12	48.05	72.22	0.43	0.60	72.64	0.23	0.31	39.35	0.186	39.22	62.69	0.26	0.42
Al_2O_3	.10	19.49	19.57	14.44	.541	3.75	14.13	.17	1.22	8.42	.030	8.51	16.55	.29	1.72
Fe_2O_3	.03	2.21	(2.21)	.94	.34	36.2	.86	.17	19.3	8.20	.072	8.36	2.86	.29	10.1
FeO	.02	5.40	(5.40)	1.00	.135	13.5	1.06	.06	5.72	8.77	.040	8.61	2.85	.20	7.12
CaO	.02	11.33	11.33	1.42	.152	10.7	1.34	.09	6.53	14.77	.064	14.68	5.19	.25	4.76
MgO	.03	10.29	10.28	.39	.135	34.6	.44	.04	9.90	13.40	.034	13.49	1.85	.10	5.62
Na_2O	.03	2.23	2.11	3.26	.284	8.71	3.43	.21	6.23	0.722	.006	.72	4.35	.19	4.32
K_2O	.02	.03	.03	5.51	.549	9.96	5.43	.21	3.83	.175	.003	.18	1.24	.09	7.04
H_2O+	.02	.11	.12	.37	.104	28.1	.31	.08	26.7	1.04	.063	1.02	1.57	.24	15.2
TiO_2	.01	.70	.70	.26	.067	25.7	.25	.03	11.2	3.84	.042	3.69	.59	.04	6.90
P_2O_5	.01	.01	.01	.10	.045	45.0	.10	.02	23.4				.13	.03	19.9
MnO	.01	.25	.23	.03	.0122	40.7	.03	.01	30.5	.177	.002	.17	.11	.02	18.8
CO_2										1.06	.013	1.04			
TOTAL		100.17	100.04	99.94			100.02			99.92			99.93		

I. Groves (Ref. 45, p. 228); allowable variations in duplicate determinations, expressed as percent of the whole rock.

II. Duplicate analyses of basalt (1091–64) by J. L. Bouvier, Analytical Chemistry Section, GSC.

III. Fairbairn et al. (Ref. 46, p. 37, Table 14); arithmetical mean ($\bar{x}$) and standard deviation (s), with added calculated relative standard deviation (C), of determinations of constituents in G-1 by 24 laboratories.

IV. Fairbairn et al. (Ref. 46, p. 38, Table 17); as in III, but by one laboratory (seven analysts).

V. Mean of 10 replicate analyses of MRG-1 by P. F. Ralph of BCM, submitted to S. Abbey for MRG-1 certification program. R-recommended values.[35]

VI. As in III, but for the determination of constituents in standard geochemical sample T-1, Msusule tonalite, by 14 laboratories as given in Supplement No. 1, Geological Survey Division, United Republic of Tanzania, Dodoma, Tanganyika, 1963.

13

TABLE 1.3 Comparison of Results Obtained for Selected Major Constituents by AAS and XRF.

SiO_2

	SY-2				SY-3				MRG-1			
	C	AAS	XRF	t	C	AAS	XRF	t	C	AAS	XRF	t
N	19	5	6	30	19	5	5	29	18	6	4	28
$\bar{x}$	59.86	60.13	60.21	59.98	59.46	59.68	59.87	59.57	39.10	39.17	39.71	39.20
s	0.483	0.222	0.591	0.485	0.472	0.479	0.409	0.468	0.495	0.325	0.494	0.497
$\%C$	0.8	0.4	1.0	0.8	0.8	0.8	0.7	0.8	1.3	0.8	1.2	1.3
R				60.07				59.71				39.22

	GA				GH				BR			
	C	AAS	XRF	T	C	AAS	XRF	T	C	AAS	XRF	T
N	26	5	12	45	24	7	13	45	26	6	10	45
$\bar{x}$	69.72	69.57	69.94	69.80	75.74	75.24	76.21	75.74	38.41	38.56	39.01	38.75
s	0.416	0.410	0.420	0.45	0.746	0.658	0.590	0.61	0.347	0.279	1.123	0.98
$\%C$	0.6	0.6	0.6	0.6	1.0	0.9	0.8	0.8	0.9	0.7	2.9	2.5
R				69.90				75.80				38.20

Al_2O_3

	SY-2				SY-3				MRG-1			
	C	AAS	XRF	t	C	AAS	XRF	t	C	AAS	XRF	t
N	6	11	6	23	7	11	5	23	8	12	4	24
$\bar{x}$	11.78	12.20	12.04	12.04	11.66	11.78	11.84	11.76	8.25	8.50	8.69	8.45
s	0.226	0.272	0.235	0.298	0.341	0.281	0.375	0.313	0.401	0.254	0.101	0.326
$\%C$	1.9	2.2	2.0	2.5	2.9	2.4	3.2	2.7	4.9	3.2	1.2	3.9
R				12.15				11.70				8.51

	GA				GH				BR			
	C	AAS	XRF	T	C	AAS	XRF	T	C	AAS	XRF	T
N	27	6	10	45	26	8	10	45	27	6	8	44
$\bar{x}$	14.70	14.71	14.59	14.67	12.64	12.67	12.86	12.69	10.30	10.19	10.25	10.28
s	0.314	0.120	0.374	0.31	0.306	0.261	0.476	0.35	0.438	0.307	0.416	0.4
$\%C$	2.1	0.8	2.6	2.1	2.4	2.1	3.7	2.8	4.2	3.0	4.1	3.9
R				14.50				12.50				10.20

K_2O	GA				GH				BR			
	C	AAS	XRF	T	C	AAS	XRF	T	C	AAS	XRF	T
N	24	7	13	46	23	8	11	44	24	8	11	46
$\bar{x}$	4.03	4.07	4.05	4.05	4.76	4.79	4.75	4.78	1.40	1.40	1.51	1.42
s	0.130	0.093	0.095	0.11	0.149	0.105	0.107	0.15	0.117	0.066	0.089	0.14
$\%C$	3.2	2.3	2.3	2.7	3.1	2.2	2.3	3.1	8.4	4.7	5.9	9.9
R				4.03				4.76				1.40

C, results obtained by classical techniques (gravimetry, titrimetry).

AAS, results obtained by atomic absorption techniques.

XRF, results obtained by X-ray fluorescence techniques.

t, calculated from only those results used for C, AAS, and XRF calculations.

T, values given in original report.

Analytical results are given in percent.

N, number of different results (some are means of replicate determinations).

$\bar{x}$, mean of values.

s, standard deviation.

$\%C$, relative standard deviation.

R, recommended value.

Values for SY-2, SY-3, and MRG-1 are from Ref. 48, Tables 4, 5, and 6, respectively.

Values for GA, GH, and BR are from Ref. 47, Table 1.

15

TABLE 1.4   Comparison of Results Obtained for Selected Trace Elements by AAS, XRF, and ES.

Ni	SY-2			SY-3			MRG-1		
	AAS	XRF	ES	AAS	XRF*	ES	AAS	XRF	ES
N	6	5	3	6	4	6	8	5	6
$\bar{x}$	15	19	7	14	20	10	176	182	211
s	6.4	11.5	1.0	6.4	13.5	3.7	35.6	43.5	29.3
$\%C$	43	61	14	46	68	37	20	24	14
R	11			11			190		

	GA			GH			GR			BR		
	AAS	XRF	ES	AAS	XRF	ES	AAS	XRF	ES	AAS	XRF	
N		3	15		3	11		4	20	2	4	1
$\bar{x}$		6	7		9	4		57	56	305	265	22
s		2.0	4.3		7.2	2.7		5.4	20.6	21.2	77.1	7
$\%C$		33	61		80	68		9	37	7	29	3
R	7			3			55			270		

Rb	SY-2			SY-3			MRG-1					
	AAS	XRF	ES	AAS	XRF	ES	AAS	XRF	ES	AAS	XRF	
N	7	11		7	10		5	6				
$\bar{x}$	234	196		219	174		8	12				
s	37.5	43.7		37.0	51.1		1.9	4.0				
$\%C$	16	22		17	29		24	34				
R	220			210			8?					

	GA			GH			GR			BR		
	AAS	XRF	ES	AAS	XRF	ES	AAS	XRF	ES	AAS	XRF	
N	3	15	2	2	13	2	2	12	3	4	13	
$\bar{x}$	192	178	166	400	390	326	164	191	137	45	57	4
s	41.0	23.0	33.0	11.0	36.0	104.0	9.2	28.6	78	14.1	27.1	
$\%C$	21	13	20	3	9	32	6	15	57	31	48	2
R	175			390			175			45		

nipulative errors. Spark source mass spectrometry is capable of providing very precise results at concentrations in the parts per billion range, but this technique is very time-consuming. Neutron activation analysis can similarly yield good precision at very low concentration levels.

For the determination of trace elements, each of the above methods has areas of application where it is the best method available and where it produces the most precise results. These aspects will be discussed in more detail in the appropriate chapters. It can be stated in summary, however, that precision in trace element determinations is not an easy value to measure nor to describe on a general basis. Nor are reliable results at the trace level readily obtained. Virtually any compilation of

TABLE 1.4 *(Continued)*

Sr	SY-2			SY-3			MRG-1			GA		
	AAS	XRF	ES	AAS	XRF	ES	AAS	XRF	ES	AAS	XRF	ES
N	11	11	8	9	9	10	7	5	7	7	18	12
$\bar{x}$	269	258	290	310	280	339	268	259	263	324	319	345
s	24.7	26.7	37.8	14.3	38.6	91.7	5.9	18.0	64.0	39.0	65.0	119.0
$\%C$	9	10	13	5	14	27	2	7	24	12	20	35
R	270			310			270			305		

Sr	GH			GR			BR		
	AAS	XRF	ES	AAS	XRF	ES	AAS	XRF	ES
N	4	14	8	7	11	13	8	17	11
$\bar{x}$	15	11	11	539	544	478	1414	1313	1309
s	3.4	4.7	5.4	48.0	99.0	122.0	195.0	133.0	658.0
$\%C$	23	43	49	9	18	26	14	10	50
R	10			550			1350		

Zn	SY-2			SY-3			MRG-1		
	AAS	XRF	ES	AAS	XRF	ES	AAS	XRF	ES
N	11	7	2	11	7	2	9	3	2
$\bar{x}$	254	249	224	247	236	216	189	205	177
s	19.2	20.9	33.2	14.2	20.7	51.6	15.1	22.8	26.9
$\%C$	8	8	15	6	9	24	8	11	15
R	250			240			185		

Zn	GA			GH			GR			BR		
	AAS	XRF	ES	AAS	XRF	ES	AAS	XRF	ES	AAS	XRF	ES
N	5	4	5	5	3	4	3	3	5	7	3	5
$\bar{x}$	94	82	53	85	86	80	77	75	59	170	131	161
s	35.0	25.0	28	12.9	29.8	33.4	30.3	14.7	31.9	34.1	15.5	17.6
$\%C$	37	31	53	15	35	42	39	20	54	20	12	11
R	75			80			60			160		

AAS, results from atomic absorption techniques.
XRF, results from X-ray fluorescence techniques.
ES, results from optical emission spectroscopy techniques.
All results are in parts per million.

N, number of different results (some are means of replicate determinations).
$\bar{x}$, mean of values.
s, standard deviation.
$\%C$, relative standard deviation.
R, recommended value (? refers to doubtful value).

Values for SY-2, SY-3, and MRG-1 are from Ref. 47, Table 2. Values for GR, GA, GH, and BR are from Ref. 48, Tables 9, 10, 11, and 12, respectively.

[a] One value of 125 ppm omitted.
[b] One value of 870 ppm omitted.

trace element results in an interlaboratory collaborative study will give a range of results for trace elements often exceeding an order of magnitude. Chromium results of 15 and 400 ppm as reported in the GR study,[48] and copper results of 2 and 40 ppm in SY-1 program,[47] are only two examples. A study of the results in Table 1.4 shows that each of the three methods will, on occasion, give results which are significantly different from the accepted value, indicating that it is subject to interference or some other cause of inaccuracy, even though results obtained by it have an overall low standard deviation. Results so far off the mean show also that sources of error can be overlooked in even the best laboratories. An interlaboratory and interinstrumental comparison study has confirmed the above observations.[49]

1.4.2   Accuracy

As stated previously, the standard deviation is a quantitative measure of the precision of an estimate. There is no guarantee, however, that a *precise* answer is an *accurate* one. The accuracy of a determination is the concordance between it and the true or most probable value; the numerical difference between the two values is the *error*. The big problem is to be able to define a *true* value. The closest we have been able to come to doing this in natural geological samples is for the all too few standard reference materials such as those used in Tables 1.3 and 1.4.

Prior to the initial distribution of standard reference samples such as W-1 and G-1, only sketchy statements of the probable precision of the separate determinations in a rock analysis were available, and no similar statement of the probable accuracy of these determinations is known to the authors. Estimates of the accuracy of an analysis (when considered!) were evidently based largely upon the degree of success with which mineralogical formulas could be derived from the analytical data. The initial stage of the G-1 and W-1 project was an exercise in the determination of the precision of rock analysis because no true values for the two rock samples were known. This was, of course, recognized at the start of the project and, partly because of the shock occasioned by the disturbingly large variations in the results for some constituents,[50] it was decided to test both precision and accuracy by the use of a material of known composition, in this case a six-component synthetic silicate glass ("haplogranite") approximating to the composition of G-1.[51] An important consideration arising out of the investigation was the recognition of a bias in the determinations of SiO_2 and Al_2O_3; the mean value for the combined $SiO_2 + Al_2O_3$ was in excellent agreement with the true value, but that for SiO_2 was 0.41% *low* and that for Al_2O_3 was 0.41% *high*. The probable sources of error giving rise to this bias have been discussed in detail,[52] and it has been shown that it can be minimized by careful attention to the techniques of dehydration and washing of precipitates when the conventional method is used, or by the use of other methods, such as the combined gravimetric-colorimetric method for silica described by Jeffery and Wilson.[53] Any correlation study of a proposed reference material has the added advantage that the results from several independent methods will help to point out any bias that is unique to any one of the methods, provided that the precision of each method is adequate. However, even the best established values for reference materials are published as recommended values, and the use of the word "true" is rigorously avoided. There has been no agreement as to what "recommended" means nor how to establish a recommended value, although a considerable effort is being made to reach a concensus on this problem.[54]

In spite of the above statement, the recommended values for some constituents are now considered to be very close to the true values, and the standard reference materials are widely accepted and used to establish the *accuracy* of a method or of a set of results. It is possible to quantify the accuracy of an estimate by defining the *confidence limits* associated with it. The statistically derived confidence limit is defined as $ts/\sqrt{N}$ where s is the standard deviation, N is the number of measurements, and t is a value drawn from t distribution tables.[41-44]

A worked example is probably the best way to describe the use of confidence limits in estimating the accuracy of a method or an analytical result. We shall assume that an analyst is requested to analyze a rock sample for SiO_2 and was asked to establish whether or not the method used had any appreciable bias and to state the reliability of the result (i.e., the precision and accuracy).

Let us further assume that the analyst selected a reference material with a recommended value for SiO_2 of 61.57% and analyzed both the sample and the standard five times, obtaining the results shown in Table 1.5. The mean, the standard deviation, and the relative standard deviation are also given, the last two being the values that establish the precision of the analysis (not very good in this case). To answer the question regarding the accuracy of the result, the analyst calculated the confidence limits and chose to do so at the 95% confidence level. To accomplish this the analyst went to a table of t values (Ref. 55, p. 280) and found for five measurements (4 degrees of freedom) and for a 95% confidence interval a t value 2.132. Thus for the standard sample,

$$95\% \text{ confidence limit} = \frac{ts}{\sqrt{N}} = \frac{2.132 \times 0.2702}{\sqrt{5}} = 0.26$$

TABLE 1.5   Statistical Data for Calculation of
Confidence Limits

Measurement	Sample	Standard
1	54.0	61.3
2	53.5	62.0
3	54.2	61.5
4	53.2	61.8
5	53.8	61.6
N	5	5
$\bar{x}$	53.74	61.64
s	0.3975	0.2702
%C	7.4	4.4
95% confidence limit	$\pm 0.38\%$	$\pm 0.26\%$

This means that the measurement of SiO_2 in the standard resulted in a value of 61.64% ± 0.26% at the 95% confidence level. That is, there are 95 chances out of 100 that the true value will come within the range of 61.38% and 61.90%. The accepted value of 61.57% does fall within this range, and the analyst can be sure that, at the 95% confidence level, the method used has no appreciable bias. The answer to give for the unknown sample would then be 53.74% ± 0.38% at the 95% confidence level.

It should be pointed out that the confidence limits are independent of the concentration in the sample, but they are a direct function of the precision of the determination. In the above example, the less precise answer for the unknown sample has a larger confidence interval (±0.38%) than does the answer for the standard material (±0.26%). The only difference between the two is the higher standard deviation for the former. The confidence that the analyst has that there is no bias observed in this method must be tempered by the observation that a bias of, say, + 0.15% would not be detected. In order to detect a bias of this magnitude the analyst would have to make many more measurements or improve the precision of the method used to obtain a standard deviation of less than 0.1573. This comes from solving for s in the equation above with N = 5, t = 2.132, and a confidence limit of 0.15. Guidelines for reporting imprecision and inaccuracy have been published.[38]

The availability now of many standard reference materials with well established values for the major and minor elements found in most rocks and minerals has helped to establish the quality of results for these elements obtained by new techniques and methods. Unfortunately the same quality of recommended values is not available for most elements found at the trace level (100 ppm or less). As a consequence of this, any analyst who hopes to set up a new method for a wide range of trace elements will be faced with the difficult problem of obtaining standards to cover adequately the variety of elements and their concentration ranges that are likely to be encountered.

The importance of standard reference materials in the daily operation of a laboratory is illustrated by the large number of new, would-be standard reference materials being introduced by workers around the world, and by the establishment, in 1977, of a new journal, *Geostandards Newsletter*, by the International Working Group of the Association Nationale de la Recherche Technique, Paris.

1.5 SUMMARY

The foregoing material is intended to set the scene, and to provide a sort of aperitif for the more solid fare to follow. In this regard, brief mention has been made of a variety of instrumental methods that find use in rock and mineral analysis, about which more will be said in Chapters 7–12. The importance and use of standard reference materials have been touched upon lightly, preparatory to more detailed coverage in Chapter 2, and the statistical approach, this time with regard to sampling and sample preparation, is encountered again in Chapter 3.

Since the appearance of the first edition of this work in 1968, a number of other texts have appeared which treat the analysis of rocks, minerals, and related materials in varying depth and with varying emphasis.[56–60] Bennett[61] has also reviewed the developments in silicate analysis, since the mid-1940s, covering both the ceramic and the geochemical fields.

References

1. A. W. Groves, *Silicate Analysis,* 2nd ed., George Allen and Unwin, London, 1951, pp. xi–xviii, 28–37.
2. W. F. Hillebrand, G. E. F. Lundell, H. A. Bright, and J. I. Hoffman, *Applied Inorganic Analysis,* 2nd ed., Wiley, New York, 1953, pp. 793–807.
3. H. S. Washington, *The Chemical Analysis of Rocks,* 2nd ed., Wiley, New York, 1910, pp. 1–31.
4. M. E. Weeks and H. M. Leicester, *Discovery of the Elements,* 7th ed., J. Chem. Educ., Easton, PA, 1968.
5. R. C. Chirnside, *J. Soc. Glass Technol.,* 43(210), 5T–29T (1959).
6. F. J. Langmyhr and P. R. Graff, *Norg. Geol. Unders.,* 230 (1965).
7. E. L. P. Mercy, *Geochim. Cosmochim. Acta,* 9, 161 (1965).

8. E. A. Vincent, "Analysis by Gravimetric and Volumetric Methods, Flame Photometry, Colorimetry and Related Techniques," in A. A. Smales and L. R. Wager, Eds., *Methods in Geochemistry,* Interscience, New York, 1960, pp. 71–78.

9. I. A. Voinovitch, J. Debras-Guedon, and J. Louvrier, *L'Analyse des Silicates,* Hermann, Paris, 1962, pp. 19–22.

10. A. D. Maynes, *Anal. Chim. Acta,* **32,** 211 (1965).

11. A. D. Maynes, *Chem. Geol.,* **1,** 61 (1966).

12. P. F. Ralph, B. C. Department of Mines and Petroleum Resources, private communication.

13. G. R. Lachance, *X-Ray Spectrom.,* **8,** 190 (1979).

14. H. J. Lucas-Tooth and B. J. Price, *Metallurgia,* **64,** 149 (1961).

15. G. R. Lachance and R. J. Traill, *Can. J. Spectrosc.,* **11,** 43 (1966).

16. F. Claisse and M. Quintin, *Proc. 13th Colloq. Spectrosc. Int.,* (Ottawa, 1967), Adam Hilger Ltd., London, 1968.

17. S. D. Raspberry and K. F. J. Heinrich, *Anal. Chem.,* **46,** 81 (1974).

18. R. Jenkins, *Adv. in X-Ray Analy.,* **22,** 281 (1979).

19. W. K. de Jongh, *Norelco Rep.,* **23,** 26 (1976).

20. K. Govindaraju, G. Mevelle, and C. Chouard, *Anal. Chem.,* **48,** 1325 (1976); M. Roubault, H. de la Roche, and K. Govindaraju, *Sci. de la Terre,* **9,** 339 (1962–1963).

21. C. O. Ingamells, *Anal. Chem.,* **38,** 1228 (1966).

22. N. H. Suhr and C. O. Ingamells, *Anal. Chem.,* **8,** 730 (1966).

23. F. W. E. Strelow, C. J. Liebenberg, and A. H. Victor, *Anal. Chem.,* **46,** 1409 (1974).

24. S. O. Agrell and J. V. P. Long, "The Application of the Scanning X-Ray Microanalyser to Mineralogy," in A. Engström, V. Cosslett, and H. Pattee, Eds., *X-Ray Microscopy and X-Ray Microanalysis,* Elsevier, Amsterdam, 1960.

25. D. G. W. Smith, *Microbeam Techniques, Short Course Handbook,* CO-OP Press, Edmonton, Alta., 1976.

26. A. G. Plant, "Applications of Microbeam Techniques to Mineralogy," in D. G. W. Smith, *Microbeam Techniques, Short Course Handbook,* CO-OP Press, Edmonton, Alta., 1976.

27. A. Råheim, *Lithos,* **8,** 221 (1975).

28. E. W. White, "Application of Soft X-Ray Spectroscopy to Chemical Bonding Studies with the Electron Microprobe," in C. A. Andersen, Ed., *Microprobe Analysis,* Wiley, New York, 1973.

29. K. Keil, "Applications of the Electron Microprobe in Geology," in C. A. Andersen, Ed., *Microprobe Analysis,* Wiley, New York, 1973.

30. G. B. Dalrymple and M. A. Lanphere, *Potassium-Argon Dating,* Freeman, San Francisco, 1969.

31. G. E. F. Lundell and J. I. Hoffman, *Outlines of Methods of Chemical Analysis*, Wiley, New York, 1938, p. 15.

32. H. S. Washington, *Am. J. Sci.*, 4th Ser., **10**, 59 (1900).

33. S. Abbey, "Studies in 'Standard Samples' of Silicate Rocks and Minerals, Part 4," 1974 ed. of "Usable" Values, Geological Survey of Canada, Paper 74-41, 1975.

34. F. J. Flanagan, *Geochim. Cosmochim. Acta,* **37,** 1189 (1973).

35. S. Abbey, A. H. Gillieson, and G. Perrault, *Can. J. Spectrosc.,* **20,** 113 (1975).

36. O. H. J. Christie, *Talanta,* **22,** 1048 (1975).

37. F. Daniels, J. W. Williams, P. Bender, and C. D. Cornwell, *Experimental Physical Chemistry,* 6th ed., McGraw-Hill, New York, 1962.

38. D. Garvin, *J. Res. Nat. Bur. Stand.,* **76A,** 67 (1972).

39. R. A. Chalmers and E. S. Page, *Geochim. Cosmochim. Acta,* **11,** 247 (1957).

40. F. Chayes, *Am. Mineral.,* **38,** 784 (1953).

41. D. A. Skoog and D. M. West, *Fundamentals of Analytical Chemistry,* 3rd ed., Holt, Rinehart & Winston, New York, 1976.

42. E. L. Bauer, *A Statistical Manual for Chemists,* Academic, New York, 1971.

43. W. J. Youden, "Accuracy and Precision: Evaluation and Interpretation of Analytical Data," in I. M. Kolthoff and P. J. Elving, Eds., *A Treatise on Analytical Chemistry,* Vol. 1, Part 1, Interscience, New York, 1959.

44. H. J. Halstead, *Introduction to Statistical Methods,* Macmillan, Toronto, 1960.

45. A. W. Groves, *Silicate Analysis,* 2nd ed., George Allen and Unwin, London, 1951, pp. 28–37, 224–236.

46. H. W. Fairbairn et al., *U.S. Geol. Surv. Bull.,* 980 (1951).

47. S. Abbey, A. H. Gillieson, and G. Perrault, "SY-2, SY-3 and MRG-1, A Report on the Collaborative Analysis of Three Canadian Rock Samples for Use as Certified Reference Materials," Canada Centre for Mineral and Energy Technology, Ottawa, MRP/MSL 75–132 (TR), 1975.

48. M. Roubault, H. de la Roche, and K. Govindaraju, *Sci. de la Terre,* **15,** 351 (1970).

49. I. B. Brenner, L. Gleit, and A. Harel, *Appl. Spectrosc.,* **30,** 335 (1976).

50. W. G. Schlecht, *Anal. Chem.,* **23,** 1568 (1951).

51. H. W. Fairbairn and J. F. Schairer, *Am. Mineral.,* **37,** 744 (1952).

52. P. G. Jeffery, *Geochim. Cosmochim. Acta,* **19,** 127 (1960).

53. P. G. Jeffery and A. D. Wilson, *Analyst,* **85,** 478 (1960).

54. S. Abbey, R. A. Meeds and P. G. Belanger, *Geostand. Newsl.,* **3,** 121 (1979).

55. W. Mandenhall, *Introduction to Statistics,* Wadsworth Publ. Co., Belmont, CA, 1965.

56. A. Volborth, *Elemental Analysis in Geochemistry, A. Major Elements*, Elsevier, Amsterdam, 1969.

57. H. Bennett, and R. A. Reed, *Chemical Methods of Silicate Analysis*, Academic, New York, 1971.

58. A. J. Easton, *Chemical Analysis of Silicate Rocks*, Elsevier, Amsterdam, 1972.

59. P. G. Jeffery, *Chemical Methods of Rock and Mineral Analysis*, 2nd ed., Pergamon, Oxford, 1975.

60. R. D. Reeves, and R. R. Brooks, *Trace Element Analysis of Geological Materials*, Vol. 51, Chemical Analysis Series, Wiley-Interscience, New York, 1978.

61. H. Bennett, *Analyst*, **102,** 153 (1977).

CHAPTER 2

THE WORKING ENVIRONMENT

Modern instrumental methods of analysis require slightly more elaborate facilities than do the standard classical methods. These requirements range from separate rooms for the protection of instruments from ubiquitous laboratory fumes to special facilities for ventilation. The requirements are not so different, however, that major structural alterations are required in any laboratory likely to house one of the several different types of instruments that might be required. It should be pointed out that even though instrumental methods are superseding more and more of the classical methods, no laboratory should place itself in the position of being unable to utilize any of the latter through lack of the basic supplies, facilities, or equipment. Thus this chapter, while dealing with the laboratory requirements for instrumental methods, should be supplemented by reference to other texts dealing with classical methods of rock and mineral analysis.[1-4]

2.1 LABORATORY AND SERVICE FACILITIES

2.1.1 The Laboratory

The amount of laboratory space required will depend on the techniques to be used, the workload, and the complexity of any special preconcentration or sample preparation requirements.[5] Large laboratories are more suited to routine analytical work, particularly when the samples are analyzed in batches, but for the more complex work of mineral analysis a smaller laboratory is preferable; traffic flow is minimized and space is provided for the specialized apparatus that is often needed.

At the GSC the chemical laboratories are located on the seventh floor of an eight story building. This has the advantage of minimizing the exposure of the laboratories to airborne dust and other contaminants and, because the fan-room is located directly above the laboratories, ensures a good exhaust in the fume hoods. Structural vibration, as it affects delicate instrumentation such as an analytical balance, has not been a serious problem.

A convenient module is one that is about 6 m wide and 9 m deep, which will provide adequate space for two or three analysts. At the GSC the more specialized work is done in single modules and half-modules; double modules are used for work of a routine nature on large numbers of samples. The balance room is placed between two double module laboratories, thus affording a convenient access for the largest number of people.

The laboratories should be well lighted, preferably without exposure to direct sunlight, and should have adequate bench and fume hood space. A fume hood should not be less than 2 m wide; a large part of the work is done in the hood, and there must be room for a steam or water bath and a hot plate, as well as the usual services supplied in fume hoods. Provision should be made for at least one special fume hood designed to permit the safe use of perchloric acid; this should include facilities for flushing the duct and walls of the fume hood with water, the use of explosion-proof electrical outlets, and the exclusion of gas outlets. The bench space should include some at desk height to accommodate pieces of equipment which are more conveniently operated at this level. Desk space should be provided away from the laboratory bench, preferably in a separate room adjacent to the laboratory. In planning the layout of the laboratory it is necessary to leave some free wall space for the storage of such items as gas cylinders, filing cabinets, and laboratory carts.

There is much to be said for the use of built-in equipment. By suspending such items as gas burners and filter stands from the back of the laboratory bench, the surface of the bench is more easily cleaned. The utilization of the floor of the fume hood for built-in steam or water baths, hot plates, and furnace wells makes for efficient use of the available space. Threaded holes in the back wall of the fume hood, into which support rods can be screwed, provide a useful means for keeping the fume hood floor free of support stands and bulky structures.

At the GSC there are two types of balance tables in use on the fifth, sixth, and seventh floors. In the one type, the balance platform is fastened to a metal plate, and the latter in turn is fastened to the top of a large steel cylinder containing about 150 lb of steel slugs. The cylinder hangs freely inside a wooden pedestal which rests upon hard rubber supports; the weight of the cylinder pulls the balance platform against hard rubber supports on the top of the pedestal. The whole is surrounded, but not touched, by a separate work table. The second type is a simple but effective slab-sided terrazo table that is cast in one piece and weighs about 400 lb; the table is strengthened internally by brass rods, and the dimensions are such as to give the maximum damping effect, aided when necessary by isolation pads placed under each of the two side slabs.

Instruments such as atomic absorption spectrophotometers, X-ray fluorescence spectrometers, emission spectrographs, and electron microprobes are best kept in separate rooms, each with its own unique ancillary equipment. When these instruments are being used for trace element analyses, especially at subparts per million levels, they should be located in rooms in which the air pressure is maintained at a positive pressure, relative to that in the rest of the laboratory. This holds true for any room in which sample decomposition, concentration, and other preparation steps are performed. The "isolation" conditions need not be as extreme as those provided in the Lunar Receiving Laboratory,[6] but reasonable precautions against contamination are an important consideration in any trace element analysis program.

2.1.2 Safety

Safety is a first priority in the planning of laboratory facilities, and several texts are available which provide detailed information,[7, 8] one or two of which should be found on the analyst's book shelf. Only a few of the more important safety precautions which should be considered in the planning stage will be mentioned here. Every laboratory room should have at least two well separated exits, and fume hoods should be placed so that they are not adjacent to either of them. Eye baths and safety showers should be easily accessible in each laboratory, and they should be checked regularly (at least once each month) to ensure that they are in working order. Ethers and other chemicals capable of gradual conversion to explosive compounds should not be stored for extended periods, and all volatile chemicals should be kept in a properly ventilated storage area. Cylinders of compressed gases should be chained to a wall or bench or otherwise secured to prevent their being knocked over, and they should not be moved about unless the regulator has first been removed and the cap replaced.

Any work which involves the use of carcinogenic compounds such as aromatic solvents and heavy liquids must be performed in an adequately ventilated area. The BCM laboratory has designed a down-draft fume hood for mineral separations involving such high density liquids as diiodomethane (specific gravity up to 3.325). The floor of the fume hood is separated from its walls by a peripheral gap through which the fume hood vapors are drawn and then passed through a set of activated charcoal filters before being vented to the outside atmosphere. This avoids the problem of trying to lift fumes that are very much heavier than air, and the efficiency of the venting is such that no trace of any heavy liquid can be detected in the mineral separation laboratory, even during periods of maximum use.[9]

2.1.3   Services

The services which are routinely required include electrical outlets of both 110 and 220 V capacity. Special requirements for large equipment such as X-ray units may be supplied by special wiring as needed, and it is advantageous to make provision for bringing in these special lines when the services are installed. Line fluctuations are a common occurrence, and for smaller instruments such as spectrophotometers it will be necessary to provide voltage stabilizers; larger instruments are generally equipped with some means of voltage stabilization. Electric hot plates, ovens, and muffle furnaces are usually provided with some type of temperature control, but for accurate work, such as in the sodium peroxide sinter procedure, it is wise to measure critical temperatures with a reliable pyrometer.

A supply of compressed air is required for many operations, from providing the air necessary for the air–acetylene mixture used in an atomic absorption spectrophotometer to the blowing out of dust from sample and grinding vials. A glass-wool filter should be inserted in the line to remove the droplets of oil and particles of dust that accompany the current of air. Vacuum outlets are also very useful for many operations, although water aspirators attached to the faucet of a cup sink will usually suffice. While electrically operated furnaces, ovens, and hotplates have many advantages, including increased safety and easily automated temperature controls, the gas burner still plays a very important part in many laboratory operations. Some municipalities, however, deliver low BTU gas at low pressures, and an independent source such as a bank of propane tanks is usually much more satisfactory.

A plentiful supply of hot and cold water is, of course, a necessity; the faucets of the cup sinks and fume hoods only need to supply cold water. The type of faucet that permits the mixing of hot and cold water is preferable, and for those operations where close control of the temperature is necessary, as in photographic darkrooms, special mixing, heating, and refrigeration devices must be provided. When water is circulated as a cooling medium, provision should be made to filter out the suspended or entrained solid matter that is commonly carried by the water as it is forced through the pipes at high pressure.

A central supply of distilled water has both advantages and disadvantages and is usually not provided, unless the scale of laboratory operation is large enough to warrant it. While it does eliminate the need for individual water stills which require separate operation and maintenance, it does not always permit the degree of quality control that is desirable. At both the GSC and the BCM laboratories, a portion of the condensate from each water-still is passed through a mixed anionic-cationic resin column (which has a very long life) to provide both distilled as well as distilled and deionized water. The latter is used for the analytical work, except when the possible presence of organic material is undesirable; the distilled water is used for the rinsing of glassware, and in water baths to minimize the formation of boiler scale. A periodic check should be made on the quality of the water used.

2.2 EQUIPMENT

The increasing application of instrumental techniques to rock and mineral analysis has resulted in a change in emphasis on the nature of the equipment used. Burettes, for example, are now employed in only one or two of the determinations commonly made in rock and mineral analysis. Platinum ware, on the other hand, is used much more frequently than before. Most of the items used in the classical silicate techniques are now employed in instrumental techniques for such applications as sample decomposition, the preparation of standard solutions, the concentration of trace elements, or the separation of interfering elements prior to the chosen measurement step. Particularly good coverage of materials, reagents, and apparatus is available in other texts,[2, 10, 11] but it will be useful here to discuss briefly a few items as a guide to proper usage.

2.2.1 Glassware

Borosilicate or chemically resistant glass, of which Pyrex and Kimax are the two most common types, is the most suitable material for beakers, flasks, funnels, and other similar items; it has a high resistance to chemical attack and to mechanical and thermal shock. Alkaline solutions will attack it readily, particularly when boiled in it for any length of time, and solutions containing free hydrofluoric acid should not, of course, be used in it unless the introduction of varying amounts of silica and other constituents is of no consequence. Other solutions, including water, will also attack the glass in time, and it is a good practice to retire glassware to less important usage when signs of wear become visible. The attack on the glass by acids other than hydrofluoric is generally negligible, and solutions that are to be left standing in glass for any length of time should be acidified. Care must be taken not to defeat the use of chemically resistant glass by the unwitting use of less resistant glass in the form of stirring rods and cover glasses. *Vycor*, a silica glass containing approximately 96% SiO_2, and *fused silica* or *quartz* glass, either transparent or translucent, are useful when contamination by alkalies must be avoided. Both of these glasses can be used at temperatures much above the working limit for Pyrex or Kimax, and the fused silica glass has a very high resistance to thermal shock; they have the disadvantage of high cost.

The same care and attention that are given to beakers and flasks should be given to burettes, pipettes, dispensers, and the other articles of glassware in common use. These should also be made of chemically resistant glass and should be treated with the respect that they deserve. Burettes and separatory funnels fitted with Teflon stopcocks are superior to those with glass stopcocks. Burettes in burette stands should be fitted with caps (small polyethylene thimbles) to keep out dust and grease when they are not in use; filling them with water results in attack on the glass over a period of time. The use of rubber policemen as guards for burette tips, particularly for those of the semimicro variety, will prevent chipping which may render the burette unfit for use. Care should be taken with the tips of pipettes; they are easily chipped, and the rate of flow may change considerably as a result.

The method used to clean glassware is always subject to personal preference. The removal of the film of grease that forms on glass walls is necessary for reasons both esthetic and practical; drainage is seriously impaired if the surface is not clean (Ref. 12, p. 502). A mixture of concentrated sulfuric and chromic acids will remove a film of grease, but will also leave an adsorbed film of chromium ion which is removed only with difficulty. A hot solution of sodium hydroxide and potassium permanganate will remove organic material resistant to the sulfuric–chromic acid mixture. Warm dilute nitric acid, hydrochloric acid, or *aqua regia* will remove stains, but the glass surface will likely become sensitized and will retain adsorbed ionic films. Detergents are also commonly used, and some laboratories find the use of automatic (and programmable) dishwashers to be efficient and satisfactory. Unless kept in sealed cupboards, glassware will accumulate a film of dust and grease on standing. It is advisable to wash glassware after use and to clean it immediately before putting it into use again.

The proper use of volumetric glassware is of critical importance in preparing reliable standards for AAS and other techniques. For example, it is mandatory to check the calibration of any new piece of volumetric glassware. A good account of glassware calibration is found in Ref. 10, pp. 135–152 and in Ref. 13, pp. 437–446. There is still a surprising number of analysts to be found who do not know the correct way to use a transfer pipette or a burette, and a useful description of these techniques is given in Ref. 14, pp. 238–243 and Ref. 10, pp. 139–144.

The need to extend the detection of trace elements to lower and lower limits has emphasized the importance of thorough cleaning of glassware and the realization that a glass surface, as mentioned above, will adsorb trace amounts of ions which are then available to contaminate subsequent analyses. It is advantageous to keep a separate stock of glassware for use in trace element determinations, and as many blank determinations as feasible should be included in each batch of analyses.

2.2.2 Porcelain

Porcelain does not find extensive use in rock and mineral analysis, but casseroles are used for evaporation and precipitates are occasionally ignited in porcelain crucibles. Porcelain is more resistant than glass to the action of basic solutions, provided that the inner glazed surface is in good condition, but it is not as resistant to acids. It can, of course, be used at much higher temperatures than glass, but care must be taken to avoid any interaction between the glazed surface and the crucible contents. Platinum crucibles are sometimes heated to a high temperature (above 1200°C) while resting in a porcelain crucible, and spot melting of the glaze can severely alter the weight of the platinum crucible. A common chemical reaction is that between the porcelain and the alkali salts in incompletely washed precipitates, producing a fused lump that is completely useless. When the inner glaze of the porcelain ware becomes dulled through chemical attack and physical abrasion, the porcelain should be discarded or reserved for less important use.

2.2.3 Plastics

The use of plastic ware in all kinds of laboratories has become very pronounced in the last couple of decades, but in spite of its many advantages, there are some disadvantages that also must be recognized. For example, some extruded plastics carry with them molybdenum contamination from the molybdenite used as a lubricant in the extruding machinery; zinc and barium have been found to leach from the walls of some types of plastic bottles, and titanium is a frequent contaminant introduced into analyses from the paper lining of plastic caps.

Polyethylene (polythene) is now used to fabricate a wide variety of chemically resistant common laboratory items that make the handling of solutions containing hydrofluoric acid much easier than was formerly possible. Ordinary polyethylene should not be heated much over 50–60°C, but an irradiated form is now available that will withstand heating to 100°C without deformation. Polyethylene wash bottles offer convenience of manipulation and close control of delivery; the fine jet is an easy means of filling semimicro burettes, for example, without the formation of trapped air bubbles. Polyethylene cover glasses, which can be easily made from the base of empty hydrofluoric acid bottles, give useful service at water bath temperatures. Concentrated nitric and sulfuric acids discolor polyethylene, and many organic solvents attack it. It also has a tendency to adsorb ions and to release them later on, and standard solutions that are stored in polyethylene for other than a short period should be restandardized before use. For other than standard solutions and those materials that attack it, polyethylene storage containers have many advantages. Polyethylene finds extensive use in providing unbreakable laboratory items, for example, graduated cylinders, beakers, and watch glasses. The beakers are widely used for hydrofluoric acid solutions such as those employed in the dissolution of borax fusion melts. Perspex and Lucite are trade names for polymethyl methacrylate, a transparent, hard, colorless plastic that is very useful for the fabrication of apparatus, and as safety shields and special hoods and boxes. It is easily machined, and the edges can be bonded together by various solvents.

Teflon has become a valuable aid to the analytical chemist because it is both strong and chemically inert to most laboratory reagents. In addition, it can be machined and used at temperatures up to 250°C. It has been found useful as a coating around magnetic stirring bars, as a liner for aluminum or stainless steel reaction bombs, as evaporating dishes, and as tubing in automatic dispensers and pipettors, to name only a few applications. It is often used as a substitute for platinum ware, but it is expensive and cannot be recycled or reformed.

2.2.4   Platinum

At one time the most expensive item in the equipping of a laboratory for rock and mineral analysis was platinum ware such as crucibles, dishes, tongs, and forceps. Today, while platinum and platinum alloys play an irreplaceable role as fusion vessels and moulds, the cost of an instrument such as an XRF spectrometer dwarfs any expenditure on ancillary platinum equipment.

The most commonly used fusion vessels and moulds are those made of a platinum–gold alloy which is nonwetting with respect to lithium metaborate and lithium tetraborate fusion melts. This characteristic facilitates the extraction of beads from moulds and the cleaning of crucibles between fusions and has the additional advantage of preventing loss of sample due to creeping.

Because of the routine use of platinum ware in rock and mineral analysis, its properties, use, and maintenance will be considered in detail.[15]

The following applies to equipment made from both pure platinum and platinum alloys.

Platinum, because of its softness, is easily distorted and damaged and requires more care than is usually given to it. It is easily scratched; glass rods should be used with caution in removing fusion cakes from crucibles, and a rubber policeman or a plastic rod is a better tool for such an operation. Any deformation caused by dropping or squeezing (especially by tongs; the sharp-nosed variety can produce flutings in the rim of a crucible or dish that are difficult to remove) should as soon as possible be corrected by reshaping; all platinum ware should be provided with a hardwood shaper of appropriate size and shape. If deformations are not corrected, they will become centers for the development of cracks. Platinum ware should always be kept clean and bright on both the inside and the outside (the cleaning process will be described later) as a protection for the considerable investment of money involved, and in order to ensure its being in proper condition for analytical use. Recrystallization resulting from prolonged heating at a high temperature will give the ware a greyish, speckled look and, unless removed by burnishing, will lead to the development of pinholes and cracks; this is particularly so on the outside of the crucible bottom which receives the full force of the heat, and which is too often ignored in the cleaning operation.

Prolonged heating of platinum at temperatures above 1200°C should be avoided because of the volatilization of platinum that will occur; when such heating is necessary, a correction for the loss in weight must be made. Oxidizing conditions should prevail at all times. At no time should the reducing portion of a gas flame be in contact with the platinum; acetylene is particularly bad in this respect, and one occasion is known in which a hole was rapidly made in the bottom of a 30 ml crucible unknowningly in contact with the sharp blue cone of an acetylene flame. Platinum ware should preferably be handled only with platinum-tipped tongs when hot (pure nickel will also serve) and should rest on supports of fused silica, clay, or platinum, never on iron or other metal supports. It is appropriate to comment here on one type of deformation that develops because of the unequal expansion of platinum during heating; expansion occurs more readily in the upper part of the crucible than in the lower part which is confined in the triangular support, leading to the development of a "cuspidor" shape that is most undesirable. Platinum basins and dishes should rest on porcelain or plastic-covered rings during the evaporation of solutions and not directly on copper or brass rings. Hot platinum ware should be cooled on a nonmetallic surface, such as a polished stone slab.

Although it is relatively chemically inert to most reagents and chemical compounds, platinum is attacked to some degree by a number of substances, and seriously by a few of them. Some precautions which should be taken during some common operations are as follows:

1. *Fusion.* Alkali borates (meta and tetra), borax, and boric acid may be fused at high temperatures with little attack on the container, but lithium carbonate cannot be so fused because it attacks and is adsorbed by the platinum,[11] alkali fluorides and bifluorides do not react, but pyrosulfate (and bisulfate) as well as sodium and potassium carbonate fusions will dissolve a small amount of platinum if the fusion is heated above about 700°C, or prolonged unduly. Nitrates, nitrites, and peroxides should not be fused alone, but a small amount mixed with sodium or potassium carbonates will do no harm. Alkali hydroxides, sulfides (tellurides and selenides), and cyanides must be avoided.

2. *Ignition.* The first rule is that an unknown substance should not be heated in platinum; the unsuspected presence of elements such as lead, antimony, or arsenic, especially if there are reducing substances present, can result in the ruination of a crucible or dish. The oxides, carbonates, sulfates, and oxalates of those elements which are not easily reduced, such as Si, Al, Ca, Mg, Mn, Ti, Zr, Th, and rare earths, may be safely ignited in platinum at reasonable temperatures (1200°C); if suitable precautions are taken to maintain an oxidizing atmosphere, the compounds of elements such as Fe, Ni, and Co may also be ignited safely. BaO and Li_2O attack platinum and should not be heated in it, nor should their compounds be heated in the presence of reducing substances. Filter papers should be burned off at a low temperature before the crucible is heated strongly; carbon, at high temperatures, will form platinum carbides and may cause embrittlement of the platinum. The paper should not be allowed to ignite; this not only creates undesirable drafts which may cause material to be lost from the crucible but also establishes a reducing atmosphere that may cause damage to the crucible, either from carbon monoxide or by reducing some substance, such as phosphate, arsenate, or antimonate, which will subsequently attack the platinum. The ignition of magnesium ammonium phosphate is particularly prone to trap carbon. Wyatt[10] points out that even silicates may be a source of trouble if heated at about 1000°C in the presence of sulfur and carbon, because brittle platinum silicide is formed. Volatile halides must also be avoided. The presence in a sample of as little as 1% sulfur as a sulfide, if fused in platinum with lithium borate salts without prior ignition, will cause serious etching of the crucible.

3. *Acid decomposition.* At normal temperatures, platinum is affected only by *aqua regia* and thus hydrochloric and nitric acids, or other

mixtures which liberate nascent chlorine, should not be used in platinum. If potassium nitrate is used as an oxidant in an alkali carbonate fusion, followed by dissolution of the cake in hydrochloric acid, the evaporation must not be done in platinum, or several milligrams of the basin will be added to the analysis; it is thus preferable to use sodium peroxide as the oxidant in the fusion so that platinum may then be used if it is desirable. At water bath temperature the only precaution necessary is to avoid the heating in platinum of those solutions that contain much ferric chloride (i.e., the HCl solution of the fusion cake from highly ferruginous material) because attack will occur. At high temperatures phosphoric and sulfuric acids will attack platinum, but only slightly, after prolonged heating; there is no appreciable attack during the relatively brief heating of sulfuric acid to fumes of SO_3 in the volatilization of silica with HF and H_2SO_4, and in the recovery of silica from the pyrosulfate fusion of the R_2O_3 mixed oxides.

The cleaning of platinum ware is accomplished by a combination of acid treatment, fusion, ignition, and burnishing. When dents and other deformations have been removed, the piece should be examined for stains. Digestion in hot concentrated HCl will remove most iron stains, although two or more such treatments, interspersed with ignitions (5–10 min) of the piece at red heat under oxidizing conditions, may be necessary. If the result is not entirely satisfactory, a brief fusion with a small amount of alkali bisulfate or pyrosulfate at a dull red heat is generally all that will be necessary, and the piece, after removal of the fusion cake, is ready for burnishing. Clean sand, not too coarse (which scratches the surface unnecessarily) nor too fine (which is not sufficiently effective) is used; it should be free from dirt and relatively free of other minerals, and the grains should be well rounded rather than angular. A well-sorted, clean sea sand is most useful. A small amount of sand is placed in the piece, moistened, and rubbed vigorously around the sides and over the bottom, holding the ware in such a way as to minimize deformation by squeezing; after 1 or 2 min of scouring, the sand is scraped into the palm of the hand and the piece is carefully rubbed over the sand, with particular attention being paid to the bottom. The lid is also scoured, again using the sand in the palm of the hand. Rubber policemen are sometimes used but should be avoided; the rubber is worn rapidly and exposure of the glass rod can lead to serious scratching of the platinum. The index finger is the most flexible tool for this work. When the burnishing is finished, the piece should be washed free from sand and then washed with soap and water to remove the film of "platinum black" that is present. Finally it is reshaped and ignited at red heat for about 5 min, cooled, and examined for the absence of stains and recrystallization. New platinum ware should be ignited and examined for iron stains, for iron is often present as a surface contamination; it is a useful precaution to digest new ware in warm, concentrated hydrochloric acid before its initial use.

Platinum-tipped tongs are cleaned by rubbing the platinum feet briskly against sand held in the palm of the hand; the metal legs are cleaned with emery paper. The tongs should be cleaned when they become discolored and always after they have been used to hold a crucible or dish during a pyrosulfate fusion; during the latter operation there is always a transfer of a trace of the fusion mixture to the tongs, and this could be a serious source of contamination in future operations.

2.2.5   Other Materials

The only other materials that have significant application in rock and mineral analysis are silver, nickel, and iron, in the form of crucibles for fusions. The melting point of silver is too low (960°C) to permit its use at high temperatures, but it is suitable for fusions with sodium and potassium hydroxides, having the advantage over nickel that there is only a slight attack on the crucible; any silver dissolved is easily precipitated and removed when the fusion cake is dissolved in dilute hydrochloric acid. There is a tendency for the contents to creep, and if the fusions are done in a muffle furnace, they must be handled with care in removal from the furnace to avoid loss of material. Shell[16] prefers to use silver crucibles for the fusion of silicates on which a precise determination of iron is to be made, because crucibles of Pt or Pt–Rh retain an indefinite amount of iron which can only be recovered by repeated ignition of the crucible under oxidizing conditions and leaching in hot acid.

Nickel crucibles have a relatively long life when used for alkali hydroxide fusions, and the amount of nickel that is introduced into the solution is not usually objectionable. They are also superior to iron crucibles for fusions with sodium peroxide. Iron crucibles have the advantage of low cost and, where the introduction of iron is of no consequence, are widely used for fusions with sodium peroxide.

2.2.6   Heating Equipment

There are several different types of heating equipment required during the course of an analysis. The sample must first be dried, usually at 105–110°C, in a drying oven equipped with a temperature control; the oven should be capable of giving temperatures up to about 250°C in order to handle certain hydrous minerals which yield their water (water of constitution as well as moisture) only at higher temperatures. Evaporations are usually carried out on a steam or water bath, and little attention need be given to them during the course of the evaporation; the bath should be capable of 24 hr operation without danger of becoming dry, so that such evaporations may be made overnight. Baths with either four or six holes can be accommodated in a 2 m wide fume hood; little use is made now of single-hole water baths. A convenient and effective substitute for steam or water baths, when solutions are to be evaporated in dishes, is provided by tall form beakers of 200 ml capacity resting on a hot plate; the beakers are half-filled with water, a few carborundum boiling chips are added, and the dishes are supported directly on the rim of the beaker, or on rings. Infrared heating lamps, either with internal reflectors or inserted in reflectors, are a rapid and efficient means of evaporation; the combination of steam bath, water bath, sand bath, or, less preferably, hot plate and the heat lamp will considerably hasten the evaporation. Surface heating is less subject to bumping than is the conventional bottom heating, and more use should be made of it. It is necessary that the lamp and reflector be wiped free of accumulated deposits before use.

Hot plates, which usually can be operated up to about 250°C if electrically heated, and to high temperatures if gas heated, find a multitude of uses. Their surface is not usually completely smooth, however, and thus uneven heating of vessels will occur, leading to the danger of loss of material by bumping. Because they tend to heat unevenly, the analyst must get to know the vagaries of a particular hot plate and be able to select areas most suited to specific needs. Squares of thin asbestos paper help to produce a finer gradation of temperature but may cause harm to the hot plate because of the development of hot spots beneath the insulating paper.

Sand baths, heated either by a gas burner or, preferably, by a hot plate, offer some advantages over the use of a hot plate. The temperature is uniformly distributed throughout the sand, and the loose nature of the sand enables crucibles and dishes to be nested securely, at the same time subjecting a larger area to the heat. In the GSC laboratories it has been found that enamel photographic trays, approximately $45 \times 24 \times 5$ cm and two-thirds filled with coarse sand, are very convenient receptacles for up to 12 50-ml platinum dishes; the tray is placed on a hot plate, and is easily removed when the hot plate is needed for another purpose. Care must be taken to remove any adhering sand from the bottom of the crucibles or dishes so heated.

The most useful and versatile source of heat in the laboratory is the gas burner, and it is preferable to a hot plate when temperature must be carefully controlled. The Tirrill burner, with its gas regulatory needle valve, is very useful for low temperature work (boiling of solutions, rapid evaporations, initial charring of papers, and some ignitions) while the Meker, or Fisher, type of burner is more suitable for higher temperatures. The Meker burner presents little hazard for the heating of platinum ware but, as pointed out previously, at no time must the inner blue cone of the Tirrill burner come into contact with the platinum. Earthenware chimneys, supported by steel tripods, serve both as a support for the triangles used to hold crucibles and as a shield against drafts. Metal evaporators,[12] supported by iron rings, are efficient air baths for evaporating the contents of crucibles and small dishes, but they offer no particular advantage over a sand bath.

A multiburner device is now available which will hold up to six crucibles and moulds for preparing fused sample discs for XRF analysis (Fig. 2.1). It is designed to gyrate in such a way that the samples are stirred vigorously and continuously during the fusion with the result that the time required for a fusion is reduced to a total of about 8 min or less. When the fusion is completed, a lever is turned which inverts the crucibles, and the melts pour into moulds which swing into place as the crucibles invert. The discs formed are durable and of a high quality and can be measured on the spectrometer without the need for polishing. In fairness it must be admitted that a certain amount of personal skill is required to ensure the production of "good" discs.

Heating at a high temperature is best done in an electrically heated muffle furnace which will give temperatures of 1200°C or more, sufficient to complete the heating of hydrous alumina to constant weight. Such furnaces, preferably large enough to take at least four 30 ml crucibles, give reasonably uniform heating conditions and freedom from reducible substances such as one encounters in the gas flame. It must be borne in mind, however, that the temperature indicated by the pyrometer scale is that existing at the tip of the thermocouple, and there may be a difference of as much as 100–200°C between the temperatures at the front and the back of the furnace. There must also be access for air in order that oxidizing conditions will prevail. The heating elements must *not* be exposed; even when covered they contribute a fine metallic dust to the inside of the heating chamber, and crucibles should not be placed on the floor of the chamber but rather on a removable refractory plate or tray which is cleaned before use. The charring of filter paper prior to the ignition of a precipitate is conveniently done by starting with a cold muffle furnace and allowing it to heat slowly; when more than one crucible is placed in the muffle, however, the crucible covers should be three-quarters in place to prevent both the contamination of the contents of one crucible by those of another, and the loss of material because of the drafts

Figure 2.1. Claisse Fluxer used for fusing samples and casting them into discs suitable for subsequent analysis by XRF. Photograph by R. Player, BCM.

created by the combustion. Many electrical muffles now available are controlled by solid state circuitry to within ± 10°C of a temperature set on a digital dial. Also available are furnaces capable of rapid heating, reaching 1100°C in under 15 min.

Blast lamps find some use in rock and mineral analysis, but are unnecessary if a muffle furnace capable of giving temperatures up to 1200°C or better is available. A glass blower's oxygen-gas hand torch is a very convenient and easily controlled source of high temperatures, such as is needed to complete the final heating of the glass bulb in the Penfield method for the determination of total water.

Some mention should be made of the types of tongs that will be found useful. For handling hot platinum ware it is essential that the tongs be equipped with platinum feet or shoes (pure nickel is an acceptable substitute but the platinum is better); by inserting suitable pieces of stainless steel between the lower and upper parts of the tongs they can be increased to 45 cm or more, a convenient length for use with a muffle furnace. Sharp-nosed tongs are very useful for gripping dishes and crucibles, but if a lid is present it must first be removed; platinum claws, capable of grasping both 25 and 30 ml crucibles and fitted to the lengthened tongs previously described, will facilitate the introduction and removal of crucibles during heating without the need for displacement of the lids. Care is necessary when the sharp-nosed tongs are used; if they are squeezed too tightly the effect is rather to open the jaws and crucibles or dishes may be dropped.

2.2.7 Filter Papers

There are various brands of filter paper on the market, each available in several grades which are usually comparable in thickness, porosity, and wet strength. Of these the Whatman and Schleicher and Schuell (S & S) papers are probably the best known, and they will be used to illustrate the size and the grade of paper when needed. Detailed discussions of these and other papers are given in several textbooks[10–12] to which reference should be made.

Qualitative papers have no application in rock and mineral analysis. The quantitative papers used must be of the "ashless" variety (ash content generally less than 0.1 mg per 9 cm circle), papers that have been washed with both hydrochloric and hydrofluoric acids; hardened quantitative papers, which have a greater wet strength and increased rapidity of filtration, find only occasional usage. For coarse precipitates, such as separated silica, the Whatman No. 41 (S & S 589 Black Ribbon) is most useful. Medium-grained precipitates such as calcium oxalate are best filtered through Whatman No. 40 (S & S 589 White Ribbon), while the fine-grained precipitates such as barium sulfate require the use of Whatman No. 42 (S & S 589 Blue Ribbon), or even of Whatman No. 44 (S & S 589 Red Ribbon).

The diameter of the filter paper for a filtration should be governed by the volume of the precipitate to be retained by it, since the precipitate should not occupy more than about one-half of the volume of the filter paper cone. If a paper larger than necessary is used, it requires additional washing to free the entrained soluble matter, but there is nothing more frustrating than a filter paper which turns out to be too small for the job! Paper circles are available in diameters ranging from 4.25 to 15 cm; the 7, 9, and 11 cm circles will be found most useful, with the 9 cm diameter paper being the most readily adaptable.

The use of filter paper pulp is recommended by some analysts as an aid in promoting more rapid and efficient filtration. In the double precipitation of the R_2O_3 group elements its addition, as the macerated paper used in the first precipitation, is unavoidable, and it does ensure a more easily filtered precipitate. The indiscriminate use of paper pulp is not justified, however, because the improved filtering and ignition qualities that it imparts to the precipitate are offset by certain disadvantages. It contributes to the final weight of the ignited precipitate and, if a correction must be made for it, the weight of the paper pulp used must be known; it also presents an additional surface for the trapping of foreign ions and necessitates additional washing, which may be undesirable if the precipitate is more than slightly soluble in the wash solution. The presence of paper pulp, intimately mixed into the precipitate, may also foster the development of the undesirable reducing conditions mentioned previously, because of the difficulty of ensuring that all of the organic material is charred at a low temperature. In hot acid solutions a degradation of the paper can occur, and some elements may, as a result, be reduced to lower valence numbers; the subsequent precipitation of these elements may be interfered with, or rendered incomplete. There are other ways of improving the filtering qualities of precipitates, and the use of paper pulp should be a last resort.

2.2.8 Specialized Items

There are many minor pieces of specialized equipment which serve a very useful purpose. Because they may prove of interest to other analysts, a few such items are described here.

Long forceps, approximately 25 cm in length and made of nickel-plated steel, are a convenient means of handling weighing dishes and crucibles without contamination from the fingers, and they are not as heavy nor as clumsy as crucible tongs. Smaller forceps, 15 cm in length and fitted with platinum tips, are used to handle filter papers for transfer to crucibles. If the semicircular jaws of light wire tongs (a variety which opens when squeezed) are bent at right angles to the legs, and the wire jaws are covered with thin rubber tubing, these will be found very useful for removing crucibles from desiccators and balance pans; regrettably such simple tongs are also very difficult to obtain.

Wooden sample trays, approximately 30 × 22 × 2.5 cm, with holes of the appropriate size drilled to a depth of about 2 cm, will keep sample vials in order and facilitate their movement in the balance room, particularly for batch analysis. Temporary gummed labels on the trays will help to identify batches of samples at a glance. Fiberglass cafeteria trays are ideal for carrying large numbers of beakers, bottles, and other items. They have the advantage of being resistant to most mineral acids (but not to hot crucibles nor organic solvents), and they are conveniently cleaned and stored. A stainless steel trolley with two shelves is ideal for transporting racks of samples to and from equipment such as dilutors and atomic absorption spectrometers.

Several new devices for pipetting both micro and macro amounts of solutions are available, and they are a very welcome addition to an analyst's equipment. They have disposable tips so that cross contamination is avoided and the delivery precision is advertised to be $\pm 1\%$. The introduction of dispensers and pipettors that can be used to dispense reagents is equally welcome. Some have the disadvantage of using glass bores and glass cylinders that need lubricating. They quite often freeze into one position when used for dispensing mineral acids and are difficult to loosen. Regular lubrication and cleaning can prevent this, however, and the facility with which they dispense preset aliquots of reagents is very time-saving. Automatic dilutors are available which take an aliquot of a sample or reagent and then discharge it together with enough diluent to deliver a solution having a preset dilution factor. The precision is about $\pm 1\%$ in most models.

In the Pratt method for the determination of ferrous iron the sample is decomposed in a platinum crucible with a mixture of sulfuric and hydrofluoric acids kept at a gentle boil for 5–10 min. This requires a very small flame to avoid undesirable bumping or too rapid boiling and, since the decomposition must be done in a fume hood, the flame must be protected from drafts. A conical chimney fastened to the burner will give some protection, but at the GSC a simple device is used which not only allows close control of the flame and protects it from drafts, but also provides a safe support for the crucible, a necessary requirement in view of the hazardous nature of its contents. The burner housing and crucible support stand, shown in Figs. 2.2 and 2.3, consists of a small but heavy plywood box, $20 \times 18 \times 18$ cm, having one side missing; it is lined with transite and covered with a sheet of transite in which is cut a 9 cm hole. A silica or vitreosil triangle, large enough to support a 45 ml tall form platinum crucible, is fastened securely over this hole. A micro gas burner, with attached chimney, is fastened to the base of the box beneath the central hole and is connected to a gas source by means of a needle valve inserted through a hole in one side of the box; a Tirrill burner, used for the initial heating of the crucible, can be connected by a Y joint to the same gas source.

A special support stand for use in the determination of total water by the Penfield method was designed by Courville.[17] The details of the sup-

Figure 2.2. The apparatus used for the determination of ferrous iron (modified Pratt method) by potentiometric titration. The decomposition of the sample is done in a fume hood, using the special burner housing and crucible support stand shown at the right. Photograph by Photographic Section, GSC.

port stand are shown in Fig. 2.4. It consists of a plywood base and tray holder which support a plastic overflow dish and metal cooling tray; the glass tube for the water determination rests, at a slight downward inclination, in slots cut into the sides and is supported just above the surface of the water–ice mixture in the cooling tray. A transite shield is fastened to one side to serve as a heat reflector.

2.3 MISCELLANEOUS INSTRUMENTATION

It is possible, given the necessary time, to do first class analytical work with little more in the way of instrumentation than a photoelectric colorimeter, but when the rapidity with which the analysis can be made become important, then the use of more complex instrumentation is indicated. In general, the capability of an instrument to perform a particular

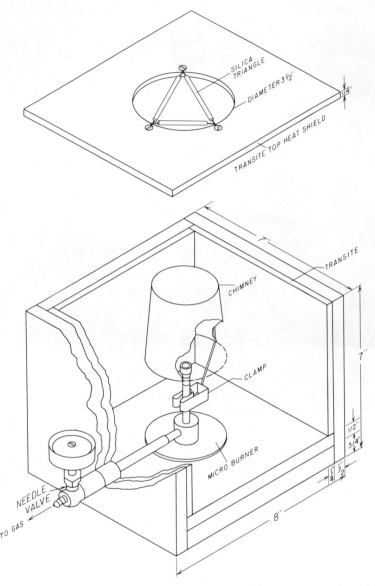

Figure 2.3. Diagram of special burner housing and crucible support stand used in the decomposition of the sample for the ferrous iron determination.

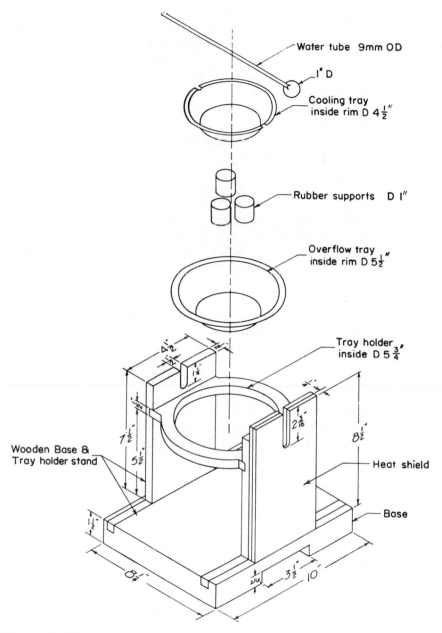

Water tube 9mm OD

1″ D

Cooling tray
inside rim D $4\frac{1}{2}$″

Rubber supports D 1″

Overflow tray
inside rim D $5\frac{1}{2}$″

Tray holder
inside D $5\frac{3}{4}$″

Wooden Base &
Tray holder stand

Heat shield

Base

Figure 2.4. Diagrammatic view of special support stand for total H_2O determination. Reproduced with permission of *The Canadian Mineralogist*.

function will exceed that of the hand or eye, and the analyst is enabled thereby to make measurements that otherwise would be impossible without lengthy separations and purifications. Provided that such instruments are recognized as being no more than laboratory tools able to do only the jobs that they are set to do, then the modern rock and mineral analyst has available a host of laboratory helpers.

The need for many more results and the increasing costs of labor have led to the development of many automatic and time-saving features.[18] This trend is evident in even the most basic analytical procedures such as weighing, pipetting, and diluting.

2.3.1 Balances

No laboratory can function without analytical balances for both rough and precision weighing. Many new models and innovations are being introduced annually,[19] and any detailed description of the latest developments is out of date almost immediately. A recent introduction is that of the single pan torsion balance having a sensitivity down to 0.1 mg. These balances are very tolerant of vibrations and for this reason have found application on oceanic survey vessels. Automation has been introduced into such features as taring and digital readout. A computer- and printer-interfacing compatability has also been introduced into balances in the last few years to permit the automation of the analytical sequence and the computer treatment of data. The cost of balances having some or all of these features is very high, but in certain situations the labor-saving advantages make them economical to use.

The balances found in the majority of laboratories involved in rock and mineral analysis are of the single pan type supplied by Sartorius, Mettler and Oertling, among others. Each laboratory should have at least one balance with a loading capacity of 100 g and a sensitivity of 0.05 mg per scale division, as well as one or more with a loading capacity of 200 g and a sensitivity of 0.2 mg per scale division. A microbalance is a necessity for semimicro- and microchemical work with samples weighing 100 mg or less. A top loading balance with a capacity of 1 kg and a sensitivity of 100 mg per scale division is useful for weighing large samples and reagents. It often transpires that, for trace element determinations, a sample need only be weighed to the nearest milligram. Thus a top loading balance of the type having a cover for protection against drafts can be used with a considerable saving in time. An additional advantage is that they are usually much less expensive than the more precise single pan analytical balances mentioned previously.

2.3.2   Spectrophotometers

For absorptiometric measurements there is need for a spectrophoto-meter which will enable measurements to be made routinely in the ultra-violet and visible regions, that is, in the range from about 200 to 1000 nm. The available instruments range from the simple to the complex in operation and capability; there is an advantage in having a less expensive instrument for routine absorptiometric measurement and one capable of greater resolution and precision for special analysis and for method development work. A recording attachment which will provide a visual spectral record is a very useful addition to this instrument. Many instruments are available now with built-in recorders. Automation has been introduced into this field through the development of the autoanalyzer which has seen widespread use in the cement industry for the determination of Si, Al, Fe, Mg, Cu, and S (W. F. Mivelaz[18]). It is also extensively used in pharmacological, environmental and agricultural laboratories. An atomic absorption spectrometer can also be used for absorptiometric measurements, and it is possible to obtain an absorption cell holder that fits into the light path in the place of the burner head.

The Pyrex or fused silica cells used for absorbance measurements require careful handling in order to avoid scratching the optical surfaces, and they should, of course, be kept free from surface films and fingerprints. Unless matched cells are available, however, it is unlikely that the absorbance of distilled water will be the same for each cell, and it is necessary for accurate work to determine a correction for each cell, relative to one cell in which the absorbance is arbitrarily considered to be zero. Because wear is inevitable, the cells should be numbered and the relative cell correction checked each time that the cells are used. When not in use, the cells should be kept immersed in a beaker of water containing a mild detergent. Before use they should be rinsed in distilled water, the outside surfaces patted dry with absorbent tissue, and the optical surfaces polished with lens paper. Once the cells are placed in the cell holder, they should not be removed until the series of absorbance measurements is finished; a pipette, connected to a water aspirator pump, can be used to remove the solutions and water rinses from the cells.

2.3.3   pH Meters

A pH meter finds limited application in rock and mineral analysis for the measurement of pH, but it can serve usefully as a potentiometer in potentiometric titrations and is directly applicable to ion-selective electrode techniques for the determination of F and Cl. There are several types available, but it is preferable to acquire one that has both pH and millivolt scales, and a scale expansion feature. Some models are equipped with a digital readout capability and can be interfaced to an automatic printer.

2.3.4   Fluorimeters and Electroanalytical Instruments

Other instruments which have been used to a varying degree in the determination of some elements are the fluorimeter, the polarograph, and the anodic stripping voltammeter. The fluorimeter is used for the determination of trace quantities of uranium and beryllium; the preparation of the fused beads for final measurement is empirical in nature and requires a special burner and quenching apparatus. The polarograph has also been used chiefly for trace element analysis, but methods for the determination of major constituents such as titanium and aluminum have been developed. Anodic stripping voltammetry is a trace element method capable of detecting concentrations as low as a few parts per billion, but there are many interferences due to the complex nature of most rock and mineral samples.

2.3.5   Minicomputers and Microprocessors

The dramatic drop in the price of computer components over the last few years has made the introduction of minicomputers and microprocessors into the laboratory increasingly common,[20, 21] either as a peripheral (Fig. 2.5) or as an integral (Fig. 2.6) component of an instrument. They are capable of monitoring and adjusting virtually any parameter of instrument operation, such as setting the slit width on an atomic absorption unit, changing the analyzing crystal on an XRF spectrometer, or collecting voltage data from any number of photosensitive detectors on a direct reading optical emission spectrometer. They can, at the same time, store calibration and measurement information, perform fairly complex calculations involving corrections (as in correcting for matrix effects in XRF analysis), and print out the concentrations of the analytes in the samples without further attention being required on the part of the analyst.

The use of computers in these various applications will be discussed in more detail in later sections, but it must be pointed out that they can be a mixed blessing. The analyst must take every precaution to ensure that the problems which are routinely encountered when operating the instruments in a manual mode are either corrected automatically or else brought to attention so that the appropriate corrections can be made.

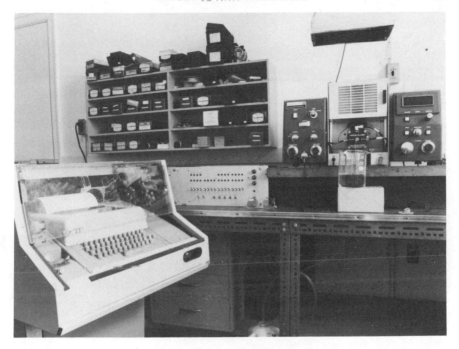

Figure 2.5. Atomic absorption spectrometer interfaced to an external microprocessor and teletype unit. Photograph by R. Player, BCM.

2.4 REAGENTS AND STANDARDS

No matter what analytical method is chosen by an analyst for a particular determination, the analyst must be concerned with the purity of the reagents to be used and the quality of the standards for the calibration of the instruments employed. The more dependent analysts become on instrumental methods, particularly those that are automated, the more critical they have to be about the reagents and standards chosen.

2.4.1 Reagents

The number of reagents used in the analysis of a rock or mineral drops significantly when the classical chemical techniques are replaced by an instrumental technique such as XRF. A knowledge of the purity of the reagents used is, however, critical in either case, and failure to give sufficient attention to such information can be the source of large errors, particularly in the determination of trace elements.

Figure 2.6. Atomic absorption spectrometer incorporating a microprocessor, keyboard, and cathode ray tube display. Reproduced with permission of Instrumentation Laboratory, Inc.

There are two grades of reagents that are of particular importance. The first, the primary standard reagents, are of invaluable service in many ways and are supplied with a complete statement of the analysis of each bulk lot; the analyst generally can accept these statements at face value. The other type is the analyzed reagent; these are supplied with either an actual analysis of the bulk lot, or with a specification giving the maximum limits of impurities that may be expected. Other grades of reagent will likely require preliminary purification, depending on the use to which they are to be put, although there are numerous occasions when the use of a technical grade chemical is entirely acceptable.

Of the common reagents needed in rock and mineral analysis, only one really requires special preparation. Aqueous ammonia (ammonium hydroxide), as commercially supplied in glass bottles, has a high reagent blank,[22] and its use should be restricted to those occasions when its impurities will either be of no consequence or will be eliminated by subsequent steps in the analysis. Aqueous ammonia, in any concentration up to 15 M, is easily and readily prepared by bubbling ammonia gas through distilled and deionized water in a polyethylene bottle immersed in an ice water bath. Saturation of the solution, indicated by a steady stream of bubbles rising to the surface, occurs when the concentration is about 15 M.

Apart from chemical impurities, one must guard against solid impurities such as wood, dust, and paper fibers introduced during the manufacturing processes, and solutions prepared from solid reagents should be filtered before use. At the same time, care should be taken in the use of spatulas and scoops to remove reagents from bottles, for hygroscopic reagents will attack glass and metal, and hard, caked reagents will abrade them.

A special warning should be given about the use of organic reagents, because their purity is usually much inferior to that of the inorganic reagents commonly used, and is often surprisingly low. It is well to check the purity of a new supply of reagent before using it (and preferably *before* the previous satisfactory supply is exhausted); because organic reagents are subject to deterioration on standing, periodic checks should be made.

The use of alkali metal borate fluxes for the fusion of samples preparatory to subsequent analysis is now widespread. Lithium metaborate and tetraborate are the compounds most commonly used. This has resulted in the marketing of fluxes of much higher purity, but it is still advisable to keep on hand a stock of reagents from a single batch, sufficient to last at least two years, especially if the flux is to be used in the determination of trace elements. Very useful guides to reagent chemicals are put out by the American Chemical Society[23] and by Analar Standards Ltd.[24]

2.4.2 Standards

Any measurement is meaningless unless it can be compared to a standard. An example of a standard is the commonly used *meter*, for which " . . . It has been internationally agreed (1961) that the m. be defined as 1,650,763.73 times the wavelength in vacuo of the orange-red radiation corresponding with the transition between the energy levels $2p_{10}$ and $5d_5$ of the Kr^{86} atom . . . "[25] Governments have recognized the importance of standards to trade and commerce and have controlled them by statutes and by other means ever since ancient times. Analytical chemists are no less dependent on adequate standards than are others, and this has led to the formation of such bodies as the independent Bureau of Analyzed Samples Ltd. in England and the government-supported National Bureau of Standards in the United States. These two organizations are concerned with the standards requirements of a wide variety of chemists, and it has become necessary for those analysts concerned with rock and mineral analysis to establish a set of reference materials adequate for their own purposes. As was discussed in Chapter 1, G-1 and W-1 were the forerunners of such programs. The lessons learned in the collecting, crushing, mixing, splitting, distributing, and data-collecting aspects of the G-1 and W-1 samples were used to refine the process for G-2 and subsequent samples.[26]

This set of samples, and others like them, are very important to any rock and mineral analyst as a source of reference materials on which to base an evaluation of a new method, an instrument, or even a new analyst. The data which have been collected in these programs are not of consistent quality, however, and the analyst must be aware of their limitations.[26-31] Because these reference materials are so important, the values recommended for them and their sources are listed in Appendix 1. In spite of the large number of results collected for the trace elements found in these samples, there remain large gaps in the commonly required concentration ranges for many elements. For some such as Se and Te, for example (Ref. 29, p. 2), only a few scattered results are available for a limited number of the samples. In addition, not all of the results are dependable, and a means of establishing the quality of published values is being sought by many workers.[32-35] A very complete compilation of reference materials available and the present status of their application has been published by Koch.[36]

Optical emission spectroscopists have other sources of useful standards for their work. One such example is the set of standard powder mixtures designed as semiquantitative reference materials which is available from Johnson Matthey Chemical Ltd. under their registered trade name Spectromel. Microprobe spectroscopists are not as fortunate, and it is difficult for them to find suitable standard materials. This is especially true in the quantitative analysis of mineral grains by XRF.

The need for concern for the tools of the measurement system, which includes standards as well as the other items discussed in this chapter, is stressed by Amore.[37] Emphasis upon quality assurance through good analytical practices is the best method for improving the reliability of analytical results.

References

1. H. Bennett and R. A. Reed, *Chemical Methods of Silicate Analysis—A Handbook,* Academic, London, 1971.

2. W. F. Hillebrand, G. E. F. Lundell, H. A. Bright, and J. I. Hoffman, *Applied Inorganic Analysis,* 2nd ed., Wiley, New York, 1953.

3. J. A. Maxwell, *Rock and Mineral Analysis,* Interscience, New York, 1968.

4. P. G. Jeffery, *Chemical Methods of Rock Analysis*, 2nd ed., Pergamon, Oxford, 1975.

5. W. R. Ferguson, *Practical Laboratory Planning*, Halsted Press, New York, 1973.

6. M. A. Reynolds, N. L. Turner, J. C. Hurgeton, M. F. Barbee, D. A. Flory, and B. R. Simoneit, *Analytical Methods Developed for Applications to Lunar Sample Analyses*, American Society for Testing and Materials, Philadelphia, PA, ASTM STP 539, 1973.

7. W. Handley, *Industrial Safety Handbook*, McGraw-Hill, Maidenhead, 1969.

8. Manufacturing Chemists Association, *Guide for Safety in the Chemical Laboratory*, Van Nostrand, New York, 1972.

9. W. M. Johnson, L. E. Sheppard, and D. N. Beaton, *Can. Jor. Earth Sci.*, **14**, 2422 (1977).

10. G. H. Wyatt (p. 13), F. E. Beamish and W. A. E. McBryde (p. 431), in C. L. Wilson and D. W. Wilson, Eds., *Comprehensive Analytical Chemistry*, Vol. 1A, Van Nostrand, New York, 1959.

11. C. R. N. Strouts, H. N. Wilson, and R. T. Parry-Jones, *Chemical Analysis; the Working Tools*, Vol. I, Clarendon Press, Oxford, 1962, pp. 78–109.

12. I. M. Kolthoff and E. B. Sandell, *Textbook of Quantitative Inorganic Analysis*, 3rd ed., Macmillan, New York, 1952, pp. 183–201, 502.

13. F. P. Treadwell and W. T. Hall, *Analytical Chemistry*, Vol. 2, 9th ed., Wiley, New York, 1951.

14. D. A. Skoog and D. M. West, *Fundamentals of Analytical Chemistry*, Holt, Rinehart & Winston, New York, 1963.

15. H. A. Foner, *Lab. Practice*, 944 (1965).

16. H. R. Shell, *Anal. Chem.*, **22**, 326 (1950).

17. S. Courville, *Can. Mineral.*, **7**, 326 (1962).

18. *Advances in Automated Analysis* (Technicon Int. Congr., 1970), Vol. II, *Industrial Analysis*, Futura Publishing Co., Mount Kisco, NY, 1972.

19. R. O. Leonard, *Anal. Chem.*, **48**, 879A (1976).

20. Q. Bristow, *J. Geochem. Explor.*, **4**, 371 (1975).

21. J. W. Frazer and F. W. Kunz, Eds., *Computerized Laboratory Systems*, American Society for Testing and Materials, ASTM STP 578, 1975.

22. F. J. Langmyhr and J. T. Håkedal, *Anal. Chim. Acta*, **83**, 127 (1976).

23. *Reagent Chemicals, American Chemical Society Specifications*, 5th ed., American Chemical Society, Washington, DC, 1974.

24. *"Analar" Standards for Laboratory Chemicals*, 6th ed., Analar Standards Ltd., London, 1967.

25. J. Grant, Ed., *Hackh's Chemical Dictionary*, McGraw-Hill, New York, 1972.

26. F. J. Flanagan, *Geochim. Cosmochim. Acta*, **33**, 81 (1969).

27. R. Sutarno and G. H. Faye, *Talanta*, **22**, 675 (1975).

28. G. H. Faye, *Chem. Geol.*, **15**, 235 (1975).
29. S. Abbey, "Studies in 'Standard Samples' for Use in the General Analysis of Silicate Rocks and Minerals, Part 6," 1979 ed. of "Usable" Values, Geological Survey of Canada, Paper 80-14, 1980.
30. F. J. Flanagan, *Geochim. Cosmochim. Acta,* **37,** 1189 (1973).
31. F. J. Flanagan, U.S. Geological Survey, Prof. Paper 1155, 1980.
32. F. Bastenaire, *Geostand. Newsl.,* **3,** 115 (1979).
33. S. Abbey, R. A. Meeds, and P. G. Bélanger, *Geostand. Newsl.,* **3,** 121 (1979).
34. M. Sankar Das, *Geostand. Newsl.,* **3,** 199 (1979).
35. A. L. Wilson, *Analyst,* **104,** 1237 (1979).
36. O. G. Koch, *Pure Appl. Chem.,* **50,** 1531 (1978).
37. F. Amore, *Anal. Chem.,* **51,** 1105 A (1979).

CHAPTER 3

SAMPLING AND SAMPLE PREPARATION

3.1   GENERAL

The observation is often made that an analysis is no better than the sample that it represents. This relationship is, unfortunately, too often ignored, and much time and labor are expended upon samples not worthy of the effort. Conversely, conclusions are often drawn from these analyses that are not warranted by the nature of the samples that they represent, thus adding an often incorrect interpretation to unnecessary work. It is incumbent upon the analyst to ensure that the sample taken is suitable for the analytical work requested; the nature of the sample is dictated in part by the expected degree of accuracy of the desired analytical information, and the analyst must be prepared to interpret this to the geologist, and to give guidance where it is needed.

Only scant attention has been paid to the importance in rock analysis of proper sampling and sample preparation, other than qualitative discussions of inherent difficulties, a state of affairs that still exists today. This is in contrast to the proper sampling and preparation of ore samples which is discussed in some detail in many publications.[1-5] The detailed discussions of ore and shipment sampling reflect the economic importance of taking a truly representative sample. Such samples will decide, for example, whether or not mining is started in a particular area, or how much is to be paid for a shipment of ore or concentrate. There is some indication that more attention is now being paid to sample taking and sample preparation for rock and mineral analysis,[6-15] but much more is needed, especially with respect to collecting samples in the field. If the geologist has not collected a sample that is truly representative of the outcrop or geological unit of interest, then even the best analytical work will give data that are open to misinterpretation.

Until recently the number of rock analyses made was dictated largely by the time and expense involved, and the samples submitted for analysis were frugally selected. The advent of more rapid methods of analysis has reduced the need for careful selection on the grounds of frugality. Instead of a single outcrop, large areas are subjected to detailed chemical study, and the scope of interpretation of the analyses has broadened in consequence. With increased opportunity for the accumulation of data has come, however, an increased need to select proper samples and to apply the analytical data in a sound manner. This has led to a notable emphasis on the application of statistics to geology, in particular to geochemical studies. Sampling, analyzing, and comparing results are considered to be three basic procedures that are governed fundamentally by statistical theory,[16, 17] and the increased accumulation of data that now appears to characterize present-day studies is a response to this trend and its need for abundant data to satisfy statistical requirements.

The intent of this chapter is to draw attention to the many problems involved in the process of securing a proper sample for analysis and to the means whereby these problems may be resolved or minimized. Discussions of sampling and sample preparation are included in most publications dealing with methods of rock and mineral analysis.[18-20] Texts on geochemistry and geochemical prospecting also give some attention to sampling.[9, 21-24] Some mathematical considerations of errors introduced when samples are taken from heterogeneous powders for both major and trace element determination have appeared in the literature, but much work remains to be done.[8, 10, 12, 15, 25, 26]

The sampling of minerals presents a different problem, except for the few occasions when large deposits of such minerals as barite, gypsum, and clays are to be examined. A mineral sample is usually obtained by concentration from a host rock, and the quantity of material available for analysis is governed by both the percentage of the mineral in the rock and the ease with which it can be concentrated and purified (see Sec. 3.5).

The rock and mineral analyst is very often confronted with samples of stream sediments and soils about which there is an entirely different philosophy of sampling. This type of sampling is designed to isolate areas of anomalous element concentration in a much larger geographical setting. For this reason, relative rather than absolute element concentrations are important and sampling techniques are designed accordingly.[9, 21, 24, 27]

3.2   SAMPLING

The process of sampling is often affected by factors that have nothing to do with the securing of a statistically sound sample. As mentioned previously, the availability of analytical services or facilities will influence the number of samples that are taken for analysis. If only a few analyses can be made, the geologist must select those samples that illustrate certain specific details, and local, rather than general, considerations will be paramount. If, on the other hand, the opportunities for analysis are favorable, then the conditions for sampling may not be so and the number of samples to be taken may have to be governed by such considerations as the availability of outcrops, the difficulty in securing them, and the distance and effort required to transport them. If we assume, however, a set of circumstances in which the influence of these factors is not significant, then the problem becomes one of where to take the sample and how many and how much of each.

3.2.1 Homogeneity and Grain Size

The degree of homogeneity of a rock mass is an obvious first consideration, but Laffitte[28] and Grillot[29] have also drawn attention to the importance of grain size as a factor in sample selection, a consideration discussed much earlier by Larsen.[30] Their conclusions as to the size of sample necessary to give significance to the analysis must be startling indeed to those who are content with a single hand specimen broken from an outcrop that is conveniently available and easily sampled. Conners and Meyers[23] suggest two sequential sampling stages, the first of which is designed to determine the extent of compositional variation, and the second is planned according to the geochemical variation found as a result of the first stage studies.

Laffitte[28] takes as an example the sampling of a gneiss containing crystals of microcline, from which a series of 100 samples of 1 kg each is taken from the top to the bottom of the mass. The samples are crushed to pieces approximately 1 cm^3 in size, mixed thoroughly, and a representative sample of 1 kg is carefully prepared, the K_2O content of which is subsequently determined to be 6%. Assuming that all mineral grains have equal weight, it is shown that the possible error (twice the standard deviation, at the 95% confidence limit) is 0.58%, and thus 5 of the 100 samples could have a K_2O content falling outside the limits of 5.42–6.58%. This inhomogeneity is not revealed in the composite sample but requires a series of samples for its statistical evaluation. A similar situation has been described by Shaw and Bankier[17] for the occurence of rubidium in 28 diabase samples from Ontario[31]; the calculated average rubidium content (m) was found to be 144 ppm, with a standard deviation (s) of 86 ppm, and thus at the 95% confidence limit the range of rubidium content ($m \pm 2s$) would be 0–316 ppm. A composite sample of the 28 diabases would probably have given an average rubidium content close to 144 ppm, but would have given no indication of the range that existed. Burks and Harpum[32] found that bulk samples of granitic rocks may give analyses that are not representative of the area sampled and which may not fall in the field of composition of any natural rock in the area. It is suggested[17] that for geochemical work the only safe rule is "the more samples the better."

The importance of grain size and its influence upon the size of the sample that should be taken has been considered by Laffitte[28] for the determination of both major and minor constituents. Consider a granodiorite in which all of the mineral grains are approximately the same size and which has an average CaO content of 2.4%; the CaO content of the quartz, mica, and potash feldspar grains can be taken as zero, while that of the plagioclase feldspar, which makes up 40% of the rock, is 6%. If it is assumed that the weight of a grain having a volume of 1 cm^3 is 2.5g, then the size of sample necessary to give desired limits of error in the determination of CaO for a given grain size is shown in Table 3.1. If the size of the largest grain present is taken as the governing size, then phenocrysts and porphyroblasts will greatly increase the size of sample needed; the percentage of the constituent sought must also be considered.

3.2.2 An Integrated Approach to Sampling Control

It will be useful to expand some of the concepts discussed in Sec. 3.2.1 in a more quantitative manner, in light of the recent work that has been done.

The first of the three sampling stages is the collection of the sample in the field. The definition of a sample given in Hackh's Chemical Dictionary[33] includes the term "systematically taken," and this is the problem addressed by Conners and Myers,[23] Visman,[34] and others. The second stage involves sample size reduction, from the sample collected in the field (or from a conveyer belt) to the sample submitted to the laboratory. Gy[35,40] has contributed the design for a sampling slide rule to assist in the process, and Ottley[36] gives a good description of its use. Finally, the sample submitted to the laboratory must be subjected to comminution

TABLE 3.1   Influence of Grain Size on the Size of
Sample Required [after Lafitte (Ref. 28, p. 729)].

Grain Size (cm^3)	Number of Grains	s (%)	$m \pm 2s$ (%)	Sample Weight (grams)
0.1	100	0.3	1.8–3.0	25
0.1	2,500	0.06	2.28–2.52	625
0.1	10,000	0.03	2.34–2.46	2500
0.01	10,000	0.03	2.34–2.46	250

and subsampling steps before the analytical work can be performed. Ingamells and co-workers have derived a laboratory sampling constant and
sampling diagrams which have been published in a series of papers; these
should be read by all involved in the analysis of geological materials.[10,
11,13,15,37,38] Because the analytical variance must be known in order to
isolate the sample reduction variance, and both are required to determine
the field sampling variance, this discussion will treat them in that order.

a.   *Ingamells' Constant* K_s

The sampling constant K_s is the weight of sample (in grams) required
to ensure, at the 68% confidence level, that the sampling uncertainty will
not exceed 1%. It has been derived in two different ways,[11,37] and an
approximation for the equation (for a two-component mixture) is

$$K_s = 10^4 \frac{(K - L)(H - L)u^3d}{K^2} \qquad (1)$$

where

K = overall percent concentration of X
H = percent concentration of X in the high-X component
L = percent concentration of X in the low-X component
u = maximum grain size, in centimetres (that is, effective linear mesh
    size)
d = density of the high-X component

This equation assumes that a precise analytical method is used.

Let us consider the derivation of K_s by analytical measurement. Nu-
merically, the constant K_s for a particular constituent X of a sample is
equal to the square of the coefficient of variation obtained by determining
X in a series of 1 g subsamples (assuming an insignificant analytical
variance). This is derived from

$$K_s = R^2w \qquad (2)$$

where R is the relative deviation in percent, and w is the sample weight
in grams. Thus K_s can be obtained from N repetitive (and precise) determinations of X in a series of w gram subsamples, so that

$$K_s = R^2w = 10^4 \frac{w\sum(X_1 - \bar{X})^2}{(N - 1)\bar{X}^2} = \frac{(100s)^2}{\bar{X}^2} w \qquad (3)$$

where s is the standard deviation, $\bar{X}$ is the mean of the N analytical
results, and s^2 is the variance.

As an example, consider a set of 10 results for Au reported as part of a certification program for the Certified Reference Material MA-1.[39] The 10 results from 15 g subsamples (in ounces Au per ton) are: 0.520, 0.530, 0.520, 0.510, 0.520, 0.520, 0.520, 0.520, 0.520, 0.530. From this, $\bar{X}$ = 0.5210, s = 0.0057, and from equation (3), K_s = 18 g. In other words, to ensure a sampling uncertainty not exceeding 1% at the 68% confidence level for MA-1, a subsample of at least 18 g must be taken.

b. Gy's Constant C

Once a decision has been made about the size of the subsample required for the analytical work (and once the analytical variance is known), it is possible to consider how best to reduce a sample that has been collected in the field (or in the mill) to a size manageable at the laboratory scale. Gy's equation and his slide rule find their application at this stage of the sampling problem. The sampling equation, as discussed by Ottley,[36] is

$$w = \frac{Cu^3}{s^2} \tag{4}$$

where

w = weight of sample required, in grams
u = largest particle size, in centimetres
s = standard deviation
C = a sampling constant for a particular material to be sampled

The sampling constant C is given as

$$C = fglm \tag{5}$$

where

f = "shape" factor, f = 0.5, except in the case of a gold ore when f = 0.2
g = particle size distribution factor, g = 0.25, except for closely sized material when g = 0.5
l = "liberation" factor, l = 0 for homogeneous materials and l = 1 for completely heterogeneous materials. Values of l are related to

$$\frac{u}{L} = \frac{\text{largest particle size}}{\text{liberation size}} \tag{6}$$

m = mineralogical composition factor,

$$m = \frac{1 - a}{a} [(1 - a)r + at] \tag{7}$$

with

r = mean density of valuable mineral component
t = mean density of gangue mineral component
a = average mineral content expressed as a decimal fraction of 1

"Average" values of r and t are 5.0 and 2.6, respectively, and can be used when calculating m.

Once enough information is known about an ore deposit, or any other unit that is to be sampled, in order to permit the calculation of Gy's constant C, this constant can and should be used as a criterion for designing the sample reduction technique and determining the sample sizes to be taken at the various reduction steps. An example given in Ottley's description of Gy's slide rule will illustrate this application.[36]

The problem is to design a sampling procedure for -1 inch ore which is 10% ZnS with a liberation size of 0.2 mm. The assaying accuracy required is 2% of the ZnS concentration, with samples to be taken at each of three crushing stages ($\frac{1}{4}$ inch, 10 mesh, and 100 mesh). Gy's equation can be used to show that the minimum weight of sample required at each stage is: 170 kg at the -1 inch size, 5.5 kg at the $\frac{1}{4}$ inch, 200 g at the 10 mesh, and 0.5 g at the 100 mesh size.

A rule of thumb used in this type of sampling problem is that of the Tyler screen scale technique, which states that a sample which is reduced to pass the next finer screen on the $\sqrt{2}$ scale may be split in half with no loss of information. This is only a general guideline, however, and a proper sampling process in any exploration program or mill should use the more quantitative approach available through Gy's sampling equation.

c. Visman's Constants A and B

The problem that remains now is how to ensure that a sample of ore in place, a run-of-mine sample, or a sample from a stockpile is going to represent the body being sampled to the required tolerance. Visman has offered a solution to this problem, his General Sampling Theory.[34] He introduces two constants A and B which he describes as the unit random variance and the segregation variance, respectively. The overall sampling variance is given by his equation

$$s^2 = \frac{A}{W} + \frac{B}{N} \tag{8}$$

where W is the gross sample size made up of N increments, A/W is the random variance term, and B/N is the segregation variance term. The values for A and B can be determined by taking a series of small (W_1)

and a series of large (W_2) samples (Visman suggests 25–30 of each). "Small" means that W_1 should be small enough so that the random variance is the dominant term in equation (8) and "large" means that the segregation term is dominant. The first approximation of the maximum estimate for A and B can then be calculated by

$$A = \frac{W_1 W_2 (s_1^2 - s_2^2)}{W_2 - W_1} \tag{9}$$

$$B = s_2^2 - \frac{A}{W_2} \tag{10}$$

where s_1^2 and s_2^2 are the variances obtained from the analyses of the series of small and large samples, respectively. This assumes, of course, the use of Gy's and Ingamells's constants to ensure proper sample reduction steps and adequate analytical subsampling sizes. Thus when A and B have been determined, the optimum gross sample size W and the number of increments N required to give a selected variance of s^2 can be calculated. The optimum field sample size for segregated material is obtained by differentiating equation (8) with respect to W and setting it equal to zero to obtain

$$W_{opt} = \frac{A}{B} \tag{11}$$

During any sampling operation the estimates of A and B should be made increasingly accurate to ensure that the value for W_{opt} can be adjusted to optimize the information obtained from sampling at minimum cost.[38]

d. Sampling Diagrams

Ingamells has extended the use of these constants to the generation of sampling and subsampling diagrams in order to provide a means of visualizing sampling errors.[13] As his diagrams show, if the analyte (X) occurs in discrete high-X mineral grains, then as the subsample size decreases, there is an increasing chance of not including a high-X mineral grain in the subsample, and the most probably result will be one that is lower than the "true" value for the sample. Conversely, if the minor constituent mineral is a low-X species, there is an increasing chance of excluding a low-X mineral grain as the sample size decreases, and the most probable result will be a higher than the "true" value. The diagrams are constructed by plotting the most probable result Y, obtained from samples of w grams, against the sample weight with the latter variable on the logarithmic scale of semilog paper. For a two-component mixture the most probable value

Y for an unsegregated material ($B = 0$) is given as

$$Y = \frac{2(K - L)^2 Kw + AL}{2(K - L)^2 w + A} \qquad (12)$$

where

A = Visman's random distribution constant
K = the true value of X in the sample
L = the X content of the major mineral of the two-component mineral mixture (usually the X content of the gangue)
w = sample weight, in grams

An example taken from Ingamells's discussion of sampling diagrams is that of a 10 mesh molybdenum sulfide ore sampled in two ways. A series of 100 g samples were crushed to pass a 150 mesh screen, and 2 g subsamples were taken from each of the pulps. Duplicate 2 g subsamples indicated that subsampling and analytical errors were negligible, but the series of 100 g samples showed significant variance between them. A series of 1000 g samples were treated in the same way, resulting in a large drop in the between sample variance. The value of K_s is, however, 14 kg (using a value of 0.242% Mo for the "gangue" material, 0.2 cm grain size for the MoS_2 grains). This represents the sample size of 10 mesh material that would have to be reduced to 150 mesh material (using proper reduction procedures) to ensure that the true Mo value is determined to $\pm$ 1% at the 68% confidence level. This is illustrated in Fig. 3.1 (taken from Ref. 13).

3.2.3   Methods of Sampling

There are four common types of sampling procedures. *Random* sampling is self-explanatory, but there are many difficulties in the way of obtaining a truly random sample. *Stratified* sampling involves the selection of samples from specific layers, as in the sampling of interbedded shale and limestone, or banded gneiss. The taking of samples at spaced intervals about a sample point is *cluster* sampling and is done to give information about variability over a small distance. *Systematic* sampling is probably the type most used, knowingly or unknowingly, to give information about a specific phase or feature of a mass. All of these types are best served by an organized sampling program which will yield the desired information, and foresight is preferable to hindsight.

The proper sampling of a rock mass requires a geological knowledge of the mass; this is usually confined to a plane surface, but for some

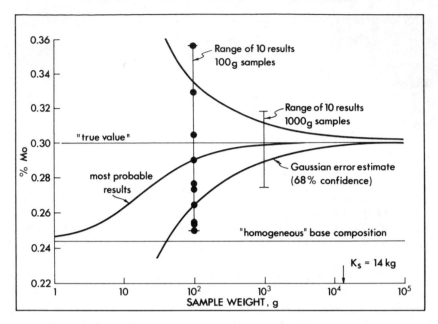

Figure 3.1. Sampling diagram for Mo ore showing the prediction of low values for the most probable result when small sample sizes are used. Reproduced with permission of *Geochimica et Cosmochimica Acta* (Ref. 13).

layered intrusions the third dimension is exposed as well.[41] If the mass is homogeneous, then as much randomness as possible should be injected into the sampling procedure; if heterogeneous, the mass should be dealt with as a series of rock types, and consideration must be given to the effect of one type upon another (e.g., hybridization of surrounding material by xenoliths). In the case of a banded gneiss, possibly the most difficult of any rock to sample meaningfully, stratified sampling should be used if the bands are sufficiently coarse to permit it; the alternative is to take a large sample and homogenize it by crushing, an obvious compromise. Washington[42] preferred to take a carefully selected hand specimen from a homogeneous mass, but favored the sampling of individual facies in a heterogeneous one, and the combining of the resultant analyses in relative proportions to give an average composition. Grout[6] obtained surprisingly good results when analyses of representative samples of banded gneiss were compared with those of larger chip samples.

There are several ways in which a sample may be obtained. Single specimens, weighing about 2 kg, are obtained by breaking them from an outcrop with a geological hammer (grab sampling); lichen and weathered material should be removed by washing and scrubbing, if necessary, and the sample thoroughly dried. This should be done *before* the sample is submitted for analysis, for only the collector knows the extent to which trimming should be done on the specimen. *Chip sampling* involves the taking of several kilos of chips, each weighing about 50–100 g, from various sites on the mass being sampled; *channel sampling*[43] is similar, except that this requires the cutting of a channel, 7–15 cm in width and 2–5 cm in depth, across the cleaned surface of the mass, using a diamond saw, hammer and chisel, or pneumatic drill. All of the material cut out is included in the sample and generally the longer the channel, the better the sample. *Core drilling* provides an excellent opportunity to obtain a series of samples, and is becoming increasingly popular; an elegant way to obtain an average sample from a diamond drill core is to grind the core lightly with a carborundum grinding wheel along one side and collect the powder. A comparison of grab sampling versus channel sampling in limestones has been made by Galle.[44]

Mention has already been made of the need to consider grain size in the selection of samples (see Sec. 3.2.1 and 3.2.2). It must be emphasized, however, that the end purpose of the sampling must govern the sampling procedure. Most of the foregoing has been concerned with the sampling of a rock mass for the determination of the major constituents, a relatively simple job compared to that required for a study of the distribution of those constituents in low concentration which may occur in scattered accessory minerals, such as sulfides, zircon, and apatite, or camouflaged in major mineral lattices. The sampling plan needed for a study such as this latter one will be more complex and more extensive than that for the estimation of major constituents. Lamar and Thomson[7] have described the sampling of limestone and dolomite deposits for the spectrographic determination of trace and minor elements.

The sampling program used by Pitcher and Sinha[45] to obtain samples for a petrochemical study of the Ardara Aureole is instructive, although Flinn,[46] who subsequently analyzed their data statistically, has pointed out certain failings in their procedure. Pelitic schists make up about 365 m of the aureole surrounding the Ardara pluton, part of the Donegal Granite Complex. This horizon was sampled inside and outside by means of eight crosstraverses, each traverse being allotted five rectangular areas measuring 45 × 135 m each. Five collecting points were selected within each rectangular area, and samples of about 300 g each were taken at each point. They were washed, split, and one-half of each was broken into small pieces. The pieces from a single rectangular area were mixed to give an aggregate sample of about 750 g. In order to show that five collecting points were sufficient, chemical analyses were made on five samples taken from one rectangular area, and on a composite of the samples prepared as described above. Some variation occurred in the content of Si, Mg, Ca, Na, and K between individual samples, but the average values of the five analyses were in good agreement with the values obtained for the composite sample.

Stream sediment sampling for geochemical purposes is determined largely by the area to be covered and the type of terrain to be sampled. The sample frequency can vary from an average of one per 10 square kilometers to several per square kilometer. The rationale for stream sediment sampling is that the products of weathering of many minerals are transported either in solution (and later precipitated or adsorbed), or mechanically as particles in the waters of a catchment basin. There exists then a relatively accessible medium by which the area drained by the stream may be sampled for anomalous concentrations of economically interesting elements.

The sediment samples are usually taken from the center of the stream, well away from any area of bank collapse. The sample should consist of the finest material available, and it should be taken from behind rocks, in backeddy pools, and in the main stream bed, with several grams being taken from different locations in each sample site to make up a 50–100 g composite sample. This is the ideal situation of course, and it is not possible at all sampling sites. The samples should be placed in new sample bags (special paper ones are available), and extreme caution should be taken to avoid contamination from such diverse causes as mine dumps, bridges, industrial effluents, or even the handling of coins just prior to sampling. The proper planning of a stream sediment survey will depend on the specific exploration problem to be solved, and reference should be made to sources in which methods of geochemical exploration[9, 21, 24, 27] are discussed.

3.3   PRELIMINARY TREATMENT OF THE SAMPLE

A preliminary examination of a rock or mineral sample received for analysis is a useful practice and may result in a saving of analytical time. When possible, the analytical work to be done should be discussed with the submitter; useful information can often be obtained about the nature of the sample, and the objective of the analysis and the analytical work may be reduced or modified as a result. The shotgun approach of a "complete" analysis is often wasteful of analytical time, and the determination of 4 or 5 constituents may provide as much information as will the determination of 12 to 15 of them.

3.3.1 Cleaning and Trimming

It is preferable that the removal of unwanted material from a specimen be done by the collector before it is submitted, so that no error will be introduced by the removal of pertinent material. If this is not done, then trimming instructions should be obtained before the sample is treated further, and lichen, encrusting dirt, and identification marks (including adhesive labels) should be noted for removal also. Wet samples should be dried at 100–105°C and allowed to come to equilibrium with their surroundings before being sampled. The disposition of the bulk of the residue after sampling should be determined, and arrangements made to save a piece for thin section preparation or other purposes *before* the whole sample is reduced to less than pea size.

3.3.2 Preliminary Examination

The presence of sulfide minerals or accessory minerals containing such elements as fluorine and boron should be noted, because these will affect the choice of analytical methods to be used. When no other information is available, a modal analysis may suggest useful modifications in the analytical approach.

Mineral samples are usually submitted as material having a grain size between 60 and 80 mesh and are often obtained by concentration with heavy liquids. If the mineral has a noticeable odor of volatile organic compounds, the sample should be dried in an oven at 100–105°C in order to expel this organic material, then allowed to come to equilibrium in air before further treatment. The degree of purity of the sample is really the responsibility of the submitter, but a precautionary examination will do no harm and may reveal the presence of unwanted impurities which make further purification of the sample necessary; the analytical data may later be used to calculate the formula of the mineral, and the analysis of an impure sample can be very misleading.

Such detailed examination of samples is not always possible nor practical, particularly where large numbers of samples are being submitted. The responsibility must then rest with the submitters, but every effort should be made to acquaint them with the problems involved. Those engaged in sample preparation should be informed of the procedure to be followed and care should be taken to see that the usefulness of the sample is not impaired by careless preparatory work.

3.3.3   Recording of Sample Data

It is a good practice to keep a record of the weight of the sample submitted for analysis, particularly of minerals, for such information is often useful at a later date. This again may not be practical when a large number of samples is submitted. Pertinent information about the sample should be recorded by the laboratory on a simple form, such as the specimen field number, the rock type, any significant special features such as the presence of appreciable sulfides or other potentially troublesome constituents, and the nature of the analytical work requested. It is assumed that the submitter will have recorded the pertinent geological and other information necessary to identify and characterize the specimen.

3.4   PREPARATION OF ROCK SAMPLES FOR ANALYSIS

Sample preparation, which includes the crushing, splitting, grinding, sieving, mixing, and, if necessary, the drying of the sample, is an important first step, and much has been written about the procedures for all types of materials.[47-50] Because the efforts to provide a proper sample are easily undone by improper sample preparation, the latter must not be left to chance. There are many ways to prepare samples, and it is necessary to choose the best procedure for the job at hand.[51]

Contamination of the sample during sample preparation is an everpresent and unavoidable danger. One cannot prepare a rock or mineral sample for analysis without changing its composition in some respect, through either the addition of foreign material, the preferential loss of material, or the oxidation of constituents. The goal is one of minimizing these undesirable effects, either by preventing their ready occurrence or by choosing a procedure that will favor a type of contamination that is the least harmful. Thiers[52] has discussed the problem of the contamination of biological samples in trace element analysis and has suggested means of controlling it; his comments are applicable to rock and mineral samples as well. Volborth[53] grinds a duplicate split of the sample in both ceramic and hardened steel pulverizers and analyzes both splits in order to determine the contamination from each pulverizer to obtain the true composition.

Stream sediment samples must be dried at low (105°C) or even ambient temperatures. If the samples are to be analyzed for volatile elements such as Hg, they must not be oven dried. Because trace elements are of interest, precautions must be taken against contamination, and these include not using brass sieves, using a sample bag only once, and isolating the samples from mineralized ore samples. These may seem to be obvious precautions, but all have been found to be sources of contamination in stream sediment programs.

3.4.1 Errors in Sample Preparation

Rock samples are usually submitted in the form of chips (about 2 × 2 × 1 cm), or as hand specimens (about 8 × 8 × 5 cm), ranging in weight from about 50 g to 40 kg or more. It is necessary to obtain from this an aliquot portion of 20–50 g that will represent the larger sample, as the latter is supposed to represent the larger mass.

There may be special circumstances which permit the selection of one or two chips as being representative, but it is generally necessary to treat the whole sample. Grout[6] obtained good agreement between the SiO_2 content of both single specimens and composite samples of banded gneiss, except for one sample in which, by error, the whole of the composite sample was not crushed and quartered; the SiO_2 content of the hand specimen was 71%, but only 66% SiO_2 was obtained for the poorly sampled composite.

Laffitte[28] has emphasized the importance of grain size (see Sec. 3.2.1) in determining the size of aliquot that should be taken and in governing the magnitude of the relative error to be expected in the determination of constituents. With care, the error introduced by the sample preparation can be made negligible; for a 100 g sample having an average grain size of 4 mm, the relative error introduced into the determination of a major constituent could amount to 8%, but if the grain size is reduced to 1 mm, the relative error is reduced to about 1%. The premature quartering of samples should thus be avoided if the precision of the sampling is to be the same as that of the chemical analysis, particularly for those elements present in the less abundant minerals.

Grillot[29] studied the effectiveness of various types of crushing and grinding apparatus in reducing samples of both soft and hard rock to − 200 mesh. He found that there is an unavoidable loss of fine dust during this reduction, particularly during the mechanical grinding; a loss of 5 g was found for 200 g of both hard and soft rock after about 15 min of grinding. A 100 g sample of hard rock powder having an average diameter of about 0.3 mm, after being twice passed through the grinder, yielded about 10 g of flour having a diameter of less than 0.08 mm (200 mesh).

The process of coning and quartering is probably the most popular method of reducing the sample to a small representative portion. It is tedious, however, and unless done carefully it can lead to gross errors from segregation as the heavier fragments tend to roll to the base of the pile and do not always distribute themselves equally about the base.

Improper sampling and dust losses are indeed major sources of error in sample preparation, but another important one is contamination of the sample. There is no way of avoiding wear on the comminution equipment and thus the material of which the equipment is made must be chosen so that the inevitable contamination will be of no consequence in subsequent analyses or the extent of the contamination is known so that it can be accounted for. For example, steel pulverizer plates should not be used to grind samples which are to be analyzed for trace concentrations of chromium or nickel.

Comminution equipment made from a wide variety of materials is now available, and a judicious selection can preclude most contamination problems. A necessary precaution with any new piece of comminution equipment is to determine the contaminants (and their extent) that its use might introduce. In the next section some specific examples of contamination from comminution equipment are given.

3.4.2   Types of Equipment

There are several types of equipment that may be used for the various stages of sample preparation,[19, 22, 54, 55] and the choice is dependent upon the subsequent work to be done on the sample. As was mentioned in the previous section, it is impossible to avoid the introduction of some foreign material during sample preparation, and the best alternative is to choose the type of contamination that will be the least objectionable. Samples must first be reduced in a jaw crusher to a particle size suitable for further reduction in a grinder of one type or another and, because of the hardness of the sample material, abrasion of the crusher or grinder will occur (Ref. 9, p. 132). The ideal arrangement would be to crush the sample to the desired particle size between blocks of similar material, but this is not practical. It is also possible to heat the sample in an electric furnace to about 600°C and then quench it in distilled water; the sample will disintegrate readily to a grain size suitable for grinding in an agate mortar.[56, 57] This method is not only undesirable because of the possible effect of the heating and quenching upon the constituents of the sample, but it may also be dangerous; explosive decomposition of the material may occur when it is withdrawn from the furnace, even with slow heating.[20]

The reduction of the sample to manageable size can be done with a rock trimmer or hydraulic rock crusher, or by a simple bucking board and muller. The latter may be made of steel or ceramic material, and Myers and Barnett[57] have shown that contamination is introduced by the bucking board, as well as by automatic crushers and grinders. Bloom and Barnett[58] ground both quartzite and massive quartz to less than 100 mesh on bucking boards made of both high alumina ceramic and steel, and also in an agate mortar. The steel bucking board added objectionable amounts of Fe, Mn, Cr, Cu, Ni, and V to the samples; only 0.003% Mg was added to the quartzite and quartz by the ceramic bucking board and muller, and 0.0001% Ti to the quartz alone.

Further crushing of the material may be done in large, hardened-steel mortars of the Plattner (or "diamond") type in which the diameter of the pestle is only slightly less than that of a collar which in turn fits snugly into a depression in the base of the mortar. Sandell[59] studied the contamination of quartz and microcline by crushing them in a Plattner mortar and found that the amount of iron introduced from the mortar was less if the collar was not used; the crushing was also more efficiently done, but a cardboard screen must be used to prevent loss of material. Because Cu, Cr, and Ni were also introduced, the steel used for the mortar should be low in these constituents. Storm and Holland[60] found no significant contamination by Ni when quartz was crushed in a steel mortar to a size between 120 and 200 mesh. The Ellis mortar[61] is larger and heavier, and the crushing may be done entirely by the pestle without the use of a hammer; the pestle is also much smaller than the collar, and abrasion is minimized.

Jaw crushers are widely used to reduce the large fragments to pea size or smaller. Grillot[29] studied the efficiency of a jaw crusher, a cylinder grinder and an automatic agate mortar for the grinding of both hard and soft rocks, and his results are given in Table 3.2.

Jaw crushers are a major source of contamination because burring of the steel plates will occur with the consequent introduction of slivers of steel (tramp iron) into the sample. Face plates should be of hardened steel (they may be vertically grooved for more efficient operation) and should be changed when they become roughened. Economy in operation at this point can be wasteful of effort at a later stage. A hydraulic rock crusher, in which the jaws used for breaking specimens are replaced by two steel blocks, will produce material of about 1 mm in diameter.[22]

There is a small jaw crusher available (Progressive Exploration Products, NV) equipped with high alumina ceramic jaws and cheek plates which will take the pea size material from a larger jaw crusher down to approximately − 10 mesh. This is only slightly more coarse than the product that can be obtained with a disc pulverizer equipped with ceramic plates, and the jaw crusher has the dual advantage of crushing a sample

TABLE 3.2 Efficiency of Crushing and Grinding [after Grillot (Ref. 29, pp. 9–10)].

Equipment Used	Nature of Rock	First Pass	Second Pass	Third Pass
Jaw crusher	Soft	50% < 1 mm		
	Hard		62% < 1 mm	
Cylinder, grinder	Soft	42% > 0.3 mm	5.5% > 0.3 mm	<1% > 0.3 mm
	Hard	30% > 0.3 mm	2.5% > 0.3 mm	1% > 0.3 mm
Automatic agate mortar	Soft (20 g, 0.3–0.08 mm)	6 g > 0.14 mm	5 g > 0.14 mm	
		5 g, 0.14–0.08 mm	3 g, 0.14–0.08mm	
		9 g < 0.08 mm	12 g < 0.08 mm	
	Hard (20 g, 0.3–0.08 mm)	2 g > 0.14 mm	2 g > 0.14 mm	
		3 g, 0.14–0.08 mm	3 g, 0.14–0.08 mm	
		15 g < 0.08 mm	15 g < 0.08 mm	

more rapidly than the latter and it is much easier to clean. This results in a much higher sample throughput, a major consideration for most laboratories.

Roller crushers are also used for material first crushed in large jaw crushers. The equipment used at the Department of Geology, University of Manchester, has been described by Smales and Wager,[22] and was designed to reduce small pieces of rock to a powder with a minimum of contamination. It consists of two sets of rollers, one set of which reduces the coarse material to about 80 mesh while the other continues the reduction to −200 mesh.

There are several devices that may be employed for the reduction of the sample to a particle size between 80 and 100 mesh. Cylindrical grinders or pulverizers use opposing plates, one of which is stationary while the other revolves at a variable distance from it. The sample is fed from a funnel into the gap between the plates, and the desired reduction is achieved by narrowing the gap. Table 3.2 shows the results that may be obtained in this way. Steel plates will introduce considerable metal contamination, and ceramic plates are much to be preferred.[62] However, the ceramic plates cannot be adjusted sufficiently close together to give a sample having a grain size much smaller than about −20 mesh without occasioning an excessive amount of wear which results in an unacceptably high level of contamination (ceramic chips have been found on occasion) and a short plate life. A cylinder mill that will handle as little as 0.5 g of sample, and is easily cleaned, has been described.[63] The cone grinder, hammer mill, and rotary beater are other forms of pulverizers that can be used; it was found at the GSC that excessive loss of fine material occurred with the rotary beater, and the metal screens used with this device are a major source of contamination. The BCM tested an ingeniously arranged, doubly gyrating ball mill utilizing ceramic containers and balls and found it to be very slow and to introduce excessive contamination. All of these machines require relatively large samples.

Further reduction of the sample to 150 mesh or finer can be achieved by various methods. The Schwingmühle (swing-mill) of Siebtechnik, Mühlheim, Germany, is used by Danielsson and Sundkvist[64] for the preparation of samples for the tape machine technique of spectrographic analysis, where uniformity of particle size is particularly important. This equipment may be obtained with a single large mill, or with as many as six smaller mills. Each mill consists of a container, a ring and/or solid grinders and a cover. These are made of hardened steel but can also be obtained with ceramic, agate, or tungsten carbide liners and grinders. The Bleuler Mill (rotary or swing mill, Willy Bleuler Company, Switzerland) and the Shatterbox (Spex Industries, MI) are similar machines. A ball mill, which is a mechanically rotated porcelain container holding about one-third of its volume of flint pebbles or porcelain balls, will produce a very fine powder and also thoroughly mix the sample. It is, however, somewhat difficult to clean and requires a large sample because of the amount that is retained on the surface of the balls and on the walls of the container. Related to the ball mill is the "paint-shaker,"[54] which uses ceramic balls to do the grinding; the sample is placed in a ceramic cylinder with one or more balls, six of these containers are clamped in a device used to mix the contents of paint tins and vigorously shaken. Ballard, et al.[65] have investigated the grinding and mixing efficiency of stainless steel ball mills for the preparation of samples for X-ray diffraction and spectrographic analysis.

Mortars are commonly used to grind the sample to the desired final particle size. There are many types—agate, mullite, tungsten carbide, glass, and porcelain (these latter two are now rarely used). They may be used manually (some laboratories prefer this, believing it to be faster than automatic grinding[66]) or operated mechanically. Grinding under water is occasionally done[67]; the fines are decanted periodically into a reservoir and when all of the sample has been satisfactorily reduced, the water is removed by decantation and evaporation. This method also reveals any "tramp iron" present as a surface scum. Allen (Ref. 61, p. 813) found that a weight loss of 0.291 g by a mortar and pestle occurred when 200 g of quartz sand were reduced to pass through a 150 mesh screen; the abrasion of the mortar and pestle is greater in automatic grinding than in manual grinding. Hempel (Ref. 61, p. 814) found that steel and glass mortars were abraded less than agate, but in rock analysis the contamination from steel and glass is undesirable. The use of a boron carbide mortar has been discussed by Boulton and Eardley.[68]

Table 3.3 shows the contaminants introduced by different types of equipment in the BCM laboratories.

There are other types of equipment which can be used, some of which are designed for specialized problems, such as the mica pulverizer described by Neumann.[69] A food blender can also be used to pulverize mica. A special container which permits the crushing of iron meteorites at liquid air temperature has been described.[70, 71]

Sieving of the ground material is a step that must be given careful consideration. Frequent sieving during the grinding removes the finer fractions and speeds up the grinding process, and also ensures that the bulk of the powder will be of the desired size. Care must be taken, however, to avoid contamination by the material of the sieve; if trace elements are to be determined, brass sieves should be avoided, and stainless steel sieves used instead. Care should be taken to ensure that the

TABLE 3.3 Degree of Contamination, as Determined by Semiquantitative dc ES, for Pulverizing Equipment Used under Varying Conditions in the BCM Laboratory

Pulverizing Equipment	Rated Capacity	Sample[a] Size	Pulverizing Time	Contaminant Level[b]
Diamonite mortar and pestle	—	10 g	—	Al(5%), Mg(<10), Ca(<10), Fe(<10), Cu(<1), Ti(<10)
Ceramic ball mill (new)	20–25 g	10 g	0.75 hr	Al(5%), Mg(<10), Ca(<5), Fe(<10), Ti(<5)
Ceramic ball mill (well used)	20–25 g	10 g	0.75 hr	Al(>10%), Mg(0.05%), Ca(<10), Fe(40), Ti(10)
Ceramic ball mill (well used)	20–25 g	25 g	1.5 hr	Al(0.4%), Mg(<10), Ca(<10), Fe(10), Cu(<2)
Ceramic disc mill	—	30 g	5 min	Al(50), Mg(<10), Fe(<5)
Ceramic disc mill	—	30 g	10 min	Al(0.25%), Mg(20), Ca(<10), Fe(10), Cu(<2), Ti(5)
Steel disc mill	10–15 g	10 g	5 min	Al(40), Mg(<10), Ca(<5), Fe(0.1%), Cu(<1), Ti(6)
Rotary disc pulverizer	—	—	—	Al(20), Mg(<10), Fe(2.2%), Pb(50), Cu(70), Zn(40), Mn(50), Ni(30), Mo(0.5), Cr(15)

[a] All samples were precrushed quartz.
[b] Contamination level in parts per million unless otherwise noted. The elements listed are the only ones detected.

stainless steel sieves are not soldered. Some commercially available ones are, and Sn and other contaminants have been detected as a consequence of their use. The type of sieve commonly used for work in which contamination must be minimized consists of silk or nylon bolting cloth stretched over a glass or plastic holder. The importance of the thorough mixing of material that has been sieved is shown by Lundell and Hoffman[72]; the composition of the material can change significantly with sieve size.

Mixing can be done manually by rolling the powder from one side to the other on a rubber cloth or sheet of glazed paper; the material must tumble over itself and must not just slide along the surface of the sheet. Mechanical mixers vary from simple to elaborate devices. O'Neil[73] places a taped sample jar in sawdust in a pebble mill container and rotates the latter for 24 hr. A spiral mixer that can be rotated on a lathe (a smaller model is turned by a stirrer motor) is used at the GSC; the sample bottles (which should be no more than half-filled) are held at a 45° angle and the rotation of the mixer imparts both a rotary and a vertical movement to the contents. This device has been expanded to a large wooden drum capable of mixing 144 samples at once. Another type employs a wooden cylinder which rotates freely on two rubber rollers turned by means of a small motor; the cylinder has an inner stainless steel container fitted with baffles which cause the powder to be thoroughly tumbled as the cylinder rotates. There are also more elaborate mixers or blenders which can be used for quantities of material from several pounds to less than one pound. Whatever method is used, however, the mixing must be thoroughly done.

There remains only the final step of reducing the bulk sample to a size that will meet the needs of the analysis. A sample of 30–50 g is usually a very ample quantity, leaving sufficient material after the analysis for storage against future needs; this quantity is not always available, however, particularly for mineral samples. Large samples (one of several kilos) can be reduced to 100 g by coning and quartering; again, there is the danger that unequal segregation of heavier material will occur during the rolling and coning of the material. A riffle, or Jones splitter, is a device having an even number of narrow sloping chutes, with alternate chutes discharging in opposite directions; these vary in size from those able to handle several kilos, to microsplitters for small mineral samples which are vibrated electromagnetically. The sample is carefully poured in a stream across the top of the riffle, and alternate portions are collected in containers placed below each set of chutes. This requires that a uniform stream of material be poured on the surface, and there is also considerable loss by dusting. A rotary splitter consists of a rotating table carrying six pie-shaped containers; the sample is poured into a conical receiver and falls slowly through a chute terminating just above the sample containers which are rotating beneath it.

3.4.3 Sample Preparation Procedures

There are many factors which govern the choice of procedure, and often the final choice must be a compromise. Some contamination is inevitable and that procedure which will keep it to a minimum should be selected. The degree of magnitude of the contamination will be directly related to the final particle size, because the longer grinding period required to produce a finer particle size will naturally result in a greater amount of wear on the grinding material. The mesh size should be fine enough to ensure that the sample will be as nearly homogeneous as possible. Broadly speaking, the sample should pass through a 100 mesh screen for chemical analysis, whereas for emission spectrographic analysis, 200–300 mesh material is desirable to reduce subsampling errors introduced by the small aliquot that is used. It has been argued that sample powder of even −200 mesh is too coarse.[74] XRF analysis also requires 200–300 mesh material in order to minimize particle size problems when pressed powder samples are used.

The size of the sample will often dictate the method to be used. A small sample is better prepared by hand; too much loss occurs when larger, automatic equipment is used. Very small samples can be ground directly in automatic pulverizers such as the Mixer-Mill or Wig-L-Bug used in the preparation of samples for ES analysis.

The effect of grinding upon such constituents as ferrous iron and water must also be considered. Mauzelius[75] showed that excessive grinding oxidized a major part of the ferrous iron, and he recommended the use of the coarsest powder permissible. This was confirmed by Hillebrand,[76] who believed that the oxidation was due more to the local heat created by the friction of the pestle than to the increased surface that is exposed to air; he also demonstrated that there is a considerable increase in the water content with increasing fineness of the powder. Hillebrand et al.[61] discussed the problem in some detail, including the possibility of doing the grinding in an organic medium such as alcohol or carbon tetrachloride. Some analysts prefer to do the ferrous iron determination on a separate, coarser portion of the sample, but this is objectionable because of the differences in moisture content, state of oxidation, and homogeneity that may exist between the portions. If two portions are used, the moisture must be determined on both in order to correct for its effect on the ferrous iron, as well as on the other constituents determined on the finer portion.

Hillebrand et al.[61] prefer to grind the whole sample in a nonoxidizing medium rather than to use a separate portion of coarser material.

Each laboratory will have its own preferred procedure which is dictated by both the type of analysis to be done and the equipment available, but such procedures should be open to modification as new equipment becomes available and as new concepts are introduced. The following procedure is used routinely at the GSC. The sample is first broken into pieces (approximately 3 cm cubes) in a crushing machine fitted with ceramic-plated jaws or, if too large for the crusher, with a hydraulic rock trimmer. These smaller pieces are then reduced in the jaw crusher to about 3 mm or less, starting with the jaws set wide apart to minimize loss of particles, and reducing the gap width on successive passes; three or four passes should be sufficient. If the size of the sample necessitates the use of a larger crusher with steel-plated jaws, the plates should be changed when they begin to show wear.

The pulverized material from the crusher is reduced to a convenient and representative sample size by coning and quartering. The material is rolled from one corner of the paper or mat to the opposite corner by raising one corner and causing the material to tumble over upon itself, and the process is repeated by raising each corner in succession until thorough mixing of the sample has been achieved. The material is then gathered as a flat cone in the center of the sheet (flatten the cone, if necessary, with a spatula; if a steel spatula is used, it should be demagnetized to prevent the removal of magnetic grains) and it is then divided into four quarters and opposite quarters are removed. The mixing and quartering are repeated and opposite quarters removed in alternating sequence, until a sample of about 30 g in weight is obtained. This is quartered and the two sets of opposite quarters are reserved for further grinding. It is the practice at both the GSC and BCM to prepare two subsamples from each sample, one of which is used for chemical analysis and the other for spectrographic analysis, whether or not both types of analysis have been requested by the submitter. Some laboratories like to keep a small amount (20–50 g) of coarse material as a reserve against future contingencies; the residues of analyzed samples should *always* be retained.

The samples (10–20 g) are reduced to a particle size between 100 and 200 mesh by shaking each in an individual small ceramic ball mill for up to 30 min; sets of six ball mills, in a special holder, are clamped in the jaws of a modified paint-shaking machine. Twelve samples can be ground at the same time if both sides of the machine are utilized. If the grinding is to be done by hand instead, transfer the sample to a 100 mesh sieve (silk bolting cloth is preferable), and sieve it gently over a piece of glazed paper. Grind the oversize portion in an agate or mullite mortar; again sieve, and repeat these operations until all of the sample has passed through the screen. At no time must a particular sieve fraction be taken for the analysis; differences in the susceptibility to grinding of various minerals make it almost certain that compositional differences will exist between sieve fractions.

Transfer the final powder to a numbered plastic vial, seal with a tight friction cap, and store it until needed. Before use, mix the sample for 20–30 min in an automatic mixer, pour the contents of the mixer vial on a piece of glazed paper, quarter the pile, and with a spatula (demagnetized steel, bone, or plastic) transfer the quarters successively to the original sample vial; do *not* pour the sample into the vial, as this causes a segregation of particles according to density. This mixing should be repeated before use whenever the sample has been stored for any length of time, in order to eliminate the segregation that may occur during storage.

The BCM follows a slightly different practice in the crushing stage. The sample is broken into pieces of about 5 cm across, and these are passed through a large jaw crusher to give pieces less than 1 cm. If the sample is large (greater than 500 g), it is split at this stage in a stainless steel Jones riffle to produce a sample of 250 g or more. If the sample is 500 g or less, it is not split at this stage but the entire sample is crushed in a small jaw crusher with ceramic plates (see Sec. 3.4.2) to pass through a 10 mesh screen. It is then reduced by cone and quartering as above, and about 25 g is ground in the "paint shaker" to less than 150 mesh.

3.5 PREPARATION OF MINERAL SAMPLES FOR AGE DATING

The separation of sulfide minerals by laboratory scale flotation techniques is chiefly a mineralogical problem and will not be dealt with here. While this section is primarily intended as a discussion of the separation of minerals for subsequent age dating determinations,[50, 77] many of the techniques can be applied to separations of minerals for other analytical purposes as well.

3.5.1 Types of Equipment

The comminution equipment needed, such as jaw crushers and pulverizers, has been discussed earlier (Sec. 3.4.2). Disc pulverizers should be adjusted so that the plates are far enough apart to allow a large portion of the sample to remain sufficiently coarse so that the desired size fraction can be obtained. A visual examination of the rock prior to crushing can often tell an experienced technician what spacing of the plates is to be used.

The size fractions are separated by means of different sized sieves stacked one above the other in a sieve shaker, and agitated to cause the powder to distribute itself on the different sieves. When the separate size fractions have been obtained, they are washed in water in a cylindrical column constructed of glass or Plexiglass® (Fig. 3.2). The column is fitted with a fritted disc near the bottom to allow air bubbles from an aspirator attachment to rise through the sample and carry light dust particles and most of the biotite by overflow into a sieve placed to catch the latter. The water flow and air pressure are adjusted to optimize the washing action, and the hornblende and heavier minerals remain in the column. A bank of infrared lamps is a very convenient way of drying the samples at a low temperature (<40°C) after this treatment.

The separation of heavy minerals may be done with a Wilfley concentration table. The bulk sample is fed by gravity to an inclined table bearing

Figure 3.2.   Cylindrical columns used for washing size fractions prior to subsequent mineral separation steps. Photograph by R. Player, BCM.

a number of riffles; the heavy minerals are caught by the riffles while the lighter ones float over them. A similar but more refined separation is achieved with the superpanner. In this device a thin film of water flows down a sloping flat surface; the surface layer of the film moves more rapidly than its bottom layer, and thus minerals are separated along the sloping surface according to their specific gravity. By varying the slope of the surface, separations can be made among minerals whose specific gravities vary only slightly.

The Frantz isodynamic separator is the most popular of the devices used to separate minerals on the basis of their magnetic susceptibility; the powder moves down a vibrating chute parallel to the pole pieces of an electromagnet which separates the more magnetic and the less magnetic particles into two streams which fall into separate collector vessels at the foot of the chute. By varying both the direct current to the electromagnet and the inclination of the chute, the forces applied to the particles may be controlled. Prior to treatment with the Frantz separator, the Carpco direct-roll magnetic separator can be used to make a preliminary coarse separation, thus reducing very considerably the bulk of the material that must be treated in the Frantz.

The use of heavy liquids requires the use of such apparatus as separatory and filtering funnels, funnel racks, and similar items. It is also essential to laboratory safety to have adequate ventilation because of the carcinogenic properties of the diiodomethane, bromoform, and other heavy liquids used. The down-draft fume hood described in Sec. 2.1.2 is one way of ensuring that the fumes are properly vented so that the dangerous compounds are absorbed by the activated charcoal filters.

The heavy liquids most commonly used include Clerici solution (100 g of thallium formate and 100 g of thallium malonate in 10 ml of water at 20°C gives a specific gravity of 4.25) (Ref. 50, p. 194), bromoform (specific gravity 2.87), tetrabromoethane (specific gravity 2.96), and diiodomethane (specific gravity 3.2). These are all very poisonous and should be handled with extreme care.

3.5.2 Sample Preparation

Most minerals exist in close proximity to other minerals, and the separation of one from another is not an easy task. A mineral may be intergrown with another mineral, often of the same grain size, and the exclusion of these "joined" grains from the sample could be a potential source of error. In order to separate minerals, by one means or another, it is necessary to crush the sample to a particle size that is smaller than the smallest mineral grain to be separated; this must not, however, be carried too far, or a product of such fineness will be produced that surface attraction will make a clean separation virtually impossible. Mackenzie and Milne[78] have shown the considerable change in mica that is caused by excessive grinding. Another possible problem is the occurrence of very small mineral grains as inclusions in larger grains, such as rutile needles or zircon grains in micas, and chlorite in feldspar; again, it is almost impossible to obtain a pure concentrate.

When the sample has been crushed, it is then separated into different size fractions by passing it through a set of sieves on a sieve shaker. There is some difference of opinion as to the optimum particle size distribution for mineral separation. The GSC generally reduces material to a -100 to $+200$ mesh (-150 to $+74$ μm) range (see Appendix 2). The BCM chooses a much coarser range of -20 to $+100$ mesh (-840 to $+150$ μm), unless mineral grain sizes necessitate the use of a finer sample. The choice is often dictated by the analyst who will be carrying out the isotopic analysis for the age dating procedure.

When the samples have been crushed and sized, the different fractions are then examined under a microscope to determine the largest grain size fraction that has the desired mineral grains free of any gangue material. This fraction is then selected for further work. If not enough material is available in that size fraction to yield a sample of sufficient weight, then coarser material, if available, is ground to the selected size. It should be pointed out that, whenever possible, only mineral separates from one size fraction should be submitted for age dating purposes. This may not be possible in many cases, but when feasible it does increase the confidence placed in any date obtained from the sample.

3.5.3 Mineral Separation

There are no two mineral separation problems which are exactly alike. The grain sizes may differ, the individual minerals may be altered to a lesser or greater extent, and the magnetic susceptibilities of the various minerals in the sample can be different. In addition, the specific gravity of a mineral may vary from one sample to the next. Hence no set procedure is going to effect a good separation of hornblende, for example, from every sample submitted. The size fraction used can differ, the initial washing step may or may not preclude the necessity of having to use a preliminary separation with a heavy liquid of low specific gravity, the settings on a magnetic separator will differ from one sample to another, and, on occasion, even the order of applying magnetic and heavy liquid separation techniques can be reversed.

Keeping in mind this problem of sample variability, however, the procedure following the comminution and the all-important sizing is generally as follows, beginning with the washing step. This step has two purposes, one being to cleanse the selected size fraction of very fine particles which will contaminate the heavy liquids in the subsequent steps. The second purpose is to make, by careful adjustment of the parameters, a separation of the lighter minerals from the denser ones. Under favorable circumstances this can be used to concentrate biotite (see Sec. 3.5.1) so that only a small cleanup step will be needed subsequently to make the biotite concentrate suitable for analytical work. The washing step can be accomplished by means of a superpanner or a water column. The superpanner has the advantage of being able to process a larger sample in a given period of time, but the selective concentration of biotite is difficult to obtain with it.

When the sample has been washed it should be dried at a temperature low enough to avoid the loss of volatile analytes (such as Ar) from the matrix of the individual mineral grains. When the sample has dried, the next step is most commonly a heavy liquid separation using bromoform to effect the separation of the lighter quartz and feldspar groups (floats) from hornblende and other heavier minerals (sinks). Biotite may either float or sink at this stage, depending on the specific gravity of the bromoform used and that of the biotite. If zircon is to be separated, a heavy liquid of higher specific gravity, such as diiodomethane, should be used initially.

The heavy liquid separation procedure is most readily performed in a separatory funnel fitted with a large-bore Teflon stopcock. The size of the funnel will depend on the amount of sample, but one of 500 ml capacity is a convenient size. Approximately 200 ml of heavy liquid is poured into the funnel, up to 25 g of the sample is then poured in and the mixture is stirred to ensure that all of the sample particles are wetted. The heavy grains are allowed to settle to the bottom and the light fraction is stirred again. When no more particles are seen to sink, the operator should look carefully at the liquid in the funnel to determine the extent of the separation between the heavy and light mineral fractions. There will usually be an area of clear liquid between the layers, but this need not be the case, especially when a highly altered mineral having a wide range of specific gravities is present. The stopcock on the separatory funnel is opened carefully, and the heavy minerals at the bottom are flushed out and caught in a coarse filter paper (10 cm squares of industrial grade paper towels are satisfactory for coarse fractions) in a filter funnel positioned below the funnel (Fig. 3.3). The heavy liquid filtrate is retained for future use. The light mineral fraction is then removed from the separatory funnel in a similar manner, and the funnel and the mineral con-

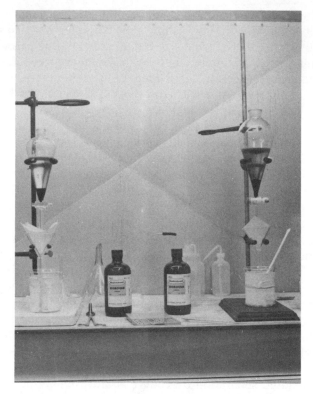

Figure 3.3. Separatory funnel and collection funnel for the heavy liquid step of a mineral separation procedure. Photograph by R. Player, BCM.

centrates are washed thoroughly with dimethyl sulfoxide, acetone, or alcohol. The washings should be collected in a separate container for future reclamation of the heavy liquid (see Sec. 3.5.4).

The light and heavy fractions may be further segregated by treatment with either bromoform or diiodomethane diluted with acetone or alcohol to give a range of specific gravities from 2.6 to 3.2. Specific gravity chips are available (Raynor Optical Co. Ltd., England) to assist in adjusting the dilutions. They range in specific gravity from about 2.3 to 3.7, in somewhat irregular steps approximating 0.04 specific gravity units each. This can be too coarse an interval in making some separations, and a specific gravity balance, although slower, can be used instead.

When a finer separation is undertaken, it is often more convenient to use smaller separatory funnels and a smaller volume of heavy liquid. Much attention to detail is required and many separations may involve painstaking adjustment of the specific gravity of the liquids. A formula, and the resulting dilution curves for adjusting the specific gravity, are available (Ref. 50, pp. 193, 197).

Magnetic separation can be used at an earlier stage to separate magnetic material from nonmagnetic, but the Frantz isodynamic separator is very useful at this stage for the separation of mineral grains of similar specific gravity but of different magnetic susceptibility. As mentioned earlier, no two mineral assemblages are similar, and the settings of current applied to the magnets, and the degree of tilt, have to be determined empirically for each new sample in order to give the optimum separation. The Frantz separator can be used for high throughput by adjusting it to a steeper incline, but the efficiency of the separation drops significantly. It is usually more convenient to decrease the sample bulk by gravity separations before attempting a magnetic separation with the Frantz.

The Carpco high-intensity, induced-roll magnetic separator has a higher sample throughput capacity than does the Frantz and can be used as a separation step preliminary to either the Frantz or the heavy liquid stage. The Eriez induced-roll magnetic separator incorporates a belt-type feed instead of a free drop, and this allows a greater control over the feed rate.

The Davis tube magnetic separator is a useful means of obtaining magnetic fractions from very fine-grained samples. This is usually difficult with either of the dry magnetic separators discussed above. The Davis tube, an inclined tube into which the sample is introduced as a slurry, has a two-dimensional motion which subjects the sample to a twisting type of agitation. There are two electromagnets on either side of the tube which attract and hold the magnetic fraction. The rate of water flow through the tube, the vigor of the agitation, and the magnitude of the current applied to the magnets enable an effective magnetic separation to be made. A mistake often made is that of not placing a screen below the outflow to catch the nonmagnetic fraction, which is then lost. Another common mishap is to turn off the magnet before removing the screen holding the nonmagnetic fraction. Needless to say, neither mistake is likely to occur more than once (or twice!) with an operator, but it seems to be a natural law that each new operator has to experience both.

Simple devices can also be used to make separations, such as a small hand magnet enclosed in a piece of paper and lightly drawn over the surface of a powder, or a centrifuge. For handpicking grains under the microscope, Murthy devised a suction apparatus in which an intravenous needle connected to a soft plastic bottle sucks the mineral grains into a collecting tube when the pressure on the bottle is released.[79] A similar arrangement, but with constant suction, utilizes a thin plastic hose fitted with a similar intravenous needle and connected to a plastic bottle which in turn is connected to a simple water aspirator.

3.5.4 Heavy Liquid Recovery

The high cost of heavy liquids makes their recovery an important aspect of any mineral separation operation. The most common technique is to utilize their insolubility in water, although in some instances distillation is necessary (Ref. 50, p. 210). The recovery of the heavy liquids from the acetone and alcohol washings is effected by adding a large excess of water to the washings and shaking the mixture vigorously. The acetone and alcohol are miscible with the water, but the heavy liquid is not and will sink to the bottom of the container. Most of the aqueous layer is then decanted and the process repeated twice more. The final separation of the heavy liquid from the water should be done in a separatory funnel and the heavy liquid filtered through paper before it is stored.

A heavy liquid reclaimed in this manner never has the maximum specific gravity of the unused compound because not all of the washing solvent is extracted. However, the specific gravity is sufficiently high for the liquid to be useful for most purposes.

Heavy liquids are unstable to light and they decompose to form I_2 (diiodomethane) and Br_2 (bromoform) which will discolor the solution to the point where it cannot be used. Shaking the heavy liquid with a 5–10% solution of Na_2CO_3 or NaOH will remove much of the discoloration.

3.6 PREPARATION OF SILT AND SOIL SAMPLES

Silt and soil samples are usually collected for mineral exploration purposes, although occasionally they are associated with an environmental monitoring program. The sample treatment considered here will be applicable to those collected for geochemical or exploration purposes (Ref. 9, pp. 254–256).

The samples as received in the laboratory are very often wet and they must be dried either at room temperature (if time is not important or if Hg or other volatile constituents are to be determined) or in an oven at 105°C. Most often the samples are collected in paper bags, and they need not be taken out of them for the drying step.

Once the samples have dried they should be crushed to break up any lumps or cakes which may have agglomerated during drying. This crushing need not be done if the sample on drying is in a finely divided state. Care should be taken to avoid base metal contamination of the sample from the steel jaws of the crusher, such as when small pieces of rock collected with the silt and soil samples are broken.

The most common sieve size used for screening soil and silt samples is 80 mesh, with the −80 fraction being used for the analysis. Other fractions are used on occasion (Ref. 9, p. 255). The screen should not be of brass or stainless steel; nylon and aluminum screens are commonly used, and bolts of nylon mesh can be obtained for use either in the field or in the laboratory. The screen can be cut to the appropriate size and fitted over the end of any plastic or nonmetallic cylinder to serve as a noncontaminating sieve.

3.7 PREPARATION OF SPECIAL SAMPLES FOR ANALYSIS

The foregoing material has been concerned chiefly with the routine treatment of rock and mineral samples. There are many other types of samples which require modification of these routine procedures; no attempt will be made to cover all of the possibilities, but a few of the more obvious procedures will be mentioned.

The handling of wet material is facilitated by first drying the material, but some samples such as soils and clays have a high water content at room temperature which changes readily in response to atmospheric changes. It is usually necessary to dry these materials to prevent caking during the grinding; the sample should then be spread out on a piece of glazed paper on the balance table, covered lightly to protect it from dust, and allowed to come to equilibrium with its surroundings before it is placed in a sample vial. This applies to other samples also having a high water content. Marshall and Jeffries[80] discuss ways of cleaning soils prior to mechanical analysis; a number of special centrifuge tubes for heavy liquid separation are also described. Further details are given by Searle and Grimshaw.[81]

Methods for the sampling of water, and the handling of the samples prior to analysis, are described in detail by Rainwater and Thatcher.[82]

The problem of obtaining a truly representative sample of a meteorite is a considerable one, and differs in magnitude from the iron to the stony types. Iron meteorites can be sampled by either drilling or milling, and the purity of the material can be checked by means of a binocular microscope. Some analysts prefer to sample the metal and inclusions separately; a bulk composition is derived by analyzing the fractions separately and combining them on the basis of an areal analysis of a large slice cut from the surface. Others prefer to use thin slabs free from visible inclusions which are totally dissolved without further treatment.

In stony meteorites the analyst is faced with the presence of three or four phases—silicate, metal, sulfide, and phosphide; drilling may be used to isolate small samples for study, but it is not helpful in obtaining a sample for bulk analysis. The sampling problem is related directly to the nature of the sample in that the amount of sample that must be taken is governed by its homogeneity. Magnetic separation is, next to handpicking, the simplest way to effect a separation, but the use of a magnet results in two fractions, a magnetic (metal) one with some silicate grains either joined or trapped, and a nonmagnetic (silicate) fraction with some metal. Separate analyses of these fractions mush each be corrected for the presence of some of the other fraction, often on the basis of certain assumptions that are not strictly valid. Various reagents such as mercuric chloride have been used to remove the metal fraction, but their use always raises the question of their effect on the other phases. A very elegant technique has been developed at the British Museum (of Natural History); dry chlorine is passed over the heated sample, resulting in removal of the metal as volatile chlorides and leaving the silicate phase in an untouched form.[83] Onishi and Sandell[84] dissolved the metal and sulfide phases of chondrites in aqua regia to separate the silicate phase, but some of the latter is decomposed in the process; troilite can be dissolved in hydrochloric acid saturated with bromine. Berry and Rudowski[70] have designed a special container for crushing chondritic meteorites at liquid air temperature; embrittlement of the metal phase (Ni < 20%) makes it possible to crush the whole sample.

References

1. G. R. Davis, *Bull. Inst. Min. Metall. (Lond.)*, No. 673, (Dec. 1962), and No. 674, p. 255 (Jan. 1963).

2. M. E. Joseph, *Proc. Australasian Inst. Mining Metall.*, **202,** 81 (1962).

3. H. S. Sichel, *Bull. Inst. Min. Metall. (Lond.)*, No. 544, p. 261 (Mar. 1952).

4. *Annual Book of ASTM Standards*, Part 32, *Proposed Method for Sampling Iron Ores*, 1973, American Society for Testing and Materials, Philadelphia, p. 1182.

5. *Kirk-Othmer Encyclopedia of Chemical Technology*, 2nd ed., Vol. 17, Wiley, New York, 1968.

6. F. F. Grout, *Am. J. Sci.*, **24,** 394 (1932).

7. J. E. Lamar and K. B. Thomson, *Illinois State Geol. Surv. Circ.*, 221, 1956, 18 pp.

8. A. W. Kleeman, *J. Geol. Soc. Aust.*, **14,** 43 (1967).

9. A. A. Levinson, *Introduction to Exploration Geochemistry*, Applied Pub. Ltd., Calgary, Alta., 1974.

10. C. O. Ingamells and P. Switzer, *Talanta*, **20,** 547 (1973).

11. C. O. Ingamells, *Talanta*, **21,** 141 (1974).

12. K. J. D. Ridley, A. Turek, and C. Riddle, *Geochim. Cosmochim. Acta*, **40,** 1375 (1976).

13. C. O. Ingamells, *Geochim. Cosmochim. Acta*, **38**, 1225 (1974).

14. A. W. England, "To Sample the Moon," in *Sampling, Standards and Homogeneity*, American Society for Testing and Materials, ASTM STP 540, 1973.

15. C. O. Ingamells, J. C. Engels, and P. Switzer, *Proc. Int. Geol. Congr.*, 24th Session, Sect. 10, 405, 1972.

16. D. M. Shaw, *Trans. R. Soc. Can.*, 3rd Ser., Sect. IV, **55**, 41 (1961).

17. D. M. Shaw and J. D. Bankier, *Geochim. Cosmochim. Acta*, **5**, 111 (1954).

18. A. W. Groves, *Silicate Analysis*, 2nd ed., Allen and Unwin, London, 1951, p. 17.

19. H. B. Milner, *Sedimentary Petrography*, Vol. 1, 4th ed., rev., Allen and Unwin, London, 1962, pp. 54, 101.

20. A. C. Oertel, "Internal Rep. Div. of Soils," C.S.I.R.O., Australia, 1961.

21. H. E. Hawkes and J. S. Webb, *Geochemistry in Mineral Exploration*, Harper & Row, New York, 1962, Chap. 3.

22. A. A. Smales and L. R. Wager, Eds., *Methods in Geochemistry*, Interscience, New York, 1960, p. 4.

23. J. J. Conners and A. T. Meyers, "How to Sample a Mountain," in *Sampling, Standards and Homogeneity*, American Society for Testing and Materials, ASTM STP 540, 1973.

24. A. H. Lang, *Prospecting in Canada*, 4th ed., Geological Survey of Canada, Economic Geology Rep. No. 7, 1970.

25. A. D. Wilson, *Analyst*, **89**, 18 (1964).

26. C. L. Grant and P. A. Pelton, "Role of Homogeneity in Powder Sampling," *Sampling, Standards and Homogeneity*, American Society for Testing and Materials, ASTM STP 540, 1973.

27. P. Bradshaw, D. R. Clews, and J. L. Walker, *Exploration Geochemistry*, Barringer Research, Rexdale, Ont., 1972.

28. P. Laffitte, *Bull. Soc. Geol. France*, 6th Ser., **3**, 723 (1953).

29. H. Grillot, Bureau de Recherches Géologiques et Géophysiques Minières, Publ. 20, 1957, 40 pp.

30. E. S. Larsen, *Am. J. Sci.*, **35**, 94 (1938).

31. H. W. Fairbairn, L. H. Ahrens, and L. G. Gorfinkle, *Geochim. Cosmochim. Acta*, **3**, 34 (1953).

32. H. G. Burks and J. R. Harpum, *Rec. Geol. Surv. Tanganyika*, **10**, 64 (1963).

33. J. Grant, Ed., *Hackh's Chemical Dictionary*, 4th ed. McGraw-Hill, New York, 1972.

34. J. Visman, *Mater. Res. and Stand.* (Nov. 8, 1969).

35. P. M. Gy, *L'Échantillonage des Minerais en Vrac: Théorie Générale*, Vol. 1, Sociéte de l'Industrie Minérale, Saint Etienne, France, 1967.

36. D. J. Ottley, *World Mining*, 40 Aug. (1966).

37. C. O. Ingamells, *Talanta*, **23**, 263 (1976).

38. C. O. Ingamells, *Talanta*, **25**, 731 (1978).

39. G. H. Faye, W. S. Bowman, and R. Sutarno, Canada Centre for Mineral and Energy Technology, Ottawa, Report MRP/MSL 75-29 (TR), 1975.

40. P. M. Gy, *Sampling of Particulate Materials: Theory and Practice*, Elsevier, Amsterdam–New York, 1979.

41. C. H. Smith, Geological Survey of Canada, Paper 61-25, 1962, 16 pp.

42. H. S. Washington, *Manual of the Chemical Analysis of Rocks*, 2nd ed., Wiley, New York, 1910, p. 43.

43. W. E. Hill, Jr., *State Geol. Surv. Kansas Bull.*, Part I, 152, 1961, 30 pp.

44. O. K. Galle, *Trans. Kansas Acad. Sci.*, **67**, 100 (1964).

45. W. S. Pitcher and R. C. Sinha, *Quart. J. Geol. Soc. London*, **113**, 393 (1957).

46. D. Flinn, *Geochim. Cosmochim. Acta*, **17**, 161 (1959).

47. C. A. Bicking, "Principles and Methods of Sampling," in I. M. Kolthoff and P. J. Elving, Eds., *Treatise on Analytical Chemistry*, 2nd. ed., Vol. 1, Part 1, Interscience, New York, 1978, Chap. 6.

48. C. R. N. Strouts, H. N. Wilson, and R. T. Parry-Jones, *Chemical Analysis— the Working Tools*, rev. ed., Clarendon P, Oxford, 1962, p. 46.

49. R. C. Tomlinson, "Sampling," in C. L. Wilson and D. W. Wilson, Eds., *Comprehensive Analytical Chemistry*, Vol. IA, Elsevier, New York, 1959, Chap. II.

50. M. Allman and D. F. Lawrence, *Geological Laboratory Techniques*, Arco Publ. Co., New York, 1972.

51. W. P. Huleatt, *Eng. and Min. J.*, **151**, 62 (1950).

52. R. E. Thiers, "Contamination in Trace Element Analysis and Its Control," in D. Glick, Ed., *Methods of Biochemical Analysis*, Vol. 5, Interscience, New York, 1957, p. 273.

53. A. Volborth, *Appl. Spectrosc.*, **19**, 1 (1965).

54. P. J. Lavergne, Geological Survey of Canada, Paper 65-18, 1965, 23 pp.

55. G. C. Lowrison, *Crushing and Grinding*, CRC Press, Cleveland, OH, 1974.

56. L. H. Ahrens and S. R. Taylor, *Spectrochemical Analysis*, 2nd ed., Addison-Wesley, Reading, MA, 1961, p. 41.

57. A. T. Myers and P. R. Barnett, *Am. J. Sci.*, **251**, 814 (1953).

58. H. Bloom and P. R. Barnett, *Anal. Chem.*, **27**, 1037 (1955).

59. E. B. Sandell, *Anal. Chem.*, **19**, 652 (1947).

60. T. W. Storm and H. D. Holland, *Geochim. Cosmochim. Acta*, **11**, 335 (1957).

61. W. F. Hillebrand, G. E. F. Lundell, H. A. Bright, and J. I. Hoffman, *Applied Inorganic Analysis*, 2nd ed., Wiley, New York, 1953, pp. 813–814.

62. P. R. Barnett, W. P. Huleatt, L. F. Rader, and A. T. Myers, *Am. J. Sci.*, **253**, 121 (1955).

63. M. D. Schlesinger, S. Nazaruk, and L. Reggel, *J. Chem. Educ.*, **40**, 546 (1963).

64. A. Danielsson and G. Sundkvist, *Spectrochim. Acta*, **15**, 126 (1959).

65. J. W. Ballard, H. I. Oshry, and H. H. Schrenk, *J. Opt. Soc. Am.*, **33**, 667 (1943).

66. A. G. Sergeant, *Internal Manual of the Geological Survey of Great Britain*, 1962.

67. W. R. Schoeller and A. R. Powell, *Analysis of Minerals and Ores of the Rarer Elements*, 3rd. ed., Hafner, New York, 1955, p. 1.

68. J. F. Boulton and R. P. Eardley, *Analyst*, **92**, 271 (1967).

69. H. Neumann, *Nor. Geol. Tidsskr.*, **36**, 52 (1956).

70. H. Berry and R. Rudowski, *Geochim. Cosmochim. Acta*, **29**, 1367 (1965).

71. A. J. Easton and J. F. Lovering, *Geochim. Cosmochim. Acta*, **27**, 753 (1963).

72. G. E. F. Lundell and J. I. Hoffman, *Outlines of Methods of Chemical Analysis*, Wiley, New York, 1938, p. 21.

73. R. L. O'Neil, *J. Sed. Pet.*, **29**, 267 (1959).

74. G. W. Ordrick and N. H. Suhr, *Chem. Geol.*, **4**, 429 (1969).

75. R. Mauzelius, *Sver. Geol. Under. Arsbok*, **1**, 3 (1907).

76. W. F. Hillebrand, *J. Am. Chem. Soc.*, **30**, 1120 (1908).

77. L. D. Muller, in J. Zussman, Ed., *Physical Methods in Determinative Mineralogy*, 2nd ed., Academic, New York, 1977.

78. R. C. Mackenzie and A. A. Milne, *Mineral. Mag.*, **30**, 178 (1953).

79. M. V. N. Murthy, *Am. Mineral.*, **42**, 694 (1957).

80. C. E. Marshall and C. D. Jeffries, *Soil Sci. Soc. Am. Proc.*, **9**, 397 (1945).

81. A. B. Searle and R. W. Grimshaw, *The Chemistry and Physics of Clays and Other Ceramic Materials*, 3rd. ed., Ernest Benn, London, 1959.

82. F. H. Rainwater and L. L. Thatcher, U. S. Geological Survey Water Supply, Paper 1454, 1961, 301 pp.

83. A. A. Moss, M. H. Hey, and D. I. Bothwell, *Mineral. Mag.*, **32**, 902 (1961).

84. H. Onishi and E. B. Sandell, *Geochim. Cosmochim. Acta*, **9**, 78 (1956).

CHAPTER 4

SAMPLE DECOMPOSITION

4.1 GENERAL

The selection of the correct means of sample decomposition is critical to the success of any analysis. The large number of possible methods available to the analyst only adds to the importance of the considerations involved in the selection. The substitution of instrumental methods for the "classical" and "rapid" techniques of analysis has dictated a corresponding modification of both the considerations involved and the method selected. The factors that influence the decision will be considered in this chapter and the major emphasis will be on sample decomposition, with some attention paid to disssolution conditions for common refractory minerals. The chapter is in no way intended as a substitute for the very useful, indeed, indispensable book on sample decomposition written by Doležal, et al.,[1] and their extensive review of recent developments (Supp. Ref. 1). A recent translation of Bock's book on the subject will also serve as a guide (Supp. Ref. 2).

The use of aspirators for the introduction of samples to the flame of atomic absorption spectrometers or the plasma of emission spectrographs imposes constraints on the amount of dissolved solids which can be tolerated in sample solutions. Another constraint is the introduction of ions which might interfere with subsequent analyses by suppressing or enhancing the signal or by forming molecular species with atoms of the analyte. In addition, the viscosity of the sample solution must be the same as that of the standard to preclude differences in the aspiration rates of the two which will result in faulty calibration. For XRF analysis the measuring surface must be homogeneous, optically flat, and the sample itself of infinite thickness with respect to the incident X-rays.

All of the above must be considered, and many other factors as well in the selection of a method for sample decomposition. As a result no general rule regarding the most advantageous choice can be given, because the latter is governed by both the nature of the sample and the nature of the analytical work that is to follow. Obviously the method must effectively decompose the sample; in dealing with a rock sample, however, it must be borne in mind that a rock is a mixture of minerals, and, depending upon the ultimate goal of the analysis, *effective* decomposition does not necessarily imply *complete* decomposition,[2] although the careful analyst will be happier when the two are synonymous. In the determination of ferrous iron, for example, the sample is heated near the boiling point with a mixture of hydrofluoric and sulfuric acids for a fixed period of time; in most cases there remains a small residue, composed chiefly of quartz, which can be safely ignored. Continuation of the heating until the residue is dissolved would not only serve no useful purpose but would, in this instance, increase the opportunity for air oxidation of the ferrous iron.

As was mentioned above, the introduction of large quantities of material, which may be troublesome in subsequent stages of the analysis, should be avoided. In this respect the use of acids has a definite advantage over the use of flux because the excess acid can usually be more readily removed. Also, fluxes are not so readily obtained in as pure a form as are acids [Supp. Ref. 3, Chap. 5, and Supp. Ref. 4 contain useful discussions of methods for the purification of acids], and this is a matter of some concern when trace element analysis is contemplated.[3] In the same vein, the decomposition medium should not seriously attack the fusion vessel, with the introduction of potentially troublesome material into the analysis and accompanied, in some cases, by the retention of other material in the vessel's walls. Acids again have an advantage over fluxes in that the attack on the vessel is usually less because of the lower temperature that is employed. The advantages and disadvantages of various types of crucibles have been discussed in an earlier section (Sec. 2.2) and need not be repeated here.

The method of decomposition used should not result in the loss of constituents by volatilization when the determination of these constituents is to follow. Careful studies have been made of this, and the results suggest that the analyst must be on guard against unsuspected losses. Chapman et al.[4] state that, as a result of their study of the volatility of elements from mixed hydrofluoric and perchloric acid solutions, prediction of the volatility of many elements from mixed acid solutions is a virtual impossibility. Under their conditions Si, B, Ge, As, Sb, Cr, Se, Os, Ru, Re, and Mn could be volatilized. Hoffman and Lundell[5] added HCl or HBr to boiling (200–220°C) perchloric or sulfuric acid solutions of various metals; quantitative distillation from $HClO_4$ of Sb, As, Cr, Ge, Os, Re, Ru, and Sn, and from H_2SO_4 of Ge, As, Sb, Se, Sn, and Re, was obtained, with a partial loss from both acid solutions of Bi, B, Sn, Mo, Te, and Tl. Gases formed by reaction are easily driven off, such as carbon dioxide, hydrogen sulfide, and phosphine (if phosphides are present). Boric acid will be lost from a boiling aqueous solution, and boiling concentrated sulfuric or perchloric acid will volatilize phosphoric acid.

The method of decomposition can be divided into three groups: (1)

decomposition by acids, both oxidizing (HNO_3, $HClO_4$, hot concentrated H_2SO_4) and nonoxidizing (HCl, HF, H_3PO_4, HBr, dilute H_2SO_4, and dilute $HClO_4$); (2) decomposition by fluxes, either by fusion with acid or alkaline compounds, or by sintering; (3) decomposition by other means, such as in bombs, in sealed tubes, and by chlorination. The acid fluxes include the bisulfates, pyrosulfates, and acid fluorides; the losses that occur during fusions with these fluxes result from the same general conditions that cause losses from acid solutions. The hydroxides, peroxides, carbonates, and borates are the most common basic fluxes, and there is less danger of loss by volatilization during fusions. At fusion temperatures, the alkaline peroxides are very powerful oxidizers. Oxidation occurs during fusion with alkali carbonates chiefly by bringing the elements into a more intimate contact with air than is possible by simple ignition; bisulfates have a more restricted oxidizing power. An additional advantage of the use of a flux is that the color of the solidified cake is often indicative of the presence of certain constituents. The elements are usually raised to their highest valence state; the most highly colored melts are those formed by the transition group elements.

The specialized methods are employed usually when other methods fail, but they also have certain advantages, such as a short period of decomposition or the production of a simple, relatively salt-free solution, that commend them to more general use and outweigh the disadvantage of the specialized equipment that is required.

Many rocks and minerals are readily acid soluble, such as carbonates, phosphates, many sulfides and sulfates, chlorides, and borates; oxides are more readily dissolved by the nonoxidizing acids. Silicates which contain a high proportion of bases such as calcium are decomposed by strong acids such as HCl, but not so those of an acidic nature with much Al or Fe^{3+} present; complete decomposition is claimed, however, with mixtures of HCl, HNO_3, H_2SO_4, and $HClO_4$.[6] Hydrofluoric acid will, of course, effectively decompose most aluminosilicates. There are some minerals, such as kyanite, beryl, zircon, and tourmaline, which are attacked only slightly by the common acids and fluxes at normal operating temperatures and pressures. For these, recourse is had to the more specialized methods listed previously. It is also possible, by preliminary treatment, to make a sample more amenable to decomposition by ordinary means; Hoffman et al.[7] found that kaolin is quickly decomposed by dilute HCl if it is roasted at 700°C for 15 min, whereas previously it was only slowly attacked by the acid.

Decomposition is generally expedited by using a very finely ground sample (-200 mesh), but preparation of this powder may not be entirely advantageous, as discussed in Sec. 3.4. Antweiler[8] has attempted to avoid the inaccuracies introduced during the crushing, grinding, and sieving of rock samples by using rock fragments. He employs three methods: decomposition with HF, fusion with a mixture of Na_2CO_3 and H_2BO_3, and fusion with Na_2O_2; the first method is generally preferred.

The sample decomposition techniques used in the analysis of stream sediments and silts collected for geochemical survey purposes form a special group (Ref. 9, pp. 247–252, and Ref. 10). They consist of attacks by both mineral and organic acids, hot and cold solutions, and dilute concentrated acids. In all cases the goal is to maintain consistency of all analytical conditions for the whole suite or series of samples in any given exploration program, and precautions are taken to ensure that it is achieved. An interesting study on the efficiency of the extraction of metals from various minerals by some of the digestion techniques used in exploration geochemical analysis was published by Foster.[11] Whitehead has described the use of dilute acetic acid as a decomposition reagent for the determination of Ca and Mg associated with the carbonate fraction of carbonate rocks (Supp. Ref. 5).

4.2 DECOMPOSITION WITH ACIDS

4.2.1 With Hydrofluoric Acid

The use of hydrofluoric acid is a time-honored procedure in silicate analysis, and there are few silicates that are not more or less decomposed by it.[12] Of these few, kyanite, axinite, beryl, zircon, and some tourmalines and garnets are the most common, but under special conditions these can also be decomposed.[13]

The decomposition of muscovite and quartz is slow, although for the latter it has been found that reducing the particle size can increase the reaction rates (Ref. 1, p. 34). The minerals spinel, rutile, graphite, cassiterite, and pyrite are not decomposed in HF although many difficultly soluble oxides are.[14] Minerals containing elements which form insoluble fluorides, such as magnesium and lead, may dissolve only slowly because of the shielding effect of the precipitated fluoride. Decomposition of rocks and minerals at room temperature by HF in the presence of an oxidizing agent is the basis of a method for the determination of ferrous iron (Ref. 2, pp. 45–55); maximum decomposition is usually reached in less than 24 hr.

The acid used in the laboratory is approximately 48% HF (w/w), with a specific gravity and molarity of approximately 1.19 and 29.0 respectively. Caution is necessary in its use at all times; it is very poisonous, and contact between the cold concentrated acid and the skin will produce irritating white spots, while the hot acid produces yellow, ulcerous sores that require immediate, special medical treatment. Vigorous flushing of the affected area with cold water is recommended as a preliminary treatment. Gloves, either rubber or plastic, should be worn when using this acid.

Langmyhr and Graff[15] have shown that no detectable loss of Si occurs when SiO_2 is decomposed by HF, even at 100°C; only a small amount is lost initially when the solution of fluorosilicic acid is evaporated to a smaller volume (0.2 mg SiO_2 is lost from 46 mg SiO_2 present when a 1:3 HF solution is evaporated from 25 to 2 ml at 100°C). This has been made the basis of an analytical scheme in which the sample is dissolved in HF alone, and weight aliquots are taken for the subsequent determination of various constituents. When solutions are taken to dryness, or to fumes of a higher boiling acid, the loss of Si is quantitative. Other elements such as B, As, Ge, and Sb form volatile fluorides also.[4] Clear solutions are seldom obtained when minerals are decomposed with HF; the residues were first thought to be undecomposed material but are now known to be precipitates formed by secondary reactions.[14]

The acid should never be added to a dry powder, such as dehydrated and ignited SiO_2, because of the danger of loss due to the rapidity of the reaction. The sample should be moistened with water, the acid added cautiously, and the covered container allowed to stand without additional heating until the initial reaction has taken place. This reaction may be accompanied by the generation of much heat, and highly siliceous material such as chert or glass sand will dissolve without the need for additional heating (5 g of powdered chert will dissolve in as many seconds). It may be necessary to slow down the reaction by the addition of cold water.

HF by itself is more effective than when mixed with another mineral acid. In spite of this, HF is seldom used alone because of the need to expel the last traces of fluorine from the solution or residue, and both sulfuric and perchloric acids are used for this purpose.[12] Perchloric acid has the advantage over sulfuric acid in that almost all of the perchlorates are easily soluble. Sulfuric acid is more effective in removing residual fluoride than is $HClO_4$, but even after repeated evaporations to SO_3 fumes, traces of fluoride ions remain. To effect as complete a removal as possible, beakers and lids should be washed down and the solution refumed. It is sometimes convenient to add boric acid to the solution to complex the residual fluoride as the slightly dissociated fluoboric acid, but this is not totally effective[16] and $AlCl_3$ has been found to be more reliable. The removal of fluoride ions can also be effected with oxalic acid, and this technique has been used in the determination of K in micas for age dating purposes (Ref. 1, pp. 34–35).

The failure to remove fluoride ions can lead to the formation of insoluble fluorides and subsequent errors due to the inclusion of ions of the analyte of interest in the insoluble matrix or the formation of insoluble salts of the analyte itself.

The effectiveness of the dissolution properties of HF is greatly enhanced by carrying out the decomposition under pressure. The technique was devised by Ito[13] who used a mixture of 1:1 H_2SO_4 and 48% HF in a steel bomb fitted with a Teflon liner to decompose a number of refractory minerals for the determination of ferrous iron and the alkalies. He used a finely powdered 0.5 g sample and heated the bomb at 240°C for 3–4 hr. Tourmaline, kornerupine, staurolite, garnet, chrysoberyl, axinite, magnetite, ilmenite, chromite, tantalite, columbite, baddeleyite, rutile, and corundum are totally decomposed; pyrite is only partly decomposed, as is zircon, but complete decomposition of the latter is achieved after 10 hr of heating with HF alone at 240°C. This technique has been used quite widely and has proven suitable for the dissolution of a wide variety of samples from silicates to coal.[17–19] One of the advantages of using this technique for dissolving samples for subsequent trace element determinations is the minimal use of reagents and the subsequent minimization of contamination. Knoop[20] took advantage of this in a slight modification of the technique. He dissolved 10 samples at a time by placing 50 mg of sample in Eppendorf® test tubes with 500 μl of HF. He put 10 test tubes in a Teflon-lined bomb and put the bomb in a drying oven at 130°C for 1 hr. Van Eenbergen and Bruninx (Supp. Ref. 6) investigated the potential loss of 22 elements using HNO_3–$HClO_4$ and HF–$HClO_4$ acid mixtures in a Parr bomb. They found that As, Ge, Hg, Se, and Te were retained, but Mo was lost due to the formation of molybdic acid, later found in the Teflon vessel. A similar loss of Ru during treatment with HF–$HClO_4$ was also traced to the vessel.

The Teflon liner is capable of withstanding temperatures of up to 250°C without deformation and at lower temperatures can be used many times with no apparent deterioration. There are precautions that must be observed when using devices of this sort, however, and one of them is apparent in the terminology "bomb" commonly used to describe them. The recommended maximum pressure and temperature for their use is 1200 psi and 150–180°C. A pressure greater than 1200 psi could lead to an explosion. A related precaution is to ensure that if the reaction evolves significant gases, this must be included in calculating the maximum attainable pressure for the conditions anticipated. In addition, any reaction using strongly oxidizing acids for dissolving organic materials should be done cautiously (if at all) in the bomb, and $HClO_4$ should not be used for any bomb decomposition. Sulfuric acid digestions should be carefully controlled because heating Teflon to greater than 300°C in the presence of H_2SO_4 can result in the evolution of toxic gases. The use of Teflon-lined bombs can offer significant advantages, but it is mandatory that the proper precautions be taken and that the manufacturer's instructions be read carefully and observed in detail.

A somewhat different type of low-pressure bomb (a capped 1 liter wide mouth Teflon jar) is described by Feldman (Supp. Ref. 7). Powdered samples are placed in 25–45 ml platinum dishes, inserted into the jar, and exposed to $HF-HNO_3$ vapor at 70°C for up to 16 hr.

There are some applications of HF which make use of its selective dissolution properties. In some powder X-ray diffraction problems, complex multimineral diffraction patterns are very difficult to analyze. M. A. Chaudhry of the BCM has found that the treatment of a portion of the finely ground sample with HF prior to recording the diffractogram results in the elimination of the quartz and some other peaks and leaves a much simpler pattern to analyze. This often leads to the identification of one or more minerals not attacked by the HF. The complete diffractogram can then be looked at with some knowledge of what is present, and this simplifies the analysis significantly. A second example is the selective dissolution of most silicates in the presence of quartz by using a solution of 32 g of boric acid dissolved in 75 ml of 48% HF (fluoboric acid), H_3PO_4, and 2N $FeCl_3$ (Ref. 2, p. 94).

4.2.2   With Hydrochloric Acid

Hydrochloric acid, as the approximately 37% (w/w), 12 M solution having a specific gravity of 1.19, is probably the most commonly used halide acid. It dissolves most sulfides (pyrite is a notable exception), phosphates, and carbonates, is the best solvent for oxides, and the resultant chlorides are, for the most part, water soluble. Volatile chlorides are formed by Ge, Hg, Sb, As, and Sn.

The chloride anion is a strong ligand and forms a series of complexes with many metals, including Au, Ag, and Pb. The formation of these complexes has been made use of in extraction and ion exchange separations and makes HCl a very useful solvent for numerous separation reactions.

In the presence of a strong oxidizing agent, HCl has significantly enhanced dissolving properties resulting from the oxidation of Cl^- to $Cl°$ in the case of HNO_3 (aqua regia) and $KClO_3$ and the formation of Cl(III) oxide (Ref. 1, p. 18) as well in the presence of the latter. Mixtures of HCl and chlorates are, for example, capable of dissolving the primary mineral arsenates of Co and Ni as well as molybdenite, cinnibar, and pyrite, none of which is appreciably affected by HCl alone. There is an additional advantage arising from the above conditions in that As and Sb are transformed into their nonvolatile pentavalent state.

HCl is most commonly used in glass equipment but Nalgene,[®] Teflon, Au, and Pt containers are also used. It should be pointed out that HCl with free chlorine in it (due to the presence of an oxidizing agent such as HNO_3, ClO_3^- ions, or H_2O_2) will attack both Au and Pt. These two metals are also corroded by prolonged contact with concentrated HCl in the presence of light or air (Ref. 1, p. 13).

HCl will, like HF, decompose some refractory minerals under high pressure and temperature conditions, but it is a method that has found relatively little use in rock and mineral analysis, although the solution and oxidation of organic compounds is often done in this way. The lack of attention to this method which, in addition to its ability to solubilize many difficultly soluble materials, yields a solution free of added salts and having only minimum contamination, is likely owing to the specialized equipment seemingly necessary and the somewhat hazardous aspect of the procedure. In point of fact, the equipment needed is not unduly specialized and the hazards can be confined to loss of sample only.

Originally the sample was placed in a platinum tube which in turn was placed inside a glass tube and the latter sealed, both tubes being filled with dilute HCl; the tubes were heated inside a steel tube which contained a volatile organic substance to equalize the initial pressure developed by the heated acid. Hillebrand et al. (Ref. 21, p. 850) have pointed out the difficulties of this procedure, the chief of which is the incomplete decomposition of the sample; a rather large amount (several milligrams) of platinum is also dissolved.

The only related work is that of Wichers et al. who applied the method first to the attack of refractory platiniferous material[22] and then to the decomposition of refractory oxides, ceramics, and minerals.[23] The special techniques and apparatus involved in the use of the sealed glass tube are discussed in detail,[24] and the whole has been summarized in detail (Ref. 25, p. 1038). Briefly, a 0.1 g sample is placed in a borosilicate tube, 4 mm i.d. and 2 mm wall thickness, 20 cm in length, containing 1 ml HCl, and heated at 250–300°C for periods of 16–48 hr, without external protection; larger samples are heated in tubes having an internal diameter of about 15 mm and enclosed in a steel tube containing CO_2 to provide an external pressure similar to the internal one generated in the glass tube. Among the minerals successfully decomposed in this way were cassiterite, muscovite, amphibole, tourmaline, cordierite, sillimanite, spinel, bauxite, and chromite; talc, diaspore, allanite, garnet, and sphene were almost completely attacked, while partial attack was obtained for beryl, topaz, and dumortierite. Of the plagioclase feldspars, only anorthite was completely decomposed; bytownite and andesine were partially attacked, albite and oligoclase not at all. Two acid rocks (andesite and rhyolite) were tried and found to be only partially attacked, even after 2–3 days of heating at 300°C. Mallett et al. (Supp. Ref. 8) have also used the sealed tube technique for the decomposition of platiniferous material.

HCl has been used for leaching soil and silt samples for geochemical survey purposes. Some of the more common digestions for which it is used are those involving attack by mixed concentrated acid and by both warm and cold dilute[9, 11] (1 M) acids.

4.2.3   With Nitric Acid

The decomposition of sulfides, with the attendant oxidation of the sulfide to sulfur or sulfate, and analyses where the presence of other mineral acids would cause interference, as in the determination of phosphates, are the most common applications of the strongly oxidizing nitric acid in rock and mineral analysis. It is also extensively used for samples containing organic material to oxidize any easily oxidizable material prior to the addition of perchloric acid. HNO_3 is generally ineffective with oxides but some thorium and uranium minerals, such as thorianite and uraninite, are dissolved with the simultaneous oxidation of U from the quadrivalent to the hexavalent oxidation state. Dissolution may be accompanied by separation, as in the formation of tungstic oxide or metastannic acid; Sb and Mo also form insoluble products. Most metals dissolve in HNO_3, the notable exceptions being Au and Pt (if the acid is free of chlorine). For this reason, HNO_3 is used extensively for parting Ag from Au in the alloy formed in the cupellation step of the fire assay process. The presence of Pt can make the parting difficult, however, unless sufficient Ag and Au are present (Ref. 26, p. 131).

HNO_3 is more often used in conjunction with other mineral acids, most commonly HCl. The HNO_3–HCl mixture in a ratio of 1:3 by volume is called aqua regia and, due to the presence of nascent chlorine and nitrosyl chloride, it has a much greater oxidizing power than does HNO_3 alone.

As was mentioned in Sec. 4.2.2, aqua regia will attack both Au and Pt. Rhodium is only partially dissolved. Iridium is impervious, and a Pt alloy containing 20% Ir or more will exhibit the same resistivity and is used for some laboratory equipment for this reason.

Aqua regia is used very frequently for the dissolution of most metal sulfides and arsenides and finds extensive use in geochemical analysis. It is effective for molybdenite, many bismuth minerals and, in the presence of chlorate or bromine, will render most arsenic salts quantitatively soluble. Bauxite and cassiterite are not dissolved effectively in aqua regia, and the addition of hydrogen peroxide is frequently necessary to effect a complete dissolution of manganese ores (Ref. 1, pp. 52–53). The oxidation of sulfides to sulfates is very rapid when aqua regia is used along with an iodide salt (Ref. 1, p. 58). The same results can be obtained using the *Lefort* or *inverted* aqua regia (Ref. 1, p. 49)—a mixture in which the HCl and HNO_3 are in a 1:3 ratio. Either solution can be used for determining S in pyrite, for example. The reactions are vigorous, however, and the reagents must be added slowly and the reaction mixture kept cool. Ice must be used in the case of the Lefort aqua regia reaction to preclude the precipitation of elemental S or the evolution of H_2S. A common technique used in mine laboratories for dissolving samples is to treat the sample with HNO_3 and $KClO_3$, and can be applied to samples containing up to 10% S.

4.2.4 With Perchloric Acid

The hazards associated with the use of perchloric acid cannot be overemphasized, but the almost universal solubility of perchlorates (with the notable and useful exception of K, Rb, and Cs perchlorates) favors the use of $HClO_4$ for decomposition, and it has replaced sulfuric acid in popularity, even though the latter is often more efficient. The perchlorate ion is a weak ligand, however, and silver will precipitate from $HClO_4$ solution if chloride ions are present. It is one of the strongest acids known and, while nonoxidizing in a diluted hot or cold state, is a powerful oxidant when hot and concentrated. Mixtures of hot concentrated sulfuric and $HClO_4$ have an even greater oxidizing power, equivalent to anhydrous $HClO_4$. Because of the oxidizing power of $HClO_4$, care must be exercised in its use[27, 28] if serious accidents are to be avoided. Contact of the boiling, concentrated acid or hot vapor with either organic matter or easily oxidized inorganic matter may lead to explosive reactions, and such reactions have been known to occur in the cold also; the use of HNO_3 (Sec. 4.2.3) as a preliminary oxidizer before the temperature is raised above 100°C is recommended. $HClO_4$ should never be used in an ordinary fume hood because of the liklihood of depositing potentially explosive perchlorate salts in the exhaust system. Stainless steel $HClO_4$ fume hoods with facilities to wash down the interior of both the ducting and the fume hood itself (see Sec. 2.1.1) must be used for any digestions involving $HClO_4$.

$HClO_4$ in combination with HF has been widely used for the decomposition of many silicates, but Rodgers[29] cautions that many of the established methods involving $HClO_4$ digestions should be carefully reevaluated, especially those applied to the decomposition of chrome spinels. The possibility of the volatilization of chromyl chloride from boiling $HClO_4$ solutions containing chloride ions is apparently not considered in many instances. Other elements which can be lost are Re, Se, Te and As. Kodama[30] and Chapman[4] and his co-workers have both considered the volatilization of various elements from $HClO_4$ solutions, and a familiarity with their work is advisable.

In spite of the above precautions, however, many silicate minerals can be decomposed with hot concentrated $HClO_4$ with subsequent separation and dehydration of the silica. A mixture of HF and $HClO_4$ is very effective; when the sample solution is evaporated to dryness, the excess fluorine expelled, and the residue ignited at about 550°C, thermal decomposition of the perchlorates occurs, resulting in a mixture of chlorides (alkalies), basic chlorides (alkaline earths), and oxides; hot water leaching of this residue, as described for $HF-H_2SO_4$ (Sec. 4.2.5), will effect a useful separation. Willard et al.[31] used a distillation procedure with $HClO_4$ to expel residual fluorine and at the same time to dehydrate the silica, prior to the determination of potassium as the perchlorate. Marvin and Woolaver[32] leached the ignited perchlorates with hot water, followed by removal of the alkaline earths with oxalate and oxine, in order to obtain the alkalies alone as chlorides; the use of beakers as weighing media for the alkali chlorides has been criticized. Edge and Ahrens[33] use an HCl leach for the residue from the $HF-HClO_4$ decomposition of silicate rocks for a combined cation exchange–spectrochemical method for the determination of rare earths.

$HClO_4$ can be used successfully in the dehydration of silica and is reputed to be better than HCl for this purpose. It is also a useful agent for dissolving sulfates such as barite and has been used to dissolve galena for subsequent Pb and Ag determination. $HClO_4$ solutions precipitate antimonic acid, and thus it is not a useful solvent for antimony ores. It also precipitates tungstic acid, and if Nb and Ta are present, their equivalent acids will coprecipitate (Ref. 1, p. 72).

One of the most widespread uses of the $HClO_4-HNO_3$ combination (usually in a 1:3 or 1:4 ratio) is in the digestion of silt and soil samples collected for geochemical survey purposes. The wide range of minerals susceptible to $HClO_4$ attack and the almost universal solubility of perchlorates make it very valuable for this purpose. The presence of HNO_3 is necessary to oxidize any organic matter before there is a danger of it reacting explosively with hot concentrated $HClO_4$ as the digestion proceeds.

In general, $HClO_4$ is a very valuable reagent in rock and mineral analysis, but it must be used with great care and all safety precautions must be observed in order to avoid the risk of a serious explosion (Supp. Ref. 9).

4.2.5 With Sulfuric Acid

Relatively little use is made of sulfuric acid alone to decompose rocks and minerals, chiefly because of the formation of insoluble sulfates which are often very difficult to redissolve. In the concentrated form it is oxidizing in nature and thus most oxides are not decomposed by it. It is, however, a good solvent for most naturally occurring halides in either a diluted or a concentrated form. Hot concentrated H_2SO_4 is also suitable for decomposing many As, Sb, Sn, Se, and Te minerals with no loss due to volatilization if it is warmed on a hot water bath and not taken to fumes. H_2SO_4 finds a preferred use, however, in the decomposition of rare earth minerals such as monazite (Ref. 34, p. 108).

Niobium and Ta minerals are efficiently decomposed by heating the sample with H_2SO_4 and $(NH_4)_2SO_4$ (Ref. 1, p. 65). Alkali sulfates have been used in this manner as well, and besides the minerals mentioned above, titanium minerals containing oxygen are also decomposed.

As a mixture with other acids such as HF it finds many uses. Its high boiling point (about 340°C for the 96% acid) enables it to expel most other volatile compounds, including fluorine. Its dehydrating effect on silica is a useful step when removal of dissolved silica is desired. Ignition of the sulfates and subsequent leaching with hot water will yield a solution containing little more than the alkalies and some magnesium.[35] A mixture of HF and H_2SO_4 is the common solvent for ferrous iron determinations, and a solution of HCl and H_2SO_4 is effective as a solvent for basic phosphates such as those of Al and Fe.

4.2.6 With Other Acids

Phosphoric acid, as the 85% (w/w) laboratory reagent (specific gravity 1.70, approximately 15 M), is frequently used to decompose oxides, and even chromite will be eventually dissolved. The formation of pyro- and metaphosphoric complexes at high temperature aids in dissolution. Ingamells[36] uses a mixture of phosphoric acid and sodium pyrophosphate to dissolve minerals such as biotite, garnet, chromite, olivine and bauxite, as well as Mn and Fe ores. A mixture of phosphoric and sulfuric acids is used to dissolve fused Al_2O_3[37] and chromium ores.[38] Phosphoric acid has been used extensively in phase analysis for free silica to dissolve silicates but not quartz (Ref. 2, p. 94). It can also serve, in mixtures with other acids and compounds, as a reagent for decomposing sulfur and oxygen-containing minerals of many base metals (Ref. 1, p. 75).

Acetic acid is used by Epshtein and Ginberg[39] to dissolve carbonate and phosphate minerals prior to the determination of niobium. Hydrobromic acid is a good solvent for oxides and, as a mixture with HF, was used by Cuttitta[40] to prepare a solution of Mn ores from which thallium is determined by extraction as bromide; hydriodic acid is used to decompose pyrite, chalcopyrite, stibnite, and arsenic sulfides with the formation of H_2S, which is then determined.[41]

A recent paper has been published in which *m*-benzene disulfonic acid is used in conjunction with HF in the place of $HClO_4$ or H_2SO_4.[42] It has the advantages of a high boiling point, virtually all of its salts are highly soluble, and it does not present the explosive hazard that $HClO_4$ does.

4.3   DECOMPOSITION BY FUSION

4.3.1   With Sodium Carbonate

Anhydrous Na_2CO_3 is the most commonly used of the alkaline fluxes. It will decompose silicates, oxides, sulfates, phosphates, fluorides, and carbonates, to mention only the very familiar sample types, either by the formation of a definite compound (silicate, vanadate, aluminate, chromate) or by rendering the material more amenable to attack by acids. Fusions, which should be made in vessels of platinum or platinum alloy, usually require a temperature of 1000–1200°C for a flux-to-sample ratio of about 5:1. It is well to swirl the molten mass at some stage of the fusion process to ensure complete exposure of the sample to the flux. It should be kept in mind that platinum is attacked by Na_2O (a decomposition product of Na_2CO_3 at 900°C) to form soluble sodium platinate (Ref. 1, p. 91). The corrosion of platinum is enhanced with increased oxidizing conditions. Crucibles of iron and nickel can also be used for carbonate fusions.

The oxidizing conditions normally present when fusing with an alkaline flux in air can be materially increased by the addition of a small amount of an oxidant to the flux (it follows that during these fusions care must be taken to exclude, or at least minimize, the entry into the crucible of reducing gases if a flame is used, or to maintain an oxidizing atmosphere if an electric muffle is used). Malhotra[43] decomposed chromite with Na_2CO_3 (flux-to-sample ratio of 10:1) in a platinum crucible with the aid of a current of oxygen impinging on the surface of the melt from a bent platinum tube (a platinum Rose crucible would serve as well). A current of air introduced through a hole in the crucible lid is equally effective; Fukasawa et al.[44] used this procedure to fuse carbonaceous material with Na_2CO_3.

A mixture of Na_2CO_3 and KNO_3 (1:2 to 1:4) is often used as an oxidizing alkaline flux, usually for the decomposition of sulfides, arsenides, and similar reducing substances; the proportion of the KNO_3 should not be great enough to damage the platinum crucible, nor to cause too rapid a reaction with such readily oxidizable substances as arsenides. One must not, however, treat the cooled mass with hydrochloric acid in platinum because of serious attack by the aqua regia formed. Na_2O_2 is an acceptable substitute for KNO_3 and creates no problems during dissolution of the fusion cake. It is not uncommon to mix a small amount of Na_2O_2 (approximately 0.1 g) with the Na_2CO_3 in all fusions as a precautionary measure. Fukasawa et al.[44] use a 1:1 mixture of Na_2O_2 and Na_2CO_3 for the fusion of ilmenite. $KClO_3$ may also be used in place of either KNO_3 or Na_2O_2.

A mixture of Na_2CO_3 and $Na_2B_4O_7$ (1:1) has been used to decompose cassiterite[45] for the determination of indium; May and Grimaldi[46] used a 3:1 mixture to fuse ores and rocks prior to the determination of beryllium. Chirnside et al.[47] used a 2:1 mixture for the decomposition of sapphire and ruby.

When fusing silicate materials it should be kept in mind that the more basic rocks (peridotite, basalt, etc.) require substantially higher flux-to-sample ratios than do acidic rocks (up to 15:1 as opposed to 5:1, respectively) (Ref. 1, p. 92). The samples should also be finely ground (to pass 200 mesh), but even with fine samples and large flux-to-sample ratios some minerals (zircon, cassiterite, and titanite, for example) decompose only very slowly and incompletely. The addition of borax in ratios varying from three or four times as much borax as carbonate to the same preponderance of carbonate over borax has been found to assist in the decomposition of many of the more refractory minerals. Cresser and Hargitt have used a mixture of $NaHSO_4$ and Na_2CO_3 in the analysis of soils and rocks containing chromite.[48]

4.3.2 With Potassium Carbonate

Most of what has been said about fusions with Na_2CO_3 also applies to K_2CO_3, but because the latter is more hygroscopic (it must be dehydrated immediately before use) it is seldom used except when the fusion products of some elements are more soluble than the corresponding sodium salts (e.g., niobium). According to Hoffman and Lundell,[49] however, the cake from a K_2CO_3 fusion of a sample containing much SiO_2 and elements which form insoluble carbonates (e.g., Ca, Pb, and Mg) disintegrates more readily in hot water than does one from a fusion with Na_2CO_3.

An intimate mixture of anhydrous sodium and potassium carbonates in the ratio of their molecular weights is called "fusion mixture" and has a lower melting point than either of the single salts. This is an advantage for the fusion of samples in which chlorine or fluorine is to be determined, but is little used otherwise because of the general need to operate at temperatures much above the melting point of the mixture in order to be sure of complete decomposition of the sample. A disadvantage in the use of K_2CO_3 is the addition to the solution of a large amount of potassium salts, which tend to contaminate precipitates more readily than do those of sodium.

Polezhaev[50] used a 1:1 mixture of $KHCO_3$ and KCl (flux-to-sample ratio of 50:1) to decompose silicates prior to the determination of free silica; the fusion was done in a steel crucible. Govindaraju[51] has used a K_2CO_3–H_3BO_3 fusion mixture followed by an ion-exchange dissolution for silicate rocks.

4.3.3 With Sodium and Potassium Hydroxides

Sodium and potassium hydroxides are very strong alkaline fluxes, but they are used only under special circumstances. Their relative unpopularity derives from several factors. One is the fact that they are very difficult to obtain in a pure state and thus are not applicable to many trace element determinations. Their use for trace element analysis is also affected by their reaction with most crucible materials. Dodson showed in his studies[52] that trace elements were either added to the sample (as in the case of commercial grade Ni crucibles) or adsorbed into the crucible walls from the sample (as in the case of graphite crucibles). He determined that silver crucibles were better than gold, but that zirconium ones were only slightly corroded and proved to be the best when the fusion was carried out in an electric muffle at fairly low temperatures (400–450°C). Crucibles made of specpure grade Ni were also satisfactory for low temperature KOH fusions. Potassium hydroxide has the further disadvantage of the addition of a large amount of potassium salts to the solution (see Sec. 4.3.2).

These compounds tend to froth and spit when they are heated because of their excessively hygroscopic nature. It is the usual practice to melt the flux first, allow it to cool, add the sample, and then continue the fusion.

Alkali hydroxide fusions (with a flux-to-sample ratio of 10:1 or more) are effective for the dissolution of numerous rock and mineral types (Ref. 1, Sec. 2.3). They decompose alkaline earth sulfates, carbides, silicides, clay, chlorides, and silicates quite readily. They are equally effective for titanium dioxide minerals, corundum, bauxite, and many phosphate minerals (monazite, for example). Their effectiveness as fluxes can be enhanced by the addition of other compounds. The decomposition of ilmenite and rutile is speeded up when boric acid is added, and a small amount of alkaline cyanide will ensure the complete dissolution of cassiterite. A combined alkali hydroxide-alkali peroxide flux will effect the dissolution of beryl and most zirconium ores and can also be used for the determination of W and Mo in complex ores.

Grimaldi[53] has discussed the use of NaOH and KOH in the determination of Nb. The use of KOH renders Nb salts soluble in the subsequent water leach, but when NaOH is used as the flux, they hydrolyze to give the insoluble sodium niobate.

4.3.4 With Sodium Peroxide

Sodium peroxide, a powerful flux and oxidant, is not used much in rock analysis, chiefly because it is difficult to find a vessel in which the fusion can be made and still not seriously contaminate the analysis. It is also, because of its very hygroscopic nature, somewhat difficult to handle, but obtaining reagent grade material is no longer a problem. Fusions are made at 600–700°C, and platinum is seriously attacked at this temperature, although it may be used if the crucible is first lined with a thin layer of fused sodium carbonate and the fusion is made rapidly at as low a temperature as is permissible. Silver, gold, and zirconium crucibles resist attack by the fused peroxide. Usually the fusion is made in crucibles of nickel or iron, with accompanying contamination of the fusion by the constituents of the crucible.

The use of Na_2O_2 as a flux in refractory and mineral analysis has been reviewed by Belcher.[54] He determined that crucibles made of zirconium were the least affected by the fusion. In a discussion of the decomposition and analysis of chrome spinel, however, Rodgers[29] suggests the use of silver crucibles.

In general, Na_2O_2 is used as a flux in the decomposition of a silicate rock or mineral sample only if the sample contains spinels, zircon, arsenides, sulfides, and like compounds. A peroxide fusion is also effective for the decomposition of rare earth phosphates, tungsten, niobium and tantalum minerals, zirconium oxides, and vanadates. Marvin and Schumb[55] studied the procedure in detail and found that the most effective mixture was Na_2O_2:C = 15:1, for a flux-to-sample ratio of 15:1. They also found that the residue remaining after dissolution of the fusion cake could be attributed to the spattering of sample on the lid of the crucible at the high temperature (1450°C) reached during the ignition; covering the mixture with a layer of Na_2O_2 reduced the amount of spattering. No loss in weight of the nickel crucible occurred and they successfully applied the method to a number of refractory minerals such as cassiterite, corundum, zircon, and ilmenite.

The oxidizing powers of Na_2O_2 can be mitigated to some extent by mixing it with Na_2CO_3 or NaOH. Fusing with a combined flux is a more common practice than using Na_2O_2 alone (Ref. 1, p. 112).

4.3.5   With Borax and Boric Oxide

Borax (sodium tetraborate) and boric oxide are an alkaline and acid flux, respectively, which are relatively little used at present, in spite of their having definite advantages to offer.

A disadvantage of borax, compared with boric oxide, is that it adds a quantity of sodium salts to the analysis, but Hillebrand et al.[21] believe it to be a better flux than boric acid if alkalies are not to be determined on the sample. Jeffery[56] has found that refractory oxides, particularly the earth acid and rare earth minerals, yield completely to borax fusion, and it has been noted as satisfactory for the decomposition of zircon and chromite. He fuses a 0.5 g sample with 5 g of fused borax (intumescence of the material makes such a preliminary fusion mandatory), usually for a period of about 2 hr; dissolution of the fusion cake in HF will separate elements such as Fe, Ti, Nb, Ta, and B, and evaporation of the filtrate with H_2SO_4 will remove fluorine and boron (as BF_3), leaving a pyrosulfate melt. The melt is a very viscous one, and a platinum rod must be kept in the crucible to permit occasional stirring of the mass to ensure complete contact between sample and flux.

Boric oxide is a high temperature flux that is effective with those minerals that succumb also to borax fusion. It is prepared by slowly fusing boric acid in a platinum crucible and chilling the crucible rapidly in order to shatter the cooled cake. Fusions with boric oxide are done at 1000–1100°C, and again the crucible must be cooled quickly in order to loosen the cake. As in the case of borax, the boron may be removed by volatilization as methyl borate.

Hillebrand et al.[21] consider the necessity for complete removal of the boron a serious liability in the use of these fluxes; they also warn that the volatilization of sodium borate is a serious possibility at the high temperature needed for fusion, and when alkalies are to be determined, this possible loss must be considered.

Mention should be made here of the disagreement over the use of a boric acid fusion for the determination of SiO_2 in fluorine-bearing samples. In contrast to previous claims, Hoffman and Lundell[49] found that losses of silica occur when the sample is fused either with boric oxide or with Na_2CO_3 and followed by evaporation of the H_2SO_4 or $HClO_4$ solution in the presence of added boric acid. Further, the silica is contaminated with boron, and while the latter is not completely expelled by treatment with methyl chloride, it is expelled as BF_3 during the HF treatment of the impure silica, giving high results. A similar unsatisfactory result was obtained when H_3AsO_3 was used as a flux.

Borax is useful for the decomposition of many refractory compounds including the oxygen-containing minerals of Ta, Nb, and Ti. These minerals require fusion times of 1 or 2 hr, however. Borax is also an effective flux for many zirconium minerals as well as most micas.

4.3.6   With Lithium Tetraborate and Metaborate

Lithium tetraborate ($Li_2O \cdot 2B_2O_3$) and lithium metaborate (Li_2O B_2O_3), or some mixture of the two, are the most common fluxes now in use for AAS[57-59] and XRF[60] analytical work on rock and mineral samples. They have also been used in the preparation of samples for colorimetric,[61] spark solution emission spectrometric[62] and dc arc spectrographic[63] techniques.

The lithium borate fluxes are admirably suited to this type of application for several reasons. They do not damage platinum ware (if oxidizing conditions are maintained); they do not wet gold or platinum–gold alloy crucibles (see Sec. 2.2.4), and thus the complete melt can be poured out of the crucible; they are capable of dissolving almost all minerals, and their melts are easily soluble in dilute acids. It is also possible to obtain durable glass buttons which are suitable for presentation to an XRF spectrometer with no surface treatment necessary after the buttons have been removed from their mould. In addition, they can be obtained in a very pure state and can thus be used for trace element analysis.

The differences between $Li_2O \cdot B_2O_3$ and $Li_2O \cdot 2B_2O_3$ lie in the relatively higher acidity of the latter. Thus the metaborate salt is a better flux for acidic rocks and minerals (high silica) and the tetraborate salt is better for basic ones (dolomite). Bennett and Oliver[60] have made a study of the use of these two fluxes for a wide variety of materials (see Table 4.1). They have proposed a "universal flux" of a 1 + 4 mixture of tetraborate and metaborate for samples with silica and/or alumina as major constituents. This is very close to the 73% B_2O_3:27% Li_2O ratio that gives a eutectic at 832°C in the $Li_2O \cdot B_2O_3$ phase diagram.

The tetraborate salt has been the one more commonly used for XRF work until recently, mainly because of the difficulty of obtaining the metaborate salt commercially. The latter is available now, however, and there are many advantages to using the metaborate flux. It fuses at a lower temperature (849°C versus 917°C) and it yields a much more fluid

TABLE 4.1   Suitable Fluxes for Various Types of Material

Material	Parts of Flux to 1 Part of Sample (Ignited)	
	Lithium Metaborate	Lithium Tetraborate
Silica–alumina range	4	1
Steatite	4	1
Bone china	4	1
Bone ash (apatite)	0	3[a]
Zircon	4	1
Zirconia	0	12
Titania	0	12
Limestone	0	5[a]
Dolomite	0	5[a]
Magnesite	0	10
Witherite	0	12
Barium titanate	0	12
Borax frit	0	1
Glaze, <20% PbO	0	1
Glaze, 20–60% PbO	0	12
Lead bisilicate	0	12
Enamel	0	12
Iron (III) oxide	0	12
Chrome-bearing refractory	10	12.5′
Silicon carbide	6	1.5
Boron carbide	3.25	0.88

Reproduced with permission of *The Analyst*.[60]
[a] These ratios of flux to sample are the minimum to produce clear cast beads. For purposes of XRF, greater dilutions may be advisable.

melt. Consequently the fusions can be done over a gas burner, and the melt can be much more easily poured into the mould with very little, if any, remaining in the crucible.

The fusion vessels most suited to these fluxes are those made of a nonwetting platinum alloy. It is imperative, however, that oxidizing conditions be maintained, or Fe, Cu, Pb, Co, and other metals can be reduced and extracted from the sample into the Pt.[59] Gold crucibles are less subject to this, but the low fusion temperature of Au makes it useless as a crucible for fusions in a flame. Gold is better than platinum, however, if the fusions are to be done in a furnace. It has been reported that the use of graphite crucibles at high fusion temperatures (1200°C) leads to the reduction and subsequent loss of Fe and Co,[64] and they should be used with caution. In addition, it is critical that a good grade of graphite be used. Some grades are too porous and the melt will leak out of them and ruin the furnace floor, whereas other grades will cause the melt to stick to the crucible when it is being poured out. It is also essential that maximum air circulation be maintained in the furnace so that enough oxygen is present to prevent the reduction of some elements but not enough that the crucibles will burn out too quickly.

The flux-to-sample ratio will depend on the nature of the sample and the subsequent analytical technique to be used, but in general it will range from 3:1 to 10:1 for silicates.[59] A 5:1 ratio is favored by many when the melt is to be dissolved in dilute acids (usually 4% HNO_3) for subsequent atomic absorption, colorimetric or other solution-based work. A mixture having a smaller flux-to-sample ratio will be desirable if buttons for trace element analysis by the XRF technique are being prepared, but the viscosity of the resulting melt is a limiting factor.

Samples fused for XRF analysis were quite often presented to the spectrometer as a pellet which had been prepared by crushing the melt to pass 200 mesh and then pressing it into a pellet. The ease with which fused beads of a high quality can now be made makes this step unnecessary, and the additional potential problems of contamination and particle size effects provide a large incentive to avoid this technique.

4.3.7   With Potassium Pyrosulfate

Potassium pyrosulfate is a more effective flux than is sodium pyrosulfate because the latter tends to lose SO_3 more readily and thus reduces its effectiveness (once the pyrosulfate has been converted to the normal sulfate it is without effect on undecomposed sample). Again, the sodium compounds tend to be more soluble than those formed by potassium, but this advantage is offset by the tendency of the melt to crust over more readily, thus making it difficult to see when decomposition of the sample is complete. Potassium pyrosulfate is much more widely used, and finds particular use for the fusion of the ignited oxides of the R_2O_3 group. The pyrosulfates are particularly effective with oxides and less so with silicates (quartz and opal are insoluble in sodium pyrosulfate and fusion with this flux is used to separate them from the silicates in a rock).

Potassium pyrosulfate does not attack porcelain or silica vessels, and fusions can be made in these when the introduction of platinum into the analysis must be avoided (this will occur even at a relatively low temperature). Pyrex "copper flasks" have often been used for pyrosulfate fusions for As and Sb determinations. Platinum is significantly attacked when the fusion is made at a high temperature (medium to bright red heat), and this flux is effectively used to remove stains from crucibles; 2–3 mg of platinum may be dissolved during a prolonged fusion. Schoeller and Powell[34] obtained complete fusion of the finely divided ignited R_2O_3 oxides in 5–10 min; it is common practice for the bulk of the ignited oxides to be transferred from the platinum crucible to one of Vycor® in which the main fusion is to be done, and a brief fusion (1–2 min) suffices to fuse the small amount of residue clinging to the platinum crucible. Prolonged fusion will result in the formation of the normal sulfate and the crusting over of the fused mass; the flux can be regenerated by the addition of a drop or two of H_2SO_4 (1:1) to the cooled mass, followed by gentle heating to a molten state. Hillebrand et al.[21] point out that this is often necessary toward the end of a prolonged fusion in order to convert certain sulfates, which may have been formed during the conversion of the pyrosulfate to normal sulfate, to a more soluble form.

It is often useful, following the decomposition of a sample with hydrofluoric acid, to add sulfuric acid and expel the fluorine by heating to fumes, and finally to fuse the residue with potassium pyrosulfate. A discussion of some of the uses of pyrosulfate fusions is given by Sill[65] who also mentions some of the differences between sodium and potassium salts.

4.3.8   With Alkaline Fluorides

These (KF, KHF_2, NH_4F) are low-temperature fluxes and are used chiefly for the opening up of refractory silicates and oxides, such as some niobium, tantalum, beryllium, and zirconium minerals, through the formation of soluble fluoride complexes. Although the temperature is lower, the time of fusion is usually shorter than that required for other fusions, and of course, platinum vessels must be used. As mentioned in Sec. 4.1, the losses that occur here and during fusion with other acid fluxes result from the same conditions that cause losses from acid solutions, that is, the formation of volatile fluorides. Such fusions should be carried out in a hood because of the loss of HF that occurs during the heating.

Zircon can be decomposed by fusion with KHF_2 alone at 800–900°C, but in some laboratories it has been found useful to begin the attack with a mixture of KF and HF; the crucible is gently heated until initial frothing has ceased, then heated at 800–900°C for about 90 sec to ensure complete decomposition of the zircon. Athavale et al.[66] used HF and KHF_2 for the decomposition of low-grade uranium ores.

NH_4F has the advantage of easy removal of the excess flux by vola-tilization after decomposition is complete.[67] It has been used to decom-pose beryl and tourmaline[68] at 300–400°C, combined with subsequent heating with oxalic acid.

Sill[65] is a strong advocate of KF as a flux and proposes the transposition of the fused cake to a pyrosulfate fusion by the addition of concentrated H_2SO_4. This step eliminates SiF_4 and HF so that subsequent steps can be carried out in glass apparatus. The fact that the sample is in a sulfate medium is recognized as a possible disadvantage.

4.3.9 With Other Fluxes

There are numerous other fluxes, generally mixtures of two or more compounds, to be found in the literature, and nothing is gained by at-tempting to catalog all of them. Mention is made of only a few to give some idea of the variety that has been used for special purposes.

A mixture of ZnO and Na_2CO_3 is used frequently[69, 70] for the decom-position of silicates prior to the separation and determination of fluorine. Hillebrand et al. (Ref. 21, p. 836) mention the use of lead oxide and carbonate, and basic bismuth nitrate, citing as advantages their ready removal after the fusion and the opportunity of determining the alkalies in the fused portion. Crystalline pyrophosphoric acid, heated at 250°C, is used to separate free and combined silica.[71] Ammonium halides and nitrate were used by Isakov[72] to decompose minerals and ores; he found that a mixture of NH_4Cl and NH_4NO_3 (solid aqua regia) was most effective.

4.4 DECOMPOSITION BY SINTERING

4.4.1 With Sodium Peroxide

Fusion with Na_2O_2, and the disadvantages thereof, has been discussed in Sec. 4.3.4. Rafter[73] has proposed a sintering procedure that will, at a comparatively low temperature, easily and completely decompose those minerals resistant to decomposition by the usual methods, without the slightest attack on a platinum crucible, and very slight attack on iron or nickel vessels. He found it particularly useful in the decomposition of earth-acid and rare-earth minerals; Ta_2O_5 and tantalum minerals have been decomposed completely and the cake dissolved in cold water. Lenc (Supp. Ref. 10) regards this technique as one that has application as a universal decomposition method.

Control of the temperature and time of the sintering is very important. The finely ground sample (0.2–1 g) is sintered for 7 min with 1.2–3.0 g Na_2O_2 in a platinum crucible at 480 ± 10°C.

4.4.2   With Sodium Carbonate

In speaking of modified classical techniques in silicate analysis, Chirnside states that "it is quite extraordinary that the most significant advance in silicate analysis which has taken place as long ago as 1930 has been ignored for years by most analysts in this field" (Ref. 74, p. 11 T). He was referring to the work of Finn and Klekotka,[75] who found that many aluminosilicates were readily decomposed by sintering them with a small amount of Na_2CO_3 (0.6 g sodium carbonate for 0.5 g sample, sintered at 875°C for 2 hr). The reduction in the amount of sodium salts thus introduced into the analysis is an obvious advantage. Hoffman[76] continued this investigation and applied it to various rocks and minerals, but used a 2:1 ratio of flux to sample and obtained a fused cake rather than a sinter. The mixture, 0.5 g 60 mesh sample and 1.0 g Na_2CO_3, is heated in a 75 ml platinum dish for 15 min at 1200°C in an electric muffle furnace. Subsequent steps in the separation of silica are carried out in the same dish. The procedure has much to recommend it to the analyst.

Ruby and sapphire powders have been sintered in a platinum crucible with a flux consisting of anhydrous Na_2CO_3 and anhydrous $Na_2B_4O_7$, prior to the determination of iron (about 0.002%). Only 50 mg of flux and 20 mg of sample were used, and the mixture was heated in an electric muffle furnace at 1050°C for 10 min. The lower temperature and shorter heating time required minimized the exchange of iron between the melt and the crucible.[47]

A mixture of ZnO and Na_2CO_3 was used as a sintering medium by Moriya[77] for the separation and determination of boron in silicates (10 min at 900°C); Peck and Tomasi[78] sintered silicates with a mixture of ZnO + Na_2CO_3 + $MgCO_3$ at 800°C for 30 min for the subsequent determination of chlorine. Other mixtures that are proposed are Na_2CO_3 + MgO + KNO_3 for the determination of ferric iron in pyritic residues,[79] Na_2CO_3 + oxalic acid as a general mixture for silicate decomposition,[80] and Na_2CO_3 + potassium oxalate for the decomposition of iron ores.[81]

4.4.3   With Calcium Carbonate and Ammonium Chloride

This mixture, usually consisting of 0.5 g NH_4Cl and 4 g $CaCO_3$ for a 0.5 g sample, is that commonly used in the J. Lawrence Smith procedure for the determination of the alkali metals in silicates. The procedure, in which the silicates are decomposed by sintering them with the calcium oxide and calcium chloride formed from the thermal decomposition of the calcium carbonate and its reaction with the ammonium chloride, replaced for some years the Berzelius procedure which began with decomposition of the sample by a mixture of hydrofluoric and sulfuric acids. In recent years there has been a return to this latter method, which lends itself readily to the flame photometric determination of the alkali metals.

The complete separation of the alkali metals as chlorides is not achieved. A small amount of magnesium will likely be present in the final residue, and under some conditions a small amount of calcium may be present also. Conversely, not all the lithium will be present, although Shneider[82] found that this sintering procedure, rather than the Berzelius decomposition, gave accurate results for Li_2O in spodumene.

There have been many changes suggested for the sintering mixture, ranging from the elimination of the NH_4Cl to the use of $BaCO_3$ instead of $CaCO_3$. Stevens[83] suggested the substitution of $BaCl_2 \cdot 2H_2O$ for the NH_4Cl, in order to remove sulfates both present and formed during the ignition and prevent them from accompanying the final chloride residue. Its use also eliminates the need for slow initial heating required to expel the NH_3 formed during the sintering reaction. Esikov et al.[84] fused those samples having more than 0.5% Rb with $CaCl_2$ alone in a graphite crucible for the flame photometric determination of Rb and Sr.

Simon and Grimaldi[85] used a mixture of $CaCl_2 \cdot H_2O + CaO + MgO$ for decomposing mineralized rocks and molybdenites prior to the determination of rhenium. A mixture of CaO and KNO_3 has been used to decompose various sulfides in which germanium was to be determined.[86]

4.5 DECOMPOSITION BY OTHER METHODS

4.5.1 By Bombs

The use of bombs is more widespread than the use of sealed glass tubes (see Sec. 4.2.2), particularly that of the Parr peroxide bomb.

Parr[87] first described the extension of his calorimeter bomb to the decomposition of samples (0.25–0.50 g) for analytical purposes in 1908; he obtained excellent decomposition of galena, sphalerite, and pyrite, and also a shale, fireclay, and titanium ores. Since then, many new applications have been described and bombs have been developed in sizes from 2.5 to 42 ml, thus accommodating micro-, semimcro-, and macrosamples. The method involves the ignition, by an electric hot wire or flame-induced hot spot, of a mixture of sample, Na_2O_2, and an accelerator (commonly $KClO_4$) in a nickel cup enclosed in a steel bomb; the reaction, which usually requires less than 1 min, results in the formation of a molten mass which quickly solidifies to a cake containing the sodium salts of most of the elements in the sample, plus excess sodium oxide and hydroxide. For a description of the apparatus and operating procedures, the appropriate instruction manual should be consulted.

Ingles[88] used an electric ignition bomb for the decomposition of refractory uranium ores, such as euxenite, in order to reduce the loss of sample during the explosive reaction and to minimize the danger inherent in this method. He found this to be a useful general method for decomposing ores. He also found that a ratio of Na_2O_2 to sugar carbon of 15:1 resulted in a higher temperature of fusion with more effective attack on the ore; even with the higher temperature no evidence of distortion or attack of the bombs was found.

The use of HF in Teflon-lined bombs has been discussed in Sec. 4.2.1.

4.5.2   By Chlorination

It is appropriate to conclude this brief summary of methods for the decomposition of samples with reference to an elegant method that is deserving of more attention than it has received. This is the use of chlorine, passed over a sample which is heated to 250–300°C, to separate the constituents by selective volatilization of their chlorides. Some early use was made of this technique for the decomposition of iron meteorites. Moss et al.[89, 90] have shown that, provided the chlorine used is dry and free from HCl, no attack of feldspar, olivine, pyroxene, apatite, magnetite, or chromite occurs, and thus it is particularly useful in the separation of the metal and sulfide phases from the silicate phase of siderites. The sample, which need not always be pulverized, is placed in a porcelain boat inside a special Pyrex reaction tube; the chlorine is passed rapidly over it and the reaction tube is heated until the reaction begins ($FeCl_3$ sublimes), the temperature finally being raised to 350°C and maintained for at least 1 hr. After cooling, the residue in the boat is leached with water and the residue recovered. The water leach of the residue will contain Ni, Co, Cr, and Cu (and traces of Fe); Fe, S, P, and Ge (Ga?) distill quantitatively. A mixture of sulfur dichloride and chlorine has been used to decompose tantalum, niobium, titanium, and rare earth multiple-oxide minerals.[91]

References

1.   J. Doležal, P. Povondra, and Z. Sulcek, *Decomposition Techniques in Inorganic Analysis*, American Elsevier, New York, 1968.

2. R. S. Young, *Chemical Phase Analysis*, Charles Griffin & Co., London, 1974.

3. L. H. Ahrens, R. A. Edge, and R. R. Brooks, *Anal. Chim. Acta*, **28**, 551 (1963).

4. F. W. Chapman, G. G. Marvin, and J. Y. Tyree, *Anal. Chem.*, **21**, 700 (1949).

5. J. I. Hoffman and G. E. F. Lundell, *J. Res. Natl. Bur. Stand.*, **22**, 465 (1939).

6. S. Rossol, *Chem. Anal. Warsaw*, **3**, 865 (1958); through *Chem. Abstr.*, **53**, 19671[a] (1959).

7. J. I. Hoffman, R. T. Leslie, H. J. Caul, L. J. Clark, and J. D. Hoffman, *J. Res. Natl. Bur. Stand.*, **37**, 409 (1946).

8. J. C. Antweiler, U. S. Geological Survey, Prof. Paper 424-B, 322, 1961.

9. A. A. Levinson, *Introduction to Exploration Geochemistry*, Applied Publ. Ltd., Calgary, Alta, 1974.

10. P. M. D. Bradshaw, D. R. Clews, and J. L. Walker, *Exploration Geochemistry*, Reprint of a series of seven articles from *Mining in Canada* (1970) and *Can. Min. J.* (1971).

11. J. R. Foster, *Can. Inst. Min. Metal. Bull.*, p. 85 (1973).

12. F. J. Langmyhr, *Anal. Chim. Acta*, **39**, 516 (1967).

13. J. Ito, *Bull. Chem. Soc. Jpn.*, **35**, 225 (1962).

14. F. J. Langmyhr and K. Kringstad, *Anal. Chim. Acta*, **35**, 131 (1966).

15. F. J. Langmyhr and P. R. Graff, *Anal. Chim. Acta*, **21**, 334 (1959).

16. P. R. Graff and F. J. Langmyhr, *Anal. Chim. Acta*, **21**, 429 (1959).

17. A. M. Hartstein, R. W. Freedman, and D. W. Platter, *Anal. Chem.*, **45**, 611 (1973).

18. B. Bernas, *Anal. Chem.*, **40**, 1682 (1968).

19. Y. Hendel, *Analyst*, **98**, 450 (1973).

20. P. Knoop, *Anal. Chem.*, **46**, 965 (1974).

21. W. F. Hillebrand, G. E. F. Lundell, H. A. Bright, and J. I. Hoffman, *Applied Inorganic Analysis,* 2nd ed., Wiley, New York, 1953, pp. 835–851.

22. E. Wichers, W. G. Schlecht, and C. L. Gordon, *J. Res. Natl. Bur. Stand.*, **33**, 363 (1944).

23. E. Wichers, W. G. Schlecht, and C. L. Gordon, *J. Res. Natl. Bur. Stand.*, **33**, 451 (1944).

24. C. L. Gordon, W. G. Schlecht, and E. Wichers, *J. Res. Natl. Bur. Stand.*, **33**, 457 (1944).

25. I. M. Kolthoff, and P. J. Elving, Eds., *Treatise on Analytical Chemistry*, Vol. 2, Part I, Interscience, New York, 1961, pp. 1027–1050.

26. E. E. Bugbee, *A Textbook of Fire Assaying*, 3rd ed., Wiley, New York, 1949.

27. Manufacturing Chemists' Assoc., Inc., General Safety Committee, *Guide*

for Safety in the Chemical Laboratory, Van Nostrand, Princeton, NJ, 1954, pp. 117–119.

28. G. F. Smith, *Analyst*, **80**, 16 (1955).

29. K. A. Rodgers, *Mineral. Mag.*, **38**, 882 (1972).

30. K. Kodama, *Quantitative Inorganic Analysis*, Interscience, London, 1963.

31. H. H. Willard, L. M. Liggett, and H. Diehl, *Ind. Eng. Chem. Anal. Ed.*, **14**, 234 (1942).

32. G. G. Marvin and L. B. Woolaver, *Ind. Eng. Chem. Anal. Ed.*, **17**, 554 (1945).

33. R. A. Edge and L. H. Ahrens, *Anal. Chim. Acta*, **26**, 355 (1962).

34. W. R. Schoeller and A. R. Powell, *The Analysis of Minerals and Ores of the Rarer Elements*, 3rd ed., rev., Hafner, New York, 1955, pp. 2–4.

35. S. Abbey and J. A. Maxwell, *Chem. Can.*, **12**, 37 (1960).

36. C. O. Ingamells, *Talanta*, **2**, 171 (1959).

37. N. G. Mendliva, A. A. Novaselova, and R. S. Rychkov, *Zavod. Lab.*, **25**, 1293 (1959).

38. B. Brezny, *Hutn. Listy*, **15**, 552 (1960).

39. R. Ya. Epshtein and G. P. Ginberg, *Tr. Nauchn. Issled. Inst. Geol. Arktiki*, **119**, 84 (1961); *Ref. Zh. Khim.*, Abstr. 19D61 (1961).

40. F. Cuttitta, U. S. Geological Survey, Prof. Paper, 424-C, 384, 1961.

41. A. R. V. Murthy and K. Sharada, *Analyst*, **85**, 299 (1960).

42. R. S. Cohen, P. Hemmes, and J. H. Puffer, *Chem. Geol.*, **16**, 307 (1975).

43. P. D. Malhotra, *Analyst*, **79**, 785 (1954).

44. T. Fukasawa, Y. Takabayashi, and S. Hirano, *Jpn. Analyst*, **8**, 292 (1959).

45. A. A. Rozbianskaya, *Tr. Inst. Mineral., Geokhim., i Kristallochim. Redk. Elem. Akad. Nauk SSSR*, 138 (1961); *Ref. Zh. Khim.*, Abstr. 15D54 (1962).

46. I. May and F. S. Grimaldi, *Anal. Chem.*, **33**, 1251 (1961).

47. R. C. Chirnside, H. J. Cluley, R. J. Powell, and P. M. C. Proffitt, *Analyst*, **88**, 851 (1963).

48. M. S. Cresser and R. Hargitt, *Anal. Chim. Acta*, **82**, 203 (1976).

49. J. I. Hoffman and G. E. F. Lundell, *J. Res. Natl. Bur. Stand.*, **3**, 581 (1929).

50. N. G. Polezhaev, *Akad. Nauk SSSR*, 33 1958, *Ref. Zh. Khim.*, 10, Abstr. 34620 (1959).

51. K. Govindaraju, *Anal. Chem.*, **40**, 24 (1968).

52. E. M. Dodson, *Anal. Chem.*, **34**, 966 (1962).

53. F. S. Grimaldi, *Anal. Chem.*, **32**, 119 (1960).

54. C. B. Belcher, *Talanta*, **10**, 75 (1963).

55. G. G. Marvin and W. C. Schumb, *J. Am. Chem. Soc.*, **52**, 574 (1930).

56. P. G. Jeffery, *Analyst*, **82**, 66 (1957).

57. F. B. Barredo and L. P. Diez, *Talanta*, **23**, 859 (1976).

58. J. C. Van Loon and C. Parissis, *Analyst*, **94**, 1057 (1969).

59. C. O. Ingamells, *Anal. Chim. Acta*, **52**, 323 (1970).

60. H. Bennett and G. J. Oliver, *Analyst*, **101**, 803 (1976).

61. L. Shapiro, U. S. Geological Survey, Prof. Paper 575-B, B 187, 1967.

62. N. H. Suhr and C. O. Ingamells, *Anal. Chem.*, **38**, 730 (1966).

63. F. J. M. J. Maesson and R. W. J. M. Boumans, *Spectrochim. Acta*, **23B**, 739 (1968).

64. H. Bennett and G. J. Oliver, *Analyst*, **96**, 427 (1971).

65. C. W. Sill, *Accuracy in Trace Analysis*, National Bureau of Standards, Spec. Publ. 422, 1976.

66. V. T. Athavale, A. J. Patkar, and B. L. Rao, *J. Sci. Ind. Res. (India)*, **B21**, 231 (1962).

67. A. C. Shead and G. F. Smith, *J. Am. Chem. Soc.*, **53**, 483 (1931).

68. Yu. A. Shukolyukov and I. I. Matveeva, *Zh. Anal. Khim.*, **16**, 544 (1961).

69. F. S. Grimaldi, B. Ingram, and F. Cuttitta, *Anal. Chem.*, **27**, 919 (1955).

70. H. R. Shell and R. L. Craig, *Anal. Chem.*, **26**, 996 (1954).

71. A. I. Bulycheva and P. A. Mil'nikova, *Akad. Nauk SSSR*, 23 (1958) *Ref. Zh. Khim.*, 11, 328, Abstr. 38, (1959).

72. P. M. Isakov, *Vestn. Leningr. Univ., Ser. Biol., Geogr. i Geol.*, 117, 1955; through *Chem. Abstr.*, **49**, 15606^i (1955).

73. T. A. Rafter, *Analyst*, **75**, 485 (1950).

74. R. C. Chirnside, *J. Soc. Glass Technol.*, **18**, 5T (1959).

75. A. N. Finn and J. F. Klekotka, *J. Res. Natl. Bur. Stand.*, **4**, 809 (1930).

76. J. I. Hoffman, *J. Res. Natl. Bur. Stand.*, **25**, 379 (1940).

77. Y. Moriya, *Jpn. Analyst*, **8**, 667 (1959).

78. L. C. Peck and E. J. Tomasi, *Anal. Chem.*, **31**, 2024 (1959).

79. S. T. Balyuk, *Ogneupory*, 576 (1960); *Ref. Zh. Khim.*, Abstr. 11D120 (1961).

80. S. M. Efros and O. Ya. Bilik, *Sb. Stud. Rabot Leningr. Tekhnol. Inst. Lensoveta* (Leningrad) 13, 1956; through *Chem. Abstr.*, **54**, 8434^a (1960).

81. O. Quadrat, *Chem. Anal. Warsaw*, **4**, 405 (1959); through *Chem. Abstr.*, **53**, 16803^a (1959).

82. L. A. Shneider, Obogashch. Rud, **3**, 31 (1958); Ref. Zh. Khim., 971, Abstr. 30, (1959).

83. R. E. Stevens, *Ind. Eng. Chem. Anal. Ed.*, **12**, 413 (1940).

84. A. D. Esikov, G. S. Beschastnova, and G. N. Yakovlev, *Byull. Komis. Opred. Absol. Vozrasta Geol. Formats. Akad. Nauk SSSR*, 76, 1962; *Ref. Zh. Khim.* 19GDE, Abstr. 3G47 (1963); through *Anal. Abstr.*, **10**, 3088 (1963).

85. F. O. Simon, and F. S. Grimaldi, *Anal. Chem.*, **34**, 1361 (1962).

86. S. A. Dekhtrikyan, *Dokl. Akad. Nauk Arm. SSSR*, **28**, 213 (1959); *Ref. Zh. Khim.*, 469, Abstr. 30 (1960).

87. S. W. Parr, *J. Am. Chem. Soc.*, **30**, 764 (1908).

88. J. C. Ingles, *Topical Rep.*, Radioactivity Division, Mines Branch (Canada), TR-131/55, 1955, 12 pp.

89. A. A. Moss, M. H. Hey, and D. I. Bothwell, *Mineral. Mag.*, **32**, 802 (1961).

90. A. A. Moss, M. H. Hey, C. J. Elliott, and A. J. Easton, *Mineral. Mag.*, **36**, 101 (1967).

91. J. R. Butler and R. A. Hall, *Analyst*, **85**, 149 (1960).

Supplementary References

1. Z. Sulcek, P. Povondra, and J. Doležal, *Decomposition Procedures in Inorganic Analysis*, CRC Crit. Rev. in Anal. Chem., 225 June (1977).

2. R. Bock, *A Handbook of Decomposition Methods in Analytical Chemistry*; J. L. Marr, Transl., Int. Textbook Co., Glasgow and London, 1979.

3. M. Zief and J. W. Mitchell, *Contamination Control in Trace Element Analysis*, Wiley, New York, 1976.

4. R. W. Dabeka, A. Mykytiuk, S. S. Berman, and D. S. Russell, *Anal. Chem.*, **48**, 1203 (1976).

5. D. Whitehead, *Chem. Geol.*, **18**, 149 (1976).

6. A. Van Eenbergen and E. Bruninx, *Anal. Chim. Acta*, **98**, 405 (1978).

7. C. Feldman, *Anal. Chem.*, **49**, 825 (1977).

8. R. C. Mallett, R. Breckenridge, I. Palmer, K. Dixon, and G. Wall, Natl. Inst. Metall. Rep. 1852, 1976.

9. A. A. Schilt, "Perchloric Acid and Perchlorates," G. Frederick Smith Chemical Co., 1979.

10. J. Lenc, *Sklar Keram.*, **26**, 197 (1976); through *Anal. Abstr.*, No. 6B2, **33** (1977).

PART II

Methods of Analysis for Analytes Requiring Individual Methods of Preparation or Detection

CHAPTER 5

The Determination of Moisture, Total Water, Loss on Ignition, Carbon, Sulfur and Phosphorus

CHAPTER 6

The Determination of Iron, Uranium, Thorium, Fluorine, Chlorine and Tungsten

THE DETERMINATION OF MOISTURE, TOTAL WATER, LOSS ON IGNITION, CARBON, SULFUR, AND PHOSPHORUS

5.1 GENERAL

There are some constituents of a "complete" or total analysis of a rock or mineral which cannot be determined by one or more of the common instrumental techniques (AAS, XRF, or ES) used in rock analysis. Methods for the determination of these constitutents are described in this chapter and the next. This will not only ensure that methods are given for all common rock constituents but will also enable a laboratory which does not have an XRF capability, for example, to determine S and P_2O_5 by alternative methods.

5.2 MOISTURE (H_2O^-)

5.2.1 General

Moisture or, as it is variously designated, hygroscopic water, minus water, nonessential water, H_2O^-, or $H_2O^{-110°C}$, should be part of any rock or mineral analysis that lays claim to completeness. It is part of the total hydrogen content of the sample, as discussed in detail by Hillebrand et al.,[1] who class it as nonessential water (excluding the possible presence of H_2O as water of crystallization). It is water held by surface forces such as adsorption and capillarity, and its magnitude is related more to physical properties and sample treatment than to composition. It is sometimes considered to be an indicator of the freshness of the sample material. Manheim[2] has discussed the problems encountered with the determination of moisture in sediments; Drysdale et al.[3] have done the same for natural glasses.

The division into "minus" and "plus" water is an arbitrary one and the determination of moisture provides the petrographer with a means for reducing analyses to a common moisture-free basis. It is essential, however, that the moisture content of the sample remain constant during the analysis, unless all portions needed for the analysis are weighed at the one time. The fineness of the powder, which governs the amount of surface exposed, has a great effect on the initial amount of adsorbed moisture, but later variations in the moisture content are caused chiefly by changes in humidity. It is thus preferable that the various portions of the sample needed in the analysis be weighed as close together in time as possible.

Grinding of the sample will cause changes in the moisture content; the latter will tend to increase because of an increase in exposed surface, but the heat and pressure developed during the grinding will have an opposing effect. The sample should be ground to the fineness which will enable the expulsion of the hygroscopic water in reasonable time, but excessive grinding should be avoided.

Heating at 100–110°C will drive off most of this nonessential water, although on occasion temperatures much above this are necessary. This is an indirect method of determination, because the loss in weight is assumed to be due to moisture alone. The actual temperature of heating is not important since the amount given up at either temperature differs little, but it is important that the temperature used should be stated. There is nothing to be gained from drying the sample before bottling; a change occurs each time that the bottle is opened to moisture-laden air, and if the sample contains hydrous minerals, serious errors may result. It is preferable to allow the sample to reach equilibrium with the atmosphere and then to weigh all portions as close together in time as possible.

The determination of moisture may be done either as a separate step, or as the initial step in a detailed analytical scheme, and the type of container used for the determination will be governed by the subsequent treatment of the sample. Crucibles of platinum, porcelain, and nickel have been used; the advantages of platinum over these others have been stated and, in addition, the sample is then ready for fusion. Porcelain or nickel crucibles, or a glass weighing bottle, will do if the sample is not to be used for further work. The use of a watch glass is not recommended because of the ease with which losses may occur.

In some minerals the moisture content may be the major part of the total water present, and failure to determine the moisture, present chiefly as adsorbed water, will give an entirely misleading significance to the values for H_2O^+. Variations in the H_2O^+ content of samples analyzed cooperatively by various laboratories may be due as much to incomplete drying of the sample as to the determination of the total water content.

5.2.2   Determination of Moisture

Weigh accurately about 1 g of sample, or less if the supply is limited, which has been ground to approximately 150 mesh and thoroughly mixed, and transfer it to a clean, weighed 30 ml platinum crucible (a porcelain or nickel crucible if this sample portion is not to be fused). Weigh the covered crucible and sample; the difference in weight of the sample should not exceed 0.2 mg. Place the uncovered crucible in an oven, cover it with a 7 cm diameter filter paper, and heat it at 105–110°C for 1 hr. Transfer the crucible to a desiccator, cover, and allow it to cool for 30 min before weighing.

If the loss in weight exceeds 1 mg, the heating, cooling, and weighing should be repeated until constant weight is obtained. If the loss in weight exceeds 5 mg, the crucible should be heated at a higher temperature, such as 125°C, to see if a further loss occurs, indicating the presence of a significant amount of hydrous material.

When a significantly hygroscopic sample is to be weighed, it is essential that it be in equilibrium with the atmosphere. Spread the sample on a clean sheet of glazed paper and allow it to stand exposed overnight (a paper canopy should be erected to protect the sample from atmospheric dust), then weigh all portions to be used in the analysis at the same time.

The same sample may be used for the determination of loss on ignition in either a platinum, fused quartz, or porcelain crucible, but those made of porcelain should first be treated to remove the inner glaze. This is accomplished by igniting the crucible, half-filled with carbonate-rich material, in an electric muffle furnace for 3–4 hr.

Little weight change should occur after this treatment.

5.3   WATER (H_2O^+)

5.3.1   General

The "water content" of a rock or mineral is generally taken to mean that amount of water which is expelled from the sample when it is heated at a temperature greater than 105–110°C and excludes the water (moisture, H_2O^-) which is driven off when the sample is heated at a temperature in or below this range (Sec. 5.2). If the sample is only air-dried, then the water obtained by the following methods is the *total water* content of the sample, assuming complete release and collection of the water. On subtraction of the H_2O^- (moisture) content of the sample, if known or subsequently determined, or if the sample used was first oven-dried (105–110°C), a value is obtained that is variously designated as H_2O^+, water of constitution, $H_2O > 110°C$, and essential or bound water. The comments made in Sec. 5.2 apply to this section as well and should be reviewed. Because of the many factors which affect the water content of a rock or mineral, it is unavoidable that the division of the total water content into temperature-controlled fractions must be an arbitrary one, and this should be borne in mind when these values are considered. It has been shown that the loss of water is dependent more on the duration than upon the temperature of heating, at least for natural glasses;[4, 5] Chalmers[6] believes that, in order to interpret the role of water released at any particular temperature, it is necessary to combine the information obtained from weight loss–time–temperature curves with that from differential thermal analysis curves. For general discussions on this subject, including the role of hydrogen in minerals, reference should be made to Hillebrand et al. (Ref. 1, pp. 814–822), Groves (Ref. 7, pp. 269–274), and to a review by Mitchell.[8]

The methods for the determination of water in silicates are not numerous, and most of them involve a gravimetric measurement of the water collected. Titrimetry, with Karl Fischer reagent,[9, 10] the measurement of dielectric constant, and electrolysis have also been used. The last method involves the use of a flow gas to carry the water vapor into a cell where it is absorbed in a hygroscopic electrolyte and quantitatively electrolyzed. The magnitude of the electric current is proportional to the number of moles of water absorbed per unit time.

The loss on ignition (usually at about 1000°C) of a sample is occasionally reported as a substitute for the determination of total water, but its use is fraught with danger, and at best it is more useful as a relative than as an absolute value. Its significance, with examples of such, has been discussed in detail;[1, 11] some analysts reject it as being completely misleading.[7] It is obvious, of course, that the value obtained by igniting the sample at 1000°C will represent the water content only in the absence of other similarly volatile constituents such as carbon dioxide, sulfur, fluorine, chlorine, and, to some extent, the alkali metals; it also requires the absence of constituents that are readily oxidized or reduced during ignition, chief among which is ferrous iron. Riley[12] has shown that the amount of FeO remaining unoxidized after the ignition of silicates is very variable (0.16–3.0%), and a reliable correction for this oxidation cannot be applied; several examples are also given by Hartwig-Bendig.[11] The use of loss on ignition for the determination of water in sediments and sedimentary rocks has been investigated by Manheim,[13] who recommends 1000°C as the most suitable ignition temperature; he also mentions the possibility that the sample may become contaminated by material from the furnace walls during the heating.

All minerals do not give up their water readily; talc, topaz, staurolite, chondrodite and phlogopite are the worst offenders but titanite[14] and some varieties of epidote are also troublesome. The chief problem lies in ensuring that all of the water is expelled, which can be done by using

very high temperatures capable of breaking down the most resistant structure, but which then requires a special furnace capable of attaining these
elevated temperatures and a sample tube capable of withstanding the
latter. Hartwig-Bendig[11] has discussed this problem in detail with abundant supporting data, and Peck (Ref. 15, pp. 17–21, 61, 83) has presented
a very practical evaluation of the factors affecting the determination.
There are some minerals that will yield all of their water on simple ignition
but, as a rule, it is better to employ a flux to break down the structure
(thus making it possible to use a lower maximum temperature); the flux
also serves to retain other volatile constituents such as fluorine and sulfur
and provides a supply of oxygen to prevent any reduction of water to
hydrogen during the ignition. Among the numerous fluxes that have been
suggested, there are three that find most use—sodium tungstate, lead
oxide (litharge),[16] and lead chromate. Peck[15] tested these three fluxes
with hornblende and found that, while no single flux was completely
satisfactory, a mixture of two parts of PbO and one part of $PbCrO_4$ was
sufficiently reactive to decompose hornblende and did not intumesce
disadvantageously with samples containing carbon or carbon dioxide. The
flux mixture is dried by heating it to 800°C and is then kept in a tightly
sealed bottle. Sodium tungstate was found to be not sufficiently reactive,
as well as being hygroscopic, and Peck does not recommend its use. It
has been found (WMJ) that litharge was successful in the decomposition
of refractory minerals having high magnesium content, such as talc and
the standard reference material (SRM) PCC-1.[16]

If a purification and absorption train is used in the determination of
water, there will be little likelihood of interference from other volatile
constituents not completely retained by the flux. When the water is to
be weighed directly, as in the Penfield method, the effect of these other
volatiles, particularly carbon dioxide, must be considered, or serious
errors may be introduced. Litharge, or the litharge–lead chromate mixture
of Peck, will retain the sulfur, chlorine, and fluorine present, unless these
latter are present in unusually large amounts. The addition of freshly
ignited calcium oxide to the flux–sample mixture will help to retain excessive amounts of fluorine. Carbon dioxide is not retained by the flux
and, unless eliminated, may cause a twofold error; (1) if present in large
amount it may carry some water vapor out of the tube, and (2) if not
removed from the tube before it is weighed, the CO_2 will be counted as
H_2O and cause results to be high. The usual method of correction for
carbon dioxide is to allow the tube to "drain," that is, the tube is supported at a 45° angle with provision for retaining the water, and a correction is applied to compensate for the water vapor lost during the
"draining" period. Peck[15] prefers to omit the center bulb of the Penfield

tube and to connect the tube horizontally to a special bulb containing water and air; after a time, the water-saturated carbon dioxide in the tube becomes so thoroughly diluted with water-saturated air that it no longer has any effect upon the results. A displacement procedure, in which the air in the tube is replaced almost completely by insertion of a glass rod, is also effective. It is not possible to correct for the presence or organic carbon which may be oxidized to water and carbon dioxide during the fusion–ignition of the sample.

The use of an absorption train, usually in conjunction with a combustion tube furnace, has been described by several authors[1, 7, 11, 12] and will not be discussed in detail here. The sample, contained in a platinum, porcelain, alumina, or silica boat and mixed with one of a variety of fluxes, is heated to about 1000°C in a silica or vitreosil tube while a current of dry air is drawn over it. The water is absorbed in a suitable agent such as calcium chloride, or anhydrone ($Mg(ClO_4)_2$). The flux will hold back most of the other volatile constituents, but purification tubes can be inserted in the train ahead of the water absorption tube. It is recommended that the drying agent used to dry the air that is drawn through the system should be the same as that used to absorb the sample water, and that all weighings should preferably be made under the same atmospheric conditions. It is advantageous to keep the blank as small as possible. Jeffery and Wilson[17] devised a closed-circulation system in which the air is recycled by means of a small pump; not only is a very low blank obtained, but it is only necessary to change the desiccant occasionally, instead of frequently as in the open system. Methods in which the macroprocedure is adapted to the microscale have also been described.[18]

The most practical method for routine use is the refreshingly simple one first suggested by G. J. Brush and developed later by Penfield.[19] This method, or a modification of it, is frequently used, and in careful hands it is capable of giving very satisfactory results. The sample, mixed with a flux, is heated in a bulb which is blown on the end of a length of borosilicate tubing; the expelled water is condensed and collected in the cooled center portion of the tube, and the fused bulb and sample are pinched off. For a gravimetric determination the tube and collected water are weighed. The tube is then dried, reweighed, and the loss in weight is that of the water expelled from the sample. An alternative method is to rinse the water out of the tube into a titration vessel for a Karl Fischer titration. The other pieces of equipment usually required include a long-stemmed funnel for the introduction of the sample–flux mixture, a capillary plug to minimize the flow of air in the tube during the ignition period, and a small cap with which to close the tube prior to the weighing step or the titration step.

The chief disadvantage of the Penfield method is that it is difficult to attain an ignition temperature sufficiently high to expel all of the water from the sample before the bulb collapses. Another disadvantage is that any carbon dioxide also present in the sample will interfere seriously unless special precautions are taken to eliminate it.

Peck[15] has designed a furnace which will heat four samples concurrently at 900°C and which requires little attention during the heating period of about 50 min. Penfield[19] also used a furnance, made of firebrick and lined with charcoal, for minerals which would not yield their water readily (no flux is used); the tube was wrapped in platinum foil to support it. The use of a propane–oxygen mixture in a glass-blower's hand torch, properly regulated, will also yield a temperature of 900°C before collapse of the tube. The objection to the use of the Penfield method on the ground that a sufficiently high temperature cannot be obtained[7] is no longer valid.

Riley[12] believes that the Penfield method has a fundamental flaw, in that not all the water that is liberated is condensed. Peck[15] determined the combined water in two amphiboles by his modification of the Penfield method and obtained results that were virtually identical with those obtained by Friedman and Smith,[20] who heated the samples under vacuum in an induction furnace (1450°C), reduced the H_2O to H_2 over hot U metal, and measured the volume of the hydrogen.

The Penfield method has been adapted to the semimicroscale by Guthrie and Miller,[21] and to microanalysis by Sandell.[22] The latter places a closely fitting helix of nichrome ribbon around the end of the tube to prevent sagging when it is heated.

Harvey (Ref. 1, pp. 834–835) proposed a method that combines features of both the Penfield and the combustion–absorption techniques in simplified form. The sample, in a platinum boat, is heated in a silica test tube supported by an asbestos heat screen; a weighing bottle containing $CaCl_2$ is connected to the test tube, and also a small $CaCl_2$ drying tube to remove moisture from air drawn into the apparatus during the cooling period. The method requires the absence of significant amounts of volatile constituents other than water and carbon dioxide, the latter being removed by repeated vertical displacement.

The advantages gained by titrating the collected water with Karl Fischer reagent instead of weighing it include a reduction in the time required for the determination, elimination of the need to "drain" any CO_2 and thus avert the loss of some water, and the possibility of doing samples in batches of 20 or more. Some disadvantages are the interference of H_2S, the necessity of having to protect the reagent and the reaction vessel from atmospheric water, and the frequent reagent standardization required (twice daily) because of the instability of the reagent. A comprehensive coverage of the use of the Karl Fischer reagent in water determination is given by Mitchell.[8]

The reactions involved are as follows:

$$C_5H_5N \cdot I_2 + C_5H_5N \cdot SO_2 + C_5H_5N + H_2O \rightarrow 2C_5H_5N \cdot HI$$
$$+ C_5H_5N \cdot SO_3$$
$$C_5H_5N \cdot SO_3 + CH_3OH \rightarrow C_5H_5N(H)SO_4CH_3$$

There is an excess of the hydroxyl-containing compound present (usually methanol) to prevent the pyridine sulfur trioxide complex formed in the water reaction from reacting with water. This latter is not desirable because the reaction is not as water-specific as is the primary one and thus there would be more possibility of interference (Ref. 23, p. 608).

5.3.2 Determination of Water by the Modified Penfield Method

This modified method is readily adapted to routine use. It makes use of a simple water tube without a center bulb and the apparatus for supporting and cooling the tube during the ignition that was devised by Courville.[24] The elimination of carbon dioxide is achieved by displacing the air in the tube, after drawing off the fused end, with a closely fitting glass rod which is then weighed with the tube; this is the procedure used by Sandell in his micromethod.[22] It is convenient to make the determinations in groups of eight; the tubes, with condensed water and inserted glass rod, are accumulated on a rack in a refrigerator until all eight are ready for the next step, that of coming to equilibrium in the balance room prior to the first weighing. A supply of the tubes, cleaned with cleaning solution and thoroughly rinsed with distilled water, is kept in an oven having a temperature of about 130°C.

Apparatus

Penfield tube. This is made from borosilicate tubing having an outside diameter of 9 mm and walls of 1 mm in thickness, overall length of 250 mm, with a bulb at one end having a diameter of approximately 2.5 cm. The walls of the bulb must be of uniform thickness, and the open end must be free of any "lip" that reduces the size of the opening.

Tube support. As shown in Fig. 2.4, it consists of a heavy plywood support having a space at one side of the base to accommodate the bottom of a Bunsen burner during the initial stages of the heating. The heat shield is made of two 6 mm pieces of asbestos board screwed to the front vertical support; a single sheet of 12 mm asbestos board would suffice as well. The overflow tray is made of plastic in order to minimize "sweating" of the dish when the cold water drips into it. The cooling tray is made of aluminum and the rubber supports are made from rubber stoppers cemented to the two trays. The height of these supports should be such that, with the overflow tray in position, the notches in the rim of the cooling tray are level with the bottoms of the two slots in the vertical supports of the tray holder. The open end of the tube should have a slight downward inclination to prevent any condensed water from running back into the hot portion of the tube.

Funnel. This is a long-stemmed funnel which, when supported, will reach nearly to the bottom of the sample tube, and which can be easily and rapidly inserted and withdrawn.

Capillary plug. Heat and draw a length of capillary tubing to a short taper such that, when cut off and covered with a piece of thin rubber tubing, it will fit tightly into the open end of the water tube.

Tube cap. Insert a short piece of glass rod into a 12 mm length of firm rubber tubing having an internal diameter such as to enable it to fit easily but firmly over the open end of the water tube.

Displacement rod. This is a 20 cm long piece of 6–7 mm o.d. glass rod, fire-polished at both ends, which will slide easily into the water tube. When not in use, these rods should be stored in used water tubes to keep them clean.

Reagents

Lead oxide flux. Place a quantity of PbO (litharge) in a 90 ml nickel crucible and heat it with a moderate flame of a Bunsen burner for 30 min, stirring frequently with a glass rod and crushing any lumps that form. Store on a watch glass in a drying oven at 110°C. Transfer the amount required for immediate use to a desiccator for cooling.

Procedure

Place a clean water tube, which has been dried for at least 2 hr at a minimum of 110°C, in a horizontal support, insert a narrow glass tube into it, and by means of suction, draw a gentle current of air through the water tube until it is in equilibrium with room conditions. Support the long-stemmed funnel (which must be dry) in a vertical stand and insert it into the water tube so that the end of the stem is at about the center of the bulb.

Weigh and transfer 1.000 g of sample (or less if the water content is likely to be greater than 1%) to a small porcelain crucible and add 2 g of flux. Mix with a spatula and transfer by means of the funnel to the bulb of the water tube. Brush any remaining powder into the funnel and brush the walls of the funnel; expedite the transfer of powder to the bulb by tapping the rim and stem of the funnel with the spatula. Add about 0.5 g of flux to the crucible, rinse the crucible with it, and add it to the funnel.* Lift the funnel so that the lower end of the stem is at the entrance to the bulb and again tap the funnel to dislodge any powder clinging to the stem walls. Withdraw the funnel and close the tube with a capillary plug.

Carefully raise the tube to a horizontal position (no powder must escape from the bulb) and wrap the center 8–10 cm of the tube spirally with a 15 cm strip of wet cloth (about 12 mm wide). Place the tube in the cooling tray of the tube support, which should contain crushed ice and water, so that the cloth wrapping touches the cold water and the bulb projects about 2 cm beyond the heat shield.

Place a Bunsen burner in position so that a small flame is beneath the bulb. Rotate the tube slowly to prevent caking of the sample, and gradually increase the burner flame until the full heat of the burner is attained and the flame envelopes the bulb (this must be done slowly over a period of 5 min in order to avoid explosive expulsion of volatiles with the possibility of some water being carried out of the tube.) Commence heating the bulb with a brushlike flame of an oxygen–propane hand torch, while steadily rotating the tube, and gradually increase the heat of the flame until the walls of the bulb begin to collapse (the sample and flux should be a molten mixture by now). Rotate the tube rapidly enough to prevent the bulb from sagging and heat it as strongly as possible; collapse the walls of the bulb about the fused sample, starting at the bottom of the bulb and working toward the orifice (this reduces the possibility of the fused material blocking the orifice and causing a rupture of the bulb before the ignition is finished). Finally, using a sharp flame, fuse the tube at the junction with the bulb, and with tongs draw off the fused end and discard it. Heat the sealed end of the tube briefly to round any sharp projections.

Keep the tube in a horizontal position and cautiously wipe the hot end with a wet cloth to cool it somewhat before pushing the wet wrapping strip over the hot end of the tube. Place the end of the tube in the ice bath, pat dry the other end with a lint-free cloth, remove the capillary plug, and quickly insert a clean, dry displacement rod; if there is a space remaining at the upper end of the tube, displace the air in it once or twice with the end of another displacement rod. Close the tube with a tube cap

---

* If appreciable fluorine is present, it is necessary to "fix" it by adding approximately 0.3 g CaO (obtained by igniting 0.5 g $CaCO_3$ at 900°C for 30 min and storing it in a desiccator until needed) as the final step.

and carefully dry the outside of the tube by patting it (do not rub) with a lint-free cloth.* Place the tube on a horizontal rack and allow it to stand for 20 min near the balance before removing the tube cap and weighing the tube. The weighing is best done by supporting the tube on a wire saddle of known weight.

Carefully remove the displacement rod and dry it by heating it cautiously with a low flame; dry the water tube in the same manner. (They may be left overnight in an oven having a temperature of at least 110°C if this is more convenient.) Cool the rod and tube, while drawing a gentle current of air through the latter. Return the rod to the tube, wipe the tube with a damp cloth, and again pat dry. Place the tube on a horizontal rack, allow to stand for 20 min near the balance, and weigh as before. The difference between the two weights is that of the water in the sample. A blank should be run at regular intervals and always when a new batch of flux is used.

Conversion Factors

$$H_2O \underset{8.9357}{\overset{0.1119}{\rightleftarrows}} H$$

5.3.3   Determination of Water by Karl Fischer Titration

This method is identical to the method described in Sec. 5.3.2 up to the end of the fusion and cooling steps, with the exception that no displacement rod is required. The water condensed in the tube during fusion is titrated instead of being weighed.

Apparatus

The apparatus for the fusion is as described in Sec. 5.3.2. The dead-stop titration technique (Ref. 23, pp. 488–491) requires apparatus for aquametry (available from most scientific supply houses) equipped with a micro reaction vessel such as that supplied by Labindustries or its equivalent. This consists of a dual burette system for dispensing large and small aliquots of titrant, a reaction vessel closed to the atmosphere, an aspiration or suction device, a magnetic stirrer and stirring bar, a meter or a Pt electrode circuit,[8] a stand, and clamps. Also needed are reagent storage vessels and pump mechanisms for filling the burettes with reagent and the butanol–methanol solution without exposing the system to the

---

* The development by friction of a charge of static electricity on the surface of the tube can cause a serious error in the determination. At the GSC it was found useful to keep a mounted ionizing unit in the balance chamber to dispel static electrical charges from the water tubes and to pass this, or an ionizing brush, lightly over the tube before weighing.

atmosphere. All openings to the atmosphere must be protected by drying tubes filled with $Mg(ClO_4)_2$ or some other desiccant.

Reagents

The PbO flux is prepared as discussed in Sec. 5.3.2. A supply of stabilized Karl Fischer reagent and its diluent (both available from most scientific supply houses), 1 : 1 butanol–methanol and methanol are also required.

Procedure

The weighing and fusion steps are the same as those described in Sec. 5.3.2 up to the point where the tube has been cooled subsequent to the fusion. At that point the tube is closed with a cap, without inserting a displacement rod, and stored until all of the samples have been fused and are ready for titration.

Before titration of the samples it is necessary to determine the blank corrections for the butanol–methanol solution and the fusion step, and to standardize the Karl Fischer reagent. The automatic burettes, which should always be drained between each series of determinations, should be rinsed by cycling the appropriate solution once into the burette and back into the reservoir. The titration vessel should be rinsed with a small quantity of methanol from a wash bottle, and the rinsings removed by means of suction.

Transfer 10 ml of the butanol–methanol solution to the titration vessel for the determination of the blank correction. Insert the meter and titrate to the desired end point. If a Labindustries aquameter is used, the end point is reached when, after the addition of one drop of titrant, the needle stays past the dark portion of the dial for 25 sec or more. Remove the solution from the titration vessel, and rinse the latter with a few milliliters of methanol. Note the blank correction to be made.

Add 10 ml of the methanol–butanol solution to the titration vessel, quickly add 50 µl (i.e., 50 mg) of water, and again titrate to the end point as before. Correct the titration reading for the blank on the alcohol solution and calculate the titre of the reagent. This standardization should be repeated during the analysis of a set of samples.

To determine the water content of a sample, add 5 ml of the alcohol solution to the glass tube from the Penfield fusion and transfer this to the titration vessel with the aid of a small funnel, rotating the tube during the transfer to ensure that the inside of the tube is thoroughly rinsed with the alcohol solution. Repeat this step with an additional 5 ml of alcohol solution, and then titrate the contents of the flask as before. Determine a blank correction for the fusion step by carrying out all of the steps of the procedure but omitting the sample. Correct the burette reading for each sample for the blank obtained in this way, and for that obtained for the methanol–butanol solution, and then calculate the percent water.

5.3.4 Determination of Total Water by Combustion–Infrared Absorption Measurement

See Sec. 5.6.4c.

5.4 LOSS ON IGNITION

5.4.1 General

The use of the loss on ignition result as a substitute for the total water determination has been discussed in Sec. 5.3.1. These considerations are important because significant interpretative errors can result from an incautious use of loss on ignition data. In spite of this, however, its determination can serve a useful purpose, particularly for carbonate rock analysis. The preferred ignition temperature is about 1000°C, and the heating is usually started in a cold muffle furnace, the temperature being gradually raised to prevent the loss of sample by too sudden expulsion of volatiles; about 30–60 min heating at 1000°C is usually sufficient to bring the sample to constant weight. Manheim[13] has discussed this determination in considerable detail.

A method for the determination of carbon dioxide in limestone and dolomite by differential loss on ignition has been described.[25] The loss on ignition for high calcium samples approaches closely the actual carbon dioxide content, but the presence of carbonaceous material and other volatiles in less pure samples introduces a large error into the result. By determining the loss on ignition after heating at 500°C for 25 min (CO_2 is retained, other volatiles are evolved), and then heating again for at least 1 hr at 1000°C, the loss in weight due to evolution of carbon dioxide can be calculated. The oxidation of pyrite present ($>0.2\%$) in the sample will introduce a further error, but a correction can be applied; the method fails when siderite and magnesite are present.

5.4.2 Determination of Loss on Ignition

Place the crucible and sample used for the determination of moisture (Sec. 5.2), uncovered, in a cold electric muffle furnace and allow the temperature to rise slowly. When the temperature reaches 500°C, cover the crucible and continue the heating to 1000°C. Heat at this temperature for 30 min, cool in a desiccator, and weigh. Repeat the ignition at 1000°C until constant weight is obtained. Calculate the weight loss at 1000°C.

5.5   CARBON

5.5.1   General

Carbon is usually present in the form of independent accessory carbonates such as calcite, dolomite, and siderite, but it may also be present as a carbonate in the structure of some silicates such as cancrinite, or in other minerals such as apatite. It may also occur as graphite, or even as carbide; in the form of diamond it is not likely to be encountered by the analyst.

The concentration of carbon dioxide, that is, of carbonate carbon, is very low in most igneous rocks unless the latter have been subjected to much alteration (e.g., by weathering). In the form of graphite or carbonaceous matter (a mixture of graphite and organic matter), carbon is usually present only in small amounts, although it may reach a concentration of several percent in rocks such as graphite schists and slates. The geochemistry of carbon in igneous rocks has been discussed in detail by Hoefs.[26]

In noncarbonate sedimentary rocks, such as shale and sandstone, the carbon content is very variable, but usually it is much higher than in igneous rocks. In the carbonate rocks it is, of course, a major constituent.

Carbon as carbonate seldom interferes in the determination of other constituents. Care must be taken to avoid mechanical loss when acid is added to a slurry containing carbonates, as in the determination of ferrous iron (Sec. 6.1.2). When present as graphite it may be more troublesome in a physical sense than in a chemical one, but as carbonaceous matter, or as organic matter alone, it will create several problems for the analyst. Fusion of such a sample with sodium carbonate in a platinum crucible may result in appreciable attack on the crucible (it should be noted that those samples that have organic matter present usually contain other elements such as sulfur and arsenic which are responsible for much of the damage done to the crucible) but this is eliminated, or reduced, by a preliminary roasting of the sample prior to fusion and by the addition of a small amount of an oxidant such as sodium peroxide to the flux. Mention has also been made of its deleterious effect upon the accurate determination of ferrous iron (Sec. 6.1.1), and of water (Sec. 5.3.1). Corbett and Mizon[27] have discussed the determination of carbon in high sulfide materials.

In the following discussion of methods for the determination of carbon it is difficult to confine the discussion to silicates alone, since most methods are primarily intended for carbonate-rich materials. In general, these methods are applicable to silicates simply by using an appropriately larger amount of sample.

"Carbonate" Carbon

The comments made about the use of loss on ignition as a means of determining water (Sec. 5.3.1) apply equally as well to the determination of carbon dioxide by this method. It can be used as a rough check on the magnitude of the amount of carbonate that may be present in a sample, provided that other volatile constituents are either absent or have already been determined, but it is only under very special circumstances that much reliance may be placed upon the results. A method involving controlled loss on ignition is described by Galle and Runnels[28] for the determination of carbon dioxide in limestone and dolomite; the sample is ignited first at 550°C to remove all noncarbonate volatile constituents, weighed, and then ignited at 1000°C to decompose the carbonate minerals. Warne[29] investigated the method in some detail and found that ankerite could be treated similarly, but not magnesite, siderite, and possibly some other carbonates as well. Waugh and Hill[30] applied it to pyritic limestones of relatively high purity, but a correction must be applied for the oxidation of sulfide to sulfate.

The most straightforward procedure is that in which the volume of carbon dioxide released by acid decomposition of the carbonates or other CO_2-bearing minerals is measured directly. Fahey[31] has described a simple and rapid procedure in which the sample is heated with HCl and NaCl in a Pyrex® test tube, and the volume of the carbon dioxide is measured by the displacement of mercury in a burette connected directly to the test tube; it has been found particularly suitable for rocks containing less than 10% carbon dioxide. Wolff[32] surrounds the reaction and measuring vessels with a thermostatically controlled water jacket and keeps the temperature and pressure constant. In both of these methods it is imperative that no gases other than carbon dioxide be released, which is a serious disadvantage; an important advantage is the relatively large volume occupied by 1 mg of carbon dioxide (0.405 cm^3, at standard temperature and pressure[32]). A quantitative manometric method for the determination of calcite and dolomite in soils and limestones has also been described.[33, 34]

Shapiro and Brannock[35] devised a simple and rapid modification of Fahey's method as part of their rapid silicate rock analysis scheme, which involves only the measurement of the volume of gas formed during treatment of the sample with hot acid. The sample is placed in a borosilicate test tube which is fitted with a side-arm tube at an angle of about 60°, and paraffin oil, $HgCl_2$ (both as a solid and as a saturated solution), and HCl are added; the tube is heated to about 120°C in an aluminum heating block, and the liberated carbon dioxide displaces the oil in the side-arm tube where its volume is measured and compared with that of a known weight of CO_2 liberated under similar conditions. The $HgCl_2$ is added to prevent the formation of hydrogen by reaction of the acid with any "tramp" iron present in the sample. The method is recommended for those samples that contain less than 6% carbon dioxide.

A further refinement of the volumetric technique is one in which the measured volume of liberated gas, which is mixed with a certain amount of CO_2-free air and water vapor, is passed through a solution that will preferentially absorb carbon dioxide; the volume of CO_2 is then found by difference when the gas volume is measured a second time. Goldich et al.[36] used this procedure to determine CO_2 in lake marls and obtained results that were comparable with those obtained by the standard gravimetric absorption train method. The volume of liberated CO_2, plus air and water vapor, is measured in a water-jacketed gas-measuring burette using acid sodium sulfate solution as the containing fluid; the CO_2 is absorbed in potassium hydroxide solution in an absorption pipette, and the gas volume is measured again to give the volume of carbon dioxide by difference. Measurement of the temperature and pressure is done in order to apply a correction for the vapor pressure of the sodium sulfate solution. Shapiro and Brannock[35] have modified the apparatus of Goldich et al. somewhat and use this procedure for the rapid determination of CO_2 in those samples in which its concentration is probably greater than 6%; the need to correct for the vapor pressure of the sodium sulfate is eliminated by standardizing the method with a sample of known CO_2 content which is carried through the procedure concurrently with the samples.

The same concept is used in the increasingly popular Leco semiautomatic carbon determinator. The sample is heated in a Leco induction furnace in a stream of purified oxygen which carries the evolved gases through a dust trap, a sulfur trap, and a catalytic converter (to convert any CO to CO_2) and then into the carbon determinator. The mixed gases displace a measured volume of dilute H_2SO_4. When the combustion is finished, the collected gases are first passed into a reservoir containing KOH solution, which absorbs the CO_2, and then back into the original H_2SO_4 burette. The difference in volume of displaced H_2SO_4 is a measure of the amount of CO_2 that was evolved from the sample in the induction furnace. This represents the total carbon content of the sample, but by liberating carbonate carbon as CO_2 from a weighed portion of the sample with HCl prior to heating the sample in the induction furnace it is possible to obtain a value for noncarbonate carbon. The determination of total carbon on a second sample portion will then allow the calculation of the content of carbonate carbon by difference.

A method that is lengthy but still popular is that in which the carbon dioxide, liberated by acid treatment or ignition of the sample and purified by passage through absorbers which will remove volatile constituents other than CO_2, is absorbed in a suitable medium, the increase in weight of which is then determined. Air or nitrogen, freed of CO_2, is usually employed as a carrier gas. This procedure has been described in detail in standard texts (Ref. 1, pp. 766–778, 934–935; Ref. 7, pp. 109–117).

The size of sample used will depend upon the carbon dioxide content, but for most silicates a sample of 2–3 g is preferable. When acid is used to liberate the CO_2, it is usually hydrochloric which is selected because of the solubility of its compounds; its high volatility is a possible source of undesirably high blanks, however, and many analysts prefer to use phosphoric acid instead. A condenser attached to the decomposition flask will remove the bulk of the HCl–water vapor formed when the acid mixture is boiled to expel the liberated CO_2; if not so removed it will necessitate frequent changing of the desiccant [$CaCl_2$, P_2O_5, anhydrous $Mg(ClO_4)_2$] in the absorption train. A variety of substances are used to remove such unwanted constituents as HCl, H_2S, and Cl_2, among which are anhydrous copper sulfate, chromic acid–phosphoric acid, silver arsenite, and manganese dioxide.

The apparatus used for this procedure varies little from analyst to analyst. It generally consists of a decomposition flask fitted with a small condenser and a separatory funnel for the addition of acid (an absorption tube containing a CO_2-absorbent, such as ascarite, removes CO_2 from the air drawn into the train through this funnel) or a combustion tube and furnace if thermal decomposition is to be used, followed by a series of absorption tubes containing absorbents to remove unwanted volatiles and vapors; one or, preferably, two weighed absorption tubes to absorb the evolved CO_2, and a final tube containing absorbents to protect the weighed absorption tubes from the accidental sucking in of air, complete the apparatus. Gas bubbler tubes at the beginning and end of the absorption train serve to indicate the rate of passage of the flow gas, and also give warning of the presence of a leak in the system; water should not be used in these bubblers, and a mixture of phosphoric and chromic acids will function doubly as a flow rate indicator and as a scavenger of hydrogen sulfide. Peck (Ref. 15, pp. 48–50, 79–80) describes a compact absorption train that features a combined reaction vessel and condenser.

Jeffery and Wilson[17] have modified the conventional absorption train into a closed circulation system (the apparatus, slightly modified, can be used for the determination of noncarbonate carbon as well). The air is recycled through the system by means of a small pump, and significantly lower blank values were obtained by Jeffery and Wilson for this method (0.8 mg CO_2 per determination for HCl, <0.1 mg for H_3PO_4) than were obtained for the conventional method (1.5–2.0 mg CO_2 per determination). Riley[12] determines carbon dioxide and water simultaneously by heating the sample to 1100–1200°C in a combustion tube through which a current of purified nitrogen is drawn, and collecting the liberated CO_2 and H_2O in suitable absorbents; an auxiliary tube containing copper wire which is heated to 700–750°C is necessary to remove acidic oxides of nitrogen that form during the combustion of the sample. The apparatus has been modified to permit the simultaneous determination of H_2O and CO_2 in as little as 10 mg of sample.[18]

The liberated CO_2 may be absorbed in a medium such as a solution of $Ba(OH)_2$ and the resulting precipitate of $BaCO_3$ either filtered off and weighed, or the excess $Ba(OH)_2$ determined by an acid–base titration. Hoefs[26] utilized a special absorption cell in which the CO_2 is absorbed in weakly alkaline $Ba(OH)_2$–$BaCl_2$ solution and determined by means of a potentiometric titration in the special cell; after the determination of carbonate carbon, the carbonaceous matter is oxidized with chromic acid and the determination finished in the same way. A nonaqueous titration has been proposed in which the liberated CO_2 is absorbed in formidimethylamide and then titrated with potassium methoxide in benzene–methanol using thymolphthalein as indicator.[37]

Bouvier et al.[38] have modified this method of nonaqueous absorption and titration to utilize an automatic Leco sulfur titrator, an adaptation of the manual titration procedure developed by Sen Gupta.[39] Acid-generated CO_2 (carbonate) or Leco induction furnace-generated CO_2 (total) is absorbed in a solution of triethanolamine in acetone and titrated with a solution of sodium methylate to a thymolphthalein end point. The precision attainable is good, and the detection limit is about 0.01% C. Another nonaqueous titration scheme is that developed by Read[40] in which the acid-liberated CO_2 is absorbed in a dimethylformamide solution and titrated to a thymolphthalein end point with a solution of tetrabutylammonium hydroxide.

Jeffery and Kipping[41, 42] have utilized gas chromatography for the determination of carbon dioxide and other volatiles in rocks and minerals. The sample (5 mg for carbonates, up to 1 g for other rocks), in a small (12 ml) reaction flask, is treated with orthophosphoric acid, and the liberated CO_2 is carried by a stream of hydrogen through a chromatographic column packed with silica gel where a separation from other volatile constituents is achieved. Detection is by means of a thermal conductivity cell, and the sensitivity for CO_2 is high, as little as 5 ppm in a 1 gm sample being detected. Calibration of the apparatus is done with pure $CaCO_3$ and must be done frequently. A dynamic sorption apparatus has been adapted by Thomas and Hieftje[43] for the determination of CO_2; a thermal conductivity cell is used for the final measurement.

"Carbonaceous" Carbon

The determination of noncarbonate carbon is similar in most respects to the determination of carbonate carbon, but oxidation, as well as decomposition, are involved. Both of these operations must be completed, particularly if the determination is to be of the total carbon, that is, both the carbonate and the noncarbonate fractions. If a combustion method is used to decompose the sample, the oxidation of the carbonaceous material is achieved by carrying out the combustion (usually at 800–1000°C) in a stream of purified oxygen or air, or by mixing an oxidizing flux such as lead chromate, vanadium pentoxide, potassium chromate, or dichromate with the sample (the flux also retains most if not all of other volatiles present), or by a mixture of both methods. Wet oxidation of the carbonaceous matter is achieved by heating the sample with an oxidant such as chromic acid; this is usually combined with a preliminary determination of the carbonate carbon by acid decomposition of the sample, the apparatus requiring little or no modification for the second step. Both combustion and wet oxidation methods require the prior removal of other volatile constituents before the final determination of the liberated CO_2; combustion of the sample in air may result in the formation of oxides of nitrogen, in addition to those of sulfur if the latter is present in the sample, and these too must be removed.

Because of the usually small amount of carbon that is present in silicates, measurement of the volume of gas formed by oxidizing it to CO_2 is usually not feasible.

The determination of carbonaceous matter by difference from the determination of total and carbonate carbon is not very satisfactory when the sample contains much carbonate and little carbonaceous matter. A preliminary acid treatment of the sample with HCl, or HCl and HF, will eliminate the carbonate (and also sulfides)[1, 44–46]; the residue is filtered onto an asbestos pad, dried at 105–110°C, and then the pad and residue are burned in a combustion tube in the usual way. Volatile or soluble components of the carbonaceous matter may be lost by this precedure; Frost reports no significant loss.[45]

The apparatus needed for the gravimetric determination of the CO_2 liberated by burning the sample is similar to that used for the determination of carbonate carbon, except for the substitution of one or more combustion furnaces for the acid decomposition flask and condenser, and the insertion of additional absorption tubes to remove volatiles not encountered in the acid-evolution procedure. Hillebrand et al.[1] have described an elaborate furnace and train in which purified air is drawn through the system and the sample is fused with a mixture of lead and potassium chromates; a heated, reduced copper spiral is used to reduce oxides of nitrogen, and granular lead chromate is packed into the exit end of the tube and heated to 300–400°C to remove sulfur and halogen compounds. A three-unit split electric furnace is very useful when different portions of the purification and absorption train must be maintained at different temperatures. Vovsi and Bal'yan[46] purify the oxygen used for the oxidation by passing it through granulated Cr_2O_3 heated to 600–650°C; nitrogen or sulfur oxides formed during combustion of the samples are removed by passing the gas through silica gel impregnated with chromic acid. (Any chromic acid carried over is caught in silica gel impregnated with a saturated solution of ferrous sulfate in 1:3 sulfuric acid.)

The carbon dioxide may also be absorbed in a solution of barium hydroxide and the determination completed titrimetrically.[1, 45]

The efficiency of the wet oxidation procedure, using a phosphoric acid–chromic acid mixture, was investigated by Dixon[47] who found that a second oxidation flask was necessary to ensure the complete oxidation of the gaseous carbon compounds which would otherwise escape absorption in the train. Mercuric oxide was found to enhance the activity of the acid mixture but was added only to the second flask because, at the final high temperature to which the first flask is heated, it shortens the active life of the oxidant. The acid mixture in the second flask also serves to trap unwanted volatile constituents. Groves[7] found that for the usually low carbon content of most rocks only one oxidation flask was necessary; he cautions against the use of rubber connections to the oxidation flask because of the high blank values that usually result. Jeffery and Wilson[17] modified Dixon's procedure by using a closed circulation system (see also Sec. 5.3.1) that gives a reagent blank of only 0.7 mg (C?) per determination compared to the 2–3 mg per determination found for Dixon's method. They also use a single oxidation flask, but after the initial oxidation appears to be complete, they recommend that the walls of the flask be rinsed down to recover graphite or other carbonaceous matter adhering to them, that more acid mixture be added, and that the oxidation be repeated; low recoveries were obtained for samples containing a large amount of graphite when this precaution was not observed. They prefer to make the determination of carbonate carbon a separate step for such samples, rather than to combine the two determinations. Hoefs[26] applies the wet oxidation procedure only to samples containing >0.1% carbon, using a single flask and finishing the determination potentiometrically.

Hillebrand et al.[1] used sulfuric and chromic acids as the oxidizing mixture, but it is difficult to avoid the concurrent liberation of large amounts of unwanted SO_3. Phosphoric acid has an added advantage over sulfuric acid in that, unlike the latter, it does not exert an oxidizing effect at high temperatures and can be used for a preliminary determination of carbonate carbon without the danger of partial oxidation of some carbonaceous matter as well. The use of a mixture of $K_2Cr_2O_7$ in H_2SO_4, with silver sulfate, for the rapid wet oxidation of organic carbon at 150°C, has been discussed by Frost[48] as a means for the rapid and simple determination of carbon in sedimentary rocks, particularly in shale; the excess potassium dichromate is determined by titration with 0.5 N ferrous sulfate, using orthophenanthroline as indicator. A correction must be applied for oxidizable sulfur compounds that are also present in the sample; other reducing agents, such as ferrous iron, will also interfere, but sulfur exerts the most effect.

Frost[45] has also investigated the applicability of three different methods for the determination of total and organic carbon in sedimentary rocks: (1) a gravimetric combustion method, (2) a gasometric method, with fusion in a Parr microbomb (the potentially high blank limits this to samples containing >0.2% organic carbon), and (3) a combustion method followed by absorption in barium hydroxide solution and titration of the excess OH^- with hydrochloric acid. The first is a control method and Frost prefers to do the routine determinations by (2) if the organic carbon content is greater than 0.5%, by (3) if it is less than this amount.

A two-stage ES procedure for carbonate carbon and "equivalent graphite" carbon (graphite plus diverse organic material) has been described.[49] An aliquot of the sample, with added quartz, is burned in a copper electrode; another aliquot is treated with HCl to remove carbonate carbon, and the dried residue, again with added quartz, is burned as before. Measurement of the density of the cyanogen bands in the spectrum of the first burning gives the total carbon content; similar measurements on the second spectrum give the "equivalent graphite."

Ferris and Jepson[50] used a nonaqueous titration similar to that of Read[40] to titrate CO_2 generated from the dry combustion of organocarbon in clay minerals. It consisted of titrating the CO_2 absorbed in a dimethylformamide solution to a thymolphthalein end point with a solution of tetrabutylammonium hydroxide in toluene–methanol as the titrant.

5.5.2   Determination of "Carbonate" Carbon

a.   By an Acid Evolution–Gravimetric Method

Apparatus

The apparatus does not usually vary significantly from one laboratory to the next. The CO_2, liberated by treatment of the sample with hot acid in a flask fitted with a condenser, is purified of unwanted volatile constituents by passage through absorbing media and is finally collected in a weighed absorption tube for the final gravimetric determination. The apparatus described here has been modified to permit the simultaneous running of two determinations, and use has been made of a commercially available purification and absorption train (Fig. 5.1), modified to operate in a vacuum system, instead of a pressure system as originally intended. The reaction vessels are 150 ml flat-bottomed wide-mouthed extraction flasks fitted with Graham condensers to remove most of the water from the gas, and 50 ml cylindrical separatory funnels for the addition of acid to the samples. Air enters the system by the way of each dropping funnel after first passing through a small tower containing soda asbestos (ascarite) which removes carbon dioxide. After leaving the condenser the gas passes through a tower containing anhydrous magnesium perchlorate and then into the Burrell Carbotrane (Burrell Corporation, Pittsburgh, PA), which carries two separate purification and absorption systems. The gas passes in order through a cartridge containing granulated MnO_2 (Sulsorber C) which removes the oxides of sulfur, a cartridge containing anhydrous magnesium perchlorate (Exsorber C) which removes the last traces of water vapor, and then through a weighed cartridge (Disorber C) containing soda asbestos and anhydrous magnesium perchlorate in which the CO_2 is absorbed. The bubble trap contains a silicone oil and serves both to indicate the rate of flow and to prevent any entry of CO_2 or moisture into the Disorber C cartridge should the systems accidentally "back up."

The glass inlet and outlet of each cartridge plug into rubber sleeves on the ends of rigid connecting copper tubing; the admittedly large mass of the Disorber C cartridge (about 100 g) is counterbalanced by the convenience of handling and ease of replacement.

Procedure

Weigh and transfer a 1.00 g sample (the sample weight used will depend on the amount of CO_2 expected to be present and will vary from 0.2 g for limestones to as much as 5 g for silicate rocks containing less than 0.1% CO_2) to the reaction vessel, and add about 35 ml of water which

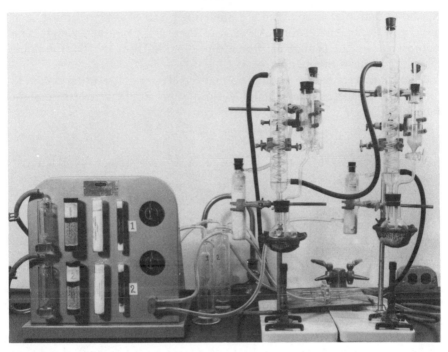

Figure 5.1. Dual apparatus for the acid evolution–gravimetric determination of "carbon-ate" carbon using a Burrell Carbotrane. Photograph by Photographic Section, GSC.

has been freshly boiled and cooled to remove dissolved carbon dioxide. Connect the flask firmly to the condenser and separatory funnel, start a flow of cooling water through the condenser, and open the stopcock of the separatory funnel to admit a slow current of CO_2-free air. Flush the system in this manner for 10 min. A glass delivery tube having a right angle bend at each end is inserted in the Carbotrane in place of the Disorber C cartridge.

Close the stopcock of the separatory funnel and add to the funnel 30 ml of HCl (1 : 1). Remove the glass delivery tube from the Carbotrane and insert the weighed Disorber C cartridge. Open the stopcock of the separatory funnel and allow the acid to flow slowly into the flask.

After all of the acid has been added, and any visible reaction has ceased, heat the contents of the flask to boiling and boil for 2 min. Remove the source of heat, reduce the flow of air to approximately one bubble per second, and continue the flow for 15 min.

Close the stopcock of the separatory funnel and also the one connecting the bubble trap to the vacuum line; remove the Disorber C cartridge, seal the outlet and inlet tubes with a connecting piece of rubber tubing, and place it in the balance room. Allow it to stand for 5 min and weigh. The increase in weight is that of the carbon dioxide from the "carbonate" carbon present in the sample.

A blank determination should be made daily, or more often if humid conditions prevail, and a suitable correction applied to subsequent determinations.

Conversion Factors

$$CO_2 \underset{3.6644}{\overset{0.2729}{\rightleftarrows}} C$$

b.   *By an Acid Evolution–Volumetric Method*

In this procedure the carbon dioxide is liberated by the action of the hot acid on the sample in a closed system, and the measured volume of air and carbon dioxide is passed through a solution of potassium hydroxide; the CO_2 is absorbed and is determined as the difference between the measured volumes before and after passage through the KOH solution. This method is essentially that described by Goldich et al.[36] and modified by Shapiro and Brannock[35] for their rapid analysis scheme; they recommend its use for samples containing more than 6% carbon dioxide. The need to correct for temperature and pressure conditions is eliminated if standards are run with each batch of samples and the factor used in calculations is based upon these. Very satisfactory results have been obtained at the GSC for samples containing as little as 0.3% CO_2; the method is simple in operation and rapid, each determination taking about 15 min. The presence of other volatile constituents which are absorbed by KOH will, unless they are removed, cause serious error in the determination.

Apparatus

The apparatus is shown in Fig. 5.2 and consists of the following. (1) A 50 ml Pyrex filtration flask is connected by a tightly fitting rubber stopper to a small (20 cm) Liebig condenser, and with the side arm connected by a length of Tygon™ tubing to a 500 ml separatory funnel. (2) The condenser is connected by a piece of right angle glass tubing to one outlet of a three-way stopcock joined to a 100 ml burette enclosed in a water jacket in which a thermometer is also suspended; a second outlet of the stopcock is joined to a gas-bubbling pipette having two compartments and a three-way stopcock, and the lower part of the burette is connected by a length of tygon tubing to a leveling bottle, supported on a leveling clamp.

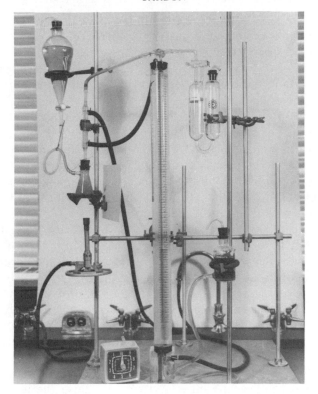

Figure 5.2. Apparatus for the volumetric determination of "carbonate" carbon. Photograph by Photographic Section, GSC.

Reagents

Hydrochloric acid–sodium chloride solution.   To 1500 ml water add 500 ml of 12 *M* HCl and saturate the dilute acid (1 : 3) with sodium chloride. Add a few milligrams of methyl red indicator to the mixture.

Potassium hydroxide solution.   (50% w/v). Dissolve 1000 g of KOH in water in a plastic beaker, cool, dilute to 2 liters, and store in a plastic bottle.

Procedure

1.   Fill the separatory funnel with the acid solution and allow it to flow into the connecting Tygon™ tubing until all air is expelled. Add sufficient acid solution to the leveling bottle to fill the burette when the bottle is raised to the level of the stopcock. Add sufficient solution (about 200 ml) to the gas pipette and adjust the level in the absorption compartment with

the leveling bottle, so that there is sufficient KOH to accommodate the passage through it of 100 ml of gas. Mark this level.

2.   Turn the stopcock of the gas pipette so that gas does *not* bubble through the solution and open the connection between the gas pipette and burette by turning the burette stopcock to the appropriate position; by raising and lowering the leveling bottle, adjust the level of the KOH solution to the reference mark on the absorption compartment. Close the stopcock of the gas pipette.

3.   Open the connection between the burette and the condenser by turning the burette stopcock to the appropriate position and by raising the leveling bottle, fill the burette to the stopcock opening and close the burette stopcock. Lower the leveling bottle to the bottom of the burette. The system is now ready for use.

4.   Weigh 0.1000 g of reagent grade $CaCO_3$ into the flask and attach the flask firmly to the condenser; connect the side arm of the flask to the tubing from the separatory funnel (the acid solution should just fill the side arm) and open the connection between the evolution system and the burette (the level of the liquid in the burette will drop slightly).

5.   Open the stopcock of the separatory funnel and allow 10 ml of the acid solution to enter the flask. Heat the contents of the flask continuously to near boiling, then boil for 3 min and discontinue the heating.

6.   Open the stopcock of the separatory funnel and allow the acid solution to flow into the evolution system until the liquid just reaches the burette stopcock; close the burette stopcock. Raise the leveling bottle until the levels in the burette and in the bottle are the same and record the burette reading. Note the temperature inside the water jacket at the time of the reading.

7.   Turn the burette and gas pipette stopcocks to their proper settings, and by raising the leveling bottle, force the gas from the burette through the KOH solution. Lower the leveling bottle, after turning the gas pipette stopcock to the appropriate setting, until the gas has been drawn into the burette again. Repeat the passage through the KOH twice more. After the third passage adjust the level of the KOH solution to the reference level marked on the absorption compartment and close the burette stopcock. Adjust the height of the leveling bottle so that the heights of the liquid in the burette and bottle are the same, and record the burette reading. The difference between the two readings is that of the volume of CO_2 released from the sample. Note the temperature, which should not have changed by more than 1°C.

Repeat the procedure (steps 1–7) with a sample of suitable size. At the end of a group of samples, run another 0.1000 g portion of $CaCO_3$ and average the results obtained for the two standards.

Compute the factor for the standard:

$$\text{factor} = \frac{44.01}{\text{average volume of } CO_2 \text{ obtained}}$$

Compute the percentage of CO_2 in the sample:

$$\%CO_2 = \frac{\text{volume of } CO_2 \times \text{factor} \times 0.1000}{\text{sample weight}}$$

5.5.3 Determination of "Total" Carbon

a. *By a Combustion–Gravimetric Method*

Apparatus

The combustion is done in a high temperature tube, $3.8 \times 3.2 \times 90$ cm and having a reduced end, which is heated to 1000°C in a split cylindrical furnace capable of continuous operation at about 1200°C. The reduced end of the combustion tube is connected to a cartridge containing granulated MnO_2 [Sulsorber C from a Burrell Carbotrane, (Sec. 5.5.2a)] for the removal of SO_2 and chlorine, which in turn is followed by a Fleming purifying jar containing a mixture of chromic and phosphoric acids (add concentrated chromic acid to 85% phosphoric acid until the solution is a deep reddish orange) which serves both as an indicator of the rate of flow of gas through the train and as a trap for SO_3 and oxides of nitrogen. A U tube containing pumice impregnated with ferrous sulfate (the pumice is soaked in 1:3 H_2SO_4 saturated with ferrous sulfate and dried at 110°C) serves as a trap for any of the chromic–phosphoric acid solution that might be carried out of the Fleming jar by the gas. A U tube packed with anhydrous magnesium perchlorate removes any water vapor from the gas, which then passes through a weighed (about 50 g) Stetser–Norton carbon dioxide absorption bulb containing a lower layer of soda asbestos (Mikhobite or Ascarite) and an upper layer of anhydrous magnesium perchlorate to absorb any water produced by the absorption of carbon dioxide; a small weighed U tube also containing soda asbestos and anhydrous magnesium perchlorate follows the absorption bulb and serves to catch any CO_2 not absorbed in the latter. A U tube filled with anhydrous magnesium perchlorate and a second purifying jar containing chromic–phosphoric acid solution protect the weighed absorption tubes from contamination in the event of a "backing up" of the system. The purifying jar also serves as a flow rate indicator and, with the first bubbler, as a check on the absence of leaks in the vacuum system.

Air is drawn through the system at a rate of about 3 bubbles per second by means of suction; it is first made CO_2-free by passing it through a U tube containing soda asbestos. The U tube is connected to the hollow stem of a high temperature, double baffle heat reflector which in turn is connected to the wide end of the combustion tube by a tightly fitting rubber stopper. The temperature of the furnace is regulated by a variable rheostat.

The sample, mixed with an oxidizing flux (reagent grade vanadium pentoxide), is burned in a ceramic boat (95 mm long) which is used only once. A small wooden stand is a useful device for carrying the weighed carbon dioxide absorption vessels to and from the balance room.

All connections in the train are made with tygon tubing, except the connections to, and between, the two weighed carbon dioxide absorption vessels which are of soft rubber for ease of handling. Closely fitting pieces of heavy rubber tubing, sealed with Teflon plugs at one end, are used to close off the outlets and inlets of the absorption vessels when they are removed from the train; they are removed during the weighing. A piece of glass tubing is inserted in their place in the train during the period of weighing so that flushing of the system may be continued in preparation for the next determination.

Procedure

Weigh a sample of suitable size such that not less than 10 mg of CO_2 will be produced during the combustion, add 0.50 g of reagent grade vanadium pentoxide and mix well. Carefully spread the mixture on the bottom of a clean ceramic boat. The boats should first be ignited in a furnace for several hours and then stored in a sealed glass jar until needed; they should be handled with forceps, or finger cots.

Insert the weighed carbon dioxide absorption vessels into the train (close off the system on either side of them during this operation), then quickly insert the boat in the mouth of the combustion tube, push it into the hot central portion with a steel rod, insert the rubber stopper, and open the system to suction. Continue the heating for 30 min.

Close off the system on either side of the absorption vessels and disconnect them from the line; close the stopcocks of the U tube and seal the outlet and inlet of the Stetser–Norton bulb with the rubber plugs. Place the tubes on a support and allow them to stand for 10 min in the balance room. Before weighing, remove the plugs from the absorption bulb and briefly open the stopcocks of the U tube. Insert the glass tube in the train in place of the absorption vessels and flush the system for 10 min before beginning another determination.

A blank determination should be made daily; because of the complexity of the train these blanks tend to be rather high but should not be much more than 1 mg.

b. *By a Combustion–Volumetric Method*

The method to be described was developed for the analysis of carbon in steel but it has been adapted to the determination of total carbon in rocks and minerals with no difficulty. It has the advantage of being much more rapid than the other methods, and of having lower blanks. The equipment used is the commercially available Leco induction furnace (No. 521-000) and the corresponding Leco semiautomatic carbon analyzer (No. 572-200) made by Laboratory Equipment Corporation, Saint Joseph, MI. The method is that supplied with the equipment.

Apparatus

A cylinder of compressed oxygen is equipped with a regulator and a Leco purification train for the oxygen flow. The train consists of an H_2SO_4 bubbler, a single glass column enclosing an ascarite layer separated with a glass wool plug from an anhydrone layer, and a rotometer to regulate the oxygen flow rate. This train scrubs any CO_2, moisture, and acid from the oxygen stream.

The furnace unit consists of the Leco induction furnace previously mentioned, a Leco quartz combustion tube with a built-in jet to direct the oxygen flow into the combustion area, and a second purifying train. This second purifying train is designed to ensure that only CO_2 reaches the carbon analyzer, and it consists of a cloth dust filter, a manganese dioxide trap for sulfur gases such as SO_2, and a Leco catalytic converter and furnace for converting any CO to CO_2.

The Leco ceramic reaction crucible used to contain the sample can be used once only, and an adequate supply should be kept on hand. Because of the high temperatures involved (the reaction in the crucible can exceed 1600°C), the various O-rings, the combustion chamber, the ceramic crucible support, and connecting tubing require frequent replacements, and a supply of these items should also be kept in stock.

The carbon analyzer consists of a burette, a leveling bottle, and a CO_2 absorption reservoir. The burette and the reservoir are enclosed in a water jacket through which water is circulated by a small pump to maintain as constant a temperature as possible in the measurement burette and absorption reservoir. The temperature of the circulating water is monitored with a thermometer set into a well in the flow. The atmospheric pressure should be monitored by a good mercury barometer to ensure that a reliable pressure correction can be applied to the volumetric CO_2 measurement.

The Leco model numbers given above for the induction furnace and the semiautomatic carbon analyzer are given for reference only. There are other furnaces (both resistance and induction models) which would serve as well, and there is also a manual carbon analyzer available, although the semiautomatic one mentioned previously is much more convenient to use.

Reagents

The KOH absorption solution is prepared by dissolving reagent grade KOH in distilled water in a ratio of 50 g KOH per 100 ml water. To prepare the red leveling solution, add 0.4 g of methyl orange to 200 ml of distilled water and heat to boiling, cool, and filter. From this prepare an H_2SO_4 solution containing 5 ml of H_2SO_4 per 100 ml of water and add 2 ml of a wetting agent. The dye is added to make it easier to read the burette. The volume of KOH and H_2SO_4 solutions that are required will depend on the particular model of carbon analyzer used. The KOH solution will have to be renewed occasionally, but the H_2SO_4 solution will last indefinitely.

In addition to the solutions listed above, granulated tin and iron chip accelerators are required. Leco products are not mandatory provided that the S and C contents of the product chosen are sufficiently low.

Procedure

The procedure used will depend on the model of the furnace and the type of analyzer available. Detailed instructions accompany the equipment, however, and in general no modifications to the methods given are required. Included in the instructions are directions for establishing a blank for the apparatus and testing it for leaks. A trouble-shooting guide is also available.

In general, the method for the equipment detailed above involves accurately weighing 0.5 g of sample (or less for limestone samples) into a crucible, adding a scoop (about 1 g) of tin accelerator and about 0.5–1 g of iron chip accelerator, and placing the uncovered crucible on the ceramic pedestal of the induction furnace. The stopcock of the burette is turned to the exhaust position briefly, and then the burette is filled with the red leveling solution, the crucible is raised into the combustion area, the oxygen flow is adjusted to 1.5 l/min, and the burette stopcock is adjusted to allow the oxygen to sweep into the burette. The combustion is initiated and allowed to go to completion, and enough oxygen is swept in to force the red leveling solution two-thirds of the way down the calibrated stem. The oxygen flow is then stopped and the hot crucible removed.

The meniscus in the burette is set at zero, and the gases collected during the combustion are forced into the absorption vessel containing the KOH solution. The residual gases are forced back into the burette and the menisci of the red leveling solution in the burette and auxiliary reading tube are adjusted to the same level. The burette reading (which will be above the zero position because of the CO_2 absorbed in the KOH solution) is corrected for the temperature of the circulating water and the barometric pressure, and the carbon content of the sample is calculated.

This method gives good results over a wide concentration range (0.02% to about 12.0%) and is very rapid (50–60 samples per working day). It is necessary to run frequent standards and blanks to ensure that the reagents and purifying trains are functioning efficiently; all tubing and the dust trap should be cleaned frequently. In spite of these drawbacks, however, it is a dependable, rapid method.

c.   By a Combustion–Infrared Absorption Method

See Sec. 5.6.4c.

5.6   SULFUR

5.6.1   General

The sulfur content of most silicate rocks is very low (<0.1%), and the sulfur is usually in the form of a sulfide such as pyrite (FeS_2) or pyrrhotite (Fe_7S_8); the sulfur-bearing silicates nosean ($Na_8(Al_6Si_6O_{24})SO_4$) and häuyne ($(Na, Ca)_{4-8}(Al_6Si_6O_{24})(SO_4, S)_{1-2}$) are seldom encountered. The rocks may have been subjected to sulfide mineralization, of course, but the nature and source of the sample will usually give ample warning of this possibility. Stony meteorites may contain much sulfur present as troilite (FeS). Among the sulfates often encountered in small amounts are barite, celestite, gypsum, and alunite ($K_2Al_6(OH)_{12}(SO_4)_4$); barite is the most common of these minerals and a spectrographic analysis will again alert the analyst to the possibility of its presence. Sulfur may also occur in the elemental (native) form or in organic compounds, but such occurrences are unusual.

The analytical problems raised by the presence of sulfur in its various guises are encountered chiefly at the beginning of a classical main portion analysis, and in the determination of ferrous iron (Sec. 6.1). If the sulfur is present as a sulfide and is to be determined, the sample must not be ground too finely or some of the sulfide may be oxidized to SO_2 and be lost; the sulfide may also be oxidized to sulfate, and this will be a source of error if the sulfide is to be determined by a method involving the evolution of hydrogen sulfide.

The form(s) in which the sulfur is present in the sample can be determined with some assurance by boiling about 1 g of the powdered sample with dilute HCl (1:4); the odor of hydrogen sulfide or the blackening of a piece of lead acetate paper will indicate the presence of an *HCl-soluble sulfide* such as pyrrhotite. If, on the addition of a few drops of 5% barium chloride solution to the filtrate from this acid digestion, a white precipitate is obtained, it is likely that *HCl-soluble sulfates* are present in the sample. If the remaining residue is again boiled with acid, this time dilute nitric acid (1:4) containing some bromine, and the filtrate is again tested with barium chloride, a white precipitate indicates the presence of *HCl-insoluble sulfides* such as pyrite, or possibly *elemental sulfur*. Fusion of the residue with sodium carbonate and potassium nitrate, followed by leaching, acidification with HCl and the addition of barium chloride, will reveal the presence of *HCl-insoluble sulfates* such as barite, or possibly of *organic sulfur*.

The determination of the total sulfur is almost always necessary (this is usually all that is needed for small amounts, 0.1% or less). A separate determination of either the sulfate or the sulfide sulfur will permit the results to be expressed in terms of the latter two forms, assuming that elemental sulfur and organic sulfur are not present. If only the total sulfur content is determined and the form in which the sulfur exists in the sample is not known, then it should be reported as sulfide sulfur only.

In a summation of the analysis of a sulfur-bearing sample it is usually necessary that a correction be made to the analysis total by subtracting the O-equivalent of the S that is present as sulfide if the other elements which form the sulfides, especially iron, are to be reported instead as oxides. No correction is necessary if the sulfur present as sulfate is reported as SO_3. The correction for the sulfide requires a knowledge of the sulfide mineralogy of the sample, unless the sulfur content is less than 0.1%, because of the differing behavior of the sulfides during the acid decomposition of the sample prior to the determination of ferrous iron. (It is highly unusual for sulfides other than pyrite and pyrrhotite to be present in significant amounts in silicates that are not ore samples.) Pyrite is scarcely attacked by the hot sulfuric and hydrofluoric acids of the modified Pratt method and not at all by the cold Wilson procedure; it is much more probable that the pyrite will be decomposed completely during the determination of total iron, and the ferrous iron of the pyrite will be counted as ferric iron.

$$2 \, FeS_2 \equiv Fe_2O_3$$

$$\frac{30}{4S} = 0.374$$

and thus the oxygen equivalent of the sulfur known to be present as pyrite is the percent sulfur multiplied by 0.374. Pyrrhotite, on the other hand, is readily attacked by acids and the iron will be reported as ferrous iron; the O-equivalent is $\% \text{ S} \times 0.437$.

$$Fe_7S_8 \equiv 7 \text{ FeO}$$

$$\frac{70}{8S} = 0.437$$

Peck (Ref. 15, pp. 51–53, 82–83) recommends that when the amount of sulfur is small and it is not practical to distinguish between individual sulfides, the correction to be subtracted should be $\% \text{ S} \times 0.5$. It is also better to report the total sulfur as S, rather than to attempt to separate the types.

The problems that may be encountered in the determination of sulfur in rocks and minerals have been treated in detail by Groves (Ref. 7, pp. 117–123, 207–208) and by Hillebrand et al. (Ref. 1, pp. 711–724, 948–951); a general coverage of the analytical chemistry of sulfur is also available.[51–53] Young has included a section dealing with the analysis of the different forms of sulfur (sulfide, sulfate, etc.) in his monograph on phase analysis,[54] and reference should also be made to the monograph by Doležal et al.[55]

"Sulfide" Sulfur

When sulfur is known to be present only as a sulfide, its determination is a relatively simple task. The sample may be fused with an oxidizing flux such as a mixture of sodium carbonate and potassium nitrate, or decomposed by digestion with a mixture of oxidizing acids and bromine. Both methods result in the oxidation of the sulfide to sulfate. The fusion method has the advantage of removing a number of elements such as iron and the alkaline earths, but does introduce a large amount of alkali salts into the solution; when there is much sulfide present (e.g., > 2%), however, the fusion method is preferred to acid digestion because of the possibility that some sulfide may escape oxidation and be lost by the latter method. The acid digestion method has the advantage of removing silica and of introducing no additional material; it is necessary, however, to remove or reduce ferric iron, if much is present, because it will contaminate the barium sulfate when the determination is finished gravimetrically. A detailed discussion of the gravimetric determination of sulfur as $BaSO_4$ can be found in several texts.[1, 53, 56]

The sample may also be heated at a high temperature (about 1400°C) in a stream of oxygen or air (in the latter case the sample is mixed with an oxidizing flux) to convert the sulfide to SO_2 or SO_3, or to a mixture of both.[7, 57-59] The gases are absorbed in a suitable solution, and the determination is finished colorimetrically, titrimetrically, or gravimetrically. This general method is the basis for the Leco induction furnace–titrimetric method developed for the determination of sulfur in steel and adapted to rock analysis,[38] and Garcia has applied it with success to zinc ores.[60]

It is not usually possible, however, to assume that the sulfur is present only as sulfide, and the methods given above could, for the most part, be used to determine the total sulfur content as well. It is common practice to so determine the total sulfur and then to determine, on a separate portion, that part of it which can be put into solution by boiling the sample with dilute HCl (so-called soluble sulfate); the difference between the two values is expressed as sulfide sulfur. It is obvious that this is not entirely satisfactory; insoluble sulfate will be counted as sulfide, and care must be taken to avoid the oxidation of soluble sulfides during the acid treatment. Vlisidis[61] does the acid digestion under an atmosphere of nitrogen, and adds $CdCl_2$ to precipitate any sulfide ion liberated at the same time; precipitated $BaSO_4$ and any barite present are removed, fused with Na_2CO_3, dissolved in HCl, and the barium is precipitated with Na_2SO_4. For the usually low sulfur content of silicates the errors so introduced will be small, but unless there is reason to suspect the presence of more than one form of sulfur, it is better to report the total sulfur as sulfide.[15]

The most reliable method is one in which the sulfide ion itself is liberated. Murthy et al.[62] digested a mixture of sulfides and sulfates with combined hydriodic and hydrochloric acids in a special apparatus; the H_2S formed by the reaction is swept into a suspension of cadmium hydroxide, the latter is poured into 2 N acetic acid containing a known amount of iodine, and the excess iodine is titrated with a standard solution of sodium thiosulfate. The reaction is slow but satisfactory for sulfides such as galena and sphalerite; for pyrite and chalcopyrite it is necessary to employ a more concentrated hydriodic acid solution and to add a small globule of mercury to the reaction vessel.[62] A similar precipitation of CdS is used by Smith et al.[63] to determine sulfide sulfur in oil shale in which sulfate and organic sulfur compounds are also present; after the separation of the soluble sulfate, the residue, in a special refluxing flask, is treated with aluminum hydride dissolved in tetrahydrofuran to decompose the sulfides, and the hydrogen sulfide formed is passed through a solution of cadmium sulfate. The sulfuric acid formed in the absorption vessel is titrated with sodium hydroxide solution. Any free sulfur present in the sample will be reduced also. Sorensen et al. employed a modification of this H_2S generation–precipitation technique for the sensitive and precise measurement of the sulfide concentration in soils and minerals, using a sulfide ion electrode–titrimetric measurement with Cd^{2+} as the titrant.[64]

Organic sulfur and free sulfur are two forms of sulfur that are not often encountered, especially in silicates. They are determined as part of the total sulfur content because they are oxidized to sulfate during decomposition of the sample, but for their separate determination such oxidation must be avoided. Groves[7] extracts the free sulfur by refluxing the sample with carbon tetrachloride in a Soxhlet extractor; the extract is evaporated to dryness in a weighed platinum dish and the extraction repeated until the weighings show that all free sulfur has been extracted. Portions of the sample, after the extraction, are used for the determination of sulfide and sulfate sulfur. Smith et al.[63] used the dried residue remaining after the removal of sulfate and sulfide sulfur from oil shale for the determination of organic sulfur by fusing it with Eschka mixture ($MgO–Na_2CO_3$, 1:2), leaching the sinter, acidifying the filtrate after the addition of bromine water, and precipitating the sulfur as barium sulfate. Steger[65] has described the determination of the elemental sulfur content of minerals and ores by vaporization in a closed tube with final determination by spectrophotometry.

"Sulfate" Sulfur

Reference has already been made to the determination of sulfate sulfur in the previous section. If it is the only form in which sulfur is present in the sample, whether as a soluble or an insoluble sulfate, its determination is a simple matter; fusion with sodium carbonate will ensure that all of it will be in solution prior to its determination by one of several possible methods. When sulfides are present, certain precautions must be taken to avoid including all, or a portion, of them in the value for sulfate.

The determination of soluble sulfate (usually reported as SO_3) is accomplished by digesting the sample for a brief period with dilute hydrochloric acid. Water-soluble sulfates, the silicates nosean and häuyne, gypsum, anhydrite, and (possibly) celestite will dissolve wholly, while some others may dissolve to a limited extent. Peck[15] boils the sample with water before adding the HCl in order to expel air that might oxidize the hydrogen sulfide formed when soluble sulfides (e.g., pyrrhotite) are also dissolved.

Some acid-insoluble sulfate minerals can be put into solution by boiling the sample with sodium carbonate (e.g., lead sulfate) or sodium hydroxide (alunite). This is useful when it is necessary to determine the truly acid-soluble sulfates. An initial treatment with sodium hydroxide will extract alunitic sulfate, the residue can then be boiled with dilute HCl to dissolve the gypsum and anhydrite. Peck[15] has pointed out, however, the uncertain state of knowledge regarding the behavior of the sulfide minerals in a boiling basic solution and suggests that, unless the mineralogy of the sample is known and a separation will serve a useful purpose, it is better to confine oneself to the simplest determinations and to allow the petrographer to distribute the sulfur as he sees fit. A simple digestion with HCl, under nonoxidizing conditions, will yield a value for acid-soluble sulfate that will not be greatly in error, provided that the total sulfur content is low.

To obtain acid-insoluble sulfate and thus total sulfate sulfur it is necessary to remove the sulfides that are soluble with difficulty, such as pyrite, from the residue left over from the determination of acid-soluble sulfate. This latter is digested with a mixture of aqua regia and bromine, and the remaining residue is then fused with sodium carbonate to decompose barite and other insoluble sulfates.

Fusion and acid treatment are not the only methods of attack that may be used to solubilize sulfate, although they are usually the simplest and most direct means. Mention has been made of the use of a high temperature combustion furnace for decomposition of the sulfates to give either SO_2, SO_3, or a mixture of both. It is also possible to reduce the sulfate to sulfide. Rafter[66] reduced sulfate by heating it with graphite in a nitrogen atmosphere and precipitated the sulfide as silver sulfide.

Total Sulfur

The decomposition of the sample by acid digestion in an oxidizing environment, to ensure the conversion of the sulfur present to sulfate, is the simplest approach, but this will include only the acid-soluble sulfate and sulfide; any acid-insoluble minerals will be lost during the determination. With this limitation in mind, however, a rapid determination can be made; Shapiro and Brannock[35] digest the sample with a mixture of aqua regia and carbon tetrachloride, precipitate the sulfate with barium chloride, and after a 10 min period for digestion, filter, ignite, and weigh the barium sulfate. Goldich et al.[36] use the filtrate from the R_2O_3 separation for the determination of total sulfur in marl; the sample was first ignited, under oxidizing conditions, to convert all sulfur to sulfate. Although some loss of sulfur occurred during the ignition, a comparison of the values obtained by this rapid method and those obtained by digesting the marls with hydrochloric acid and bromine favored the rapid method in reproducibility and completeness of recovery. Wilson et al.[67] prefer an acid digestion to fusion because the latter approach usually results in the crucible being attacked; they recommend a mixture of aqua regia, HF, and $HClO_4$ (in a Teflon dish) with V_2O_5 added to expedite the oxidation of pyrite and any organic matter that may be present. Additional calcium is added when necessary to give at least a two-fold excess over that needed to combine with the sulfate present in order to prevent a loss of SO_3 during the evaporation with perchloric acid.[1] These authors also investigated the effect of any barium present in the sample on the recovery of sulfate and found that a serious loss of sulfate occurs both when the BaO content is around 1% and when the BaO content is low but the sulfur content is high; when both the BaO and the S contents do not exceed 0.2% each, as in most silicate rocks, the loss of sulfate is negligible.

Murphy and Sergeant[68] have determined total sulfur in silicate rocks by heating the sample in a mixture of $NaClO_3$ and concentrated HCl to convert sulfide sulfur (including pyritic sulfur) to sulfate. The residue remaining after evaporation of the acid is transferred to a refluxing apparatus where H_2S is generated in a reflux reaction. The refluxing reagents are sodium iodide, red phosphorus, hypophosphorus acid, orthophosphoric acid, and propionic acid. The H_2S which is generated is absorbed in a KOH solution and titrated with a 2-(hydroxymercuri) benzoic acid solution to a dithizone end point. The method is applicable to a concentration range suitable for most silicate rocks, approximately 5–2000 ppm sulfur.

The fusion of the sample with an alkaline flux, to which an oxidant is added if necessary, is the most widely used procedure for decomposing the sample and rendering the sulfur in the form of a soluble alkaline sulfate. The flux is almost always anhydrous Na_2CO_3, with a small amount of KNO_3 or Na_2O_2 added unless sulfates are known to be the only form of sulfur present in the sample. Na_2O_2 and $NaOH$[56] are also used; these two fluxes put more aluminum into solution than does Na_2CO_3, and because of the predilection of $BaSO_4$ for adsorbing and occluding foreign ions, the latter flux is preferred. The fusions are better done in an electric muffle than over a gas flame because of the danger of the sulfur from the gas being picked up by the alkaline flux, and the subsequent leaching of the cake should be done on an electrically heated water or steam bath, or hot plate. Because of the presence of oxides of sulfur in the normal laboratory air, as well as in the reagents used, a determination of the reagent blank should be made at the same time as the sample determination.

The determination of total sulfur by the fusion method can be combined conveniently with the determination, on the same portion of sample, of zirconium, barium, chromium, and rare earths.

As was mentioned in the discussion of sulfide sulfur, heating the sample to 1400°C or higher in a stream of oxygen (or air if an oxidizing flux is present) has been used for the determination of sulfur. It results in the evolution of total sulfur as SO_2 or SO_3 which can be absorbed in a suitable medium with the determination finished gravimetrically, titrimetrically, or colorimetrically.[38, 50, 59, 64] The Leco method has been studied by Harrison and Spikings[69] who reported that yield tests gave an average recovery of 93.9% of the sulfur present and that there was an average loss of 5% of the sulfur on the walls of the delivery tubes leading from the furnace to the titrator. This was reduced to a loss of 2% when the tubes were heated to 350°C by means of a heating coil. A further improvement in the yield resulted from the use of iron-free vanadium pentoxide to obtain a more even burn with less surging and sputtering. In most silicate rocks, however, the sulfur content is so low that only the precaution of heating the delivery tubes results in a significant improvement. The Leco method is suitable for sulfur contents ranging from less than 10 ppm to the high levels found in sulfide concentrates when appropriate sample sizes and titrant dilutions are used.

Arikawa et al.[70] have described a method for determining sulfur in igneous rocks in amounts as low as 10 ppm by heating the sample, mixed with V_2O_5, at 1000°C in a stream of nitrogen. Any SO_3 generated is reduced to SO_2 on copper wire at 950°C. The SO_2 is absorbed in a solution of 0.5 mM sodium tetrachloromercurate(II), and the measurement is made with a UV spectrophotometer at 228 nm. A calibration curve can be prepared using a solution of Na_2SO_3.

Total sulfur can also be determined by XRF.[71, 72] In this case the X-rays are of very low energy (2.307 keV or a wavelength of 0.53731 nm), and either flushing of the system with helium or the use of vacuum conditions is required (see Sec. 11.2.29). There are some problems, such as a decrease in total intensity when both free sulfur and sulfide sulfur are exposed to X-rays but not when sulfate sulfur is similarly exposed. Also, samples with sulfate sulfur give higher intensities than do sulfide samples with an equivalent sulfur content when the samples are presented to the spectrometer in a ground, pelletized form.[73] In fusing rock samples to prepare a glass disc for analysis by XRF, there is some loss of sulfur even from sulfate sulfur species.

5.6.2 Gravimetric Determination of Sulfide Sulfur by Acid Decomposition and Oxidation

This procedure is used when the sample is known to contain an appreciable amount of sulfur in the form of sulfide minerals, so that the determination of sulfide sulfur as the difference between total sulfur and sulfate sulfur is not likely to be satisfactory. As discussed in Sec. 5.6.1, the procedure will determine the sulfur also present as acid-soluble sulfate (and as elemental and, to some extent, as organic sulfur as well, although this combination will rarely be encountered), but for most silicates the error will be negligible if all of the sulfur is reported as sulfide sulfur. The correction to be made for equivalent oxygen in the summation of the analysis will depend upon whether the sulfide mineral, if known, is pyrite or pyrrhotite (see Sec. 5.6.1).

Reagents

Barium chloride solution (5%). Dissolve 25 g of $BaCl_2 \cdot 2H_2O$ in water, filter, and dilute to 500 ml in a glass-stoppered bottle.

Procedure

Transfer a 1.00 g sample (80–100 mesh) to a dry 250 ml beaker, cover, and cautiously add (in a fume hood) 5 ml of a mixture of bromine and carbon tetrachloride (2 parts by volume of bromine to 3 parts of carbon tetrachloride). Allow to stand for 15 min at room temperature, with occasional swirling to ensure complete wetting of the sample by the mixture.

Add, through the lip of the beaker, 10 ml of concentrated nitric acid and let stand at room temperature for 15 min, again with occasional swirling of the contents. Place the beaker on a low temperature (<100°C) hot plate or on the covered surface of a water bath and heat until all visible reaction has ceased and most of the bromine has been expelled. Increase the temperature of the hot plate, or heat the beaker directly on the water bath, remove the cover, and evaporate the contents to dryness. Cool, add 10 ml of 12 N HCl, and again evaporate to dryness. Heat the dry residue at 100°C for 30 min to dehydrate any dissolved silica.

Drench the residue with 2 ml 12 N HCl, let it stand for a minute or so, and then add 50 ml of water. Heat to boiling and boil for 5 min. Cool for 5 min and then cautiously add about 0.5 g of reagent grade powdered zinc or aluminum to reduce the ferric iron; warm the mixture if the reaction is too slow, and stir from time to time. Filter the solution through a 9 cm Whatman No. 40 paper into a 600 ml beaker and wash the paper and residue thoroughly with hot water. Dilute the filtrate to at least 300 ml in volume, and add 2 ml of 12 N HCl.

Heat to boiling and add dropwise, by pipette, 25 ml of hot 5% $BaCl_2$ solution. Stir vigorously during the addition of the precipitant. Digest the solution and precipitate on the water bath, with occasional stirring, for 1 hr and then allow to stand overnight.

Filter the solution through a 9 cm Whatman No. 42 paper by first decanting as much of the supernatant liquid as possible through the paper, keeping most of the precipitate in the beaker. Discard this filtrate before transferring the precipitate to the paper with a stream of hot water. If the filtrate appears cloudy, pour it again through the paper and repeat this procedure until filtrate is perfectly clear. Police the beaker and stirring rod carefully and pour all rinsings through the paper. Finally, wash the paper and precipitate 10 times with small amounts of hot water.

Fold the paper around the precipitate with platinum-tipped forceps and place it in a weighed 25 ml covered platinum crucible. Wipe out the funnel and the tips of the forceps with a small piece of filter paper and add it to the crucible. Loosely cover the crucible, place it in a cold electric muffle furnace (or over the low flame of a Tirrill burner), and then burn off the paper at the lowest temperature possible (<600°C). Do not allow the crucible to become red, and do not allow the contents to catch fire. When the paper has been completely charred, displace the cover to one side, raise the temperature until the crucible is dull red, and burn off the carbon. When the residue is white, ignite it at about 900°C for 15 min. Cool in a desiccator for 30 min and weigh. Repeat the ignition, cooling, and weighing to constant weight.

If the residue is colored it should be fused with a small amount of Na_2CO_3, the leachate filtered and made acid with HCl, and the precipitation repeated. This will, unfortunately, not remove any chromium which will be present as chromate and will coprecipitate as barium chromate (the ignited residue will show a green coloration). If a significant error is likely to be introduced by the presence of chromium, it should be removed by some means such as solvent extraction, mercury cathode electrolysis, by passage through a cation-exchange resin, or, alternatively, it can be eliminated as CrO_2Cl_2 at a low temperature by heating the sample to $HClO_4$ fumes in the presence of HCl.

If it is suspected that some silica may have separated with the barium sulfate, the residue should be treated with a drop of sulfuric acid and 1 or 2 ml of hydrofluoric acid, evaporated to dryness, reignited, and reweighed.

$$\% \text{ S} = \frac{\text{weight of BaSO}_4 \text{ (g)}}{\text{sample weight (g)}} \times 0.1374 \times 100$$

Conversion Factors

$$SO_2 \underset{1.9981}{\overset{0.5005}{\rightleftarrows}} S \qquad\qquad BaSO_4 \underset{7.2807}{\overset{0.1374}{\rightleftarrows}} S$$

$$SO_3 \underset{2.4971}{\overset{0.4004}{\rightleftarrows}} S \qquad\qquad BaSO_4 \underset{3.6438}{\overset{0.2744}{\rightleftarrows}} SO_2$$

$$BaSO_4 \underset{2.9156}{\overset{0.3430}{\rightleftarrows}} SO_3$$

$$BaSO_4 \underset{2.4299}{\overset{0.4115}{\rightleftarrows}} SO_4$$

5.6.3   Gravimetric Determination of Acid-Soluble Sulfate Sulfur

This procedure presupposes that any hydrogen sulfide formed by decomposition of acid-soluble sulfide minerals also present in the sample is expelled from the solution without being oxidized; the sample is first boiled with water to remove dissolved oxygen before the addition of nonoxidizing acid. Any sulfur that is precipitated during the decomposition of sulfides will either be filtered off before precipitation of the sulfate, or will be burned off during the ignition step.

Procedure

Transfer a 1.00 g sample (80–100 mesh) to a 250 ml beaker. Add about 75 ml of water to the beaker, swirl to make a slurry, cover, and heat to a gentle boil on an electric hot plate. Boil gently for 1 or 2 min and then add, though the lip of a beaker, 20 ml of dilute HCl (1:1); the solution should not stop boiling. Continue the gentle boiling for 15 min.

Allow to cool slightly and filter the solution through a 9 cm Whatman No. 42 paper (paper pulp added to the solution in the beaker will help to ensure a clear filtrate) into a 600 ml beaker. Wash the beaker and residue several times with hot 5% HCl, pouring the washings through the filter paper, and wash the paper and residue several times with the hot dilute HCl. Discard the paper and residue, and rinse the inside of the funnel, and the outside of the tip, into the beaker.

Add 2 or 3 drops of bromcresol purple (0.04% aqueous solution) or methyl red (0.2% ethanolic solution) to the filtrate and neutralize the solution with pure aqueous ammonia. Make the solution just acid with dilute HCl (1:1), then add 4 ml in excess. Dilute the solution to 300 ml at least.

Continue with the precipitation, filtration, ignition, and weighing of the barium sulfate as described in Sec. 5.6.2..

$$\% \ SO_3 = \frac{\text{weight of } BaSO_4 \ (g)}{\text{sample weight (g)}} \times 0.3430 \times 100$$

5.6.4   Determination of Total Sulfur

a.   Gravimetrically, Following Fusion with Sodium Carbonate

Transfer 1.00 g of sample (80–100 mesh) to a 25 ml platinum crucible. Add 3 g of anhydrous Na_2CO_3 and 0.2 g Na_2O_2, and mix thoroughly with a small glass rod. Wipe the rod free of particles by rubbing it in about 1 g of Na_2CO_3 on a filter paper, and use this flux to cover the mixture in the crucible.

Place the covered crucible in a cold electric muffle furnace (or at the edge of a heated one, with the door open) and rapidly raise the temperature of the muffle to 1000–1050°C (or gradually insert the crucible into the hot furnace). Allow to heat at this temperature for 15 min.

Remove the crucible from the furnace, quickly remove the cover (caution: there may be a blob of molten material on the underside), and swirl the contents of the crucible, if possible, to distribute the cooling melt around the crucible walls. Place the crucible on a silica triangle supported on a tripod and replace the cover.

Heat the crucible cover to redness for 1 min with a Tirrill burner, to fuse any sample material that may have been spattered on the underside of the cover. Heat the crucible to dull redness for 15 sec and allow it to cool thoroughly.

Add a few ml of H_2O to the crucible and loosen the cake with a stirring rod (warm the crucible on the water bath if necessary). Transfer the cake and rinsings to a 250 ml beaker and police the inside of the crucible and cover with a rubber policeman. It is not necessary to remove any stain from the crucible, except to clean it for future use. Dilute to 50 ml, add a few milliters of ethyl alcohol, and digest on an electrically heated water bath until all soluble material is in solution, breaking up lumps with a stirring rod when necessary.

Filter the solution into a 600 ml beaker through a 9 cm Whatman No. 40 paper. Wash the residue and beaker several times with cold 1% Na_2CO_3 solution and pour the rinsings through the paper. It is not necessary to transfer the residue quantitatively to the paper unless the determination of such constituents of the residue as barium, zirconium, or the rare earths is planned. Wash the paper and contents five times with a sodium carbonate solution. The filtrate can be used for the determination of chromium by colorimetric comparison with standard potassium chromate prior to the determination of sulfate.

Add 5 ml of saturated bromine water to the filtrate and dilute to 300 ml. Cover, and add dilute HCl (1:1) cautiously through the lip of the beaker until the solution is acid (a drop of indicator will last long enough to show when this stage has been reached). Heat to boiling and boil for a few minutes to expel carbon dioxide and bromine. Cool and neutralize the solution with aqueous ammonia, than add 2 ml of 12 N HCl.

Continue with the precipitation, filtration, ignition, and weighing of the barium sulfate as previously described (see Sec. 5.6.2). The total sulfur is usually reported as % S.

b. *Titrimetrically, Following Combustion in the Leco Induction Furnace*

The procedure to be followed will depend on the type of induction furnace and titrator employed. Complete instructions for the use of these instruments are usually provided, and only the method describing the use of the specified furnace and automatic titrator will be given here.

Apparatus

The apparatus consists of a Leco induction furnace (No. 521-000) and an automatic titrator (Leco No. 532-000), Leco ceramic crucibles and covers, and a combustion tube (Leco No. 550-120); a cylinder of compressed oxygen and a purifying train (see Sec. 5.5.3b), a glass delivery tube leading from the combustion tube to the titration vessel (with joints only at either end), and a heating tape (with rheostatic control) for heating the delivery tube. Reservoir bottles for the KIO_3 and HCl solutions, a dispenser for the starch solution, and a waste storage vessel or drain are also required. The ceramic crucibles can only be used once, but the covers can be used several times. As in the case of CO_2 determination (see Sec. 5.5.3b), the high temperatures involved necessitate frequent changes of various O-rings, the combustion tube, and other small items so it is advisable to keep a good supply of these items, and of crucibles and covers, in stock.

Reagents

An iron chip accelerator (low sulfur), a KIO_3 solution (0.444 g KIO_3 per liter of distilled water), an HCl solution (15 ml of reagent grade HCl diluted to 1 liter with distilled water), and a starch solution are required. Because the photocells of the Leco titrators are designed to operate best at the particular wavelength of the colored complex formed with iodine and arrowroot starch, the solution should be prepared from this starch as follows. Make a paste of 2 g of the starch with 50 ml of distilled water; bring 150 ml of distilled water to a boil, and slowly add the starch solution with constant stirring; cool, dissolve 6 g of KI in it, and transfer the solution to a dispenser (a polyethylene one that dispenses 5 ml reliably is suitable). The starch solution should be made fresh each day, and the dispenser must be cleaned thoroughly before adding fresh solution.

Procedure

Weigh approximately 1 g of sample and mix it with 1 g of iron accelerator in a ceramic crucible. The crucible is covered and placed on the ceramic pedestal of the induction furnace, which is turned on and allowed to warm up for the required period of time. The oxygen flow is set at 1 l/min, sufficient fresh HCl solution is added to reach a prescribed level near the center of the bell-shaped portion of the titration vessel, and 5 ml of the starch solution is then added. If the titrator has not been used recently, it is advisable to first flush it out with two or three portions of the HCl solution. The sample is raised to its position without starting the combustion in order to direct the oxygen flow into the titration vessel. The end point setting of the automatic titrator is adjusted to obtain the desired color density, the KIO_3 automatic burette is set at zero, and combustion is begun. The reading on the plate-current meter must reach 400 mA to ensure that all sulfur is evolved. As the combustion proceeds, the various forms of sulfur in the sample are converted to SO_2 which is swept into the titration vessel in the oxygen stream. The SO_2 bleaches the preset blue color by reducing the iodine, and this causes the photocell-activated microswitch to open and allow more KIO_3 solution into the reaction vessel. The process proceeds automatically until no more SO_2 is evolved and the combustion is complete. The sample is removed from the furnace and the reading on the burette is recorded. If a 1 g sample and a 0.444 g/l KIO_3 solution are used, the percent sulfur in the sample can be read directly from the burette. Diluting the titrant can increase the sensitivity of the method beyond the 0.001% nominally obtained, and by taking a smaller sample the concentration range of the burette can be increased above the 0.200% inscribed on it.

c. By a Combustion-Infrared Absorption Method

A new method for the determination of total sulfur, as well as of total water and total carbon dioxide, has recently been developed at the GSC by Bouvier and Abbey.[74] It involves mixing the sample (200 mg) with 1:1 V_2O_5–WO_3 in a small nickel boat, heating the mixture in a combustion tube furnace at 950°C, and passing the effluent successively through three infrared absorption cells. The effluent emerging from the first cell is stripped of its water vapor by passage through $Mg(ClO_4)_2$, and of sulfur dioxide by MnO_2 when leaving the second cell. The output signals of all three infrared cells are channeled to a three-channel digital integrator which reads out the integrated absorbance for the three constituents. It is reported that the analytical productivity of this procedure is several times that obtainable by more conventional methods because the three determinations are made by one analyst at the same time.

Terashima[75] has described a similar method, but one employing combustion in a high-frequency induction furnace, for the determination of total sulfur and total carbon.

5.7 PHOSPHORUS

5.7.1 General

In silicate rocks and minerals phosphorus is usually present in the range of 0.1–0.5% P_2O_5 and generally occurs as orthophosphate (apatite is the most common mineral). Acid decomposition[76] or alkali fusion[55] attacks are usually effective in dissolving phosphorus minerals, although the rare earth phosphates (xenotime and monazite, for example) require an alkaline hydroxide attack[55] (see Sec. 4.3.3). Lithium metaborate fusion has been used effectively and the metaborate fusion–nitric acid dissolution technique[77] requires only the elimination of silicon before a colorimetric determination of phosphorus is possible.[78]

Phosphorus may be partially volatilized when a solution containing it is fumed strongly with H_2SO_4, although neither Bodkin[78] nor Bennett and Reed[79] have mentioned any loss even after fuming to dryness. Jeffery advocates taking $HClO_4$ solutions of phosphorus to near dryness under an infrared heating lamp (Ref. 80, p. 384). Solutions may be evaporated to dryness with other acids (HCl, HNO_3, and HF) without loss. Nitric acid is preferable, but hydrofluoric acid is usually needed to decompose the silicate; the excess hydrofluoric acid should be removed by repeated evaporation of the solution to dryness with nitric acid, and the interference of fluoride prevented by the addition of boric acid to the solution.

For most silicate rocks and minerals the determination of phosphorus is most conveniently made by measurement of the intensity of the yellow complex formed by orthophosphate, vanadate, and molybdate, or of the blue color formed by the selective reduction of the orthophosphate–molybdate complex. A preliminary separation of the phosphorus is sometimes made to eliminate the interference of silicate, arsenate, and molybdate (which form similar heteropoly complexes) and of elements such as copper, nickel, and chromium which form colored solutions, but usually it is only necessary to remove silicate by volatilization of the silicon with hydrofluoric acid. An occasional sample containing more than the usual amount of phosphorus (e.g., phosphatic sandstone) may be encountered, and extreme dilution of the sample solution will be necessary to bring it into the proper concentration range for colorimetric measurements. In this instance the determination is better made gravimetrically as magnesium ammonium phosphate (with ignition to the pyrophosphate), either after a preliminary separation of the phosphorus, such as with ammonium molybdate, or by making the precipitation in the presence of citric acid. For small amounts of phosphorus it is possible to weigh the yellow ammonium phosphomolybdate directly, or the precipitate may be dissolved in a measured amount of standard alkali solution and the excess backtitrated with a standard acid solution.

Gravimetric and colorimetric methods will be covered in some detail in the following, but because they do not have much application in silicate analysis, the titrimetric methods will not be discussed. The analytical chemistry of phosphorus has been extensively reviewed[81]; other texts treat in detail the gravimetric and titrimetric[1, 82] and the colorimetric[83] procedures.

Instrumental techniques such as XRF[84, 85] (see Sec. 11.2.22) and ES[86] have also been applied to the determination of phosphorus in rocks and minerals and, using argon-plasma emission spectrophotometry, in coal ash.[87] Mr. R. J. Hibberson of the BCM has developed a technique for determining phosphorus using graphite electrodes with a 10 A dc arc. He used standards prepared from mixtures of the NBS Phosphate Rock No. 120(a) and a granite base containing 0.005% P. For a 20 mg sample the range of concentrations that can be determined is 0.06–1% P. Neutron activation has also been applied to the determination of P in rocks.[88, 89]

5.7.2   Gravimetric Methods

The methods most often employed for the gravimetric determination of phosphorus involve its precipitation as either ammonium phospho-molybdate or magnesium ammonium phosphate, or as a combination of these two precipitation forms. Precipitation as yellow ammonium phos-phomolybdate is usually employed for the separation of phosphorus from interfering elements, with subsequent dissolving of the precipitate in aqueous ammonia and reprecipitation of the phosphorus as magnesium ammonium phosphate using magnesia reagent and aqueous ammonia. It is possible, however, to employ the yellow precipitate or its ignition product as the final weighing form, just as it is possible, with suitable precautions, to precipitate the phosphorus directly as magnesium ammonium phosphate without preliminary separation.

The difficulties attending the use of ammonium phosphomolybdate as the final weighing form are connected with the uncertain composition of the final product, this latter being affected by the conditions of the precipitation. Much has been written about this,[1, 81] and a study of the conditions favoring the quantitative precipitation of ammonium 12-molyb-dophosphate[90] using the radioisotope phosphorus-32 has eliminated much of the previous uncertainty. It is shown that quantitative precipitation is obtained after heating for 30 min at any temperature between 50 and 80°C, followed by digestion for 30 min at room temperature, with stirring at 15 min intervals; at 90°C, the precipitation of molybdic acid is noted. The precipitation is subject to interference from a number of ions, including fluoride, but the inhibiting effect of these may generally be circumvented by the use of a large excess of ammonium molybdate (twice the stoichiometric amount is recommended).[90] Vanadium(V) will be precipitated at elevated temperatures, but if reduced to V(IV) and if the precipitation is made at room temperature, it will not interfere; in general its concentration in the sample will be negligible. Heslop and Pearson[91] studied the effect of arsenate and some transition-metal ions [Fe(III), Cr(III), Mn(II), Ni(II)] on the precipitation; under the conditions necessary for the quantitative precipitation of phosphorus, arsenic is also precipitated by ammonium molybdate. Organic matter will retard the precipitation, and Peck (Ref. 15, pp. 45–48, 78–79) prefers to ignite the sample before starting to decompose it, if the presence of organic substances is suspected. Large amounts of silica must be removed, and this is readily accomplished by decomposing the sample with a mixture of nitric and hydrofluoric acids; the fluorides are largely expelled by evaporation with nitric acid alone. In the procedure recommended by Peck[15] no more than 600 mg of fluorine or 40 mg of SiO_2 remain in the solution, and these concentrations can be tolerated. The precipitation is preferably made from a nitric acid solution (perchloric acid is acceptable but hydrochloric and sulfuric acids interfere) with a mixed reagent consisting of ammonium molybdate and nitric acid or ammonium nitrate. The presence of nitric acid is necessary for the formation of the precipitate, and ammonium nitrate (as a 5–15% solution) has been used to reduce the solubility of the precipitate and to speed up the precipitation; the study by Archer et al.[90] shows, however, that ammonium nitrate has no effect upon the efficiency of the precipitation. The filtered precipitate must be washed with a dilute solution of an electrolyte (ammonium nitrate, or potassium nitrate if the precipitate is to be dissolved and titrated) to prevent peptization; it is reported that the use of a mixed ammonium molybdate–nitric acid reagent yields a precipitate that will not peptize.[90]

The composition of the yellow precipitate is ideally $(NH_4)_3PO_4 \cdot 12MoO_3 \cdot 2HNO_3 \cdot H_2O$ which, on drying at 110°C, loses nitric acid and water to give $(NH_4)_3P(Mo_3O_{10})_4$, containing 3.78% P_2O_5. This percentage may drop to as little as 3.73, depending upon the conditions of the precipitation, and it is this uncertainty of composition that militates against its use for the determination of large amounts of phosphate; for the small amount present in most rocks, however, the error thus introduced would be negligible. The precipitate may also be heated at 400–500°C for 30 min to give a residue approximating $P_2O_5 \cdot 24MoO_3$. If the precipitate is to be dissolved and an alkalimetric titration used as the finish, then it is important that the ratio of $PO_4:MoO_3$ be maintained at 1:12; if the phosphorus is to reprecipitated as the magnesium ammonium phosphate, the exact composition of the yellow precipitate is unimportant as long as the precipitation is quantitative.

The problem inherent in the precipitation of magnesium ammonium phosphate have been discussed in detail in various texts,[1, 92] and similar problems may be encountered in the determination of phosphorus, even though the precipitation is made in the reverse manner. Many elements, if present, will interfere in the precipitation, and it is often necessary to make a preliminary separation of the phosphorus with ammonium molybdate as previously discussed, or to complex these interfering elements so that they will not interfere. This latter is achieved by the addition of ammonium citrate (10–15%) to the solution, and the procedure is as accurate as that involving preliminary separation, and is more rapid. For accurate work, regardless of the manner in which the precipitation is made, a second precipitation is mandatory to ensure that the precipitate has the proper composition ($MgNH_4PO_4 \cdot 6H_2O$); for most rocks, however, the phosphorus content is so low that the increased accuracy is not worth the time required for a double precipitation and a single precipitation will suffice. To minimize errors, the precipitation should be made from an acid solution containing magnesia reagent, and also sufficient ammonium chloride to buffer the solution at approximately pH 10.5 when the solution is first slowly neutralized with aqueous ammonia and a few milliliters added in excess.

a.   *Gravimetric Determination of Phosphorus as Magnesium Ammonium Phosphate, Following Preliminary Separation as Ammonium Phosphomolybdate*

Reagents

Ammonium molybdate (5%).   Dissolve 50 g of $(NH_4)_6Mo_7O_{24} \cdot 4H_2O$ in approximately 500 ml of warm water, let stand for several hours, filter through a Whatman No. 42 filter paper, dilute to 1 liter, and store in a polyethylene bottle.

Ammonium nitrate (50%).   Dissolve 500 g of NH_4NO_3 in about 500 ml of water, let stand for several hours, filter through a Whatman No. 42 filter paper, dilute to 1 liter, and store in a glass bottle.

Magnesia reagent.   Dissolve 50 g of $MgCl_2 \cdot 6H_2O$ and 100 g of NH_4Cl in 500 ml of water, make ammoniacal and let stand overnight. Filter if necessary through a Whatman No. 42 filter paper, make just acid with hydrochloric acid, and dilute to 1 liter. Store in a glass bottle.

Procedure

Transfer an appropriately sized sample (≈ 0.5 g) to a 50 to 100 ml platinum dish and moisten with water; stir the mixture with a platinum stirring rod. Cover the dish and add 10 ml concentrated HNO_3; when any effervescence has ceased, remove and rinse off the cover, and add 10 ml of 48% hydrofluoric acid. Place the dish on a sand bath and slowly evaporate the contents to near dryness, with occasional stirring. Cool, moisten with water, and add 5 ml HNO_3 and 5 ml HF; mix and again evaporate to dryness on the sand bath. Cool, add 20 ml HNO_3 (1:1), and again evaporate to dryness; heat the contents of the dish for a further 30 min after the salts appear to be dry.

Cool, add 20 ml (1:1) HNO_3 (previously boiled and cooled to remove oxides of nitrogen), and 10 ml of 5% (saturated) boric acid solution to the dish and break up the residue with the platinum rod. Cover the dish with a plastic cover and digest the contents on a water bath until solution appears to be complete (1–2 hr); a gritty residue of quartz is often encountered at this stage. If a brown precipitate is observed (MnO_2), add a few grains of sodium sulfite and stir; the brown precipitate will disappear.

Filter through a 7 cm Whatman No. 40 paper into a 100 ml volumetric flask, and police the dish with water containing a few drops of HNO_3. Wash the dish and paper several times with this wash solution. If the residue on the paper is dark, it may contain appreciable manganese, and a white nongritty residue may be titanium phosphate. In either case, if the residue amounts to more than a few grains, it should be ignited in a platinum crucible and fused with 0.5 g anhydrous Na_2CO_3, the cake leached with water, filtered, the filtrate made acid with HNO_3, and added to the solution in the volumetric flask. Dilute to volume and mix thoroughly.

To an aliquot (usually 50 ml of the sample solution) in a 150 ml beaker, add 5 ml concentrated nitric acid and 15 ml of ammonium nitrate solution (50%). Heat to boiling, then allow to cool to about 60°C. Add 20 ml of 6% ammonium molybdate solution and stir vigorously until a turbidity appears; if none appears within a few minutes, scratch the bottom of the beaker occasionally with the glass rod until precipitation begins. Keep the solution at about 60°C for 15 min and if the precipitation is heavy, add additional ammonium molybdate reagent. Allow to stand overnight at room temperature.

Filter the solution through a 7 cm Whatman No. 42 filter paper keeping most of the precipitate in the beaker. Wash the precipitate and beaker five times by decantation with 2% ammonium nitrate solution made slightly acid with nitric acid, pouring the wash solution through the paper, and finally wash the paper five times also. Discard the filtrate and expel excess wash solution from the stem of the funnel. Wipe the underlip of the beaker with a small piece of filter paper and add it to the funnel. Place the precipitation beaker under the stem of the funnel and dissolve the precipitate on the paper by washing it with a 5% solution of aqueous ammonia containing 2 g of ammonium citrate per 100 ml. Wash the paper thoroughly with HCl, again with water, and once more with the aqueous ammonia–ammonium citrate wash solution. Discard the paper and rinse the funnel into the beaker with the ammoniacal solution. Wash down the sides of the beaker with the wash solution and stir until all of the yellow precipitate is dissolved. The volume should be less than 50 ml.

Add dropwise dilute HCl (1:1) until the solution is just acid to methyl red (0.2% in 60% ethanolic solution) and add 10 ml in excess. Add 10 ml of magnesia reagent and then aqueous ammonia dropwise, with vigorous stirring, until the solution is ammoniacal. After a few minutes add 10 ml of the aqueous ammonia, stir vigorously, and allow to stand overnight, preferably in a fume hood.

Filter the solution through a Whatman No. 42 filter paper of appropriate size, keeping as much of the precipitate as possible in the beaker, and wash the beaker and precipitate twice with 5% HCl, add 1–2 ml of magnesia reagent, and precipitate the magnesium ammonium phosphate as before. Filter, wash, ignite and weigh the $Mg_2P_2O_7$.

$$\% \ P_2O_5 = Mg_2P_2O_7 \times 0.6377 \times \frac{100}{\text{sample weight}}$$

Conversion Factors

0.4365

$P_2O_5 \rightleftarrows P$

2.2912

5.7.3 Colorimetric Methods

Two methods are in general use for the colorimetric determination of phosphorus. The one employs the yellow color of the mixed heteropoly acid formed when molybdate solution is added to an acidic solution (at least 0.7 M) containing orthophosphate and vanadate ions; the other utilizes the blue color which results from the selective reduction of molybdophosphoric acid, the intensity of the blue color being proportional to the concentration of the phosphate present in the latter. Use is also made of the yellow color of this molybdophosphoric acid complex[81, 83] after its selective extraction, but it will not be considered here. Most of the rapid analysis schemes employ the colorimetric determination of phosphorus.

The yellow molybdovanadophosphoric acid complex is the basis of the method considered to be the most specific for phosphorus. There are many conflicting data available to show the influence of various factors on the formation of the complex, such as the stability of the reagents and the use of separate reagents versus one combined reagent, the choice of acid and acid concentration, changes in temperature of the solution, and the wavelength of measurement; these have been summarized (Ref. 81, pp. 351–353). Phosphorus is not the only element that forms a colored complex of this nature, however, and the methods in use eliminate the interference of ions such as silicate, arsenate, and germanate, as well as of elements such as Fe, Ni, and Cu which form colored compounds, by preliminary separation of the phosphorus, by choice of pH, by removal of the interferences through selective volatilization or mercury cathode electrolysis, and by choice of the wavelength of measurement. Nitric acid is used most frequently, usually about 0.5 N, but HCl, $HClO_4$, or H_2SO_4 may also be used. The molybdate and vanadate reagents, about 5% and 0.25% solutions, respectively, may be added separately or as one solution, but the order of addition, if added separately, must be acid, vanadate, and molybdate; the final concentrations recommended for vanadate and molybdate are 0.002 M and 0.04 M, respectively (Ref. 81, p. 353). These authors also recommend the addition of Na_2SO_4 solution to minimize the effect of temperature changes on the intensity of the colored complex. The absorbance of the latter is measured at 400–460 nm; at the higher wavelength the interference of iron(III), copper, and nickel is minimized. The interference of these ions can also be compensated for by use of an aliquot of the sample solution, without the addition of reagents, as the blank. Fluoride ion interferes in the color development, and it is customary to bake the decomposed samples, after evaporation to dryness with nitric acid, to expel as much of the fluorine as possible; the baking should not be done at temperatures greater than 250°C or some phosphorus will be lost by volatilization. According to Boltz and Lueck[83] the optimum concentration range is 5–40 ppm phosphorus, for a 1 cm cell.

The blue color formed by the reduction of the molybdophosphoric acid is more intense than the yellow complex discussed previously, and the method is thus more sensitive, although there is at the same time a loss of color stability and an increased need for a more rigorous control of operating conditions. Boltz and Lueck (Ref. 83, p. 32) discuss the formation of two blue complexes; that which forms at an acidity of about

1 N and has an absorbance maximum at 830 nm is called "heteropoly blue" to distinguish it from "molybdenum blue," which forms at lower acidities and which has an absorbance maximum at 650–700 nm. The choice of reducing agent seriously affects the intensity and stability of the color, and the literature is replete with recommendations for particular reagents (Ref. 81, pp. 348–351). A further modification involves the extraction of the molybdophosphoric acid into an organic solvent such as isobutyl alcohol before reduction to heteropoly blue; this is one means of eliminating the interference of many other ions. Boltz and Lueck[83] use hydrazine sulfate as the reducing agent and add it separately from the molybdate; the solution is then heated for 10 min at about 100°C. Riley[93] prefers ascorbic acid for the reduction and mixes it with the molybdate as a single reagent; the addition of potassium antimonyl tartrate to the reagent was later found to give increased sensitivity when applied to the determination of phosphorus in natural waters.[94] This method has been modified for use with soils[95] in which the interference from arsenic must be eliminated. The As(V) is reduced with a mixed reducing agent ($Na_2S_2O_3$ and $NaHSO_3$); the interference of tin may be avoided but vanadium and tungsten still introduce slight errors. Chalmers[96] used the reduced form to determine phosphorus in silicate samples weighing only a few milligrams; ferrous ammonium sulfate is the reductant, and the interference of vanadate and Fe(III) is eliminated by first passing the solution through a semimicro silver reductor. The silica is not removed but is prevented from interfering by having a final acid concentration of 0.76 N H_2SO_4.

The introduction of lithium borate fusion–acid dissolution decomposition techniques to rock and mineral analysis has led to the development of colorimetric methods using aliquots for phosphorus determinations from solutions prepared for major, minor, and trace element determinations by AAS.[77, 78, 97, 98]

Ingamells[97] has reported a method for the rapid colorimetric determination of phosphorus in the presence of silica, but Bodkin[78] preferred an adaptation of the method of Fogg and Wilkinson[99] in which HF and H_2SO_4 were used to remove the silica. In both cases the absorbance of the phosphomolybdenum blue complex was measured.

a. As the Yellow Molybdovanadophosphoric Acid Complex (Method of Baadsgaard and Sandell)

The sample is decomposed by a mixture of nitric and hydrofluoric acids, followed by evaporation to dryness with HNO_3 to expel most of the fluoride retained in the residue. The residue is then dissolved in colorless HNO_3, and boric acid solution is added to complex any fluoride ion still present. An aliquot is used for the colorimetric determination of phosphorus, based upon the procedure of Baadsgaard and Sandell.[100]

Reagents

Ammonium molybdate (5%). Prepare as described in Sec. 5.7.2 (Reagents).

Ammonium metavanadate (0.25%). Dissolve 2.5 g of NH_4VO_3 in 500 ml of hot water, cool, and add 20 ml concentrated HNO_3. Stand several hours, filter if not clear, and dilute to 1 liter. Store in a glass bottle.

Standard phosphate solution (1.00 mg P_2O_5/ml). Dissolve 0.959 g of KH_2PO_4 (recrystallized and dried at 100°C) in water and dilute to 500 ml. Store in a tightly capped polyethylene bottle. This solution may be further diluted as desired.

Procedure

To the solution of the sample prepared as described in Sec. 5.7.2, containing 5 ml concentrated nitric acid, not more than 9 mg P_2O_5, and in a 100 ml volumetric flask,* add by pipette 10.0 ml of ammonium vanadate solution and 20.0 ml of ammonium molybdate solution, in that order. Mix, dilute to volume, and mix thoroughly, and then allow to stand at room temperature for 30 min. Measure the absorbance of the yellow complex at 460 nm in a 1 cm cell against a blank consisting of the reagents and concentrated nitric acid in the proportions given previously. Correct the absorbance for the effect of other colored substances, such as ferric iron (as described in the footnote), and obtain the concentration of P_2O_5 per 100 ml of solution from a calibration curve.

Preparation of Calibration Curve

To a series of 100 ml volumetric flasks add, from a 10 ml semimicro-burette, a series of aliquots of the standard phosphate solution (1.00 mg P_2O_5/ml) to give 0, 0.50, 1.00, 2.00, 3.00, 4.00, 5.00, 6.00, 7.00, 8.00, and 9.00 mg P_2O_5/100 ml. Add 5 ml colorless concentrated HNO_3, dilute to about 50 ml, and continue with the determination as previously described. Measure the absorbance of each solution against the solution containing the reagents only as blank. Prepare a calibration curve based upon the average of several readings made on each solution.

---

* If a 1.00 g sample is used for the determination, a 50 ml aliquot will enable the determination of up to 1.8% P_2O_5; such a high content would be unusual and the determination should be repeated using a smaller sample weight. The solution remaining in the flask may be used as a means of correcting for the absorbance due to the presence of other colored substances (measure the absorbance at 460 nm against water as a blank).

b.   *As the Heteropoly Blue Complex (Method of Boltz and Lueck[83])*

Reagents

Ammonium molybdate (in 10 N H_2SO_4).   Dissolve 5 g $(NH_4)_6Mo_7O_{24}$ · $4H_2O$ in about 200 ml 10 N H_2SO_4 (dilute 70 ml 36 N H_2SO_4 to 250 ml) and dilute to 250 ml with additional 10 N H_2 SO_4. Store in a polyethylene bottle.

Hydrazine sulfate (0.15%).   Dissolve 0.15 g of hydrazine sulfate $(N_2H_6SO_4)$ in water and dilute to 100 ml.

Procedure

To a 0.100 g sample (about 150 mesh) in a 50 ml platinum (or 100 ml Teflon) dish add 1 ml H_2O and 1 ml concentrated HNO_3. Evaporate to dryness on a medium temperature sand bath (about 150°C).

Add 5 ml 12 N HCl and 5 ml HF (48%) and break up the residue with a platinum or Teflon stirring rod. Evaporate to dryness.

Add 5 ml HCl, break up the residue as before, and again evaporate to dryness.

Add 6 drops of 12 N HCl to the residue, break up with the rod, and add 20 ml H_2O; heat until solution is complete, adding more H_2O if necessary.

Transfer (filter, if necessary) the solution to a 100 ml volumetric flask and dilute to volume.

Remove a 10 ml aliquot by pipette and transfer it to a 50 ml volumetric flask. Blow out any liquid remaining in the pipette into the 100 ml flask and rinse the outside of the stem and the interior of the pipette with a fine jet of water, catching the rinsings in the 100 ml flask.

Add to the 50 ml flask 5 ml of the ammonium molybdate solution and 2 ml of hydrazine sulfate solution and mix.

Dilute to approximately 1 cm from the graduation mark with water, stopper the flask, and mix thoroughly. Immerse the unstoppered flask in a boiling water bath for 10 min, remove, and cool rapidly in running water. Allow to come to room temperature, dilute to volume, and mix thoroughly.

Measure the absorbance of the heteropoly blue complex at 830 nm in a 1 cm cell, using either distilled water or a reagent blank as the reference blank solution. Obtain the concentration of P_2O_5, in micrograms, from a standard curve prepared as described below.

$$\% \, P_2O_5 = \frac{\mu g \, P_2O_5}{100}$$

Preparation of Standard Curve

Transfer 1.00 ml of the standard phosphate solution (Sec. 5.7.3a) containing 1.00 mg P_2O_5/ml to a 100 ml volumetric flask, dilute to volume, and mix thoroughly (1 ml $= 10$ μg P_2O_5). To a series of 50 ml volumetric flasks add 0, 1, 2, 3, 4, 5, 6, 7, 8, 9, and 10 ml aliquots of this solution to give a range of P_2O_5 concentrations of 0–100 μg and continue with the addition of reagents, heating, and final measurement of the absorbance at 830 nm as previously described. Plot the average of several readings for each solution against the concentration of P_2O_5 in micrograms per 50 ml.

c. Colorimetric Determination, Following Fusion with Lithium Metaborate

The following unpublished method, referred to briefly in GSC paper 74-19 by S. Abbey, N. J. Lee, and J. L. Bouvier, has been in use at the GSC since 1972 as part of the neoclassical analysis scheme in which a number of elements, both major and minor, are determined by a combination of colorimetric and AAS methods.

Reagents

Standard phosphate solution (1.00 mg P_2O_5/ml). Dissolve 0.959 g of KH_2PO_4 (dried at 110°C) in water and dilute to 500 ml. Store in a polyethylene bottle with tightly fitting cap and use this to prepare a solution containing 0.01 mg P_2O_5/ml.

Ammonium molybdate solution (in 10 N H_2SO_4). Prepare 250 ml of 10 N H_2SO_4 by diluting 70 ml concentrated H_2SO_4 to 250 ml with H_2O. Dissolve 5 g $(NH_4)_6Mo_7O_{24} \cdot 4H_2O$ in this and store in a polyethylene bottle.

Hydrazine sulfate solution (0.15%). Dissolve 0.15 g of hydrazine sulfate $(N_2H_6SO_4)$ in water and dilute to 100 ml.

Sample Decomposition

The procedure described here is designed to yield a solution which is suitable for the determination of several elements when appropriate aliquots are taken. If only phosphorus is to be determined, a smaller sample and less rigorous procedure may be used.

Weigh 0.5000 g of sample into a clean, weighed Pt crucible, cover, and determine the moisture content as described in Sec. 5.2.2. Heat the crucible (with lid partially uncovered) and sample over a Bunsen burner until crucible bottom is a dull red, and continue the heating for 5 min to ensure oxidation of all constituents.

Weigh 1 g of anhydrous $LiBO_2$ into a graphite crucible which has been ignited at 1000°C for 10 min (to provide a layer of loose carbon on the inner surface) and cooled. Transfer the ignited sample to the graphite crucible, rinse the Pt crucible with a further 0.5 g $LiBO_2$, and add it to the graphite crucible also. Mix the contents carefully, in order to not disturb the inner surface covering of the crucible, and cover with a graphite lid. Place the crucible in a muffle furnace at 1000°C for 30 min to fuse the contents.

Pour the molten bead into 30 ml H_2O in a Teflon dish, in which is also set a polyethylene collar (of a diameter such that it is not in direct contact with the liquid in the dish) to act as splashguard, and rinse the walls of the collar with a further 10 ml H_2O before removing it. Add 10 ml concentrated HCl and break up the bead with a stirring rod. Evaporate the contents of the dish to near dryness on a sand bath (approximately $1\frac{1}{2}$ hr), stirring the gelatinous mass (SiO_2) continuously near the end of the evaporation to avoid loss of sample because of spattering. To the cool mass add 10 ml concentrated HCl and 50 ml methyl alcohol, and evaporate the contents to dryness overnight on a steam bath. Add 5 ml concentrated HCl and 25 ml methyl alcohol to the dry residue and again evaporate to dryness on a steam bath. Add 25 ml concentrated HCl, then 25 ml H_2O and digest on the steam bath for approximately 30 min to dissolve soluble salts.

Filter the contents of the dish through an 11 cm Whatman No. 40 filter paper into a 250 ml volumetric flask, and quantitatively transfer the silica residue to the paper as well, rinsing the dish, cover, and rod, and washing the paper with warm 0.5 N HCl to remove soluble salts. Finally wash the filter paper with H_2O; the paper and contents may be used for the gravimetric determination of SiO_2 and the small amount of soluble SiO_2 in the filtrate determined by the colorimetric molybdenum blue procedure.

Dilute the filtrate in the flask to volume with H_2O and mix well.

Determination

Pipette 5.0 ml of the filtrate into a 50 ml volumetric flask. Similarly, pipette 3.0 ml of the standard solution containing 0.01 mg P_2O_5/ml (see below) into another 50 ml flask to serve as a control, provide a third flask to serve as a blank, and to each add 1.25 ml 1:1 HCl (about 25 drops). Pipette 5 ml of ammonium molybdate solution and 2 ml of hydrazine sulfate solution into each of the three flasks, dilute the contents to within 2 cm of the mark with H_2O, swirl the contents to mix them, and immerse the flasks in boiling water for 20 min. Cool, dilute to volume with H_2O, and measure the absorbance at 830 nm. Obtain the concentration of P_2O_5 from the calibration curve prepared as given below, applying a correction for the blank if necessary, and calculate the % P_2O_5:

$$\% \ P_2O_5 = \frac{\mu g \ P_2O_5}{100}$$

Preparation of Calibration Curve

Pipette 1.00 ml of the standard phosphate solution (1.00 mg P_2O_5/ml) into a 100 ml volumetric flask, dilute to volume with H_2O, and mix thoroughly.

$$1 \ ml = 10 \ \mu g \ P_2O_5$$

To a series of 50 ml volumetric flasks add aliquots of this solution ranging from 0 to 10 ml to give a calibration curve for P_2O_5 concentrations ranging from 0 to 100 μg/50 ml. Add HCl, ammonium molybdate, and hydrazine solutions, and treat as given above for the sample. Measure the absorbance of each solution at 830 nm and prepare a calibration curve from the average of several measurements of each.

References

1. W. F. Hillebrand, G. E. F. Lundell, H. A. Bright, and J. I. Hoffman, *Applied Inorganic Analysis*, 2nd ed., Wiley, New York, 1953.
2. F. Manheim, *Acta Univ. Stockholm*, **6**, 127 (1960).
3. D. J. Drysdale, E. D. Lacy, and J. Tarney, *Analyst*, **88**, 131 (1963).
4. B. C. M. Butler, *Mineral. Mag.*, **32**, 866 (1961).
5. I. S. E. Carmichael, *J. Petrol.*, **1**, 309 (1960).
6. R. A. Chalmers, "Chemical Analysis of Silicates," in H. F. W. Taylor, Ed., *The Chemistry of Cements*, Vol. 2, Academic, London, 1964, pp. 171–189.
7. A. W. Groves, *Silicate Analysis*, 2nd ed., George Allen and Unwin, London, 1951, pp. 95–104.
8. J. Mitchell, Jr., "Water," in I. M. Kolthoff and P. J. Elving, Eds., *Treatise on Analytical Chemistry*, Vol. 1, Part II, Interscience, New York, 1961, pp. 69–206.
9. S. Courville, Geological Survey of Canada Report of Scientific Activities, Paper 74-1, Part B, 1974.
10. A. Turek, C. Riddle, B. J. Cozens, and N. W. Tetley, *Chem. Geol.*, **17**, 261 (1976).
11. M. Hartwig-Bendig, *Z. Angew. Mineral.*, **2–3**, 195 (1939–1941).
12. J. P. Riley, *Analyst*, **83**, 42 (1958).

13. F. Manheim, *Acta Univ. Stockholm*, **6**, 127 (1960).

14. Th. G. Sahama, *Bull. Comm. Géol. Finlande*, **138**, 102 (1946).

15. L. C. Peck, *U.S. Geol. Surv. Bull.*, 1170, 89 pp. (1964).

16. W. M. Johnson and L. E. Sheppard, *Chem. Geol.* **22**, 341 (1978).

17. P. G. Jeffery and A. D. Wilson, *Analyst*, **85**, 749 (1960).

18. J. P. Riley and H. P. Williams, *Mikrochim. Acta*, 526 (1959).

19. S. L. Penfield, *Am. J. Sci.*, 3rd Ser., **48**, 31 (1894).

20. I. Friedman and R. L. Smith, *Geochim. Cosmochim. Acta*, **15**, 218 (1958).

21. W. C. A. Guthrie and C. C. Miller, *Mineral. Mag.*, **23**, 405 (1933).

22. E. B. Sandell, *Mikrochim. Acta*, **38**, 487 (1951).

23. D. A. Skoog and D. M. West, *Fundamentals of Analytical Chemistry*, 3rd ed., Holt, Rinehart & Winston, New York, 1976.

24. S. Courville, *Can. Mineral.*, **7**, 326 (1962).

25. W. E. Hill, Jr., W. N. Waugh, O. K. Galle, and R. T. Runnels, *State Geol. Surv. Kansas Bull.*, **152**, Part I, 9 (1961).

26. J. Hoefs, *Geochim. Cosmochim. Acta*, **29**, 399 (1965).

27. J. A. Corbett and K. J. Mizon, *Chem. Geol.*, **17**, 155 (1976).

28. O. K. Galle and R. T. Runnels, *J. Sed. Petrol.*, **30**, 613 (1960).

29. S. St. J. Warne, *J. Sed. Petrol.*, **32**, 877 (1962).

30. W. N. Waugh and W. E. Hill, Jr., *J. Sed. Petrol.*, **30**, 144 (1960).

31. J. J. Fahey, *U.S. Geol. Surv. Bull.*, 950, 139 (1946).

32. G. Wolff, *Z. Angew. Geol.*, **10**, 320 (1964).

33. S. I. M. Skinner, R. L. Halstead, and J. E. Brydon, *Can. J. Soil. Sci.*, **39**, 197 (1959).

34. N. Iordanov, *Talanta*, **13**, 563 (1966).

35. L. Shapiro and W. W. Brannock, *U.S. Geol. Surv. Bull.*, 1144-A, A14, A49, A53 (1962).

36. S. S. Goldich, C. O. Ingamells, and D. Thaemlitz, *Econ. Geol.*, **54**, 285 (1959).

37. J. A. Grant, J. A. Hunter, and W. H. S. Massie, *Analyst*, **88**, 134 (1963).

38. J. L. Bouvier, J. G. Sen Gupta, and S. Abbey, Geological Survey of Canada, Paper 72-31, 1972.

39. J. G. Sen Gupta, *Anal. Chim. Acta*, **51**, 437 (1970).

40. J. I. Read, *Analyst*, **97**, 134 (1972).

41. P. G. Jeffery and P. J. Kipping, *Analyst*, **87**, 379 (1962).

42. P. G. Jeffery and P. J. Kipping, *Gas Analysis by Gas Chromatography*, Macmillan, New York, 1964, pp. 107–115, 175–177.

43. J. Thomas, Jr., and G. M. Hieftje, *Anal. Chem.*, **38**, 500 (1966).

44. J. L. Ellingboe and J. E. Wilson, *Anal. Chem.*, **36**, 434 (1964).

45. I. C. Frost, U. S. Geological Survey, Prof. Paper 400-B, B480, 1960.

46. B. A. Vovsi and Kh. V. Bal'yan, *Zavod. Lab.*, **25**, 437 (1959).

47. B. E. Dixon, *Analyst,* **59,** 739 (1934).

48. I. C. Frost, U.S. Geological Survey, Prof. Paper, 424-C, C376, 1961.

49. W. H. Dennen, *Spectrochim. Acta,* **9,** 89 (1957).

50. A. P. Ferris and W. B. Jepson, *Analyst,* **97,** 940 (1972).

51. G. D. Patterson, Jr., "Sulfur," in D. F. Boltz, Ed., *Colorimetric Determination of Nonmetals,* Interscience, New York, 1958, pp. 261–308.

52. B. J. Heinrich, M. D. Grimes, and J. E. Puckett, "Sulfur," in I. M. Kolthoff and P. J. Elving, Eds., *Treatise on Analytical Chemistry,* Vol. 7, Part II, Interscience, New York, 1961, pp. 1–135.

53. L. A. Haddock, "Sulfur," in C. L. Wilson and D. W. Wilson, Eds., *Comprehensive Analytical Chemistry,* Vol. 1C, Elsevier, Amsterdam, 1962, pp. 282–295.

54. R. S. Young, *Chemical Phase Analysis,* Charles Griffin & Co. Ltd., Bucks, England, 1974.

55. J. Doležal, P. Povandra, and Z. Sulcek, *Decomposition Techniques in Inorganic Analysis,* American Elsevier, New York, 1968.

56. H. Bennett and W. G. Hawley, *Methods of Silicate Analysis,* 2nd ed., Academic, London, 1965, pp. 99–100, 286–287.

57. C. Bloomfield, *Analyst,* **87,** 586 (1962).

58. K. E. Burke and C. M. Davis, *Anal. Chem.,* **34,** 1747 (1962).

59. J. G. Sen Gupta, *Anal. Chim. Acta,* **49,** 519 (1970).

60. D. J. Garcia, *Quim. Analit.,* **28,** 112 (1974).

61. A. C. Vlisidis, *U.S. Geol. Surv. Bull.,* 1214-D, D1 (1966).

62. A. R. V. Murthy and K. Sharada, *Analyst,* **85,** 299 (1960).

63. J. W. Smith, N. B. Young, and D. L. Lawlor, *Anal. Chem.,* **36,** 618 (1964).

64. D. L. Sorensen, W. A. Kneib, and D. B. Porcella, *Anal. Chem.,* **51,** 1870 (1979).

65. H. Steger, *Talanta,* **23,** 395 (1976).

66. T. A. Rafter, *J. Sci. Tech. New Zealand,* **38,** 849 (1957).

67. A. D. Wilson, G. A. Sergeant, and L. J. Lionnel, *Analyst,* **88,** 138 (1963).

68. J. M. Murphy and G. A. Sergeant, *Analyst,* **99,** 515 (1974).

69. T. S. Harrison and R. J. Spikings, *Anal. Chim. Acta,* **67,** 145 (1973).

70. Y. Arikawa, T. Ozawa, and I. Iwasaki, *Jpn. Analyst,* **24,** 497 (1975).

71. B. P. Fabbi and W. J. Moore, *Appl. Spectrosc.,* **24,** 426 (1970).

72. W. B. Stern, *X-Ray Spectrom.,* **5,** 56 (1976).

73. H. N. Elsheimer and B. P. Fabbi, "A Versatile X-Ray Fluorescence Method for the Analysis of Sulfur in Geologic Materials," in C. L. Grant, C. S. Barrett, J. B. Newkirk, and C. O. Ruud, Eds., *Advances in X-Ray Analysis,* Vol. 17, Plenum, New York, 1974.

74. J. L. Bouvier and S. Abbey, Geological Survey of Canada, Paper 79-1B, 417, 1979.

75. S. Terashima, *Anal. Chim. Acta,* **101,** 25 (1978).

76. R. Cioni, F. Innocenti, and R. Mazzwoli, *Chem. Geol.*, **7**, 19 (1971).
77. N. H. Suhr and C. O. Ingamells, *Anal. Chem.*, **38**, 730 (1966).
78. J. B. Bodkin, *Analyst*, **101**, 44 (1976).
79. H. Bennett and R. A. Reed, *Chemical Methods of Silicate Analysis*, Academic, London, 1971.
80. P. G. Jeffery, *Chemical Methods of Rock Analysis*, 2nd ed., Pergamon, Oxford, 1975.
81. W. Rieman, III, and J. Beukenkamp, "Phosphorus," in I. M. Kolthoff and P. J. Elving, Eds., *Treatise on Analytical Chemistry*, Vol. 5, Part II, Interscience, New York, 1961, pp. 317–402.
82. S. Greenfield, "Phosphorus," in C. L. Wilson and D. W. Wilson, Eds., *Comprehensive Analytical Chemistry*, Vol. 1C, Elsevier, Amsterdam, 1962, pp. 220–236.
83. D. F. Boltz and C. H. Lueck, "Phosphorus," in D. F. Boltz, Ed., *Colorimetric Determination of Nonmetals*, Interscience, New York, 1958, pp. 29–46.
84. B. E. Leake, G. L. Hendry, A. Kemp, A. G. Plant, P. K. Harrey, J. R. Wilson, J. S. Coats, J. W. Aucott, T. Lünel, and R. J. Howarth, *Chem. Geol.*, **5**, 7 (1970).
85. S. Fregerslev and J. R. Wilson, *X-Ray Spectrom.*, **1**, 51 (1972).
86. C. V. Dutra, *Rev. Brasil. Tecnol.*, **3**, 147 (1972); through *Anal. Abstr.*, No. 3424, **24** (1973).
87. S. S. Karacki and F. L. Corcoran, *Appl. Spectrosc.*, **27**, 41 (1973).
88. A. O. Brunfelt and E. Steinnes, *Anal. Chim. Acta*, **41**, 155 (1968).
89. P. Henderson, *Anal. Chim. Acta*, **39**, 512 (1967).
90. D. W. Archer, R. B. Heslop, and R. Kirby, *Anal. Chim. Acta*, **30**, 450 (1964).
91. R. B. Heslop and E. F. Pearson, *Anal. Chim. Acta*, **33**, 522 (1965).
92. J. A. Maxwell, *Rock and Mineral Analysis*, Interscience, New York, 1968.
93. J. P. Riley, *Anal. Chim. Acta*, **19**, 425 (1958).
94. J. Murphy and J. P. Riley, *Anal. Chim. Acta*, **27**, 31 (1962).
95. J. C. van Schouwenburg and I. Walinga, *Anal. Chim. Acta*, **37**, 271 (1967).
96. R. A. Chalmers, *Analyst*, **78**, 32 (1953).
97. C. O. Ingamells, *Anal. Chem.*, **38**, 1228 (1966).
98. P. J. Watkins, *Analyst*, **104**, 1124 (1979).
99. D. N. Fogg and N. T. Wilkinson, *Analyst*, **83**, 406 (1958).
100. H. Baadsgaard and E. B. Sandell, *Anal. Chim. Acta*, **11**, 183 (1954).

CHAPTER 6

THE DETERMINATION OF IRON, URANIUM, THORIUM, FLUORINE, CHLORINE, AND TUNGSTEN

6.1 IRON

6.1.1 General

Iron, in geological materials, exists in the ferrous, ferric, and (rarely) metallic forms; it is very likely that it will be present as a mixture of two or more of these valence states. Its determination, particularly of the valence state(s) in which it occurs, is a matter of much importance; it is to be regretted that the methods available for the determination of these valence states still give cause for doubting the reliability of the results obtained. To the petrographer and mineralogist the ferric–ferrous ratio is an important parameter, and the correct determination of this ratio is still a challenge to the analyst. It is to be supposed that in a mixture as complex as a rock or most minerals the chemical equilibrium that exists in the solid state will undergo adjustment when the sample is put into solution. The best that the analyst can do is to minimize those external influences that will change this equilibrium and seek to obtain values for the different valence states that are as near to the true values as possible.

The problems associated with the determination of the valence state(s) of iron begin with the preparation of the sample. The effect of grinding on the state of oxidation of iron has long been under consideration (Ref. 1, pp. 384–403, 907–923),[2] and this has already been discussed (see Sec. 3.4.3) in some detail. Although less desirable, it is common practice now to grind all of the sample to pass through a 100 mesh screen without reserving a coarser portion for the determination of ferrous iron as is recommended by some analysts.

The total iron is usually determined now by methods involving AAS or XRF, and this will be discussed in later chapters. The ferrous iron determination must be done on a separate portion of the sample, and titrimetric or spectrophotometric methods are most frequently used. Ferric iron is seldom determined directly, but is usually obtained by difference from the separate determinations of the ferrous and the total iron contents of the sample. Metallic iron is determined either colorimetrically or titrimetrically after selective solution of the metallic portion of the sample[3], (Ref. 4, pp. 42–45).

Ferrous Iron

The methods for the determination of ferrous iron can be divided (Ref. 4, pp. 48–55) into two types, those in which the released Fe(II) is determined in a later step and those in which the Fe(II) ions are oxidized upon release by an oxidizing agent present in excess during the decomposition of the sample, the excess then being determined. Other variations in method or technique stem from efforts to ensure complete decomposition of sample, and range from simple acid attack without further precautions to fusion or acid attack in sealed tubes or bombs. Previous work of this nature, including the sealed tube method of Mitscherlich, the fusion method of Rowledge, Cooke's acid decomposition procedure, and Pratt's modification of it, have been described in detail by Hillebrand et al.,[1] Washington (Ref. 5, pp. 134–141), and more recently by Schafer.[6]

Both Groves (Ref. 7, pp. 88–94, 181–186) and Juurinen[8] describe modifications of the fusion method which was developed to ensure the complete breakdown of refractory silicates such as staurolite. Groves mixes the sample with sodium metafluoborate, $(NaF)_2B_2O_3$, in a platinum boat and fuses it in a silica tube at 950°C under an atmosphere of CO_2; the boat and the fused melt are dropped into a solution of boric and sulfuric acids, and melt is dissolved by prolonged boiling under CO_2, and the Fe(II) titrated with $KMnO_4$. The long period of boiling needed to dissolve the glassy melt can be avoided by dissolving the coarsely crushed material in an excess of oxidant (Groves used the iodine monochloride method developed by Hey[9]). Juurinen substituted a "turbid" silica tube for one of Pyrex glass used by Rowledge and modified the dissolution procedure to exclude air; he found, however, that the small amount of oxygen present in the sealed tube during the fusion does not appreciably oxidize ferrous iron. Mikhailova et al.[10] fuse the sample with sodium fluoborate in a platinum crucible under an atmosphere of CO_2 and polarograph the solution of the melt, using sodium oxalate as supporting electrolyte; polarographic waves are obtained for both Fe(II) and Fe(III); a similar procedure was used by Bien and Goldberg,[11] who fused the sample with sodium metafluoborate under nitrogen, and dissolved the pulverized sample in sodium citrate–citric acid–potassium nitrate solution before polarographing it ($+0.15$ to -0.35 V versus saturated calomel electrode). Beyer et al.[12] reported on the simultaneous polarographic determination of ferrous, ferric, and total iron in rocks following an HF–H_2SO_4 dissolution of the sample in a platinum crucible; a combination of ac and dc methods is used.

The method of decomposition by acid attack under pressure in a sealed tube (Mitscherlich) is seldom used now, but modifications of the procedure have recently been described. Riley and Williams[13] decomposed a 5 mg sample in a stoppered Teflon tube having a capacity of about 1 ml, with a mixture of HF and H_2SO_4 previously flushed free of oxygen with nitrogen. The acid mixture completely fills the stoppered tube which is heated to approximately 100°C in a boiling water bath; the determination of the ferrous iron is made colorimetrically using 2,2'-dipyridyl. The procedure is part of the microanalytical scheme proposed by Riley for the analysis of silicates. Complete decomposition of such refractory minerals as staurolite, tourmaline, and sapphirine was achieved by Ito,[14] who heated 100–200 mg samples with HF and H_2SO_4 in a Teflon-lined steel bomb at 240°C for about 4 hr; the solution is washed into a boric acid solution and the determination finished titrimetrically with potassium permanganate. Contamination of the solution by the steel of the bomb is very slight (approximately 0.04 mg Fe_2O_3 after 4 hr), but the blank consumption of $KMnO_4$ was found to increase with time of heating, and the latter should be limited to 4 hr at the most, if possible. No appreciable oxidation by the oxygen present in the air trapped inside the bomb was found; pyrite is partially decomposed by this procedure, and high results for ferrous iron are likely when it is present. Whitehead and Malik[15] advocate the use of a sealed Teflon bomb heated at 65°C for 18 hr for the determination of Fe(II) in the more refractory minerals.

Decomposition of the sample by a boiling mixture of hydrofluoric and sulfuric acids at atmospheric pressure, with air excluded either by steam or by a stream of carbon dioxide, is probably the most widely used technique; the crucible and contents are then immersed beneath the surface of a sulfuric acid–boric acid solution, and the ferrous iron is titrated with a standard solution of $K_2Cr_2O_7$, either potentiometrically or visually using diphenylamine sulfonic acid (sodium or barium salt) as indicator. Some minor modifications of this procedure have been described, such as the addition of a "spike" of standard ferrous ammonium sulfate solution to the titration beaker in order to ensure a satisfactory end point when the FeO content of the sample is small,[16] and the addition of a few milliliters of ferrous ammonium sulfate to the titration beaker, followed by titration as usual with $K_2Cr_2O_7$ to remove oxidizing agents in the $H_2SO_4–H_3BO_3$ solution prior to the addition of the crucible contents (this eliminates the need to boil and cool the water). Pyrite and some refractory minerals (e.g., tourmaline, chromite) are not decomposed; it is sometimes possible for some of these difficulty soluble minerals to be dissolved by repeated digestion of the residue (with or without further grinding). Schafer[6] decomposes the sample in H_2SO_4 and HF at 80°C in a special polyethylene vessel, purged before and during the decomposition with nitrogen, and titrates the Fe(II) potentiometrically in the decomposition vessel with

$K_2Cr_2O_7$ and diphenylamine sulfonate. French and Adams[17] have reported a method in which the sample is digested with a hot solution of equal volumes of concentrated H_2SO_4 and 40% m/v HF in a closed polypropylene bottle which is floated in boiling water for 10 min (longer if refractory minerals are present). They also include a graph of Fe(II) found as a function of digestion time in an $HF-H_2SO_4$ mixture in a covered platinum crucible. The graph shows a maximum [Fe(II)] at 20 min, after which atmospheric oxidation becomes predominant and the amount of Fe(II) found decreases steadily. Even at the maximum, however, the amount of Fe(II) found when using the platinum crucibles for digestion was less than when the closed bottle was used. Fahey[18] dissolves magnetite and ilmenite in the presence of amphibole and pyroxene without decomposing the latter silicates, by heating them overnight at steambath temperature with 1:1 HCl in a tightly stoppered 50 ml Erlenmeyer flask filled with carbon dioxide; some ilmenites require a second overnight acid treatment. Vincent and Phillips[19] used a semimicro version of the conventional procedure (15–20 mg samples in a 5 ml crucible) for iron-titanium oxide minerals. Langmyhr and Graff[20, 21] attempted to use an aliquot of a solution prepared by dissolving the sample in hydrofluoric acid in a covered Teflon vessel, with the final measurement of the Fe(II) being made spectrophotometrically; good results were obtained for low values of Fe(II), but for high values a systematic negative error was found.

The determination of Fe(II) in the presence of Fe(III) by formation of a colored (Fe(II) complex has been proposed. Walker and Sherman[22] decomposed a 0.1 g sample of soil with HF and H_2SO_4, added the mixture to a boric acid-sulfuric acid solution, filtered it, and complexed the Fe(II) in a small aliquot of the solution with bathophenanthroline (4,7-diphenyl-1, 10-phenanthroline); the red complex was extracted into nitrobenzene for spectrophotometric measurement. When the organic content of the soil exceeds 10%, high results are obtained for ferrous iron. Bathophenanthroline has also been used for the determination of Fe(II) in HCl-soluble iron oxides, but isoamyl acetate was employed for the extraction.[23] Shapiro[24] heated a 10-mg sample (200 mesh) of rock in a capped 1 oz. plastic bottle with $HF-H_2SO_4$ in the presence of about 20 mg of o-phenanthroline, and then after buffering the solution with sodium citrate, measured the intensity of the ferrous complex formed; because of the progressive fading of the color with time, the length of the heating period must be the same for samples and standards. Begheijn (Supp. Ref. 1) has reported the use of the Fe(II)-1,10-phenanthroline complex for the colorimetric determination of Fe(I), and the subsequent reduction of Fe(III), using hydroquinone, for the determination of total Fe, and of Fe(III) by difference. Many reagents have been proposed for the colorimetric determination of Fe(II); among the best known are thiocyanate, mercaptoacetic acid, and 2,2-bipyridyl. Kiss[25] has reported the synthesis and use of an Fe(II) chromogen which is stable in high temperature and high acid concentration conditions. It is the diammonium disulfonate derivative of 3-(4-phenyl-2-pyridyl)-5, 6-diphenyl-1, 2,4-triazine. Atmospheric oxidation was minimized by adding the chromogen to the acids to complex the Fe(II) as it was liberated during the dissolution stage. Whitehead and Malik[15] have adapted the AutoAnalyzer to the colorimetric determination of ferrous and total iron in rocks using 2,2-bipyridyl. Their study included the effects of pH and they found the maximum sensitivity for the 2,2-bipyridyl complex to be in the 4–5 pH range.

Mention has been made of the method in which the sample is fused with NaF and B_2O_3 in a sealed Pyrex tube. The difficulty experienced in dissolving this fused cake without oxidation of the Fe(II) led Hey[9] to carry out the dissolution in a mixture of hydrochloric acid and iodine monochloride; the I_2 formed during the oxidation of the Fe(II) was titrated with a solution of KIO_3. Nicholls[26] applied this technique to carbonaceous shales, following dissolution of the sample in a mixture of HF and H_2SO_4; the solution is first poured onto solid boric acid, then into a bottle containing HCl and ICl, after which the titration with KIO_3 solution is made as before. Unlike the Hey fusion method, however, the acid decomposition does not seriously attack any pyrite present. Banerjee[27] has described a method in which the sample is decomposed with HF and HCl in the presence of ICl and CCl_4 in a closed vessel. The liberated I_2 is titrated with a KIO_3 solution until the pink color of the CCl_4 layer due to dissolved I_2 has disappeared. Banerjee reported that there is no interference from Mn or Cr but that sulfides which decompose under these reaction conditions, and nongraphitic organic materials, do interfere.

A semimicro procedure, in which the sample solution is added to $K_2Cr_2O_7$ solution and the excess $Cr_2O_7^{2-}$ is titrated with a standard solution of ferrous ammonium sulfate, has also been described.[28] Wilson[29] chose to eliminate the possibility of aerial oxidation of the ferrous iron by decomposing the sample in the presence of an oxidizing agent, in this case ammonium metavanadate (NH_4VO_3); the sample, in an $HF–H_2SO_4–NH_4VO_3$ mixture, is allowed to stand as long as is necessary for complete decomposition to take place, and is then washed into a beaker containing boric acid solution, and the excess V(V) is titrated with a standard solution of ferrous ammonium sulfate, with barium diphenylamine sulfonate as indicator. A study of this method was published by Whipple[30] in which he concludes that, in general, the method gives excellent results but that several precautions must be observed. For example, it is better to use a V(V) solution 5 M in H_2SO_4 than one which is 1 M because the higher acid concentration impedes the oxidation of V(V) by atmospheric oxygen, and, because F^- complexes V(V), it is best to keep the concentration of fluoride ions to a minimum. Langmyhr and Graff[21] also point out that slowly soluble fluorides must be *completely* dissolved before the titration because some precipitates formed in this way contain ferrous iron. At the GSC this procedure has been modified by the addition of an excess of ferrous ammonium sulfate to the sample solution, in order to be able to make the final titration with potassium dichromate. Wilson[31] scaled his procedure down to handle samples of 3–20 mg, omitting the boric acid in order to ensure a sharp end point during the titration (made with a micrometer-syringe burette in a volume of 7 ml). He also used a colorimetric finish in which the V(IV) formed during the reaction at low pH is used to regenerate the Fe(II) when the pH is increased to 5; the Fe(II) reacts with 2,2'-bipyridyl which is added, together with beryllium sulfate, to complex the free fluoride ions. Oxidizing and reducing substances, such as organic carbon, sulfides, and the oxides of vanadium, will interfere. These procedures, with some modifications, have been extended to the determination of the oxidizing capacity of manganese compounds.[32] Reichen and Fahey[33] decompose the sample with HF and H_2SO_4 in the presence of $K_2Cr_2O_7$ and titrate the excess of the latter with standard Fe(II) solution; boric acid is omitted because it decreases the effective action of Fe(III) during decomposition of the sample, and a correction, proportional to the amount of excess $Cr_2O_7^{2-}$ found, must be made for the $Cr_2O_7^{2-}$ destroyed by the hydrofluoric acid. Garnet was completely decomposed by this procedure, but tourmaline and staurolite were unaffected. Van Loon[34] uses KIO_3 as oxidant but boils the $HF–H_2SO_4$ mixture for about 15 min to speed up the decomposition and also to volatilize the iodine formed by oxidation of the Fe(II): KI is then added, and the liberated I_2 is titrated with sodium thiosulfate. Chlorides must be absent. A different approach is described by Ungethüm;[35] the sample is decomposed in HF alone at room temperature in the presence of a measured amount of $AgClO_4$, and the excess Ag(I) is titrated potentiometrically with a standard solution of potassium bromide. Halogens will interfere; up to 5% of pyrite is not significantly attacked [and the Fe(II) of the pyrite is thus not determined].

It has been suggested previously that the value found for Fe(II) may not be the true value, because of the presence of oxidizing or reducing substances which are released to react upon decomposition of the sample. Ingamells[36] suggests that it is better to report only a value which expresses the oxygen excess or deficiency in the sample (see also Wilson[32]) and describes a procedure in which the sample is dissolved in a mixture of phosphoric acid and sodium pyrophosphate in the presence of Mn(VII) or Cr(VI); the excess of the oxidant is then titrated with a standard Fe(II) solution. The method is limited to those samples that dissolve directly in the acid mixture, do not contain sulfur or organic carbon, and do not yield peroxides when dissolved. A similar procedure was applied to ferrites by Cheng,[37] who dissolved the sample in a 0.1 N phosphatocerate solution at 280–300°C.

In addition to the possible errors associated with the sample preparation and decomposition, there are other sources of error that must be considered:

1. Some iron-bearing minerals are refractory, and even prolonged boiling with H_2SO_4 and HF will not decompose them (tourmaline, staurolite, ilmenite, magnetite). The boiling should not be too prolonged, incidentally, because of the oxidizing nature of hot, concentrated sulfuric acid. Fine grinding of the initial sample will usually increase its vulnerability to acid attack, but it is better first to repeat the attack on the insoluble residue than to risk oxidation by excessive grinding; chromite will not usually succumb and Peck[38] has warned about the surprisingly resistant behavior of siderite.

2. The introduction of "tramp" iron from the crushing and grinding equipment must be avoided; not only will this iron be counted as ferrous iron, it may reduce some of the Fe(III) present as well.

3. Aerial oxidation has been generally considered the chief source of error in the determination of ferrous iron, particularly during the process of acid decomposition and, to a lesser extent, during the titration; if the determination is being done on the semimicro- or microscale, this error can be considerable.[13] Clemency and Hagner,[39] who determine total iron and ferric iron by coulometric generation of titanous ion and obtain the ferrous iron by difference, have shown experimentally, however, that immediately following the decomposition of the sample there is a marked reduction of the ferric iron with a subsequent slow reoxidation; they suggest that this reduction, rather than aerial oxidation, may be responsible for erratic results and recommend that the period of decomposition be kept as short as possible. Various means of preventing aerial oxidation during decomposition have been suggested, usually involving the flushing of the decomposition vessel in some way with an inert gas, but if the acid mixture is heated before addition, is brought rapidly to boiling after addition and maintained at this temperature during a period of decomposition that is kept as short as possible, aerial oxidation of Fe(II) will be negligible and the use of an inert atmosphere is unnecessary. Similarly, it is customary to use water that has been boiled and cooled, or treated in some other way, to remove dissolved oxygen; Peck[38] has shown that

such a precaution is unnecessary, and Kiss[41] has demonstrated that when air is bubbled through the boiling $HF-H_2SO_4$ sample solutions, the HF–Fe system is "remarkably insensitive to aerial oxidation."

Hydrofluoric acid alone is considered to favor aerial oxidation of ferrous iron because of the formation of slightly ionized ferric fluoride with a consequent lowering of the oxidation potential of the Fe(III)–Fe(II) system; Kiss,[41] however, considers that the fluoro complexes formed by Fe on dissolution of the sample will exhibit an appreciable degree of resistance to a mild oxidant such as air. Aerial oxidation proceeds much more slowly in HCl and H_2SO_4 solutions at room temperature. The harmful effect of fluoride is overcome by complexing it with boric acid, beryllium, or aluminum[20] prior to the titration.

4.   The presence of sulfide minerals in the sample makes an accurate determination of the ferrous iron an impossible task (Ref. 4, pp. 57–58). If the sulfides are decomposable, whether by acid decomposition or by fusion, there is the strong possibility that the S(II) will reduce some of the Fe(III) present, thus giving a high value for Fe(II) and a correspondingly low one for Fe(III). The magnitude of the error depends upon the method of decomposition because, while most of the S(II) is lost harmlessly during the acid decomposition in a crucible, it will exert an effect proportionately much greater when the decomposition is carried out in a sealed tube at a high temperature and pressure; the oxidation of the S(II) to SO_3 has the capacity to reduce any Fe(III) present to give an apparent FeO equivalent to about 14 times the weight of S(II) present. Thus even 0.01% of sulfur could cause a serious error if the decomposition is done in a closed system.

Pyrite is not usually decomposed except in a bomb or sealed tube, and its presence will usually mean a low value for the FeO and a correspondingly high one for the Fe_2O_3. The presence of Fe(III) is thought to increase the solubility of pyrite. If pyrite is the only sulfide mineral present in appreciable quantity, a correction may be applied to the Fe(II) and Fe(III) values on the basis of the known sulfur content of the sample.

5.   Carbon, if present as organic matter (graphite is without effect), will give high results for Fe(II) because it will tend to be oxidized during the titration, or will reduce an added oxidizing agent. The effect is greater when $KMnO_4$ is used as the titrant.

6.   Trivalent vanadium will be oxidized during the titration with, or on addition of, an oxidant and thus will give a high value for Fe(II), but in most rocks and minerals it will be negligibly low. Pentavalent vanadium will oxidize ferrous iron.

In summary, incomplete decomposition will affect the results adversely, no matter what method is used, but the error caused by the presence of S(II) will be greater when decomposition is carried out in a closed system. The possibility of aerial oxidation can be kept to the minimum by careful and uncomplicated handling of the decomposition step, and rendered negligible by carrying out the decomposition in the presence of an oxidant. The presence of other reducing substances, however, will still cause errors under these conditions, as well as in the direct titration of the ferrous iron by an oxidant. The substitution for the oxidant of a reagent that will form a colored complex with ferrous ions during the decomposition seems to offer the best opportunity for future investigation.

Both potassium permanganate and dichromate are commonly used as the oxidant. Permanganate, which does not require a separate end point indicator, is more affected by the presence of organic matter. A fading end point is also experienced in the presence of excess hydrofluoric acid, because of the formation of slightly ionizing manganic fluoride which encourages the oxidation by $KMnO_4$ of the Mn(II) formed during the titration; the addition of boric acid to the solution removes the fluoride ions from effective action. When a chloride solution is to be titrated with $KMnO_4$, Mn(II) (as $MnSO_4$) should be added to prevent the oxidation of chloride ion to chlorine.

Bouvier et al.[40] have reported a novel method that involves the digestion of the sample at a high temperature in a mixture of concentrated H_2SO_4 and H_3PO_4. The ferrous iron reduces a stoichiometric quantity of H_2SO_4 to produce SO_2 which is measured using a modified Leco sulfur titrator (see Sec. 5.6.4b). There are, of course, serious interferences if any oxidizable matter is present, sulfides being the most common source of error in rock samples.

Kiss[41] has described a constant-current potentiometric method for the rapid determination of the oxidation state of iron in silicates subsequent to an $HF-H_2SO_4$ decomposition of the sample. The method is reported to have the advantage of not being affected by sulfides and aerial oxidation.

Because carbonates are much more readily acid-soluble than are silicates, the determination of ferrous iron in the former encounters few of the problems discussed above for the determination of ferrous iron in silicates. For a pure carbonate the method involves little more than mere decomposition of the sample in acid in an Erlenmeyer flask and immediate tritration with either $KMnO_4$ or $K_2Cr_2O_7$.

If the sample contains an appreciable amount of silicates, it may be desirable to determine the ferrous (and ferric) iron in the acid-soluble and acid-insoluble portions. The sample is attacked by dilute H_2SO_4 (or dilute HCl if $K_2Cr_2O_7$ is to be used in the titration, and the higher oxides of manganese are absent) in an Erlenmeyer flask from which air has been expelled by steam (or by a current of CO_2), cooled rapidly, and titrated immediately. The solution is filtered, the residue is washed thoroughly with water and then washed from the paper into a platinum crucible. The bulk of the water is evaporated and the determination continued as described in Sec. 6.1.2a. Total iron is determined in each portion; the titration of both the ferrous and the total iron may be done potentiometrically (see Sec. 6.1.2b), but this would involve transferring the contents of the flask, in the case of the acid-soluble portion, to a suitable beaker and thus increasing the possibility of some air oxidation of the ferrous iron.

Hillebrand et al. (Ref. 1, pp. 975–976) have emphasized the virtual impossibility of obtaining a reliable value for ferrous iron in the presence of carbonaceous matter so often found in limestones. They recommend that the decomposition be done with cold acid in a carbon dioxide atmosphere (for dolomite some heating will be necessary), that the solution be filtered through an asbestos pad rapidly (and preferably again in a carbon dioxide atmosphere), washed and titrated rapidly ($K_2Cr_2O_7$ is preferable as titrant because it is less affected by the presence of organic matter than is $KMnO_4$). The decomposition time should be kept to a minimum.

Ferric Iron

The ferric iron value is, as a rule, determined as the difference between the total iron, expressed as Fe_2O_3, and the Fe_2O_3 equivalent of the FeO value. It is thus a repository for errors inherent in either or both of the FeO and total iron determinations.

$$\% \text{ total Fe (as } Fe_2O_3) - (\% \text{ FeO} \times 1.1113) = \% Fe_2O_3$$

The work of Clemency and Hagner,[39] who used the coulometric generation of titanous ions to determine the ferric iron directly, has been mentioned previously in the discussion of the possible errors affecting the ferrous iron determination. They note that there is an initial reduction of ferric iron following decomposition by, as seems most likely, some minor constituent of the sample, and recommend the use of a shorter period of decomposition, that is, 5 min instead of the usual 10. A method has been described in which the sample is decomposed by HF and HCl in an inert atmosphere, and the Fe(III) titrated with EDTA solution, using xylenol orange as the indicator.[42] Reference has been made previously to the use of polarographic methods, and Fe(III)–Fe(II) ratios have been determined in bulk rock samples by Mössbauer spectroscopy.[43] These approaches offer hope that the direct determination of Fe(III) will eventually become part of the analytical scheme for silicate analysis.

Metallic Iron

It is unlikely that the rock and mineral analyst will need to determine the metallic iron content of a sample unless he is involved in the analysis of a meteorite. Not all meteorites contain a metal phase but when it is present, it introduces further complications into an analysis that is already sufficiently complex, and a reliable determination of the metal content becomes of first importance. It is a problem that has also been of much concern in the analysis of slags and ores (Ref. 4, pp. 42–48).*

The oldest method is that in which the sample is digested with mercuric chloride, and sometimes with ammonium chloride as well, to dissolve the metallic phase; the mercurous chloride is removed and the Fe(II) is titrated as usual. Particle size plays an important part in the dissolution and the sample should be -200 mesh to ensure complete extraction of the metal. Because of the presence of mercury salts, it is possible to determine only the iron in the metal phase. A similar procedure involves the displacement of copper from cupric sulfate by metallic iron.

Both of these methods fail when iron sulfide and phosphide are present, and Riott[44] has introduced the use of cupric potassium chloride to overcome this drawback; a very dilute acetic acid solution of the reagent is used under an atmosphere of carbon dioxide, and no attack of the iron oxides and sulfides takes place. The precipitated copper reduces the excess $CuCl_2$ to $CuCl$, from which copper is removed by displacement with aluminum; the Fe(II) is titrated as before. Only 0.1 g of metallic iron can be safely handled, and iron carbides, if present, will cause the results to be high.

Habashy[45] made a comparative study of existing methods for the determination of metallic iron in the presence of iron oxides and found that the four older methods which were considered gave results equally as good as those obtained by his proposed procedure when the particle size was 250 mesh or finer; he found that his procedure alone was reliable when the particle size was coarser than 100 mesh. He uses a variation of the $CuSO_4$ method in which the sample is digested with $CuSO_4$ and mercury; the precipitated copper amalgamates with the mercury, which is then separated, dissolved in nitric acid, and the copper determined electrolytically.

A very different approach is proposed by Easton and Lovering[46] for

---

* If it is only the removal of the metallic iron that is required, in order to permit the subsequent determination of Fe(II) and Fe(III) in the nonmetallic phase, this may be accomplished by shaking the sample with a nearly neutral solution of ferric chloride and allowing it to stand for several hours before filtering and washing the insoluble material.[42]

the separation not only of the iron of the metallic phase but of the nickel and cobalt as well. The sample is digested with $HgCl_2$–NH_4Cl and the extract, made 10 M in HCl, is placed on an anion exchange column from which the nickel is eluted first with 10 M HCl, the cobalt with 6 M HCl and the iron with 0.6 M HCl; the mercuric salts are retained on the column. Final determination of the metals is made colorimetrically, o-phenanthroline being used for the iron.

6.1.2   Determination of Ferrous Iron

a.   *Visual Titration Following Hot Acid Decomposition of the Sample (Modified Pratt Method)*

Reagents

Boric acid solution, saturated (5%).   Dissolve, with stirring, 100 g of boric acid in 1 liter of hot water, cool, and dilute to 2 liters.

Standard potassium dichromate solution (0.1 N).   Weigh 9.810 g of finely ground primary standard grade $K_2Cr_2O_7$ into a weighing bottle, and dry in an oven at 110°C for 2 hr or more. Weigh the bottle and contents and pour the reagent into a small funnel placed in the neck of a 2 liter volumetric flask; reweigh the bottle to obtain the true weight of $K_2Cr_2O_7$ used. Wash the powder into the flask with a jet of water, rinse, and remove the funnel, and wash down the neck of the flask. When the reagent is dissolved, dilute to volume at 20°C and mix thoroughly. This solution will serve as a primary standard,

$$N = \frac{\text{wt } K_2Cr_2O_7 \text{ (g)}}{2 \times 4.9035 \text{ g}}$$

but a check on the normality may be made by using an aliquot of the standard iron solution (see below) and following the procedure given in this section or Sec 6.1.2b; 10 ml of the Fe solution requires approximately 10 ml of 0.1 N $K_2Cr_2O_7$. For convenience in use, the oxidizing capacity of the potassium dichromate solution may be expressed as milliequivalent weights of FeO and Fe_2O_3, that is, for 0.1000 N $K_2Cr_2O_7$,

$$FeO \equiv 0.007184 \text{ g/ml}$$

$$Fe_2O_3 \equiv 0.007985 \text{ g/ml}$$

Standard iron solution (approximately 5 mg Fe/ml).   Weigh accurately about 1.3 g of pure iron, as wire or chips (wash in ether to remove oily film and allow to dry in air), and transfer to a 150 ml beaker. Add 50 ml

HCl (1:1), cover, and warm on a steam bath to complete solution. Cool and transfer quantitatively to a 250 ml volumetric flask. Dilute to volume at 20°C, mix thoroughly, and transfer to a 250 ml glass bottle with a greased stopper. Take aliquots only with a pipette.

Procedure

Weigh 0.500 g of sample (100 mesh) and transfer it quantitatively to a 45 ml platinum crucible having a tightly fitting cover, in the center of which is a hole 2 mm in diameter. Add about 1 ml of water, swirl the crucible to distribute the sample over the bottom to prevent caking, and add 2 or 3 drops of H_2SO_4 (1:1) to decompose any carbonates present; cover and allow to stand until all reaction has ceased.

To 10 ml of water in a 50 ml Pt dish add 5 ml H_2SO_4 (18 M) and 5 ml HF. Place the covered crucible on a silica triangle suspended firmly over a Bunsen burner having a low flame (see description of special stand in Sec. 2.2.8 and Figs. 2.2 and 2.3), slide the cover to one side, quickly (and carefully) add the hot acid mixture from the Pt dish, replace the cover, and immediately begin brushing the sides and cover of the crucible with the flame of a second burner until the contents are boiling and steam escapes through the hole in the cover. Adjust the height of the flame of the first burner (about 1 cm) so that the contents of the crucible boil gently, and continue the heating for 10 min; the heating period should not be prolonged (use an automatic timer) nor should the temperature be sufficiently high to evaporate the water to the point at which the hot concentrated H_2SO_4 begins to oxidize the ferrous iron.

To 200 ml of water in a 600 ml beaker add 50 ml of 5% (saturated) boric acid solution, 5 ml H_2SO_4 (18 M), and 5 ml 85% H_3PO_4, and mix well.

At the conclusion of the heating period at once grasp the crucible firmly with platinum-shod crucible tongs, press the cover firmly on the crucible with a glass stirring rod, and swiftly submerge the crucible below the surface of the acid solution in the 600 ml beaker; do not allow more than the Pt shoes of the tongs to touch the acid solution. At once dislodge the cover from the crucible with the stirring rod and stir to mix the contents of the crucible and beaker; do not stir so vigorously as to draw air into the liquid, but make sure that all of the crucible contents have been washed out into the beaker and that all soluble material has dissolved. Peck[38] prefers to remove and rinse off the crucible and cover at this point.

Titrate the solution immediately with $N/10$ $KMnO_4$ to the first appearance of a permanent pink tinge, or with $N/10$ $K_2Cr_2O_7$ (add 6 drops of 0.2% w/v barium diphenylamine sulfonate to serve as indicator and titrate to the first appearance of a permanent blue-violet color). When the end point is reached, swirl the crucible in the beaker to ensure that all of the sample solution has been fully titrated. For accurate work, an indicator blank should be determined and a correction made.

$$\% \text{ FeO} = \frac{\text{ml of titrant} \times \text{normality}}{1000} \times \frac{71.85}{1} \times \frac{100}{\text{sample (g)}}$$

Carefully remove the crucible cover with platinum-tipped forceps and rinse it into the beaker. By means of the stirring rod, remove the crucible and rinse it, inside and out, into the beaker. Examine the residue, if any, in the beaker with a hand lens; usually this residue will consist of white or greyish white grains of quartz which resist decomposition and which can be ignored.

The presence of brassy yellow grains of pyrite in the residue should be noted so that a correction to the FeO value, based upon the amount of sulfur present as sulfide, may be made later on if desired. This will also indicate the potentially unreliable nature of the determination, because of the possible presence in the sample of sulfides such as pyrrhotite which are decomposed under these conditions.

If there are more than a few grains of red to black undecomposed material in the residue, a second decomposition and titration should be made as follows. Decant carefully the supernatant liquid from the beaker, retaining as much of the residue as possible. Wash the beaker and residue once with water and again decant the liquid. Rinse the residue into a small agate mortar, decant the liquid, and grind the residue until no grittiness remains. Wash the slurry into the original crucible and repeat the decomposition, transfer, and titration. If some dark unattacked material remains still undecomposed, it is likely chromite, and further acid attack is of no avail. Unfortunately, because of the variable nature of chromite, it is not feasible to add a correction based on the known chromium content of the sample.

Conversion Factors

$$\text{FeO} \underset{1.2865}{\overset{0.7773}{\rightleftarrows}} \text{Fe}$$

$$\text{Fe}_2\text{O}_3 \underset{1.4297}{\overset{0.6994}{\rightleftarrows}} \text{Fe} \qquad\qquad \text{Fe}_2\text{O}_3 \underset{1.1113}{\overset{0.8998}{\rightleftarrows}} \text{FeO}$$

b. *Potentiometric Titration Following Hot Acid Decomposition of the Sample (Modified Pratt Method)*

The procedure up to the point of titration is the same as that described in Sec. 6.1.2a, except that it is not necessary to add the 5 ml of 85% phosphoric acid to the solution in the beaker.

Immerse the platinum and saturated calomel reference electrodes of a potentiometric apparatus (see Sec. 2.3 and Fig. 2.2) in the solution, add a plastic covered stirring bar, and place the beaker on a magnetic stirrer (be careful that the electrodes are kept clear of the stirring bar and avoid the formation of a vortex by too rapid stirring). The initial potential of the system will usually be in the neighborhood of 400 mV. Titrate with $N/10$ $K_2Cr_2O_7$ until a sharp increase in the potential indicates that all of the Fe(II) has been oxidized. The changes in potential will be gradual and steady until a reading of about 500 mV is reached, after which they become more pronounced and the titrant should be added at the rate of 1 or 2 drops at one time, allowing sufficient time between the additions for equilibrium to be reached; the abrupt rise in the potential (50–100 mV), marking the end point, usually occurs at 600–650 mV.

Remove the crucible and cover, rinse them into the solution, and examine the residue for undecomposed material, as previously described.

c. Visual Titration Following the Modified Cold Acid Decomposition Method of Wilson[29]

Reagents

Ammonium metavanadate solution (approximately 0.1 N).   Dissolve 10 g reagent grade NH_4VO_3 in 110 ml H_2SO_4 (1:1) and dilute to 1 liter. Store in a glass bottle or in the reservoir of an automatic pipet.

Acid mixture (H_3PO_4–H_2SO_4–H_2O = 1:2:2).   Add cautiously, with stirring, 400 ml of concentrated H_2SO_4 to 400 ml of water in a 1500 ml beaker which is cooled in a water bath. Add 200 ml of concentrated H_3PO_4 and mix well.

Boric acid solution.   See Sec. 6.1.2a (*Reagents*) for preparation.

Ferrous ammonium sulfate solution (0.05 N).

Barium diphenylamine sulfonate solution (0.2% w/v).

Standard potassium dichromate solution (0.05 N).   Prepare as described in Sec. 6.1.2a (*Reagents*), but use only 4.905 g of $K_2Cr_2O_7$.

Procedure

1.   Weigh 0.200 g of sample (if FeO is likely to be greater than 10%, use a 0.100 g sample) and transfer it to a 60 ml plastic vial.

2.   To each vial, and to four additional vials which serve as blanks, add from an automatic pipette 5 ml ammonium vanadate solution.

3.   Swirl gently to form a uniform slurry and add, from a plastic graduated cylinder, 10 ± 1 ml of concentrated hydrofluoric acid.

4.   Cover the vials (not tightly) and allow to stand overnight in a fume hood, or until the sample is completely decomposed (absence of gritty particles).

5.   To each vial add 10 ml of the sulfuric acid–phosphoric acid mixture.

6.   Pour the contents of the vial into 100 ml of boric acid solution in a 400 ml beaker; rinse out the vial, using a rubber policeman if necessary, with an additional 100 ml of boric acid solution and add the rinsings to the 400 ml beaker.

7.   Stir the contents of the beaker to effect complete solution of the contents of the vial and add, from an automatic pipette, 10 ml of ferrous ammonium sulfate solution.

8.   Add 1 ml of barium dephenylamine sulfonate indicator and titrate to a grey end point with standard potassium dichromate solution, using a 10 ml semimicro or equivalent type of burette.

9. $\% \text{ FeO} = \dfrac{\text{ml for sample } - \text{ml for blank}}{1000}$
$$\times \text{ normality} \times \frac{71.85}{1} \times \frac{100}{\text{sample weight (g)}}$$

6.2   URANIUM AND THORIUM

6.2.1   General

Because of the emphasis on uranium as an energy source, the need for the determination of both uranium and thorium has assumed greater importance. These actinide elements may be present in rocks occasionally in concentrations of 0.01–0.1% or higher, but it is more likely that it will be necessary to determine them in trace quantities for geochemical studies (Ref. 47, Chaps. 90 and 92). The initial flurry of interest in uranium as an energy source toward the end of the first half of this century led to the development of many techniques of separation and analysis which have been reviewed by several authors[48, 49] and (Ref. 50, Chaps. 35 and 39). Rybach[51] has reviewed gamma-ray spectrometry methods for simultaneous U and Th determinations and Adams and Gasparini[52] have discussed it in some detail. It is a method which has been in use for some time,[53, 54] but the development of more efficient NaI(Tl) phosphor crystal detectors and better multichannel analyzers have increased its usefulness in both the field and the laboratory. The method described in Sec. 6.2.2c was developed in the BCM laboratory by Mr. John Davies and is based somewhat on work by Hurley,[53] but it also distinguishes between "young" and "old" uranium as does the method of Domingos and Melo.[55]

These elements are probably more abundant and widespread than is commonly supposed, but unfortunately a preliminary ES analysis is not likely to indicate their presence below 0.01% because of the relative insensitivity of their arc and spark spectra. Their overall abundance in the lithosphere (Clarke value) is about 10 ppm for Th and 3 ppm for U. Granitic rocks average about 18 ppm Th and 3.5 ppm U, while basalts average about 3–4 ppm Th and 0.5–1 ppm U (Ref. 56, pp. 229–236.) Both elements are commonly associated with Nb, Ta, Zr, Ti, Pb, and the rare earths.

Neutron activation analysis is a popular method of determination for trace levels of U.[57] Bowie et al. have reported that, with delayed neutron counting, there is only a slight interference from Th and a detection limit of 0.03 μg of U is possible with a flux of 5×10^{12} n/cm^2.[58] Cumming has reported that, under similar conditions, it is not possible to reach as low a detection limit for Th as it is for U, but parts per billion levels can be attained.[59]

The method of fission track activation is also generating significant interest among analysts, and a detection limit of 1 ng/g in a flux of 10^{16} n/cm^2 has been reported by one set of workers.[60] The technique used by Weiss and Haines involves irradiation of the polished surface of a thin section of the sample with thermal neutrons to induce the fission of U atoms in the sample. The integrated fission track density is measured and compared with calibrated measurements following the application of corrections for geometric factors. Other analysts use powdered samples,[61] but this spoils the sample for further petrographic work. Matsuda et al.[62] have used this technique for the determination of U retained after chemical separation on Dowex 1 × 8 resin to achieve parts per billion detection limits.

Gaillochet et al.[63] have described a method of determining U in samples with a carbon paste electrode formed by thoroughly mixing the powdered sample with graphite and an electrolyte. A measurement of current versus voltage passed through the electrode gives a measure of the U content and also of the relative concentrations of U(IV) and U(VI).

The most widespread analytical technique used for the determination of trace levels of U is the fluorimetric method, and that for Th is by the spectrophotometric measurement of its Arsenazo III complex. Both methods are susceptible to many interferences, and it is not uncommon to do an ion-exchange separation (often with the strongly basic Dowex 1 × 8 anion-exchange resin) prior to the measurement.[64–66] Adam and Stulikova have reported the separation of U from Fe(III) and Mn(II),[67] common interferences in the fluorimetric determination of U. They extract UO_2^{2+}

selectively from an aqeuous alkaline solution in the presence of 0.5 M nitrilotriacetic acid in M NaOH using a solution of 0.5 M diphenylacetic acid in benzene as the extractant.

The fluorimetric method given in this chapter for the determination of traces of U (up to 10 ppm) is based upon the method of Ingles[68] and has been used extensively at the GSC for the determination of *total* uranium following complete decomposition of the sample. The sample, in a nitric acid solution after complete decomposition by a combination of acid treatment (HF and HNO_3) and Na_2CO_3 fusion, is extracted with ethyl acetate in the presence of a high concentration of aluminum nitrate, to separate and concentrate the uranium. The latter is transferred to an aqueous phase, and an aliquot is fused with sodium fluoride under controlled conditions; the fluorescence of the solidified bead, and that of standards and a blank treated in a similar way, is measured with a fluorimeter.

Arsenazo III is one of a series of reagents that are similar in structure to thorin but which differs from the other members in that it forms stable blue complexes with thorium at acidities up to 8 N, and is more sensitive. Much work has been done with the Arsenazo reagents by S. B. Savvin at the V. I. Vernadsky Institute of Geochemistry and Analytical Chemistry in Moscow, USSR; the method used at the GSC was developed by Abbey,[69] and it is based upon that described by Savvin. By substituting a perchloric acid medium for the hydrochloric acid one of Savvin, greater sensitivity and stability of the colored complex are achieved. Beer's law is generally effective over the range of 0–25 μg thorium in a final volume of 25 ml.

The sample (usually 0.5 g) is decomposed by sintering with Na_2O_2 in an iron crucible (carbonates are first dissolved in acid and the residue is either decomposed by HF and $HClO_4$ or by sintering it with Na_2O_2). The sinter is leached, filtered, and the water-insoluble residue dissolved in HNO_3 and HCl. Calcium is added, and the oxalates are precipitated homogeneously with methanolic methyl oxalate; the filtered precipitate and paper are digested with nitric acid to destroy the paper and finally fumed with perchloric acid. If the TiO_2 content of the sample is greater than 1%, a second oxalate precipitation is made. Excess $HClO_4$ is removed, the residue is dissolved in a controlled quantity of HCl, a small amount of ascorbic acid is added to reduce any Fe(III) that may be present, and controlled quantities of $HClO_4$, oxalic acid (to mask Zr and Hf), and reagent are added.

The absorbance of the colored complex (a mixture of the blue thorium complex and the red excess reagent) is measured at 660 nm. Calcium was found to affect the range of thorium concentration over which Beer's law is effective, as well as positively affecting the absorbance of the thorium complex; it is important that the calcium contents of the sample, standard, and blank be as close together as possible.

May and Jenkins[70] have described a similar method using Arsenazo III, but they prefer to use the HCl medium (2.5–3 N) of Savvin, and a series of precipitation steps is used to separate the thorium from other elements, including the rare earths (iodate separation of thorium).

The solvent extraction of thorium is often done with 8-quinolinol (chloroform solution); Goto et al.[71] have applied it also as a reagent for the spectrophotometric determination of thorium in monazite sand.

Sills[72] has recently described a scheme for determining the distribution of uranium and thorium isotopes, by alpha spectrometry following a pyrosulfate fusion and Aliquot-336 (tricaprylylmethyl ammonium chloride) extraction from a strongly acidic $Al(NO_3)_3$ solution.

6.2.2 Determination of Uranium and Thorium

a. *Fluorimetric Determination of Traces of Uranium Using Sodium Fluoride Fusion*

This method was developed by Sydney Abbey for the determination of 0–10 ppm uranium in silicate rocks, for which it is required that the entire sample be decomposed. It will be necessary to modify the procedure when the sample is not a silicate rock, when only the HF-soluble or acid leachable portion of the sample is of interest, or when the expected uranium content is not in the 0–10 ppm range.

Special Equipment[68]

Fluorimetry dishes.   Platinum, approximately 18 mm wide and 3 mm deep. Several sets, of 24 dishes each, will be needed. Store under distilled water.

Pipette (for measuring 0.10 ml aliquots).   *a* 100 μl microchemical pipette equipped with syringe attachment.

Aluminum trays.   14 cm × 9 cm × 4 mm, containing 24 holes (6 rows of 4 each) each 1.5 cm in diameter.

Burner.   A Fletcher radial flame burner, modified by the insertion of a loose roll of bronze screen wire (16 mesh) into the burner barrel (to diffuse gas) and the placing of a nichrome wire screen over the burner cap (to support the dishes).

Hood.   It is necessary to modify a standard fume hood by lining it with firebrick and installing a flame baffle, consisting of two sheets of heavy 6 mm mesh wire screening in the upper portion (to protect exhaust fan). The hood window is replaced by a panel of asbestos board having a mica window (20 × 20 cm).

Fluorimeter.   Jarrell-Ash Galvanek-Morrison Fluorimeter Mark V or equivalent, equipped for reflectance measurement.

Reagents

Aluminum nitrate salting solution.   Transfer 5 lb of reagent grade aluminum nitrate, having a low uranium content, to a 4 liter beaker. Add 150 ml H_2O, cover, place the beaker on a combined hot plate and magnetic stirrer, and heat, with stirring, until solution is complete. Adjust the concentration by evaporation or dilution to give a solution which boils at 130°C, and transfer the hot solution to a three-neck round-bottom 2 liter flask. Place the flask in a heating mantle controlled by a Variac, insert a reflux condenser in the middle neck, a thermometer in a well in one side neck, and stopper the other neck (solution is withdrawn from this neck when required). Store the solution at 80°C and heat to 100°C before using.

Aluminum nitrate wash solution.   Dilute 100 ml of the above solution with 73 ml H_2O and 4 ml concentrated HNO_3.

Standard uranium solutions.   Dissolve 211 mg $UO_2(NO_3)_2 \cdot 6H_2O$ in water, add 50 ml concentrated nitric acid, and dilute to volume in a 1 liter volumetric flask to give a solution containing 1 mg U/ml. Dilute 1.0 ml of this solution to 1 liter with dilute HNO_3 (1:19) to give a solution containing 1 μg U/ml.

Sodium Fluoride pellets (98% NaF, 2% LiF).   Prepared pellets, each containing 0.588 g NaF and 0.012 g LiF, can be obtained. Those supplied by Analoids (Ridsdale and Co., Middlesbrough, England), in 1000-pellet lots, have been found satisfactory.

Procedure

Decomposition.   Weigh 1.00 g of sample into a porcelain crucible and ignite over a Meker burner for 15–20 min. Cool, break up with a spatula, and repeat ignition for about 10 min. Cool. Transfer to a 100 ml Teflon dish, rinsing the crucible with a little water.

With the first batch of samples prepare a *blank*, containing 10 ml of HNO_3 (1:19) and a *standard*, containing 10 μg uranium, in separate 100 ml Teflon dishes and process these in the same way as the samples. Aliquots of the final solutions prepared from these may be used with subsequent batches of samples, provided that there is no change in the reagents and procedure used.

To blank, standard, and samples, add 5 ml concentrated nitric acid and 10 ml HF. Cover and warm on a steam bath for several hours (preferably overnight).

Rinse off and remove the cover. Add a little more HF and evaporate the contents to dryness. Take up with 5 ml HNO_3 and 5–10 ml H_2O, break up the residue with a rubber-tipped rod, and again evaporate to dryness. Repeat the dissolution and evaporation twice more.

Add 5 ml nitric acid and about 25 ml water. Break up the residue with a rubber-tipped rod, cover the dish, and warm on a steam bath to dissolve soluble matter.

Filter hot on a 9 cm Whatman No. 42 paper, receiving the filtrate in a 250 ml beaker. Wash with warm nitric acid (1:19).

Evaporate the filtrate to a low volume, transfer to a 30 ml beaker, and evaporate to <10 ml. Set aside for eventual combination with dissolved residue.

Transfer the filter and residue to a 10 ml platinum crucible. Place the loosely covered crucible in a cold muffle furnace and raise the temperature slowly until all paper is burned off.

To the cooled crucible add 0.5 ml H_2SO_4 (1:1) and about 2 ml HF. Evaporate until SO_3 fumes are no longer evolved. Ignite briefly to expel traces of acid. Add 0.5 g sodium carbonate and fuse for 15–20 min on a meker flame. Allow to cool.

Add about 5 ml water and 2–3 drops ethanol to the crucible. Cover and warm gently to dissolve the sodium carbonate. Transfer the suspension to a 50 ml beaker and set aside. Add 1 ml nitric acid and 4 ml water to the crucible, cover, and warm to dissolve soluble matter. Cool.

Carefully add the contents of the crucible to the contents of the 50 ml beaker and warm to dissolve soluble matter and expel carbon dioxide. Evaporate to less than 20 ml if necessary. Transfer this mixture to the 30 ml beaker containing the evaporated filtrate.

Evaporate to less than 10 ml, add one drop hydrogen peroxide (30%), cover and warm until the color of the titanium complex is destroyed and no more oxygen is evolved.

Extraction. Add 13 ml of the salting solution and swirl to mix. Pour into a 25 ml graduated cylinder and measure volume (V ml). Pour into a 60 ml separatory funnel, and rinse graduate and beaker into funnel with (23-V) ml of water. Allow to cool.

Add 20 ml ethyl acetate, close the funnel with a polyethylene stopper, and shake for 45–60 sec. Allow phases to separate.

Remove and discard aqueous phase, leaving any intersurficial turbidity in the funnel. Add 10 ml aluminum nitrate wash solution, shake, and separate as before.

Rinse out the funnel stem with water, discarding the rinsings. Drain the ethyl acetate phase into a 50 ml beaker. Rinse out the funnel with 7 ml H_2O and drain into the same 50 ml beaker.

Allow the ethyl acetate to evaporate at room temperature (preferably overnight). If the volume, after evaporation, is still well over 10 ml, warm gently to evaporate further.* Cool and dilute to volume in a 10 ml volumetric flask.

Fusion. Place 19 platinum fluorimetry dishes on the perforated aluminum tray and dry under an infrared lamp.

For three samples (1, 2, 3), a standard (S), and blank (B) pipette 0.10 ml aliquots onto the dishes in the following order (0 represents an empty dish):

```
1   2   3   3
S   B   S   2
S   0   1   B
1   3   2   1
S   3   2
```

Evaporate the aliquots to dryness under an infrared lamp, and transfer to the grid of the fusion burner in the following pattern:

```
      1       2       3
  3       S       B       S
2     S       0       1       B
  1       3       2       1
      S       3       2
```

Place a flux pellet on each dish except that designated as zero. Ignite the burner and adjust conditions to give fusion within 90 sec, and keep the beads molten for 120 sec longer. Turn off the burner and spray steam over the cooling beads until they no longer glow red.

Return the dishes to the aluminum tray and measure fluorescence of

---

* With some samples a reaction may take place in which the solution turns a brownish color and nitrogen oxides are evolved. *On rare occasions this reaction can be quite vigorous, to the point of causing an explosion.* The beaker should therefore be removed from the source of heat as soon as the solution begins to darken. The reaction will usually sustain itself without further heating. After it subsides, the beaker should be *cautiously* warmed to ensure that the reaction is complete and, if necessary, to reduce the volume.

all beads between 15 and 60 min after the end of the fusion. Calculate the mean fluorescence reading for each sample, for the standard, and for the blank.

$$\text{ppm U in sample} = 10 \cdot \frac{x - B}{S - B}$$

where x is the mean fluorescence reading for the sample, S the mean fluorescence reading for the standard, and B the mean fluorescence reading for the blank.

After fluorimetry, rinse the fusion dishes in warm, running water to remove the beads. Boil the dishes for 30 min in fresh concentrated HCl and rinse in cold water. Repeat boiling and rinsing, finally rinsing with distilled water. Return each set of dishes to the appropriate storage bottle of distilled water.

Conversion Factors

$$U_3O_8 \underset{1.1792}{\overset{0.8480}{\rightleftharpoons}} U$$

b. *Colorimetric Determination of Traces of Thorium with Arsenazo III in Perchlorate Medium*

This method was developed by Abbey[69] for use in the GSC, and the following outline is taken directly from the published reference.

Reagents

Sodium hydroxide (500 g/l). Dissolve 250 g in about 325 ml of water. Allow to stand overnight. Decant and dilute to 500 ml. Store in polyethylene. For the *alkaline wash solution* dilute 5 ml of this solution to 500 ml.

Methyloxalate solution. Heat oxalic acid crystals at 100°C overnight and cool in a desiccator. Break up the crust, heat at the same temperature 1 hr longer, and again cool in a desiccator. Dissolve 100 g of the anhydrous oxalic acid in 250 ml of methanol. Allow to stand at least 3 days and filter immediately before using.

Calcium nitrate solution (10 mg CaO/ml). Weigh 9 g of calcium carbonate into a large beaker. Cover with water and dissolve by adding a slight excess of nitric acid. Evaporate to dryness. Take up with a small volume of water and again evaporate to dryness. Dissolve in about 450 ml of water and adjust the pH to the green color of bromophenol blue indicator. Dilute to 500 ml.

Oxalic acid complexing solution.   Dissolve 40 g of oxalic acid crystals in about 40 ml of hot water. Filter and dilute to 500 ml. For the oxalic acid wash solution, dilute 60 ml of this solution to 480 ml.

Perchloric acid (4:1).   Mix 400 ml of perchloric acid (70%) with 100 ml of water.

Arsenazo III solution.   Dissolve 50 mg of Arsenazo III in about 90 ml of water and dilute to 100 ml.

Standard thorium solution (5 µg Th/ml).   Prepare a concentrated solution, containing about 100 µg of thorium per ml, by dissolving thorium nitrate in water, adding sufficient nitric acid to give a final normality of 1 N. Standardize gravimetrically by evaporating aliquots to dryness and carefully igniting to oxide. Dilute an aliquot to give a working solution of 5 µg of thorium per ml, in 1 N nitric acid.

Procedure

Samples should be finely ground, preferably to −200 mesh. Up to four samples can be conveniently handled in one batch.

For *silicate samples*, weigh 0.5 g of sample into a porcelain crucible. Ignite over a Meker burner for 15 min, allow to cool, break up lumps with a small spatula, and ignite again for 10 min. Allow to cool.

Set up two blanks by weighing 5 g of sodium peroxide into each of two 50 ml iron crucibles. Mix each sample with 4 g of sodium peroxide in the porcelain crucible. Line the bottom of a 50 ml iron crucible with about 0.5 g of sodium peroxide, and add the sample–peroxide mixture. "Rinse" the porcelain crucible with about 0.5 g of sodium peroxide, and use the "rinsings" to cover the charge in the iron crucible. Cover all of the iron crucibles and heat in a muffle furnace at 460°C for 1 hr. Remove from the heat and allow to cool to room temperature.

Remove the crucible covers, wipe off the outside of each crucible with clean tissue, and place in a 250 ml beaker. Add 50 ml of water to the crucible and cover the beaker immediately with a watch glass. When the reaction subsides, rinse down the material spattered on the watch glass and the walls of the beaker. Rinse off and remove the crucible. Pipette 5 ml of standard thorium solution into one of the blanks, which then serves as a check standard. Dilute all of the suspensions to about 125 ml and boil for 15 min.

Filter on a 9cm Whatman No. 42 filter paper and wash with hot alkaline wash solution. Discard the filtrate. Place the original beaker under the filter funnel, and dissolve the residue on the paper with four 5 ml portions of hot 2 N nitric acid, followed by three similar portions of hot 2 N

hydrochloric acid, and two more of hot 2 N nitric acid. Finally, wash the paper three or four times with water.

If any silica remains on the paper, ignite it in a small platinum crucible, treat with a few drops of perchloric acid and a few milliliters of hydrofluoric acid, and evaporate three times to perchloric fumes. Take up with a little water and combine with the main solution in the 250 ml beaker. (If any insoluble residue remains at this point, it should be separated and sintered with sodium peroxide.) The solution is now ready for the oxalate precipitation.

If the sample is a *carbonate*, raise the temperature slowly in the initial ignition. After the second ignition, rinse the cooled sample into a 150 ml beaker with a little water. Add 5 ml of hydrochloric acid, cover, and warm to dissolve, adding a little nitric acid if necessary. Evaporate to dryness and bake to separate silica as usual. Take up the dry residue with hot hydrochloric acid (1:1), warm to dissolve soluble matter, and filter into a 250 ml beaker. Wash with hot hydrochloric acid (1:10). Ignite the silica and treat with perchloric and hydrofluoric acids as described above. While the silica is being treated separately, evaporate the filtrate to dryness. Add 10 ml of nitric acid, boil, and again take to dryness. Add 5 ml of nitric acid and a little water, and warm to dissolve. The solution is now ready for oxalate precipitation.

Add sufficient calcium nitrate solution to bring the CaO content of the solution to about 150 mg, 10 ml of hydrogen peroxide (30%) and sufficient water to bring the volume to 125 ml. Cover the beaker and heat on a hot plate set at low heat. Remove the cover and add sodium hydroxide solution (500 g/1) dropwise from a polypropylene pipette, with stirring, until the pH is just greater than 3.8 (this can be checked with short-range pH paper; it comes quite close to the point where hydrous ferric oxide begins to precipitate). Add 15 ml of freshly filtered methyl oxalate solution, stir, and allow to digest uncovered on the low temperature hot plate for 30 min. Remove from the heat, add a little paper pulp, and again add the concentrated sodium hydroxide solution to pH 2. Allow to stand 1 hr longer. Add 5 ml of the calcium nitrate solution, dropwise, with stirring, and allow the solution to stand for another hour. Filter on a 9 cm Whatman No. 42 filter paper, covered with a thin layer of filter paper pulp. Wash with cold oxalic acid wash solution.

Place the filter and precipitate in the original beaker and add 25 ml of nitric acid. Break up the paper, rinse down the walls of the beaker, and rinse off and remove the stirring rod. Cover the beaker with a watch glass and digest on a hot plate set at low heat until a clear solution is obtained (preferably overnight).

(If there is any reason to suspect the presence of more than 1% TiO_2

in the original sample, the solution must be evaporated to dryness and taken up with dilute nitric acid. The entire oxalate precipitation procedure is then repeated, but no calcium is added.)

Rinse off and remove the watch glass, add 5 ml of perchloric acid, cover with a ribbed watch glass, and heat on a sand bath until perchloric fumes are evolved. Allow to cool, rinse down the watch glass and the walls of the beaker, and again heat on the sand bath, this time until all perchloric fumes are expelled. Allow to cool, add 1 ml of hydrochloric acid, rinse down the watch glass and the walls of the beaker, and evaporate to dryness on the sand bath. Allow to cool, add 1 ml of hydrochloric acid and 5 ml of water. Swirl to dissolve. Add a few crystals of ascorbic acid and swirl to decolorize. Pour into a dry 25 ml volumetric flask. Rinse the beaker with 10 ml of perchloric acid (4:1) and pour the rinsings into the volumetric flask. Add 5 ml of oxalic acid complexing solution directly to the flask, and swirl to mix. Rinse the beaker with 2 ml of water and add the rinsings to the flask. Add 1 ml of Arsenazo III solution and dilute to volume. Mix well before measuring the absorbance in a 1 cm absorption cell at 660 nm.

Calibration

Into six 150 ml beakers, measure 0, 1, 2, 3, 4, 5 ml of standard thorium solution (i.e., 0–25 μg of thorium). Add 20 ml of calcium nitrate solution and 5 ml of perchloric acid to each. Cover with ribbed watch glasses and continue as in the preceding paragraph. Plot absorbance against micrograms of thorium.

Conversion Factors

$$ThO_2 \underset{1.1379}{\overset{0.8788}{\rightleftarrows}} Th$$

c.   Determination of Uranium and Thorium Using Gamma-Ray Spectrometry

Apparatus

The scintillation detector is an Ortec 905-3, 5 × 5 cm NaI(Tl) phosphor well crystal mounted integrally on an RCA-6342A 5 cm photomultiplier tube. This is connected to an Ortec 276 photomultiplier tube base and preamplifier. The detector system is housed in a lead castle having walls that are a minimum of 6 cm in thickness. The high voltage power supply and the amplifier are Ortec models 456 and 485, respectively, and the multichannel analyzer (MCA) is an Ino-Tech model 5200 equipped with a dead time corrector, a region-of-interest integrate mode, a teletype output, and a Tektronix 5110 cathode ray tube display. The NaI(Tl) crystal has a resolution of 8.1% at the 0.661 MeV ^{137}Cs gamma-ray energy level.

The sample containers are plastic centrifuge vials (13 mm i.d. × 65 mm long) equipped with plastic stoppers. A balance incorporating a taring mode and capable of weighing to the nearest milligram is sufficient for most applications.

Screening Procedure

A fast method for determining the equivalent uranium in a sample involves the use of a system to measure gamma radiation. The samples and standards, in pulverized form, are packed into containers to a predetermined level, and the weight of material added is noted. The container is placed in the well of the NaI (Tl) crystal, and the gamma-ray radiation is counted for a predetermined time. The counting system is set to count over a gamma energy range from 40 to 800 keV (Band 3, Fig. 6.1a), and it is recommended that once this range is fixed, the hardware not be further adjusted. The measurable concentration range is 10 ppm to 1%, and approximately 5 g of sample is required.

At the start of the screening procedure, an empty container is counted for 1000 sec, and the background count rate is obtained from this. A standard equilibrium uranium ore of known concentration and mass is counted for 1000 sec. The count rate is calculated, the background count rate removed, and the true count rate per unit mass is calculated. This yields the relationship between number of counts per second per gram versus parts per million equivalent uranium.

The sample is then counted for 100 sec. The count rate is calculated, the background count rate removed, and the true count rate per gram calculated. Using the above relationship, the amount of equivalent uranium in the unknown can be calculated.

Assay Procedure

Again, in this procedure the gamma radiation emitted by the uranium and thorium series is counted for all standards and samples, of known weight, placed in identical containers at a constant level of material. The containers are placed in the well of a NaI (Tl) crystal, and the gamma spectra are displayed and counted on a multichannel analyzer.

The equipment is first calibrated by using standard equilibrium uranium sources. ^{60}Co is used to obtain the linear relationship of channel number versus energy levels. Modern electronic hardware is highly stable, and recalibration should be an infrequent operation.

The following two major energy bands are used:

1.   40 – 120 keV
2.   215 – 270 keV

Band 1 (Fig. 6.1a) contains the ^{234}Th peak of the uranium series and the ^{228}Th peak of the thorium series. Band 2 encompasses the ^{214}Pb peak of the uranium series and the ^{224}Ra and ^{212}Pb peaks of the thorium series. It should be noted that, for "young" uranium which has not built up any post- ^{234}Th peaks, the peaks in band 1 will be well separated, while the peaks in the post-radon region will be diminished. However, in most ores containing "old" uranium (>10^6 years), heavy elements are present causing the two peaks in band 1 to degenerate into a single peak due to natural X-ray fluorescence. Typical U and Th gamma-ray spectra are shown in Fig. 6.1

In the assay procedure, an empty container is first counted for 1000 sec, or as long as is necessary to satisfy statistical requirements, and the background count rate is calculated for each band. Each standard, of known mass and constant level of material in the container, is then counted for 1000 sec, or as long as is necessary to satisfy statistical requirements. The count rate is calculated, the background count rate subtracted, and the true count rate per gram determined for the known uranium and thorium concentrations. This operation is repeated for each standard in order to cover the whole range of possible sample concentrations. A table is constructed to relate the count rate per gram in each band to concentration. Band ratios are also calculated.

The sample, of known weight and at the same constant level in its container, is related in a similar manner to the reference samples. The true count rate per gram is calculated as before in each band.

The ratio of band 1 to band 2 is calculated. If this ratio lies between those ratios calculated for equilibrium uranium, the sample is assumed to be in equilibrium. The ratio is determined by subtracting from the net counts in bands 1 and 2 those counts attributable to Th and ratioing the resulting figures. If the ratio is 4.2 or less the sample is assumed to be in secular equilibrium with respect to uranium.

The concentration of uranium and thorium is calculated as follows:

C_{11} be count rate per gram in band 1 for standard 1
C_{12} be count rate per gram in band 1 for standard 2
C_{21} be count rate per gram in band 2 for standard 1
C_{22} be count rate per gram in band 2 for standard 2
r_1 be count rate per gram in band 1 for sample
r_2 be count rate per gram in band 2 for sample

The coefficients for the equations required to calculate the concentration of uranium and thorium in the sample are given by:

$$x_1 = \frac{(r_1 \cdot C_{22} - r_2 \cdot C_{12})}{(C_{11} \cdot C_{22} - C_{21} \cdot C_{12})}$$

and

$$x_2 = \frac{(r_2 \cdot C_{11} - r_1 \cdot C_{21})}{(C_{11} \cdot C_{22} - C_{21} \cdot C_{12})}$$

If U_1, T_1 is the amount of uranium and thorium in standard 1 and U_2, T_2 the amount of uranium and thorium in standard 2, then the amount of uranium in sample is

$$U = x_1 \cdot T_1 + x_2 \cdot U_2$$

and amount of thorium in sample is

$$Th = x_1 \cdot T_1 + x_2 \cdot T_2$$

Note: x_1 and x_2 may be negative.

6.3 FLUORINE

6.3.1 General

Fluorine seldom occurs in rocks in an amount exceeding a few tenths of 1%, and is generally much less; it is more abundant in the acid rocks. In such minerals as fluorite and cryolite it is a major constituent, and it is an essential component of fluorapatite and tourmaline and an important minor constituent of micas and amphiboles. Its distribution in rocks and minerals has been summarized by Bailey[73] and in the *Handbook of Geochemistry* (Ref. 47, Chapter 9).

The determination of fluorine is of importance not only to the petrographer but also to the rock analyst. It was once a lengthy and difficult procedure, but there is no reason now for not determining it if mineralogical or petrographical evidence suggests that it is necessary, and failure to do so can lead to serious error in the results of an analysis.

Qualitative tests for the presence of fluorine are not very satisfactory. The etching or hanging-drop test, which is specific for fluoride ion, requires some experience to be able to estimate the amount involved; the sample is heated with concentrated sulfuric acid, and the liberated hydrogen fluoride will etch a wetted glass slide exposed to the vapors, or will dissolve in a drop of water, and in the presence of silica (e.g., glass) will form silicon(IV) fluoride and hydrolyze to gelatinous silica, causing

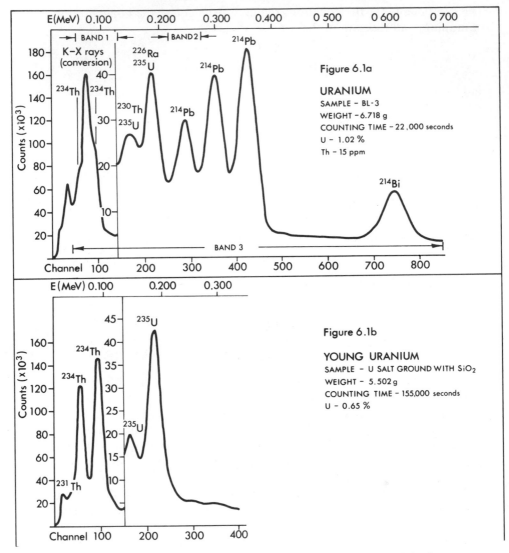

Figure 6.1. Gamma-ray spectra of selected samples of uranium and thorium.

the drop to become cloudy. If the sample is not decomposable by sulfuric acid, it must be fused with Na_2CO_3, leached in a small amount of water, and the excess SiO_2 separated with ammonium carbonate and/or zinc oxide; a mixture of calcium carbonate and fluoride is then precipitated and used for the test for fluoride. There are also several tests which

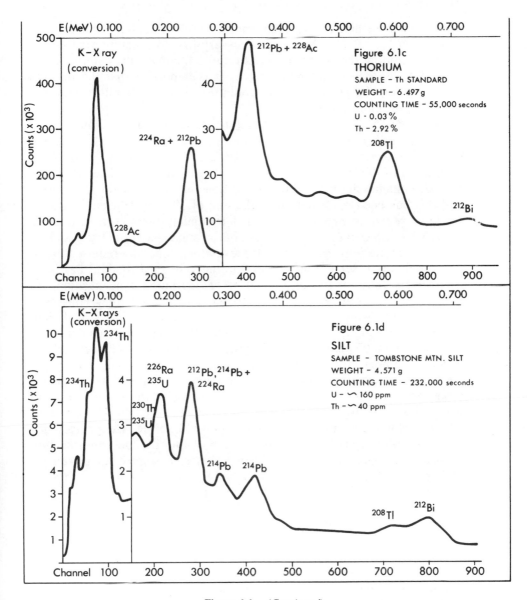

Figure 6.1. (*Continued*)

involve the bleaching of a colored complex of fluoride, but because many anions and cations interfere, it is necessary first to separate the fluoride by volatilization; the additional work required to make the determination a quantitative one is not much more. It is important that the fluorine determinations be protected from contamination by laboratory reagents such as hydrofluoric acid.

Until 1966,[74] when an ion-selective electrode became available for the measurement of fluoride content, almost all methods for the determination of fluorine were subject to many interferences and, of necessity, the separation of the fluorine was the first step in the determination. There has been much written about the separation and determination of fluorine and this has been reviewed up to the end of 1959 by Horton.[75] Among the more recent techniques used for determining trace levels of fluorine are NAA[76], ES (Supp. Ref. 2), spark source mass spectrometry,[77] and, to a limited extent, ion-selective electrodes, which generally require little, if any, preliminary separation of the fluorine. This last method has been used successfully for the determination of fluorine in concentrations up to several percent, for example, in phosphate rock,[78] but unlike the other three methods, which are capable of detecting fluorine at parts per billion levels, the sensitivity of the electrode limits this technique to a detection limit of about 0.02 ppm in solution, which translates into a detection limit of about 40 ppm in the original rock sample.[79, 80] Selig has described a method for determining fluorine which employs Gran's plots for identifying the end point of a potentiometric titration of the sample solution using a fluoride ion-selective electrode.[81] Jagner and Pavlova[82] have used a Gran's plot in a standard addition titration of fluoride. The interference of Al and other ions in the determination of fluoride ion concentration using ion-selective electrodes has been discussed in some detail by several authors,[79, 81, 83] many of whom have used the citrate ion to buffer the Al present and pH control to avoid hydroxyl ion interference.

A method for the determination of fluorine in silicates by use of an ion-selective electrode following fusion with lithium metaborate has been described by Bodkin.[84] A detailed discussion of this method will be found in Sec. 6.3.3b.

Kirsten has developed a method for the determination of fluorine at the ultramicro level.[85] Following the decomposition of the sample with WO_3 and H_3PO_4, the liberated fluorine is hydrogenated, absorbed in water, and measured spectrophotometrically. The method can be used to determine as little as 0.2 μg of fluorine and has been applied to the analysis of silica gel, trisodium phosphate, slaked lime, and other inorganic compounds. A method for the automated colorimetric determination of fluorine and chlorine in geological samples, using a Technicon

AutoAnalyser, has been described by Fuge.[86] In this method the sample is fused with Na_2CO_3, the interfering elements are retained in the residue from the water leach of the fusion cake, and the fluoride ions are measured in the filtrate after it has been acidified. The measurement is based on the bleaching effect of the fluoride ion on the zirconium–xylenol orange complex.

An indirect AAS method for the determination of fluorine has been described[87] which used the inhibiting effect of fluorine on the atomic absorption signal of magnesium. The pH characteristics of the effect permit it to be distinguished from the same effect due to other anions such as sulfate, phosphate, and silicate. Another indirect method is that of Auffarth and Klockow[88] for the routine determination of nanogram amounts of fluoride in geochemical materials. The fluoride is separated by a microdiffusion method based on hexamethyldisiloxane and then determined kinetically by its inhibiting effect on the zirconium-catalyzed reaction between perborate and iodide. Only milligram quantities of sample are used, thus making it useful for the study of inclusions.

Many manufacturers of both wavelength dispersive and energy dispersive XRF spectrometers claim that their instruments are capable of determining fluorine. It is necessary to irradiate the sample under vacuum conditions and the windows of the flow proportional counter must be very thin. Even with these precautions, however, the detection limits attainable are not low enough to be able to classify the method as being suitable for trace analysis.

6.3.2   Separation of Fluorine

In some instances it is still advantageous to separate fluorine prior to its determination, and three common methods for this will be discussed.

a.   By Steam Distillation

Fluorine is most probably[89] evolved as a mixture of SiF_4 + HF when the sample, in one form or another, is heated with one or more acids while steam is passed through the mixture, and is collected in the condensate.

1.   Some samples may be used without prior treatment, but for most silicates it is preferable to fuse the sample with an alkaline flux, remove the bulk of the silica (and the alumina) from the resultant solution, and steam distill the filtrate directly or after concentrating it to a desirable volume. The gelatinous silica that separates on acidification, if the bulk of it is not first removed, will not only retard the recovery of fluorine,

but will cause undesirable "bumping" of the contents of the distillation flask; removal is usually achieved by precipitating the silica with Zn(II), either by adding it after an alkaline carbonate fusion or, better, by fusing the sample with a mixed alkaline carbonate–zinc oxide flux (see Sec. 6.3.3b). However, Blake[90] has reported no difficulty with SiO_2 when a sodium carbonate fusion alone was used; Moyzhess[91] found no interference if the SiO_2 content of the sample was less than 40 mg.

2. There is much disagreement over the choice of acid for the decomposition, as there is also over other procedural details.[92] Sulfuric and perchloric acids are most frequently used, with perchloric being the least troublesome. A higher distillation temperature can be used with sulfuric acid (and with phosphoric acid also), however, thus decreasing the distillation time required for complete recovery of the fluorine, but the presence of these acids in the distillate is undesirable (indeed, the use of sulfuric acid has been definitely not recommended[75]). Because the higher distillation temperature is necessary to ensure the complete recovery of fluorine from some samples, such as aluminiferous ones,[89, 90, 93] some analysts prefer to eliminate the harmful volatiles from the distillate by a second distillation at a lower temperature.[94] Fox and Jackson[89] have described a simultaneous double distillation in which the uncondensed gas from the first still, at 160°C, is passed through perchloric acid, at 125°C, in a second distilling flask. Other precautions include the use of stills with special splash traps to remove entrained acid,[93, 95] and passing the distillate through an anion-exchange column from which the fluoride is later eluted preferentially.[96] Phosphoric acid is sometimes used alone, but more often as a mixture with perchloric or sulfuric acids, usually for the purpose of eliminating the adverse effect of aluminum in solution on the recovery of fluorine. It is thought that a strong complex is formed between the aluminum and fluoride which reduces the partial pressure of fluorine in the gas phase,[89] but opinions differ on the efficacy of phosphoric acid as a releasing agent. Blake[90] investigated the use of both H_2SO_4 and $HClO_4$, alone and mixed with H_3PO_4, for samples rich in aluminum and obtained the best separation of fluorine with an H_2SO_4–H_3PO_4 mixture; Moyzhess[91] tested the evolution of fluorine from a similar acid mixture in the presence of a variety of elements which form stable fluoride complexes, including zirconium and thorium, and found no interference. Others[93-95] recommend the use of H_3PO_4 in the distilling flask, usually as a complexing agent for aluminum, but Fox and Jackson[88, 89] are opposed to its use.

3. The optimum temperature for distillation[75, 94, 97] ranges from 130 to 150°C, although there is considerable disagreement on the exact temperature to be used. A higher temperature than 150°C is recommended

for aluminiferous samples,[90] up to 160–170°C if a second distillation is to be made.[94] Ingamells[93] determines the optimum temperature for his special still by experiment, which is that temperature at which no harmful amount of distillation acid is also distilled. Moyzhess[91] distilled solutions containing H_2SO_4 and H_3PO_4 at 140–190°C and found only an insignificant amount of these anions in the distillates collected at 140–150°C, and so chose 140 ± 5°C as his optimum temperature.

4.   While most of the fluorine is usually collected in the first 50 ml or so of the distillate, the presence of such elements as aluminum, boron, zirconium, or thorium, to mention only a few, dictates that a larger volume must be collected in order to be certain of complete recovery. Thus it is usual to collect 150–250 ml; Ingamells[93] recommends that a second 100 ml of distillate be collected if the fluorine content exceeds 0.5–1%. Peck and Smith[97] collect about 200 ml, but recommend that a 3% (relative) correction be added to the results to allow for retention of some fluorine by interfering compounds and adsorption on glass surfaces.

5.   Suffice it to say that the apparatus recommended by various analysts ranges from the relatively simple to the very complex and the references cited should be consulted for further details. The apparatus should be thoroughly cleaned between distillations to remove traces of fluoride adsorbed on glass surfaces; hot 10% Na_2CO_3 solution is an effective cleaning agent.

b.   By Pyrohydrolysis

The pyrohydrolytic or pyrolytic method for the evolution of fluorine (and chlorine) is a beautifully simple procedure. The sample, mixed with an accelerator (catalyst) and flux to break down the structure and facilitate the release of fluorine, is placed in a combustion boat and heated to an elevated temperature in a reactor tube through which passes a fairly rapid stream of either superheated steam or oxygen saturated with water vapor. The gaseous products are swept through either a dilute sodium hydroxide solution or water, the hydrated fluorine compounds are condensed and absorbed, and the determination is finished titrimetrically or colorimetrically. The boat and reactor tube may be of platinum, nickel, quartz, or fused silica, and the operating temperature ranges from 800 to 1000°C. Uranium(IV, VI) oxide, WO_3, and V_2O_5 serve as accelerators, and sodium tungstate is often used as a flux. Horton has discussed the various aspects of the procedure in his review.[75]

Bennett and Hawley[98] have described their procedure for aluminosilicates in some detail, including the apparatus used. They found that V_2O_5 was the best catalyst and that the minimum temperature for complete recovery of the fluorine was 850°C. Cluley[99] used U_3O_8 as accelerator and heated samples of opal glass to 1000–1050°C to ensure complete release of the fluorine; his apparatus is very similar to that used by Bennett and Hawley.

While no interference is found from silica and phosphate, which remain in the combustion boat, other volatile anions present, such as SO_4^{2-}, NO_3^-, and Cl^-, will be carried over and condensed with the fluorine; the method used for the final determination of the fluorine must take into consideration their possible presence, but may be as simple as an acid–base titration or an ion-selective electrode measurement.

c. By Ion Exchange

Mention has been made of the use of ion exchange by Sergeant[96] to eliminate anionic interferences in the distillate prior to the colorimetric determination of fluorine. The resin, Deacidite FF (chloride form, 100–200 mesh), is placed in a glass column positioned to receive the distillate as it leaves the condenser of the distillation apparatus. The absorbed fluoride is eluted with N/10 sodium acetate, and almost 100% recovery is found for up to 600 μg F^-. Glasö[100] preferred ion exchange to distillation as a means for the separation of fluoride, but the procedure as described is applicable only to those samples in which the fluoride is present in acid-soluble form, such as iron ore, apatite, and phosphate rock. The sample, dissolved in warm hydrochloric acid, is passed over Dowex 2-X 10 resin (chloride form, 200–400 mesh) and the fluoride eluted with 10 M HCl.

6.3.3   Determination of Fluorine

This procedure, which involves the fusion of the sample with sodium carbonate, the leaching of the cake, and separation of the bulk of the silica with acid zinc perchlorate, the steam distillation of the fluoride-bearing filtrate from a perchloric acid solution, and titration of aliquots of the distillate with thorium nitrate, using sodium alizarin sulfonate as indicator, was adapted by R. B. Ellestad for use in the Rock Analysis Laboratory at the University of Minnesota, from where it was introduced into the laboratories of the GSC.

Apparatus

The two-unit distillation apparatus used here is based upon that described by Shell and Craig (Ref. 101, p. 5) and is available commercially.

A three-necked 2 liter round-bottomed boiling flask serves as the steam generator for two sets of distillation apparatus; the third neck is fitted with an adapter to which is joined a piece of rubber tubing closed by a screw clamp, and this is used to regulate the supply of steam to the distillation flasks. The latter are round-bottomed flasks of 250 ml capacity having a side arm; the center neck is fitted with an adapter which holds the steam delivery tube and the thermometer. The neck of the side arm is connected by adapters to a Liebig condenser; an 8 oz polyethylene bottle is used to catch the distillate. An "anti-bump" rod is placed in each distillation flask and serves also to reduce the attack on the steam delivery tube by the acid mixture. Heating of the steam generating and distillation flasks is done by electric heating mantles or by Tirrill burners; the latter make possible a closer control of the temperature, provided that the flame is protected by a flame guard.

Reagents

Standard thorium nitrate solution (0.02 N). Dissolve 2.76 g of reagent grade $Th(NO_3)_4 \cdot 4H_2O$ in water, add 0.4 ml concentrated HNO_3, and dilute to 1 liter.

Standard fluoride solution (0.000339 g F/ml). Dry about 0.2 g of reagent grade NaF by heating it in a small platinum crucible at a low red heat; dissolve 0.1500 g of the dried salt in water and dilute to 200 ml in a volumetric flask.

Sodium alizarin sulfonate (0.05%). Dissolve 0.05 g of the indicator in 100 ml of water.

Sodium monochloroacetate–monochloroacetic acid buffer solution (2 M). Dissolve 18.9 g of reagent grade monochloroacetic acid in water and dilute to 50 ml; neutralize 25 ml of this solution by adding dropwise concentrated sodium hydroxide solution, using phenolphthalein as indicator. Add the remaining 25 ml of monochloroacetic acid and dilute to 100 ml. This solution should not be kept longer than 3–4 weeks; beyond this it will cause deterioration of the quality of the end point of the titration.

Acid zinc perchlorate solution. Dissolve 10 g of zinc oxide in 112 ml of 70% perchloric acid and add 50 ml of water.

Standardization of Thorium Nitrate Solution

Pipette several pairs of various volumes (e.g., 0.50, 1.00, and 2.00 ml) of the standard fluoride solution into 50 ml beakers. Add sufficient water to bring the total volume of each to 15 ml, and add 15 ml of 95% ethyl alcohol. Add 3 drops of the sodium alizarin sulfonate indicator and 0.30

ml of the buffer solution. Titrate each with the standard thorium nitrate solution, using a 5 or 10 ml burette, to a faint pink end point matching a reference solution prepared as follows. Dilute 2.0 ml of the 0.02 N standard thorium nitrate solution to 100 ml and add 0.50 ml of this 0.0004 N solution (equivalent to 0.01 ml of the 0.02 N solution) to 15.0 ml of water, 15.0 ml of 95% ethyl alcohol, 3 drops of indicator, and 0.3 ml of buffer solution in a 50 ml beaker. The color of this reference solution is equivalent to a 0.01 ml excess of the 0.02 N thorium nitrate solution, and this reagent blank should be deducted from each titration whose end point is matched to this color. Calculate the F equivalent of 1 ml of the standard thorium nitrate.

Procedure for Silicates

Mix 0.5 g of sample (finely ground) with 3 g of anhydrous sodium carbonate in a 25 ml platinum crucible, and fuse as quickly as possible, with only a few minutes over the Meker burner. Transfer the cake to a 250 ml beaker and leach it in 40 ml of water; *do not add alcohol to reduce manganate ion.* Filter through a 9 cm Whatman No. 40 filter paper into a 250 ml beaker and wash paper and residue with 2% sodium carbonate solution. Rinse the residue back into the original beaker, boil briefly with 20–30 ml of sodium carbonate solution, filter through the same paper, and continue washing the paper and residue until the total volume is about 100 ml. Discard the residue.

To this cold solution add, with stirring, 8 ml of the acid zinc perchlorate solution. Heat to boiling and boil for 1 min. Filter through an 11 cm Whatman No. 40 paper and wash well with hot water. Rinse the precipitate back into the original beaker, stir well with a small quantity of hot water, and again filter. Discard the precipitate. Concentrate the combined filtrates to 10–15 ml by heating the uncovered beaker on a low temperature hot plate (do *not* boil); care is needed during this evaporation to avoid loss of sample by "bumping."

Transfer the solution to the distillation flask by means of a long-stemmed funnel, using a minimum amount of water for rinsing. Place a "boiling tube" in the flask, and connect the flask into the apparatus, but use a separatory funnel temporarily in place of the steam inlet tube. Apply heat to the steam-generating flask, close the steam inlet tubes with screw clamps, but leave the vent open to the atmosphere. Start cooling water circulating through the condenser and place an 8 oz polyethylene bottle at the condenser outlet.

Close the stopcock of the separatory funnel and add to it 25 ml of 70% perchloric acid. Allow the acid to flow steadily into the flask, remove the funnel, and immediately insert the steam inlet tube.

Heat the distillation flask with the full flame of a Bunsen burner. When the temperature in the flask reaches 140°C, open the clamp on the steam inlet tube and close the vent of the steam generating flask, the contents of which should be boiling vigorously (boiling chips should be placed in the flask to maintain steady boiling).

Maintain the temperature in the distillation flask at 135–140°C by regulation of the rate of steam generation and the height of the flame under the distillation flask, and continue the distillation until about 150 ml of distillate have collected; this includes the 15–20 ml of excess water that is distilled before the maximum distillation temperature is reached. Keep the distillate alkaline to phenolphthalein (2 drops) during the distillation by adding 0.1 N sodium hydroxide solution as needed (about 2–3 ml).

When the distillation is complete, remove the flame from the distillation flask and allow the temperature to fall below 130°C. Carefully close the steam inlet tube so that no "suck-back" occurs in the steam generation flask. Disconnect the condenser, hold it vertical, and rinse the center tube with a small amount of water, combining the rinsings with the distillate.

Transfer the alkaline distillate to a 250 ml platinum dish and evaporate the solution on a water bath to about 15 ml. Carefully transfer the solution to a 25 ml volumetric flask, cool to room temperature, and dilute to volume. Pipette a 10 ml aliquot into a 50 ml beaker and add 0.2 N HCl dropwise until the pink color disappears. Add 3 drops of the sodium alizarin sulfonate indicator (which will show its reddish violet alkaline color) and continue to add the drops of 0.2 N HCl until the indicator turns yellow. Add 0.3 ml of the buffer solution, dilute to 15 ml with water, and add 15 ml of 95% ethyl alcohol. Titrate with the standard thorium nitrate solution to a shade of pink that matches the color of the reference solution, and deduct 0.01 ml as titration blank. Repeat the titration on a second aliquot as a check; if the first titration was unusually large, it is preferable to use a smaller aliquot for the second titration and, because the result of the first titration is known, it is possible to add the proper amount of water to give a final volume of 30 ml, thus increasing the accuracy of the titration. Calculate the percent fluorine in the sample.

Clean the glass apparatus thoroughly with hot 10% Na_2CO_3 between distillations.

b. *With Ion-Selective Electrode*

This method is that reported by Bodkin,[84] and it has been used extensively and successfully in the BCM laboratories. It involves fusing the sample with $LiBO_2$, dissolving the melt in dilute HNO_3, adding a buffer, and determining the fluoride ion concentration with an ion-selective electrode, using the method of standard addition.

Apparatus

An ion-selective electrode (Orion 94-09), a standard reference electrode (Orion 90-02), and a suitable voltmeter are used for the measurement of the fluoride ion concentration. Pre-ignited graphite crucibles (9 ml capacity, Spex No. 7152), a muffle furnace, polypropylene beakers and lids, Teflon-coated stirring bars, a magnetic stirrer, and a 50 ml burette are also required.

Reagents

Anhydrous LiBO$_2$

4% v/v HNO$_3$

Standard fluoride solution (100 μg/ml).   Dissolve 0.2210 g NaF, ignited at 640°C for 1–2 hr, in water, dilute to 1 liter, and transfer to a polyethylene bottle. Prepare a standard solution containing 10 μg/ml as required.

Complexing and buffer solution.   Mix 18.2 g of DCTA (1,2-diamino-cyclohexane-N,N,N^1,N^1-tetraacetic acid) with 1.5 liters of water and add sufficient 40% m/v NaOH solution, dropwise, to dissolve the DCTA. Add 300 g sodium citrate dihydrate and 60 g NaCl, adjust to pH 6.85 with HCl, and dilute to 2 liters.

Procedure

Mix 200 mg of −200 mesh sample with 800 mg LiBO$_2$, transfer to a graphite crucible, fuse at 1050°C for 10–15 min, and pour the molten material into a 150 ml polypropylene beaker containing 80 ml of 4% v/v HNO$_3$ and a stirring bar. Place the covered beaker on a magnetic stirrer and stir until the melt has dissolved (5–10 min), transfer the solution to a 100 ml volumetric flask, and dilute to volume. Transfer a 40 ml aliquot into a 150 ml beaker, add 10 ml water, 50 ml of the buffer solution, and a magnetic stirrer bar.

Insert the electrodes into the beaker, stir for 5 min to establish equilibrium, and record the potential reading. Add 10 ml of standard fluoride solution to the beaker, stir for 5 min, again record the potential reading. Calculate the fluoride concentration using the method of standard addition.

The method is applicable to a concentration range of 40 to about 4000 ppm. Higher concentrations can be measured by using a smaller aliquot of the sample solution. An experienced operator can perform more than 80 determinations per day.

6.4 CHLORINE

6.4.1 General

The chlorine content of most rocks is <0.1% and rarely exceeds 0.2%; the higher concentrations will be found in highly sodic nepheline-bearing rocks and in those containing much sodalite [$Na_8[Al_6Si_6O_{24}]Cl_2$, $Cl \simeq 7\%$] and members of the scapolite group [(Na, K, Ca)$_4$[Al$_3$(Al, Si)$_3$ Si$_6$O$_{24}$] (Cl, F, CO$_3$, SO$_4$), Cl $\simeq$ 2%], minerals in which it is an essential constituent. It is present in many other minerals as well, such as chlorapatite, and as a constituent in fluid inclusions, such as those occurring in quartz. It may be present also as sodium chloride, from contamination of the sample by seawater or brine, but in this form it is easily removed by leaching the sample with water. Because of the generally low chlorine content of most rocks it is important that the sample, at all stages of the preparation and analysis, be protected from contamination by chlorides present in the laboratory environment, including the atmosphere. The ubiquity of chloride in chemical laboratories makes it essential that a blank determination be run at the same time, and under the same conditions, as the sample. Bromine and iodine, if present, seldom exceed 0.01% and, as a rule, seldom cause significant errors in a chlorine determination.

Chlorine may be extracted from silicates by various methods, the choice depending upon the form in which it is present in the sample. As a water-soluble chloride, such as NaCl, it is readily extracted by boiling the sample with water and filtering the solution through a fine-textured single or double paper; the residue is used for the determination of the chlorine present in a water-insoluble form. Brief boiling of the sample with dilute nitric acid will extract the chlorine from acid-soluble minerals such as sodalite and chlorapatite; the addition of a few drops of hydrofluoric acid may be necessary to obtain all the acid-soluble chlorine. There is no danger of losing chlorine by volatilization if the boiling is not unduly prolonged [Groves (Ref. 7, pp. 145–147, 277–278) recommends not more than 2 min], nor will gelatinous silica be formed, at least not in sufficient amount to be troublesome. The chlorine present in the scapolite group of minerals is, however, not so easily extracted, and it is necessary to fuse the sample with sodium or potassium carbonate and extract the chlorine by aqueous leaching of the cake. The fusion approach is the most reliable one, unless it is desirable to know the extent to which chlorine is present in an acid-soluble form. Robinson[102] heated the sample, wetted with H$_2$SO$_4$ and HNO$_3$, at 310°C in a combustion tube; the evolved hydrogen chloride is swept out by nitrogen and absorbed either in a dilute solution of sodium carbonate or in deionized water. He applied the method to chloride contents ranging from several percent down to a few parts per million; the majority of the interfering elements are retained in the combustion boat as insoluble sulfates. A similar separation of chlorine can be achieved by steam distillation; Geijer[103] fused amphiboles with NaOH and then distilled the neutralized solution of the cake with H_2SO_4; the condensate is evaporated to small volume and again distilled, this time in a smaller apparatus. Pyrohydrolysis has been used by Caldwell[104] and Gillberg[105]; Greenland and Lovering[106] distilled chlorine from chondritic meteorites at 1700°C under an argon atmosphere, and finally recovered it by diffusion in a special polythene microdiffusion cell. Farzaneh and Troll (Supp. Ref. 3) passed steam over geological samples heated in an induction furnace. They measured the chlorine concentration colorimetrically at 450 nm as a mercury thiocyanate complex.

The gravimetric determination of chlorine as silver chloride is the most widely used procedure, although it is not too reliable for chlorine concentrations of < 0.05%. Only bromine and iodine interfere significantly in the determination, and in most rocks this interference is negligible. Precipitation is done by the addition of a small excess of 5% $AgNO_3$ solution to the chloride solution, acidified with HNO_3 and containing about 0.1 ml in excess per 100 ml. The precipitate is coagulated by heating it near the boiling point for 1 hr and then is allowed to stand overnight. Throughout these steps the precipitated AgCl must be shielded from light, or decomposition to metallic silver will occur (the precipitate develops a purple coloration). The precipitate is collected on a fritted glass crucible or filter paper and the AgCl is dissolved in warm, dilute aqueous ammonia to separate it from any coprecipitated siliceous matter; the filtrate is acidified with nitric acid and the precipitation procedure repeated. The precipitate is collected on a fritted glass crucible, dried at 130–150°C, and weighed. The BCM has used a modification of this in which the sample is fused with K_2CO_3, leached with hot water, filtered, the filtrate is acidified with HNO_3, and $AgNO_3$ is added. The Ag concentration is measured by AAS after the AgCl has been collected by filtration and dissolved in dilute aqueous ammonia solution. This same technique has been reported by Reichel and Acs[107] and by Westerlund-Helmerson.[108]

Groves[7] recommends a turbidimetric finish for samples which yield a precipitate that is obviously not weighable (e.g., <0.05%). The turbid solution is *not* heated but is compared, as quickly as possible, with a solution of $AgNO_3$ to which is added a standard solution of NaCl.

Peck and Tomasi[109] have described a procedure for the routine and rapid determination of chlorine up to 0.2% which utilizes a turbidimetric titration also, but with mercuric nitrate as titrant and sodium nitroprusside as the indicator. The sample is sintered with a mixed flux (Na_2CO_3–ZnO–$MgCO_3$, 7:2:1), leached with water, and acidified with nitric acid. Sodium nitroprusside is added, and the solution is titrated in the dark with 0.01 N $Hg(NO_3)_2$, in an apparatus which permits comparison of twin beams of light passed through the sample solution and through a standard turbid solution (0.4 μg/ml AgCl dispersed in a gelatin solution). When a chloride solution containing sodium nitroprusside is titrated with $Hg(NO_3)_2$, all chloride must be complexed by the mercury before a near colloidal precipitate of mercuric nitroprusside will form. This near colloidal precipitate scatters the light, and the end point is marked by brightening of the sample light beam relative to the reference beam in the standard turbid solution. The results obtained are lower than those obtained by the gravimetric procedure for chlorine contents >0.2%, if an 80 mesh powder is used for the fusion; for the occasional sample having Cl > 0.2% a more finely ground sample should be used.

Other titrimetric procedures (Ref. 1, pp. 725–736, 936–937),[110] which can be used include Volhard's method, in which the excess silver from the precipitation is titrated with ammonium or potassium thiocyanate in the presence of ferric ion as indicator, and Mohr's method in which the chloride solution, containing K_2CrO_4, is titrated with $AgNO_3$, the end point being indicated by the first appearance of red silver chromate. Geijer[103] used dichlorofluorescein as indicator for a $AgNO_3$ titration; in the Greenland and Lovering method[106] the diffusing Cl^- is absorbed in KI solution, and the final titration is made with sodium thiosulfate. Robinson[102] completed his combustion method by titrating the absorbed hydrogen chloride potentiometrically with $AgNO_3$ solution; the titration may also be done amperometrically.[111]

There are numerous procedures available for the colorimetric determination of chlorine,[110] with Fuge's automated method being of special interest.[86] Some use has been made of the red complex formed when mercury(II) thiocyanate is added to a chloride solution in the presence of Fe(III), but the most direct procedure is that of Kuroda and Sandell[112] for the determination of chlorine contents < 0.05% on as little as 0.1 g of sample, with a precision of ±0.005%. The sample is fused with Na_2CO_3, leached with water, the filtrate made acid with HNO_3, and the chlorine precipitated as AgCl; the filtered precipitate is dissolved in dilute aqueous ammonia, and sodium sulfide is added to form a yellow-brown, stable silver sulfide sol. No protective colloid is needed if the silver concentration <0.1 mg/ml. The transmittance is measured at about 415 nm, and the relation between silver concentration and extinction is linear up to 100 ppm. Caldwell[104] added $Hg(NO_3)_2$ to the condensate from pyrohydrolysis (weakly ionizing $HgCl_2$ is formed) and determined the excess mercury present colorimetrically with diphenylcarbazone in order to obtain the chlorine by difference.

A method has been developed[113] for the determination of chlorine in silicate rocks involving separation of the chlorine by ion-exchange chromatography and direct potentiometric measurement with an ion-selective electrode. The sample is fused, leached, neutralized with HNO_3, and an aliquot of the filtrate transferred to the chromatographic column.

It is possible to determine chlorine by flame photometry[114] either directly as $CuCl_2$ or $InCl$,[115] or indirectly be measuring the decrease in luminosity of a standard silver solution due to precipitation of the silver as silver chloride. $CaCl$ bands are suitable for the detection of chlorine in rocks and minerals by emission spectrography.[116] Karyakin et al. have reported the use of a pulsed vacuum excitation source for spectrographic determination of the Cl in geological samples from measurement of the 479.45 nm ionic line.[117]

Unni and Schilling (Supp. Ref. 4) have reported a radiochemical NAA determination of chlorine in silicate rocks which gives a relative precision of $\pm 5\%$ (2σ) at the 100 ppm level. The determination of chlorine by XRF is a common practice (see Sec. 11.2.11).

As in the case of sulfur (Sec. 5.6.1), it is necessary to apply a correction to the analysis summation for the oxygen equivalent of the chlorine present as chloride, if the constituents are reported as oxides.

$$2KCl \equiv K_2O$$

$$\frac{0}{2\,Cl} = 0.22$$

6.4.2 Determination of Chlorine

a. *Gravimetric*

ACID-SOLUBLE CHLORINE

Reagents

Silver nitrate solution (0.2 *N*).   Dissolve 3.4 g $AgNO_3$ in 100 ml of water containing a few drops of nitric acid and store in an amber glass-stoppered bottle.

Procedure

To a 1.0 g sample (80 mesh), in a covered 250 ml beaker, slowly add 50 ml of dilute nitric acid (1:19). Heat to boiling and boil for 2–5 min but no longer. Allow to settle and then decant the supernatant liquid through a 9 cm No. 42 Whatman filter paper, prewashed with dilute HNO_3 (1:99), into a 150 ml beaker; if the supernatant liquid appears turbid, it may be necessary to use double filter paper to ensure a clear filtrate. Wash the residue several times with the dilute HNO_3 (1:99), decanting the wash water through the paper, and finally wash the paper five times with the wash solution; the final volume should be about 75 ml. The residue may be transferred quantitatively to the paper (after removing the beaker containing the filtrate) and used for the determination of acid-insoluble chlorine as described under Determination of Total Chlorine.

Add 3 ml of 0.2 N $AgNO_3$ solution and heat to boiling, keeping the beaker out of direct sunlight as much as possible. For most samples the solution will become opalescent at this stage, without visible coagulation of the AgCl. Surround the beaker with a shield of dark paper and allow to stand overnight in a cupboard (away from containers of HCl or NH_4Cl).

Decant the supernatant liquid through a 7 cm Whatman No. 42 filter paper or a fine porosity fritted-glass filter crucible, both washed free of chloride with dilute HNO_3 (1:99), and wash the precipitate, if possible, two or three times by decantation. Wash down the sides of the beaker with 2–3 ml of dilute aqueous ammonia (1:1) and pour the solution through the filter to dissolve any silver chloride present, catching the filtrate in a 150 ml beaker. Wash the original beaker and the paper once with the dilute aqueous ammonia, and then with about 10 ml of water. Dilute to about 25 ml with water and acidify with 1:1 HNO_3, using methyl orange (0.02% aqueous solution) as indicator; add 1 ml in excess.

Add a few drops of 0.2 N $AgNO_3$ solution, heat to boiling, cover with a dark shield, and allow to stand for 1–2 hr. Filter on a weighed fine porosity fritted-glass filter crucible, wash five times with dilute HNO_3 (1:99), and dry at 130–150°C to constant weight.

A blank determination should be made at the same time as the sample analysis, following all steps of the procedure. The precipitation step will seldom yield much beyond a faint opalescence, the final value of which does not usually exceed 0.01% Cl.

Conversion Factors

$$AgCl \underset{4.0426}{\overset{0.2474}{\rightleftarrows}} Cl$$

TOTAL CHLORINE

Mix 1.0 g of sample (80 mesh), or use the residue from the acid-soluble determination, with 5 g of anhydrous Na_2CO_3 (low in chlorine) in a 25 ml platinum crucible and fuse carefully over a low flame (Sec. 4.3.1). Transfer the fusion cake to a 150 ml beaker and digest near the boiling point with 50 ml of water until the cake has completely disintegrated and all soluble matter has dissolved. Add 2–3 ml of ethyl alcohol to reduce any manganate present.

Decant the solution through a 9 cm No. 42 Whatman filter paper, previously washed with H_2O, and wash the residue several times with small portions of 1% Na_2CO_3 solution, decanting the wash solutions through the paper. Finally wash the paper five times with the wash solution and discard the paper and residue.

Cover the beaker and carefully add dilute HNO_3 (1:1) until the solution is acid to methyl orange, and add 1 ml in excess. Continue with the determination of the chlorine as silver chloride as described in the method for the acid-soluble portion. The value obtained is that of the total chlorine content of the sample, unless the residue alone from the acid-soluble chlorine determination was used for the fusion, in which case the final figure represents the acid-insoluble chloride only.

b. *Colorimetric*

This method is currently in use at the GSC. The same sample fusion and leach is used for fluorine and chlorine determinations with an aliquot being taken for each.

Apparatus

Platinum crucibles (cleaned in 1:1 HNO_3, rinsed well with distilled water and ignited for 10 min on a burner); miscellaneous glassware including 50 ml volumetric flasks, 25 ml volumetric flasks, centrifuge tubes which have also been cleaned in 1:1 HNO_3 and rinsed with distilled water; a centrifuge and a spectrophotometer.

Reagents

Standard 1000 ppm Cl solution.   Weigh 0.4125 g of reagent grade NaCl (dried at 105–110°C) into a 250 ml volumetric flask and dilute to 250 ml with distilled water. Store this in a polyethylene bottle. This is to be used for preparing the 10 ppm standard required for calibration.

Ferric ammonium sulfate solution.   Dissolve 30.14 g of ferric ammonium sulfate ($FeNH_4(SO_4)_2$) in 250 ml of 9 M HNO_3 (143 ml HNO_3 + 107 ml H_2O) and filter if necessary.

Saturated mercuric thiocyanate solution.   Add 0.88 g of mercuric thiocyanate ($Hg(SCN)_2$) to 250 ml of 95% ethanol, allow to stand overnight and filter through a Whatman No. 42 filter paper.

Weigh 0.1000 g of sample into a platinum crucible containing 1.0 g of Na_2CO_3, mix and fuse at 1000°C in a muffle furnace for 20 min. Cool the crucible, add 20 ml of water, one drop of ethanol, and allow to stand overnight or until the cake is soft. Transfer the contents of the crucible to a 50 ml volumetric flask, dilute to 50 ml, mix and let stand for 5 min. Transfer 20 ml of the turbid solution to a centrifuge tube and centrifuge to obtain a clear solution. Decant the clear portion into a 25 ml volumetric flask, acidify (*carefully*) with two or three drops of concentrated HNO_3, and wait until the effervescence has subsided.

At this point prepare a blank using distilled water and a high Cl standard (5 ml of a 10 ppm standard Cl solution) in 25 ml volumetric flasks. Add 2 ml of ferric ammonium sulfate solution, 2 ml of mercuric thiocyanate solution, dilute to 25 ml, mix cautiously (CO_2 is evolved), and allow to stand for 10 min. Pour into cuvettes and read at 460 nm on the spectrophotometer. Determine the concentration of Cl from the calibration graph, subtract the blank, and calculate the %Cl in the sample as follows:

$$\mu g\ Cl = \frac{50 \times \text{sample reading}}{\text{high standard reading}}$$

$$\%\ Cl = \mu g \times 0.0025$$

6.5 TUNGSTEN

6.5.1 General

Tungsten is a relatively rare element with an average crustal abundance in the order of 1 ppm, as compared with 2.5 ppm for U and 0.1 ppm for Ag, for example (Ref. 56, pp. 230–231). Because it is difficult to keep in solution and because its determination by AAS suffers from poor sensitivity (Sec. 9.4.18), this method is seldom used for its determination. Gravimetric methods are usually used for the determination of W at the higher concentrations and colorimetric methods for the lower ones although, as Topping has pointed out in his review,[118] titrimetric, electrolytic, XRF (Sec. 11.2.35), ES, and other methods are becoming much more common, especially when used in conjunction with solvent extraction and ion-exchange preconcentration techniques.

6.5.2 Colorimetric Determination of Tungsten

This method is used in the BCM laboratories and several commercial laboratories, in one modified form or another, for the determination of trace concentrations of W in geological samples.[119]

Apparatus

Pyrex test tubes (16 × 150 mm) and test tube racks, a hot water bath, a vortex mixer, and a centrifuge are required, as well as an analytical balance sensitive to 1 mg in 500 mg.

Reagents

10 M HCl. Dilute 835 ml conc. HCl to 1000 ml with distilled, deionized water.

SnCl₂ solution (19%). Dissolve 10 g of $SnCl_2$ in 100 ml of 10M HCl.

Dithiol solution. Weigh 1 g of zinc dithiol into a 100 ml volumetric flask, add 1 ml conc. HCl, shake for 10 sec, and dilute to 100 ml with isoamyl alcohol.

Petroleum spirit. b.p. 100–120°C.

Potassium bisulfate. Fused powder.

Standard W solutions. Dissolve 897 mg of $Na_2WO_4 \cdot 2H_2O$ in distilled, deionized water and dilute to 500 ml to give a stock solution containing 1000 μg/ml. Prepare fresh each month. More dilute solutions, prepared from this stock solution, should be made up as follows: weekly for one containing 100 μg/ml, daily for one containing 10 μg/ml, and just before use for one containing 1 μg/ml.

Procedure

Weigh 0.25 g of sample into a clean, dry test tube, add approximately 1.25 g of $KHSO_4$, mix, and heat the tube in a flame to fuse the contents, continuing the heating until effervescence has ceased and the melt is quiescent. Allow the melt to cool, add 5 ml of 10 M HCl, and leach in a hot water bath for 15 min (this leaching step can be done for one or more trays of samples at one time). Mix the sample thoroughly, cool, centrifuge, and decant the solution into another test tube.

Pipette a 4 ml aliquot of the clear sample solution into a clean, dry test tube, add 4 ml of the $SnCl_2$ solution, and mix. Heat the test tube(s) in a hot water bath preheated to 80°C for 5 min, add 1 ml of dithiol solution to each test tube without removing them from the water bath, and continue heating, with occasional shaking, until the organic phase has been reduced in volume to a small globule (*Note.* If the test tube is allowed to remain in the water bath for too long after the globule has formed, the globule will sink, and the determination will be ruined.)

Remove the test tubes from the water bath, allow to cool, add 1 ml of petroleum spirit, and shake gently. Compare the color of the petroleum spirit layer with a series of standards prepared in the same way. If the color intensity exceeds that of the highest standard, add more petroleum spirit in 1 ml portions, mix, and compare again.

Prepare a series of standard solutions containing 0, 0.2, 0.4, 0.6, 0.8, 1.0, 2.0, 3.0, .4.0, 5.0, 7.0, and 10.0 μg W by pipetting appropriate aliquots of the 1 μg/ml and the 10 μg/ml solutions into a series of 12 test tubes, each containing 4 ml of $SnCl_2$ solution, and continue with the determination as described above, beginning with the heating of the test tubes in the 80°C water bath.

This method is capable of achieving a detection limit of 2 ppm W in geological samples and has been successfully used in the BCM laboratories over a concentration range of 2–300 ppm.

References

1. W. F. Hillebrand, G. E. F. Lundell, H. A. Bright, and J. I. Hoffman, *Applied Inorganic Analysis*, 2nd ed., Wiley, New York, 1953.

2. L. M. Melnick, "Iron," in I. M. Kolthoff and P. J. Elving, Eds., *Treatise on Analytical Chemistry*, Vol. 2, Part II, Interscience, New York, 1962, pp. 247–310.

3. B. R. Sant and T. P. Prasad, *Talanta*, **15**, 1483 (1968).

4. R. S. Young, *Chemical Phase Analysis*, Charles Griffin & Co. Ltd., London, 1974.

5. H. S. Washington, *The Chemical Analysis of Rocks*, 2nd ed., Wiley, New York, 1910.

6. H. N. S. Schafer, *Analyst*, **91**, 755, 763 (1966).

7. A. W. Groves, *Silicate Analysis*, 2nd ed., George Allen and Unwin, London, 1951.

8. A. Juurinen, *Ann. Acad. Sci. Fenn.* Ser. A. Part III, **47**, 19 (1956).

9. M. H. Hey, *Mineral. Mag.*, **26**, 116 (1941).

10. Z. M. Mikhailova, A. A. Yarushkina, R. V. Mirskii, and E. A. Shil'dkrot, *Ref. Zh. Khim.*, 19GDE, Abstr. 4G122 (1964); through *British Abstr.*, **12**, 114 (1965).

11. G. S. Bien and E. D. Goldberg, *Anal. Chem.*, **28**, 97 (1956).

12. M. E. Beyer, A. M. Bond, and R. J. W. McLaughlin, *Anal. Chem.*, **47**, 479 (1975).

13. J. P. Riley and H. P. Williams, *Mikrochim. Acta,* **4**, 516 (1959).

14. J. Ito, *Bull. Chem. Soc. Jpn.*, **35**, 225 (1962).

15. D. Whitehead and S. A. Malik, *Anal. Chem.*, **47**, 554 (1975).

16. L. Shapiro and W. W. Brannock, *U.S. Geol. Surv. Bull.*, 1144-A, A48 (1962).

17. W. J. French and S. J. Adams, *Analyst*, **97**, 828 (1972).

18. J. J. Fahey, U.S. Geological Survey, Prof. Paper 424, C386, 1961.

19. E. A. Vincent and R. Phillips, *Geochim. Cosmochim. Acta*, **6**, 1 (1954).

20. P. R. Graff and F. J. Langmyhr, *Anal. Chim. Acta*, **21**, 429 (1959).

21. F. J. Langmyhr and P. R. Graff, *Nor. Geol. Unders.*, 230, 128 pp. (1965).

22. J. L. Walker and G. D. Sherman, *Soil Sci.*, **93**, 325 (1962).

23. L. J. Clark, *Anal. Chem.*, **34**, 348 (1962).

24. L. Shapiro, U.S. Geological Survey, Prof. Paper 400-B, B 496, 1960.

25. E. Kiss, *Anal. Chim. Acta*, **72**, 127 (1974).

26. G. D. Nicholls, *J. Sed. Pet.*, **30**, 603 (1960).

27. S. Banerjee, *Anal. Chem.*, **46**, 782 (1974).

28. R. Meyrowitz, *Am. Mineral.*, **48**, 340 (1963).

29. A. D. Wilson, *Bull. Geol. Surv. Gr. Brit.*, **9**, 56 (1955).

30. E. R. Whipple, *Chem. Geol.*, **14**, 223 (1974).

31. A. D. Wilson, *Analyst*, **85**, 823 (1960).

32. A. D. Wilson, *Analyst*, **89**, 571 (1964).

33. L. E. Reichen and J. J. Fahey, *U.S. Geol. Surv. Bull.*, 1144-B, 1 (1962).

34. J. C. Van Loon, *Talanta*, **12**, 599 (1965).

35. H. Ungethüm, *Z. Angew. Geol.*, **11**, 500 (1965).

36. C. O. Ingamells, *Talanta*, **4**, 268 (1960).

37. K. L. Cheng, *Anal. Chem.*, **36**, 1666 (1964).

38. L. C. Peck, *U.S. Geol. Surv. Bull.*, 1170, 39, 72 (1964).

39. C. V. Clemency and A. F. Hagner, *Anal. Chem.*, **33**, 888 (1961).

40. J. L. Bouvier, J. G. Sen Gupta, and S. Abbey, Geological Survey of Canada, Paper 72-31, 1972.

41. E. Kiss, *Anal. Chim. Acta*, **89**, 303 (1977).

42. F. Vydra and J. Vorlicek, *Chemist-Analyst*, **53**, 103 (1964).

43. G. M. Bancroft, T. K. Sham, C. Riddle, T. E. Smith, and A. Turek, *Chem Geol.*, **19**, 277 (1977).

44. J. P. Riott, *Ind. Eng. Chem. Anal. Ed.*, **13**, 546 (1941).

45. M. G. Habashy, *Anal. Chem.*, **33**, 586 (1961).

46. A. J. Easton and J. F. Lovering, *Geochim. Cosmochim. Acta*, **27**, 753 (1963).

47. K. H. Wedepohl, Ed., *Handbook of Geochemistry*, Vol II/4, Springer-Verlag, New York, 1974.

48. F. S. Grimaldi, "Thorium," in, I. M. Kolthoff and P. J. Elving, Eds., *Treatise on Analytical Chemistry*, Vol. 5, Part II, Interscience, New York, 1961, pp. 139–216; G. L. Booman and J. E. Rein, "Uranium," *idem.*, Vol. 9, 1962, pp. 1–188.

49. D. I. Ryabchikov and F K. Korchemnaya, "The Present State of the Analytical Chemistry of Thorium," in A. P. Vinogradov and D. I. Ryabchikov, Eds., *Detection and Analysis of Rare Elements,* Izdatel'stvo Adademii Nauk SSSR, Moscow, 1961, pp. 415–444.

50. R. S. Young, *Chemical Analysis in Extractive Metallurgy,* Charles Griffin & Co. Ltd., London, 1971.

51. L. Rybach, "Radiometric Technique," in R. E. Wainerdi and E. A. Uken, Eds., *Modern Methods of Geochemical Analysis,* Plenum, New York, 1971.

52. J. A. S. Adams and P. Gasparini, *Gamma-Ray Spectrometry of Rocks,* Elsevier, New York, 1970.

53. P. M. Hurley, *Bull. Geol. Soc. Am.,* **67,** 395 (1956).

54. R. D. Cherry, J. B. M. Hobbs, A. J. Erlank, and J. P. Willis, *Can. Spectrosc.,* **15,** 3 (1970).

55. J. M. Domingos and A. A. Melo, *Nucl. Instrum. Methods,* **48,** 28 (1967).

56. H. J. Rösler and H. Lang, *Geochemical Tables,* Elsevier, New York, 1972.

57. D. M. Bibby, R. P. Chaix, and A. H. Andeweg, Nat. Inst. Metall., Rep. 1625, 1974.

58. S. Bowie, M. Davis, and D. Ostla, Eds., *Uranium Prospecting Handbook,* Institute of Mining and Metallurgy, London, 1972, p. 95.

59. G. L. Cumming, *Chem. Geol.,* **13,** 257 (1974).

60. J. R. Weiss and E. L. Haines, *Rev. Sci. Instrum.,* **45,** 606 (1974).

61. D. E. Fisher, *Anal. Chem.,* **42,** 414 (1970).

62. H. Matsuda, Y. Tsutsui, S. Nakano, and S. Umimoto. *Talanta,* **19,** 851 (1972).

63. P. Gaillochet, D. Bauer, and M. C. Hennion, *Analusis,* **3,** 513 (1975).

64. J. Korkisch and H. Hübner, *Talanta,* **23,** 283 (1976).

65. J. Korkisch, I. Steffan, and H. Gross, *Mikrochim, Acta,* I(4–5), 503 (1976).

66. T. Kiriyama and R. Kuroda, *Anal. Chim. Acta,* **71,** 375 (1974).

67. J. Adam and M. Stulikova, *Collect. Czech. Chem. Commun.,* **39,** 2576 (1974); through Anal. Abstr., 28, No. 4B66 (1975).

68. J. C. Ingles, *Mines Branch Monograph,* Part II, Method U-1, Department of Mines and Technical Survey No. 866, Queen's Printer Cat. No. M32-866, Ottawa, 1959, pp. 1–37.

69. S. Abbey, *Anal. Chim. Acta,* **30,** 176 (1964).

70. I. May and L. B. Jenkins, U.S. Geological Survey, Prof. Paper 525D, D192, 1965.

71. K. Goto, D. S. Russell, and S. S. Berman, *Anal. Chem.,* **38,** 493 (1966).

72. C. W. Sills, *Anal. Chem.,* **49,** 618 (1977).

73. J. C. Bailey, *Chem. Geol.,* **19,** 1 (1977).

74. M. S. Frant and J. W. Ross, Jr., *Science,* **154,** 1553 (1966).

75. C. A. Horton, "Fluorine," in I. M. Kolthoff and P. J. Elving, Eds., *Treatise on Analytical Chemistry*, Vol. 7, Part II, Interscience, New York, 1961, pp. 207–334.

76. L. E. Fite, E. A. Schweikert, R. E. Wainerdi, and E. A. Uken, "Nuclear Activation Analysis," in R. E. Wainerdi and E. A. Uken, Eds., *Modern Methods of Geochemical Analysis*, Plenum, New York, 1971, Chap. 11.

77. J. N. Weber and P. Deines, "Mass Spectrometry," in R. E. Wainerdi and E. A. Uken, Eds., *Modern Methods of Geochemical Analysis*, Plenum, New York, 1971, Chap. 12.

78. B. L. Ingram and I. May, U.S. Geological Survey, Prof. Paper 750-B, B180, 1971.

79. B. L. Ingram, *Anal. Chem.*, **42**, 1825 (1970).

80. W. H. Ficklin, U.S. Geological Survey, Prof. Paper 700-C, C186, 1970.

81. W. Selig, *Mikrochim. Acta* (Wien), **87**, (1973).

82. D. Jagner and V. Pavlova, *Anal. Chim. Acta*, **60**, 153 (1972).

83. N. Shiroishi, Y. Murata, G. Nakagawa, and K. Kodama, *Jpn. Analyst*, **23**, 176 (1974).

84. J. B. Bodkin, *Analyst*, **102**, 409 (1977).

85. W. B. Kirsten, *Anal. Chem.*, **48**, 84 (1976).

86. R. Fuge, *Chem. Geol.*, **17**, 37 (1976).

87. C. C. Fong and C. O. Huber, *Spectrochim. Acta*, **31B**, 113 (1976).

88. J. Auffarth and D. Klockow, *Anal. Chim. Acta*, **111**, 89 (1979).

89. E. J. Fox and W. A. Jackson, *Anal. Chem.*, **31**, 1657 (1959).

90. H. E. Blake, Jr., U.S. Bureau of Mines, Rep. Invest., 6314, 29, 1963.

91. I. B. Moyzhess, *All-Union Sci. Res. Geol. Inst.*, Leningrad, **117**, 27 (1964).

92. A. M. G. Macdonald, "Fluorine," in C. L. Wilson and D. W. Wilson, Eds., *Comprehensive Analytical Chemistry*, Vol. 1C, Elsevier, Amsterdam, 1962, pp. 319–340.

93. C. O. Ingamells, *Talanta*, **9**, 507 (1962).

94. S. Megregian, "Fluorine," in D. F. Boltz, Ed., *Colorimetric Determination of Nonmetals*, Interscience, New York, 1958, pp. 231–259.

95. M. A. Wade and S. S. Yamamura, *Anal. Chem.*, **37**, 1276 (1965).

96. G. A. Sergeant, Department of Scientific and Industrial Research, London, Rep. of Govern. Chemist, 1963, 61, 1964.

97. L. C. Peck and V. C. Smith, *Talanta*, **11**, 1343 (1964).

98. H. Bennett and W. G. Hawley, *Methods of Silicate Analysis*, 2nd ed., Academic, London, 1965, pp. 97–98, 288–294.

99. H. J. Cluley, *Glass Technol.*, **2**, 74 (1961).

100. O. S. Glasö, *Anal. Chim. Acta*, **28**, 543 (1963).

101. H. R. Shell and R. L. Craig, U.S. Bureau of Mines, Rep. Invest., 5158, 30, 1956.

102. J. W. Robinson, *Anal. Chim. Acta*, **20**, 256 (1959).

103. P. Geijer, *Arkiv. Mineral. Geol.*, **2**, 482 (1960).

104. V. E. Caldwell, *Anal. Chem.*, **38**, 1249 (1966).

105. M. Gillberg, *Geochim. Cosmochim. Acta*, **28**, 495 (1964).

106. L. Greenland and J. F. Lovering, *Geochim. Cosmochim. Acta*, **29**, 848 (1965).

107. W. Reichel and L. Acs, *Anal. Chem.*, **41**, 1886 (1969).

108. U. Westerlund-Helmerson, *At. Absorpt. Newsl.* **5**, 97 (1966).

109. L. C. Peck and E. J. Tomasi, *Anal. Chem.*, **31**, 2024 (1959).

110. G. W. Armstrong, H. H. Gill, and R. F. Rolf, "The Halogens," in I. M. Kolthoff and P. J. Elving, Eds., *Treatise on Analytical Chemistry*, Vol. 7, Part II, Interscience, New York, 1961, pp. 335–424.

111. J. T. Stock, *Amperometric Titrations*, Interscience, New York, 1956, pp. 200–208.

112. P. K. Kuroda and E. B. Sandell, *Anal. Chem.*, **22**, 1144 (1950).

113. H. Akaiwa, H. Kawamoto, and K. Hasegawa, *Talanta*, **26**, 1027 (1979).

114. J. A. Dean, *Flame Photometry*, McGraw-Hill, New York, 1960, pp. 262–265.

115. A. Syty, "Halogens," in J. A. Dean and T. C. Rains, Eds., *Flame Emission and Atomic Absorption Spectrometry*, Vol. 3, Dekker, New York, 1975.

116. L. H. Ahrens and S. R. Taylor, *Spectrochemical Analysis*, 2nd ed., Addison-Wesley, Reading, MA., 1961, pp. 282–283.

117. A. V. Karyakin, E. N. Savinova, and T. P. Andreeva, *Zh. Anal. Khim.*, **24**, 468 (1969); through *Anal. Abstr.*, **19**, No. 2227 (1970).

118. J. J. Topping, *Talanta*, **25**, 61 (1978).

119. P. Aruscavage and E. Y. Campbell, *J. Res. U.S. Geol. Surv.*, **6**, 697 (1978).

Supplementary References

1. L. Th. Begheijn, *Analyst*, **104**, 1055 (1979).

2. A. Sugimae and R. K. Skogerboe, *Anal. Chem.*, **51**, 884 (1979).

3. A. Farzaneh and G. Troll, *Fresenius' Z. Anal. Chem.*, **292**, 293 (1978); through *Anal. Abstr.*, **36**, No. 3B 134 (1979).

4. C. K. Unni and J. G. Schilling, *Anal. Chim. Acta*, **96**, 107 (1978).

PART III

Methods for Determination of Major, Minor, and Trace Elements in Rocks and Minerals Using Atomic Absorption Spectroscopy

CHAPTER 7

Atomic Absorption Spectroscopy—General Considerations

CHAPTER 8

The Determination of Major and Minor Constituents in Rocks and Minerals by Atomic Absorption Spectroscopy

CHAPTER 9

The Determination of Trace Constituents in Rocks and Minerals by Atomic Absorption Spectroscopy

ATOMIC ABSORPTION SPECTROSCOPY—GENERAL CONSIDERATIONS

No attempt will be made to discuss the theoretical aspects of atomic absorption spectroscopy (AAS) since they have been very adequately covered elsewhere.[1-3] Emphasis will be placed instead upon a discussion of the applications of AAS to rock and mineral analysis, in particular on methods, common sources of error, and the preparation and use of standards. This will encompass the determination of major, minor, and trace constituents in rocks and minerals.

There are two reviews of AAS which any laboratory scientist who is working in the field would be well advised to follow as an aid in keeping up to date in this rapidly changing field. The first is the review issue (Number 5) of *Analytical Chemistry* and the second is *Annual Reports on Analytical Atomic Spectroscopy* published by the Chemical Society, London, England.

7.1 SAMPLE DECOMPOSITION

This topic has been discussed in some detail in Chapter 4 in terms of individual dissolution agents and samples. It will be useful here to repeat some of the points which relate specifically to AAS.

As was stated previously (Sec. 4.1), a constraint that must be considered when preparing a solution of a rock or mineral for analysis by AAS is the total concentration of dissolved solids. Preferably this should not exceed 1 or 2% to avoid the possibility of evaporites forming in the slot of the burner head. The gradual buildup of evaporites in the slot can cause a loss of both precision and accuracy due to changes in the aspiration rate and flame shape, and because of memory effects as well.

Another constraint which must be considered is the introduction of ions which might interfere in subsequent measurements by suppressing or enhancing a signal due to molecular, ionization, or scattering effects. The last effect is most evident when trace element determinations are attempted. Mention will be made of a possible problem of this type in the discussion of the specific element(s) likely to be affected (Sec. 7.5).

The introduction of impurities which are potential analytes from reagents used in various dissolution techniques is a common problem, and each batch of analyses must include a reagent blank to ensure that any such occurrence is detected. Fluxes are often a major source of contaminants but the lithium borates can be obtained at a level of purity suitable for the determination of most elements at the parts per million level. The parts per billion level of determination presents one with a very serious problem in keeping contamination to sufficiently low levels. It is often necessary to redistill commercial reagent grade acids,[4] and to provide "positive pressure" work areas and special air filters in the laboratory in which parts per billion level determinations are to be made.[5]

The most common dissolution techniques used in preparing rocks and minerals for analysis by AAS are fusion with lithium borate salts[6] (Sec. 4.3.6) and digestion with a combination of acids, usually including HF[7,8] (Sec. 4.2). The concept behind the fusion technique is to break down the matrix of the sample and make it soluble in an aqueous medium for subsequent aspiration into a flame or other atomizing cell. The same can be said for the mixed acid attack, but certain precautions must be taken to ensure that a complete liberation of the elements of interest is achieved.[9] That this can be a serious problem is illustrated by the many examples of refractory minerals mentioned in Sec. 4.2. One advantage of acid digestion over fusion with a flux is the ability of the former to provide a solution having a lower total content of dissolved solids for the dissolution of a given amount of sample. This advantage is most important in the determination of trace constituents when sample dilution must be kept to a minimum.

A problem which is not always obvious immediately but which can give rise to serious errors is that of sample inhomogeneity.[10] When preparing a sample for trace element determination, the analyst must choose a large enough subsample to ensure that the material to be processed is representative and not subject to variations due to sample inhomogeneity. The size of the subsample required can influence the choice of the dissolution technique. It may require, for example, a preliminary acid leach followed by fusion of the insoluble residue, and perhaps a concentration step such as a nonaqueous extraction on the combined solutions from the acid leach and the dissolution of the fusion melt. In general, the type of dissolution procedure chosen will be determined by the problem presented to the analyst. It should be apparent from the above discussion that sample dissolution is not a trivial matter but is the important first step on which will be based the remainder of the analytical process.

7.2 PRECONCENTRATION TECHNIQUES

There are several preconcentration techniques that have been used to increase the sensitivity of AAS when it is applied to trace element determinations. Among the more common ones are solvent extraction, ion exchange, and fire assay. Coprecipitation on a carrier has also been used.

7.2.1 Solvent Extraction

The most common method used in rock and mineral analysis is the extraction of the complexed metal ion into an organic solvent. The selectivity inherent in the AAS method simplifies the procedure because, in most instances, it is not necessary to exclude other metal ions from accompanying the analyte ions into the organic layer. A glance at the tabulated data for the analysis of minerals found in *Annual Reports on Analytical Atomic Spectroscopy* (Ref. 11, pp. 114–123) will show that several different organic solvents are regularly used. As Kirkbright and Sargent point out (Ref. 3, pp. 491–497); solvents which have commonly been used include methyl isobutyl ketone (MIBK), ethyl acetate, isobutyl acetate, amyl acetate, propyl acetate, and butanol. The choice of a solvent can be critical. It should be immiscible with water, directly combustible in the flame, with no toxic combustion products (such as are produced by chlorinated solvents) and with little or no smoke (which often results from the use of aromatic solvents such as toluene).

There are numerous complexing agents from which to choose. One of the more broad-spectrum reagents which is also not overly pH dependent (as is often the case) is ammonium pyrrolidine dithiocarbamate (APDC). It is active over a pH range of 2–14 with only a minor pH-dependent selectivity (Ref. 3, p. 493), and, as a 1% aqueous solution, is usually prepared daily as needed. One of its many applications has been in the determination of copper and zinc in soils for geochemical exploration purposes.[12] Many workers have used the halogens as complexing agents. For example, HBr/Br_2 has been used to concentrate Tl into isopropyl ether[13] and to extract Te into MIBK.[14] Golembeski has described the extraction of Se into a 1% solution of phenol in benzene from an HCl solution.[15] Rubeska et al.[16] have reported that the extraction of Au from an HCl solution into dibutyl sulfide is very selective and has a high coefficient of extraction. Tri-*n*-octyl-phosphine oxide (TOPO) is a complexing agent which has been used, for example, in determining trace levels of Sb in rocks.[17] Kuz'min et al. have published a review of solvent extraction techniques (Supp. Ref. 1) and Hannaker and Hughes have developed a stepwise solvent extraction scheme for 14 elements with detection limits of < 0.2 ppm (Supp. Ref. 2).

7.2.2 Ion-Exchange Separation and Concentration

The concentration of trace amounts of analyte on an ion-exchange resin or the separation of major constituents by ion-exchange chromatography followed by their AAS measurement in the elutant has also been used in many instances. Strelow et al.[18] have developed an analytical scheme for 10 major and minor constituents of silicate rocks, the basis of which is their separation by ion-exchange chromatography on a single column. The AAS determination of major, minor, and trace elements in silicates following an ion-exchange separation has also been reported.[19, 20] Anion-exchange separation in conjunction with AAS has been used for the determination of Cd in sulfide ores[21] and for the determination of seven major and trace constituents of manganese nodules collected on the floor of the Pacific Ocean.[22] The most common anion-exchange resin used is Dowex 1X8, but Amberlite IR-120 is also effective. As mentioned for solvent extraction, it is usually not necessary to obtain a complete separation of the analyte of interest from other elements when an AAS measurement is anticipated. In such cases the analyte can be eluted from the resin along with many other elements, using a minimum quantity of a strong elutant which will permit the greatest possible concentration factor. An alternative approach is to ash the resin and to leach the analyte from the ash in a minimum volume of solvent.

7.2.3   Fire Assay Concentration

Fire assay techniques (Sec. 12.3) have been used for the concentration of Au, Ag, and several of the Pt metals into a Pb button,[23] into a Ag bead[24] (for Au and some of the platinum metals), and into a tin button.[25] Diamantatos[26] has reported the collection of iridium in a lead button by fire assay prior to an AAS finish. The button or bead is dissolved in a minimum quantity of acid(s) and the analyte determined by AAS either in the acid solution or subsequent to an extraction step. In either case it is thus possible to increase the sensitivity of determination significantly and extend the detection limits to the parts per billion range.

7.3   STANDARDS

7.3.1   Synthetic Standards

A critical part of any analytical method is its calibration, and AAS is no exception to the rule. Consideration must be given to the purity of the reagents used, the matching of the standard matrix to the matrix of the analyte, the stability or useful lifetime of the standard, and many other factors.

It is often very convenient to make up multielement standards so that one set of standards can be used for a variety of determinations on different groups of samples of similar matrices. An obvious precaution in this case is to avoid using mixtures of elements or salts which are not compatible. For example, it is not advisable to use the sulfate salt of an element in making up a multielement standard which will contain Pb or Ba. Nor can a compound which must be dissolved in HCl be used as a source of an element if Ag is to be one of the elements in the standard.

The standard solution should be slightly acidic to prevent hydrolysis and the addition of a crystal of thymol will prevent bacterial growth in those solutions where it is likely to occur. A solution containing 1000 μg of element per milliliter (1000 ppm) stored in a linear polyethylene bottle will usually remain stable indefinitely if precautions are taken to avoid evaporation. Any standard solution which is more than a few months old or which is not made up by the analyst personally (and this includes commercially available standard solutions) should be considered suspect and its concentration verified, either by comparison with a previously analyzed sample (preferably an in-house reference material of similar composition) or in some other way.

When more dilute standard solutions are prepared from the 1000 ppm solution, every precaution must be taken to ensure that the dilutions are as exact as possible. It is better, for example, to make a tenfold dilution by pipetting 10 ml into a 100 ml flask and diluting to volume than by pipetting 1 ml into a 10 ml flask for similar treatment. The stability of dilute standard solutions will vary from one element to the next, but solutions having a concentration of 10 ppm and greater will usually remain stable for one week or more. Solutions having concentrations less than 10 ppm, however, should be prepared daily. The stability of solutions of some elements (such as Au) is affected by light, and they must be stored under light-free conditions.

The choice of the source of a particular element for the preparation of a standard solution is critical. The most satisfactory source for most metals is shot, bars, or wire of the element itself, of a purity suitable to the analytical application. For those elements which are not stable in the elemental state or for which, for some other reason, a compound must be used as a source, it is best to choose one that is not hygroscopic, has no affinity for atmospheric CO_2, is stable and easy to weigh, has a well-established molecular weight, and is available at a known level of purity. The source, whatever it is, should be soluble in water or in common acids or bases, should have a high molecular weight and a purity of at least 99.5%. Both elements and compounds can be obtained in reagent grade quality from most chemical supply companies and in a purer state ("5-9's" or better) from such manufacturers as SPEX Industries, Inc. (HI-PURE® materials) and Johnson Matthey Chemicals Ltd. (Specpure® materials). Tables of appropriate source materials for individual elements[27] are available (Ref. 28, pp. 609–616; Ref. 29, pp. 43–50; Ref. 30, pp. 332–336). One such table has been included in Appendix 3.

Date (Supp. Ref. 3) has described a method for the preparation of trace element reference materials using a coprecipitated gel technique.

The set of multielement standards developed by P. Ralph of the BCM laboratory, for whole rock analysis by AAS, will serve as an example. The major and minor elements determined by AAS include Si, Al, Fe, Ca, Mg, Na, K, Ti, and Mn, and the preparation of a set of standards which will cover the concentration ranges likely to be encountered for all of the elements of interest requires some very careful planning. Further, because the determination of minor or trace concentrations of some elements will be done in the presence of major concentrations of other elements such as Si, Al, and Fe, it is desirable that the source materials for these elements be of high purity, the degree of purity being defined by the specific application.

The procedure for the preparation of these standards for whole rock analysis is as follows. A set of master stock solutions is prepared for each element as given in Table 7.1 by dissolving the appropriate weight of source material and diluting it to 1000 ml. It is important that the correct amount of HNO_3 be added to give a final acid strength of 4% HNO_3 (v/v) in the 1000 ml of solution. The next step in the procedure is the preparation of the multielement standard solutions, including SiO_2 which

TABLE 7.1   Master Stock Solutions for Preparing Standards for Major Element Analysis.[a]

Oxide of Analyte	Source Used	Dissolution	Weight of Source Used (grams)	Element (ppm)	Oxide (ppm)
Al_2O_3	Al	1:1 HNO_3, 1 drop Hg	10.585	10,585	20,000
Fe_2O_3	Fe	10 ml aqua regia	13.989	13,989	20,000
MgO	Mg	1:1 HNO_3	12.063	12,063	20,000
CaO	$CaCO_3$	dilute HNO_3	35.696	14,294	20,000
Na_2O	NaCl	1:1 HNO_3	11.3150	4,451	6,000
K_2O	KCl	1:1HNO_3	9.4972	4,981	6,000
TiO_2	Ti	1:1 HNO_3, 1 ml HF	2.398	2,398	4,000
MnO	Mn	1:1 HNO_3	0.775	775	1,000

[a] The above quantities are for the preparation of 1000 ml of each stock solution. See text for details.

TABLE 7.2   Matrix Used for the Preparation of Multielement Standard Solutions for Major Element Analysis.[a]

	Al_2O_3	FeO_3	MgO	CaO	Na_2O	K_2O	TiO_2	MnO	SiO_2 (g)[b]
MS-1	45	45	45	45	50	45	40	0	0
MS-2	40	50	40	50	45	50	45	25	0.04
MS-3	35	30	35	40	40	30	35	30	0.08
MS-4[c]	50	40	30	35	35	40	30	35	0.12
MS-5	30	25	45	45	30	25	25	20	0.16
MS-6	50	10	50	15	20	50	50	0	0.21
MS-7	5	50	5	0	50	0	10	45	0.24
MS-8	20	35	20	25	10	20	0	15	0.28
MS-9	0	20	25	30	15	35	15	10	0.33
MS-10	25	15	0	20	25	5	20	40	0.36
MS-11	15	5	10	10	0	15	5	50	0.40
MS-12	10	0	15	5	5	10	0	5	0.46

[a] Matrix showing aliquots in milliliters taken from master stock solutions as prepared in Table 7.1 and diluted to 1000 ml.

[b] The SiO_2 is not added to the aliquots shown; 100 ml of a multielement solution is combined with the appropriate weight of SiO_2 (in solution) and diluted to 500 ml. See text for details.

[c] For example, this multielement solution will contain the following parts per million of each constituent: SiO_2 (2400), Al_2O_3 (200), Fe_2O_3 (160), MgO (120), CaO (140), Na_2O (42), K_2O (48), TiO_2 (24), MnO (7).

is not prepared as a master stock solution because of the problem of keeping it in solution at high concentrations. Twelve multielement standard solutions are designed to cover the range of concentrations expected in the majority of the samples to be analyzed, and the matrix for the preparation of these multielement standards is given in Table 7.2. The given volumes correspond to aliquots taken from the master stock solutions prepared as in Table 7.1. SiO_2 is not added at this stage, however, so that the multielement solutions contain all of the elements except Si. For example, MS-1 consists of a 45 ml aliquot of Al solution prepared as in Table 7.1, a 45 ml aliquot of Fe solution prepared as in Table 7.1, and similarly through to 0 ml of the Mn solution. All of these aliquots are pipetted into a 1000 ml flask, diluted to 1000 ml, and the solution transferred to a polyethylene bottle for storage.

The next and final stage involves the dissolution of the weighed portions of SiO_2 (see Table 7.2) by fusing each of them with 2.5 g of $LiBO_2$ at 1050°C for 45 min and dissolving the melt in 50 ml of 4% HNO_3 (v/v). Pipette 100 ml of each multielement solution (see Table 7.2) into 500 ml volumetric flasks, transfer the SiO_2 solution into the appropriate flask as shown in Table 7.2, add 200 ml of 4% HNO_3 (v/v), 25 ml of a 20,000 ppm CsCl solution, 50 ml of a saturated solution of H_3BO_3, and 7.5 ml of HF to each, dilute to 500 ml, mix, and transfer immediately to polyethylene bottles for storage. Do not allow the solutions containing HF to sit in glassware any longer than necessary because, even in the presence of H_3BO_3, enough Si, Na, and other elements can be leached from the glass to cause significant errors.

7.3.2   Standard Reference Materials as Standards

The use of standard reference materials (SRM) as standards for the analysis of rocks and minerals is acceptable in some instances, but the analyst must be very cautious when using them.[31] The standards that are currently available cover the concentration ranges of the major and minor elements found in most of the samples to be encountered by a rock and mineral analyst. The recommended values for these major and minor elements are generally reliable,[32] and judicious use of these SRMs as standards for these elements should not raise many problems.

The story for trace elements (<0.01%) is, however, a very different one.[33] The trace element results reported for SRMs vary over wide ranges, even for common elements such as Cu and Zn. Furthermore, although there are some 100 rock, mineral, ore, slag, and clay SRMs currently obtainable, there are several elements which are commonly requested of the analyst for which values are available for only a few of these SRMs. Thus for these elements at least, calibration based upon SRMs is not possible. Examples of such elements found in a recent compilation[32] include W, for which only 16 SRMs have "usable" reported values, Ag (12), Bi (9), Au (7), Se (9), and Te (0). In addition, there are large gaps in the concentration ranges of many other elements which decrease the utility of SRMs in preparing calibration curves for these elements, and thus it would appear that SRMs are of limited use as standards for trace element analysis.

One further point should be stressed concerning SRMs. In spite of the drawbacks mentioned, they are of invaluable aid to the analyst. The use of SRMs has the advantage that they experience the same chemical treatment as the unknowns, and any associated problems will be more readily detected. However, because most of them are in short supply, they should not be used indiscriminately nor as a routine standard on a daily basis. They should be used only to calibrate in-house standards, new methods, or instruments, and as occasional bench marks for quality control. This caution is in recognition of the cost required to collect, prepare, distribute, and analyze the sample, and to correlate and interpret the results so that a useful SRM is produced.

7.4 INSTRUMENTATION

The development of instrumentation and technology for AAS is advancing on many fronts, from burner assemblies to microprocessor-controlled instrument parameters and automated gas control attachments. New designs for hollow cathode lamps (HCL) have been introduced, as have techniques for sampling the signal more frequently to improve the monitoring of signals from flameless attachments with simultaneous background correction. Double beam and dual channel instruments are available, as are automatic sample changers, and ancillary data capture attachments. The theory and detailed description of all this equipment will not be convered here, but the application of it to rock and mineral analysis will. Ingle[34] has developed a set of equations which predict the effect various instrumental parameters will have on the precision of AAS measurements. The application of this to the improvement of the precision of flame AAS measurements (Supp. Ref. 4) and to the optimization of instrumental variables (Supp. Ref. 5) has also been discussed.

7.4.1 Light Sources and Analytical Lines

a. *Light Sources*

There are numerous light sources that have been used in AAS, including HCLs, electrodeless discharge lamps (EDL), lasers, continuum sources,[35] and, more recently, Zeeman-split sources.[36] Of these, the HCL is by far the most commonly used, although the EDL is being used with increasing frequency in applications such as the determination of As and Sb,[37] where HCL sources are not sufficiently intense or stable to give adequate detection limits. Most of these sources have been discussed in some detail in other texts,[2, 3, 30, 38] and this section will cover only some of the practical aspects as they relate to the use of HCL sources in rock and mineral analysis.

For routine control work in a concentrator or a mill, a multielement HCL is often useful for doing a series of determinations on one set of samples. Multielement lamps have the disadvantage, however, that the intensity of an individual element in such a lamp is significantly less than it is for that element in a single-element HCL. In addition, one element will often be depleted more quickly than the others in the cathode. There is also a greater possibility of spectral interference. If Cu and Pb are present in a multielement HCL, for example, the 216.5 nm Cu resonance line will interfere with Pb determinations using the 217.0 nm Pb line, and the Pb 283.3 nm resonance line should be used instead.[39]

It is usually necessary to warm up an HCL lamp for 20 min or more before it becomes operationally stable. For a double beam instrument this is not as critical as for a single beam unit, but it is not advisable to start a determination as soon as the lamp is turned on. Most instruments have a multilamp turret which allows the operator to have one or more lamps warming up while another is being used.

b. *Analytical Lines*

Each HCL gives a series of emission lines, some of which are useful analytically and some of which are not. The choice of the resonance line for a particular use is not always as straightforward as is often assumed. Margoshes discussed the theoretical prediction of analysis lines,[40] and Parsons et al.[41] extended the study to compare the theoretical predictions with experimental observations, tabulating the relative absorbance of resonance lines and, when significant, the variation of the relative absorbance with temperature. The resonance lines which correspond to a transition from the zero energy state show little or no variation in relative absorbance with temperature, but when upper energy levels are significantly populated (and this is not an unusual situation), resonance transitions can show a significant temperature dependence. Thus for some elements, such as Sn, W, Ge, Nb, and Zr, the choice of the most sensitive line may differ depending on the flame used.[41] The 247.6 nm line of Pd is the most sensitive line at flame temperatures achieved by an air–acetylene flame (2500°K), whereas the 340.5 nm line is the most sensitive in a nitrous oxide–acetylene flame (3000°K).

Other factors which can affect the choice of resonance line are the presence of interfering elements in the analyte solution, the spectral range of the instrument, the material from which the window of the HCL is made, the magnitude of the background intensity at the wavelength of interest, the characteristic wavelength response of the photomultiplier, and a high background in the flame (usually a problem below 250 nm). For example, Se has its most sensitive line at 196.0 nm, and the HCL must have a window that is transparent to radiation of that wavelength (Corning 9741 UV transmitting glass for example, is not adequate) and the photomultiplier must be sensitive to energy of that wavelength.

Interfering emission lines (from the analyte atom in the HCL, the filler gas, or the shell of the cathode) can preclude the selection of the strongest resonance line of an element (Ref. 3, pp. 442–446). In the case of Ni, for example, the 232.00 nm resonance line is the most sensitive, but it is very close to several other Ni emission lines which are either nonabsorbing or only weakly absorbing (231.98 and 232.14 nm) and to the 231.6 nm line which is emitted strongly from most HCL sources. It is difficult to isolate the 232.00 nm line from these others, even with a narrow bandpass. This leads to low sensitivity and nonlinear calibration, and the Ni 341.48 nm line is often used instead for the determination of Ni. The problem caused by emission lines from filler gas or other elements in the cathode has become well recognized, and modern lamps are usually free of this hindrance. At one time Ar was used as a filler gas in the Pb HCL, for example, and the Ar emission spectrum interfered with the strongest Pb line at 217.0 nm. The use of neon as a filler gas has circumvented this problem, and the analyst is not likely to encounter it in the future.

Norris and West[42] have discussed the advantage of using special cases of spectral overlap for the determination of major constituents by AAS. The use of a nonresonance line of one element which is located close to the resonance line of another element will permit a wide linearity range, and the need for large dilution of the sample solution can be avoided. Examples of this are the Pb 247.638 nm nonresonance line, the Pd 247.643 nm resonance line, and the Sb 217.023 nm nonresonance–Pb 216.999 nm resonance pair. Eleven such spectral overlaps have been identified.

Lovett and Parsons[43] have reported the use of the Ne 364.053 nm line from any neon-filled HCL for the determination of Rh in place of its own absorption line at 364.046 nm. The detection limit for Rh using this technique is only about one order of magnitude greater than that obtained with a Rh HCL, and its use may make it unnecessary for a laboratory to buy a Rh lamp for a special "one time" determination. In addition, many HCLs contain elements other than those on the label, and it may be possible to use them as multielement lamps.

7.4.2 Atom Cells

There are three types of atom cells which are used almost exclusively for the application of AAS to rock and mineral analysis. These are the flame cell, the thermal atomizer, and the cold vapor cell. The inductively coupled plasma source developed for ES instruments has, however, also been applied to AAS.[44] Historically the flame cell was associated with the development of AAS, and the theory, technology, and application of flames to a wide variety of analytical problems are well documented (Ref. 2, pp. 135–192; Ref. 3, pp. 197–286; Ref. 30, pp. 57–93). The nonflame thermal atom cell, such as the tantalum strip or carbon-rod atomizer, has received much attention during recent years, and it offers some potential advantages over the flame cell resulting from the more efficient atomization of the sample, such as low detection limits, direct atomization, and small sample size requirements. Rains and Menis have made a comparison of the flame and thermal atomizer cells (Ref. 45, pp. 1045–1061). The application of nonflame AAS techniques to gelogical materials has been reviewed by L'Vov[46] and Langmyhr[47] (Supp. Ref. 6). There are some disadvantages associated with carbon-rod and similar thermal atomizers. If the solid sample is to be atomized directly, for example, the small sample size required can lead to subsampling errors, and it is necessary to use a separate subsample for each determination. If the rock or mineral sample is in solution, the high solute concentration frequently encountered in such solutions can give backgrounds beyond the capacity of the instrument to correct. Also, there is an inherent problem in obtaining high precision because of the very short duration of the signal as opposed to the relatively long flow time obtainable with a flame cell.

A flameless technique which does not fit into the category of thermal atomizers is that used for Hg determinations where chemically evolved Hg vapor is measured in a flameless atom cell.[48, 49] The chemical evolution of the hydrides of several elements (e.g., arsine) and their subsequent atomization in a heated cell[37] is also a "flameless" technique which is not in the same category as the carbon-rod, tantalum strip, or similar atomizers.

The flame cell can use a variety of fuels and oxidants, but the most common combination involves acetylene as the fuel and either compressed air or nitrous oxide as the oxidant. There are many elements for which the air–acetylene flame is a satisfactory absorption cell, one providing a low noise level, high stability, and good atomizing conditions with few interferences. For the determination of some elements, however, the successful application of AAS had to await the introduction of the nitrous oxide–acetylene flame.[50] This latter flame produces a suitably high temperature, and the oxidant-to-fuel ratio can be adjusted to create a reducing region of the flame in which a sufficiently high free atom population of such problem elements as Al, Ti, Si, and V exists to enable them to be determined successfully. For example, the Al free atom fraction in an air–acetylene flame has been found to be <0.0001, whereas in a nitrous oxide–acetylene flame it can range from 0.13 to 0.29 (Ref. 3, pp. 243–255), an improvement of more than three orders of magnitude.

The generation of an efficient atom cell is critical to any AAS measurement, and the choice of oxidant, the fuel-to-oxidant ratio, the aspiration rate, and other factors must be optimized for each individual circumstance. The free atom concentration in the flame will vary with distance above the burner head, and the best region of the flame in which to make a measurement must be established for each element. In many instances these parameters will change to some extent between one make of burner and another. Thus each laboratory will have to determine the optimum operating conditions for the determination of individual elements with its own equipment, using published information as a guide for a starting point.

7.4.3   Gas Supplies and Burner Systems

A detailed description of gas supplies and burner systems is given elsewhere (Ref. 43, pp. 317–343, 433–438; Ref. 30, pp. 57–93), and this section will discuss only a few matters of practical significance, including some necesary safety precautions.

Air for the air–acetylene flame system can be obtained from commercially available compressed gas supplies, but this can require frequent replacement of cylinders when the flame source is used for an extended period of time. More preferable is a small air compressor capable of delivering air at a pressure of about 4.2–5.6 kg/cm^2 (60–80 psi), for subsequent reduction to the usual 2.8 kg/cm^2 (40 psi) required for the operation of most commercially available instruments. The compressor should be of the "oil-free" type, and a filter to remove condensed water must be inserted between the compressed air delivery line and the AAS instrument.

Cylinders of compressed gases such as acetylene and nitrous oxide should be firmly secured to a wall or laboratory bench to prevent accidental tipping and should not be placed near a radiator or other heat sources. The contents are under pressure, and this must be reduced before delivery of the gas to the instrument. A two-stage regulator, obtainable from all distributors of the compressed gases, is best suited to this purpose, and the manufacturer's instructions accompanying the instrument will indicate the delivery pressure required for the individual gases.

The compressed acetylene is dissolved in acetone for storage, and the cylinder should be changed when the gas pressure is reduced to 5.3 kg/cm^2 (75 psi) to prevent the acetone solvent from distilling over with the acetylene and affecting the flame characteristics. Acetylene should *never* be passed through copper tubing because of the formation of the explosive copper acetylide Cu_2C_2. Plastic tubing that has been certified for safe acetylene delivery is available.

Nitrous oxide is supplied in liquid form, and the cylinders are often equipped with a heated regulator to overcome the tendency for the latter to "freeze up" because of heat absorption by the vaporizing liquid on its passage through the regulator. Unless prevented, the "freeze up" will cause the gas flow rate to vary and create the hazard of a "flashback" if the flow rate is interrupted significantly. Regulators with a built-in heating element are available, but an ordinary laboratory heating tape, wrapped around the diaphragm of the regulator and connected to a rheostat, is an adequate substitute and is much less expensive. An ordinary reading lamp with its (60 W) bulb touching the regulator is also adequate.

It is important that the gas flow into the burner assembly be suitably monitored so that selected flame conditions can be reproduced from day to day (although all conditions must be optimized for each set of determinations). This is usually done with rotameters which are built into the instrument and positioned on the burner side of the final pressure regulator. The flow scale of the rotameter is often an arbitrary one, and it is then necessary to calibrate it using the graphs for the various gases which are usually provided by the instrument manufacturer.

The operation of the air–acetylene or nitrous oxide–acetylene burner is subject to the potential hazards of "flashback," but observation of the correct operating procedure (including a check of the drain line) will preclude this. The proper ignition sequence for the air–acetylene flame is first to begin the flow of the air, followed by that of the acetylene, and then ignition. For a nitrous oxide–acetylene flame, the regulator on the N_2O cylinder is opened, the air–acetylene mixture is ignited as above, the acetylene flow is increased to give a sooty, fuel-rich flame, the support gas supply is switched from air to N_2O, and the fuel and oxidant flows are adjusted to give the desired flame quality. The procedure for flame shutdown is the reverse of that for ignition. Many manufacturers now have available, as options, automatic gas control and ignition systems which will ensure that the correct sequence is followed. These have the added advantage that sensors on the burner assembly can detect changes in the gas flows due to line blockage or low tank pressure and can automatically initiate a flame shutdown before a "flashback" can occur.

The nebulizer–burner assembly is potentially the major source of "noise" in an AAS measurement, and thus its design is critical. The sample is usually introduced into the flame with the oxidant (or a small fraction of it), and a reproducible and steady aspiration rate is mandatory. This rate can vary from one instrument to another depending on the design of the assembly. In the BCM laboratory, for example, the aspiration rate for an air–acetylene burner on a Perkin Elmer model 107 instrument is about 9.5 ml/min, whereas on an Instrumentation Laboratory 351 instrument it is about 3.5 ml/min. The aspiration rate is most affected by the viscosity of the solution being aspirated, but the hydrostatic head (i.e., the length of the column of solution lifted through the capillary tube), the surface tension, and other physical factors do play some part. A change in the viscosity of the solution because of a temperature change is not usually significant unless the temperature change is greater than about ± 4°C.

The nebulization of the sample solution is a topic which is not well understood, although Kirkbright and Sargent have discussed it to some extent (Ref. 3, Secs. 8.2 and 12.1). The drop-size distribution, like the aspiration rate, is a function of the viscosity and surface tension of the sample solution with the viscosity being the predominant factor. The nebulization efficiency is affected by the mean droplet size, as is solute evaporation both inside the burner and in the flame. It is evident that the viscosity of the sample solution plays a complex and integral role in the atomization of the analyte and underlines the need to ensure that the solution matrix and acid concentrations of the samples and standards are as similar as possible. Cresser (Supp. Ref. 7) has suggested the use of an impact cup as an aid in the aspiration of solutions with high analyte concentrations, and Fry and Denton (Supp. Ref. 8) have described the characteristics of a high solids nebulizer. This subject has been covered in considerable detail in a review by Langmyhr (Supp. Ref. 6).

The most common burner is the laminar flow type (i.e., the oxidant and fuel are premixed and reach the flame in a laminar flow which produces a low-"noise", stable, reproducible, and high temperature flame). Burners which are commonly used have a 10 cm slot for air–acetylene flames and a 5 cm slot for nitrous oxide–acetylene flames. They are most often made of titanium, and modifications to the initial design have improved their flow stability and resistance to the build up of carbon around the slot. Many different burner designs have been tried, each with its advantages and disadvantages. These have been discussed by Kirkbright and Sargent (Ref. 3, Sec. 6.3.1) and by Herrmann (Ref. 30, Chapter 3). Because the flame is part of the optical path, the operator must be able to adjust the vertical and horizontal positions of the burner accurately, reproducibly, and rapidly, as it is one of the instrument parameters that must be optimized for each element.

7.4.4 Double Beam Operation

The early AAS units were usually of the single beam type with the detector being exposed only to that light originating in the HCL and passing through the atom cell and monochromator. These units were subject to drift due to fluctuations in the output of the HCL either during its warm-up period (which could last for 20 min or more) or for other reasons of instability in the lamp output. The double beam concept used in other types of spectrometers was introduced to circumvent these problems, although single beam units are still widely used.

In those instruments designed to operate in the double beam mode, the light beam is split, and part of it, the analytical beam, passes through the atom cell. The second part, often called the reference beam, passes around the atom cell and is either recombined with the analytical beam or the two are kept separate in the monochromator. The beams can then be measured by a single photomultiplier controlled by a lock-in amplifier or by separate photomultipliers, depending on the particular optical arrangement used. The measured intensity of the reference beam is used to correct that of the analytical beam for any fluctuation in intensity not due to conditions in the atom cell. Descriptions of several single and double beam instruments can be found in the textbooks on AAS previously referred to in this chapter (Ref. 2, pp. 112–113; Ref. 3, pp. 426–430; Ref. 51, pp. 359–361). One drawback inherent in the use of a double beam mode is a loss in the radiant energy from the HCL which reduces the sensitivity of the measurement.

7.4.5 Background Correction

Not all of the absorption phenomena observed in the atom cell during an AAS determination need be related only to analyte atoms of interest. At some wavelengths the flame of the atom cell may produce a strong absorption or emission and thus a high background measurement (Ref. 3, pp. 445–446). The Bi 306.77 nm line, for example, is subject to interference from the OH 306.8 nm band head which both absorbs part of the radiation from the HCL and is emitted with sufficient intensity by the flame to cause high signal noise. Most of the problems associated with flame emission can be overcome by modulating the source and the detector so that unmodulated emission from the flame is rejected by the ac amplifier. Excessive flame emission can saturate the detector, however, and damage it or give rise to a high signal noise (Ref. 1, p. 291).

Anions, organic solvent, and other species have absorbance bands which overlap analyte resonance lines in some instances and this, and light-scattering from particles in the atom cell, can give rise to background errors. The use of the carbon rod furnace has introduced background problems due to absorption and scattering from the sample matrix as it is atomized which are much more severe than those encountered when using a flame. All of these problems have led to the introduction of background correction provisions in AAS instrumentation.

One means of correcting for background (and there are many variations on this theme although the principle remains the same) is to pass a light beam from a continuum source (such as a deuterium lamp) through the flame and into the monochromator and detection system at the same time as the analyte beam is passing through the same optical path. Each light source is moderated at a different fequency so that the signals can be electronically separated. The signal from the analyte beam is made up of components representing both the analyte and the background, whereas that of the continuum source has only a background component. When the latter is subtracted (electronically) from the former, the corrected absorption measurement of the analyte is displayed.

It should be pointed out that background correction is not required in the case of most flame AAS determinations. Some of the more commonly requested elements which do require its use in the analysis of rocks and minerals at the trace level are Ag, Zn, and Cd. Background correction should not be used in situations where it is not required because a loss of sensitivity is a direct result of its use. This is due to the increased "noise" level of a deuterium lamp as compared to that of a regular HCL and to the increased uncertainty of the value of the difference when one measurement is subtracted from another. The problems attendant upon the use of a deuterium lamp can be avoided in some instances by the selection of a neighboring (with respect to wavelength) nonresonance line of another element which exhibits appreciable intensity. Such lines are not often available, however, and AAS determinations involving background correction most often involve the use of a continuum source.

7.5 INTERFERENCES

The interferences associated with the AAS technique are commonly categorized as spectral, chemical, and physical, with no firm agreement as to the precise meaning of each term. Each presents its own set of problems, some of which have been mentioned earlier, and each has been discussed thoroughly, although not always correctly, in the literature. Most texts on AAS give a good discussion of these three categories of interferences, and the subject is also covered in the review publications mentioned at the beginning of this chapter. Interference effects will be discussed briefly here from a practical point of view.

The danger of an interference effect lies in not recognizing that such an interference exists. The analysis of a series of SRMs can often reveal an interference problem, and the method of standard addition can be used to confirm a suspected spectral interference (Ref. 30, p. 321). A second method is to prepare solutions of the unknown in two concentration levels, preferably in a 2:1 ratio, and if an interference exists, the results obtained for the two concentration levels will not be in the same ratio of 2:1. The analyst must either be cognizant of potential problems from one or more of the three types of interference arising out of sample composition, or confirm the absence of interferences by means of a test as described above.

7.5.1   Spectral Interferences

Spectral interferences result when absorption and emission lines of other species present in the light source, the sample, and the sample cell interfere with the analyte absorption line. An example of this is the mutual interference between Si 250.6899 nm and V 250.6905 nm lines.[52] Fassel et al. discussed this and other spectral line interference in some detail in 1968[53] and many more observations have been reported since then. One useful source for the identification of spectral interferences is the *Handbook of Flame Spectroscopy* (Supp. Ref. 9). The simple solution to the problem, once it has been identified, is to select another resonance line of the analyte. The absorption of the analyte line can be due to a molecule formed in the flame having an absorption band close to or including the analyte line, such as the absorption of the Ba 553.55 nm radiation by CaOH (Ref. 3, p. 532). The use of background correction can overcome this type of interference, but bandpass adjustment and a change in acid concentration have been reported to be effective[54] in certain circumstances. The use of a hotter $N_2O–C_2H_2$ flame in which such interfering species as the CaOH radical decompose is also effective.

Another source of spectral interference can originate in the HCL (Sec. 7.4.1), either from the filler gas, from contaminants, or from an injudicious use of certain alloys for the metallic parts of the lamp. These problems are seldom experienced with modern lamps, but the analyst should be aware of them, especially if he is using multielement lamps, and often the use of another resonance line of the analyte will circumvent any such interference.

Absorption and emission by the flame itself may constitute a spectral interference (Sec. 7.4.5), and background correction, the use of another analyte resonance line or modulation of the source, can be used to overcome the problem.

7.5.2   Chemical Interferences

Chemical interferences are those in which chemical reactions in the analyte solution or in the atom cell have an effect on the free atom population of the analyte in the absorption zone. These interferences are commonly subcategorized into those occurring in the *condensed* phase and those in the *vapor* phase, and they have been discussed at some length in several texts.[3, 30, 51] It should be stressed that chemical interferences *cannot* be corrected by standard addition techniques.

a.   Condensed Phase Chemical Interferences

Interferences in the condensed phase include the formation of salts with a high-melting point which have been stripped of the evaporated solvent in the aerosol. They are difficult to decompose and atomize in order to obtain a sufficiently high atom concentration in the absorption zone. An example of this problem is the depressant effect of anions such as silicate, aluminate and phosphate which form compounds with the alkaline earth elements. The extent of this type of interference is dependent on the flame composition, the region of the flame in which the absorption measurement is made, and the size distribution of the nebulized sample droplets. This is because the ratio of the rate of evaporation of the difficultly atomized compound (formed between the analyte and the interfering species) and the rate of evaporation of the uncomplexed analyte particles is the determining factor in the degree of interference observed (Ref. 3, p. 515). Any factor (such as flame conditions or nebulizer efficiency) which affects this ratio will affect the degree of interference encountered.

The problem can be alleviated to some extent by optimizing conditions such as nebulizer rate, height of observation, flame conditions, and flame type. The use of a nitrous oxide–acetylene flame rather than an air–acetylene flame eliminates most chemical interferences of the type associated with the formation of a high-melting and refractory compound. If these precautions fail to resolve the problem, the addition of a releasing or chelating agent to the analyte solution can often eliminate the interference. The releasing agent (often La) combines with the interfering species, either preferentially or by mass action, if a sufficient amount is added, and prevents it from forming the refractory or difficultly atomized compound with the analyte. The chelating agent (EDTA, for example) combines instead with the analyte and thus prevents it from forming the refractory compound with the interfering species.

b.   Vapor Phase Chemical Interferences

The equilibrium established in the flame for the purpose of producing the requisite absorbing analyte atoms can be influenced by the presence of interfering species which will affect the population of the analyte atoms able to absorb energy from the hollow cathode light beam. One mechanism postulated for these observed interferences is the formation of gaseous compounds (oxides) of the analyte atom and concomitant atoms in the flame. In a fuel-rich flame the partial pressure of oxygen is low, and because of the concomitant atoms reacting with oxygen to form oxides, the dissociation of the gaseous oxide molecules of the analyte ion is greater, thus leading to an enhanced signal for the analyte in the presence of the concomitant species. Examples of this are the mutual enhancement effect of Ti and Al, and of V on Ti and Al (Ref. 30, p. 528). The use of a nitrous oxide–acetylene flame will overcome most of the problems caused by this form of interference.

Another mechanism for vapor phase chemical interference is found in the ionization equilibrium, an effect restricted mostly to the alkali metals in an air–acetylene flame and to the alkaline and rare earths, Al, V, and Ti in a nitrous oxide–acetylene flame. The effect is due to the significant influence exerted by the metals on each other's ionization equilibrium because of their low ionization potentials. An excess of K in a Na solution will, for example, suppress the ionization of the Na and thus enhance its absorption signal.

This same effect can be observed for Ba in the presence of alkali metals, but because the Ba is more highly ionized in a nitrous oxide–acetylene flame than in an air–acetylene flame, the ionization enhancement is much more effective in the former than it is in the latter.[55]

The most effective way of overcoming the analytical problems encountered as a result of this effect is to add an ionization suppressant (i.e., an easily ionized element) at a concentration high enough to overcome or "swamp" the ionization effect of the elements in the sample on one another. An example of this is the addition of Cs to the synthetic standards described earlier (Sec. 7.3.1), although Na, K, and Sr are also effective and are much cheaper. Many workers use releasing agents such as La, Ba, and Sr (which in certain instances can also act as ionization suppressants) to relieve chemical interferences.

7.5.3 Physical Interferences

Physical interferences are a result of those physical properties of the sample and sample solution which can affect the atom concentration in the absorption cell and the absorption signal generated by it (Ref. 3, pp. 507–514; Ref. 30, pp. 226–228).

The scattering of light from particulate matter in the flame (usually undissociated refractory salt particles) can lead to an apparent absorption which will suggest a higher concentration of analyte than is actually present. This problem can be overcome by background correction. It is one of the more significant encountered in the use of carbon rod and similar flameless atom cells and makes background correction mandatory for most applications of these techniques to rock and mineral analysis. In some instances, however, even this is not sufficient.

The viscosity of the analyte solution can affect the aspiration rate and hence the atom concentration in the flame. The density and surface tension of most solutions do not usually vary enough to make an observable difference in the aspiration rate unless the sample solutions are very concentrated. High salt concentrations can also lead to encrustations of solid material in the burner slot, and this affects gas flow, flame stability and nebulization efficiency, all of which affect the absorption signal. Variations in the salt and acid concentrations of the sample solution can result in similar variations in the viscosity of the solution, and for this reason the sample and standard solutions must have acid and solute concentrations which are as closely matched as possible.

7.6   OPTIMIZATION OF ANALYZING CONDITIONS

An AA spectrometer has many different parameters which must be adjusted for any specific determination. The HCL current, flame conditions, burner height, slit width, wavelength, aspiration rate, and the operating mode (single or double beam operation, integration, background correction, etc.) are all variables, many of them interdependent, which must be selected and optimized (Ref. 30, pp. 187–190; Supp. Ref. 5).

A key word in the above paragraph is *interdependent*. In the determination of Si, Al, and Ti for example, it has been the experience of the BCM that the burner height and flame stoichiometry must be adjusted together to give optimum absorption. Each HCL source is unique, and the optimum lamp current must be established by monitoring absorption and noise as the lamp current is increased. The HCL current must be high enough to avoid an excessive noise level and low enough so that the emission lines are not broadened and the sensitivity decreased. The correct support gas for the flame must be chosen (N_2O is necessary for Al determinations, for example), and, as mentioned earlier, the flame stoichiometry and burner height must be optimized.

The slit width must be chosen to optimize the signal-to-noise ratio, not just the signal alone. In AAS the bandpass should be only as narrow as is required to eliminate source emission lines other than the analytical line of interest. Only if the source has a complex spectrum or has emission lines very close to the analytical line is a narrow slit width required. When optimizing conditions for a particular determination, it is not always best to select the most sensitive line. A less sensitive line can increase the dynamic range of the analysis with no recourse to dilution being required. Alternatively, a particular spectral interference may be avoided by using an analytical line different from the one listed in the literature supplied by the manufacturer. Whatever the case, the analyst should make himself aware of alternate analytical lines which might be used.

The operating mode of the instrument can be selected so as to significantly improve a determination, but the analyst should be aware that double beam operation, for example, time-shares the signal and ratio portions of the source and thus increases optical noise. Background correction decreases the signal-to-noise ratio to a large extent and, in the case of some trace level determinations, raises the detection limits above the concentration level of the analyte.

Another important part of optimizing the instrumental conditions is the day-to-day adjustments required of the optical alignment and focusing as lamps are changed and different operators use the instrument. Before any determination is made, the instrument should be allowed to reach thermal equilibrium after the flame has been ignited, and the instrumental response should be checked with a known test solution to ensure that the expected absorption is attained, that there is no decrease in the linear range, and that the stability and precision of readings for both the base aspirant and the test solution absorptions are satisfactory. Only when these conditions have been satisfied should the analyst proceed with the determination.

References

1. J. A. Dean and T. C. Rains, Eds., *Flame Emission and Atomic Absorption Spectrometry*, Vol. 1, *Theory*, Dekker, New York, 1969.
2. B. V. L'Vov, *Atomic Absorption Spectrochemical Analysis*, American Elsevier, New York, 1970.
3. G. F. Kirkbright and M. Sargent, *Atomic Absorption and Fluorescence Spectroscopy*, Academic, New York, 1974.
4. R. W. Dabeka, A. Mykytiuk, S. S. Berman and D. S. Russell, *Anal. Chem.*, **48**, 1203 (1976).
5. J. W. Mitchell, *Anal. Chem.*, **45**, 492A (1973).
6. C. O. Ingamells, *Anal. Chem.*, **42**, 323 (1970).
7. F. J. Langmyhr and S. Sveen, *Anal. Chim. Acta*, **32**, 1 (1965).
8. F. J. Langmyhr and K. Kringstad, *Anal. Chim. Acta*, **35**, 135 (1966).
9. J. R. Foster, *Can. Inst. of Min. Bull.*, 85 (Aug. 1973).
10. C. O. Ingamells, *Talanta*, **21**, 141 (1974).
11. *Annual Reports on Analytical Atomic Spectroscopy*, Vol. 8, The Chemical Society, 1979.

12. R. Horton and J. J. Lynch, Geological Survey of Canada, Paper 75-1, Part A, 213, 1975.

13. M. Fratta, *Can. J. Spectrosc.*, **19**, 33 (1974).

14. J. R. Watterson and G. J. Neuerberg, *J. Res. U.S. Geol. Surv.*, **3**, 191 (1975).

15. T. Golembeski, *Talanta*, **22**, 547 (1975).

16. I. Rubeska, J. Korechkova, and D. Weiss, *At. Absorpt. Newsl.*, **16**, 1 (1977).

17. E. P. Welsch and T. T. Chao, *Anal. Chim. Acta*, **76**, 65 (1975).

18. F. W. E. Strelow, C. I. Lieberg, and A. H. Victor, *Anal. Chem.*, **46**, 1409 (1974).

19. A. Mazucotelli, R. Frache, A. Dadone, and F. Baffi, *Talanta*, **23**, 879 (1976).

20. R. Frache, A. Mazucotelli, A. Dadone, F. Baffi, and P. Crescon, *Analusis*, **6**, 294 (1978).

21. E. Steinners, *At. Absorp. Newsl.*, **15**, 102 (1976).

22. J. Korkisch, H. Bübner, I. Steffan, G. Arrhenius, M. Fisk, and J. Frazer, *Anal. Chim. Acta*, **83**, 83 (1976).

23. P. E. Moloughney, *Talanta*, **24**, 135 (1977).

24. J. C. Van Loon, *Z. Anal. Chim.*, **246**, 122 (1969).

25. P. E. Moloughney and G. H. Faye, *Talanta*, **23**, 377 (1976).

26. A. Diamantalos, *Anal. Chim. Acta*, **90**, 179 (1977).

27. B. W. Smith and M. L. Parsons, *J. Chem. Educ.*, **50**, 679 (1973).

28. *"Analar" Standards for Laboratory Chemicals*, 6th ed., Analar Standards Ltd. London, 1967.

29. *Reagent Chemicals*, 5th ed., American Chemical Society Specifications, Washington, 1974.

30. J. A. Dean and T. C. Rains, Eds., *Flame Emission and Atomic Absorption Spectrometry*, Vol. 2, *Components and Techniques*, Dekker, New York, 1971.

31. S. Abbey, *Geostand. Newsl.*, **1**, 39 (1977).

32. S. Abbey, "Studies in 'Standard Samples' for Use in the General Analysis of Silicate Rocks and Minerals, Part 6," 1979 ed. of "Usable" Values, Geological Survey of Canada, Paper 80-14, 1980.

33. I. Rubeska, *Geostand. Newsl.* **1**, 15 (1977).

34. J. D. Ingle, Jr., *Anal. Chem.*, **46**, 2161 (1974).

35. A. T. Zander, T. C. O'Haver, and P. N. Keliher, *Anal. Chem.*, **48**, 1166 (1976).

36. R. Stephens, *Talanta*, **24**, 233 (1977).

37. G. E. M. Aslin, *J. Geochem. Explor.*, **6**, 321 (1976).

38. G. M. Hieftje, T. R. Copeland, and D. R. Olivares, *Anal. Chem.*, **48**, 142R (1976).

39. *Analytical Methods for Atomic Absorption Spectrophotometry*, Perkin-Elmer Corp., Mar. 1971.

40. M. Margoshes, *Anal. Chem.*, **39**, 1093 (1967).

41. M. L. Parsons, B. W. Smith, and P. M. McElfresh, *Applied Spectroscopy*, **27**, 471 (1973).

42. J. D. Norris and T. S. West, *Anal. Chem.*, **46**, 1423 (1974).

43. R. J. Lovett and M. L. Parsons, *Anal. Chem.*, **46**, 2241 (1974).

44. J. M. Mermet and C. Trassy, *Appl. Spectrosc.*, **31**, 237 (1977).

45. T. C. Rains and O. Menis, in *Accuracy in Trace Analysis: Sampling, Sample Handling and Analysis*, National Bureau of Standards Special Publication **422**, 1976.

46. B. V. L'Vov, *Talanta*, **23**, 109 (1976).

47. F. J. Langmyhr, *Talanta*, **24**, 277 (1977).

48. S. Chilov, *Talanta*, **22**, 205 (1975).

49. A. M. Ure, *Anal. Chim. Acta*, **76**, 1 (1975).

50. J. B. Willis, *Nature*, **207**, 715 (1965).

51. H. H. Willard, L. L. Merritt, and J. A. Dean, *Instrumental Methods of Analysis*, 5th ed., Van Nostrand, Princeton. NJ, 1974.

52. H. Urbain and G. Carret, *Analusis*, **3**, 110 (1975).

53. V. A. Fassel, J. O. Rasmuson, and T. G. Cowley, *Spectrochim. Acta*, **23B**, 579 (1968).

54. J. E. Poldoski, *At. Absorpt. Newsl.*, **16**, 70 (1977).

55. R. Cioni, A. Mazzucotelli, and G. Ottonello, *Analyst*, **101**, 956 (1976).

Supplementary References

1. N. M. Kuz'min, V. S. Vlasov, V. K. Krasil'shchik, and V. G. Lambren, *Zavod. Lab.*, **43**, 1 (1977).

2. P. Hannaker and T. C. Hughes, *J. Geochem. Explor.* **10**, 169 (1978).

3. A. R. Date, *Analyst*, **103**, 48 (1978).

4. N. W. Bower and J. D. Ingle, Jr., *Anal. Chem.*, **51**, 72 (1979).

5. N. W. Bower and J. D. Ingle, Jr., *Anal. Chim. Acta*, **105**, 199 (1979).

6. F. J. Langmyhr, *Analyst*, **104**, 993 (1979).

7. M. S. Cresser, *Analyst*, **104**, 792 (1979).

8. R. C. Fry and M. B. Denton, *Appl. Spectrosc.*, **33**, 393 (1979).

9. M. L. Parsons, B. W. Smith, and G. E. Bentley, *Handbook of Flame Spectroscopy*, Plenum, New York, 1975.

CHAPTER 8

THE DETERMINATION OF MAJOR AND MINOR CONSTITUENTS IN ROCKS AND MINERALS BY ATOMIC ABSORPTION SPECTROSCOPY

8.1 GENERAL

AAS, although a relatively new analytical tool, has simplified the analytical chemist's life enormously. Nowhere is it more evident than in the determination of the major and minor constituents of rocks and minerals. As with other so-called "rapid" methods, it is no longer necessary to perform multiple precipitations, weighings, washings, filterings, separations, and other time-consuming steps which require much patience and experience in order to obtain a "complete" rock analysis. Instead, AAS can be used to measure virtually all of the major and minor metallic elements commonly requested.[1] It must be pointed out, however, that AAS is not a magical trick. The same degree of patience, experience, and knowledge which was demanded of an analyst in a classical analytical procedure is necessary for the successful utilization of AAS.

8.2. SAMPLE DECOMPOSITION

Sample decomposition has been discussed at some length in Chapter 4 and some general considerations were mentioned in Sec. 7.1 of Chapter 7. The two methods most commonly used for decomposing a rock or mineral sample in preparation for the determination of its major constituents by AAS are HF digestion using a Teflon bomb and some form of a lithium borate fusion.[1] The former has the disadvantage that some resistant minerals such as tourmaline, cassiterite, zircon, and titanium oxides are not always completely dissolved,[2] and some problems are encountered with pyrite and galena.[3] The judicious selection of the appropriate lithium borate flux (Table 4.1) will result in the satisfactory decomposition of most of the common rock-forming minerals, including those high in Al_2O_3.[4] A study of some 47 minerals has shown, however, that difficulties will be encountered with sulfides, some metal oxides and rare earth phosphates, and zircon and chromite.[5] A sample containing a high percentage of sulfides should first be treated with aqua regia, the solution filtered, and the residue washed prior to fusion in a Pt or Au–Pt

263

alloy crucible in order to prevent serious attack of the crucible by the sulfide and to ensure complete decomposition; this is not a concern when graphite crucibles are used for the fusion.

8.2.1   Procedure for Decomposition with Teflon Bomb

The method given here is based on that described by Bernas[6] for the determination of the major elements in silicate minerals. It is slightly modified as described by Buckley and Cranston.[7]

Apparatus

A Teflon-lined acid digestion bomb equivalent to the Parr 4745 model (Sec. 4.5.1), 100 ml volumetric flasks, funnels, 100 ml polyethylene bottles, a magnetic stirrer, Teflon stirring bars, and a drying oven.

Reagents

48% HF, freshly prepared aqua regia (HCl–HNO$_3$, 3:1), analyzed reagent grade H$_3$BO$_3$, and distilled H$_2$O.

Procedure

Transfer the carefully weighed sample (about 100 mg) to the bomb and wet it with 1 ml of aqua regia. Add 6 ml of HF, screw the cap into place by hand (but do not tighten excessively), and put the bomb in a drying oven at 100°C for 45 min, or more if clay or refractory minerals are known to be present. Remove the bomb but *do not* unscrew the cap until the bomb has cooled completely. Wash the contents of the bomb into a polypropylene bottle containing 5.6 g of H$_3$BO$_3$ in 20 ml of water, and a Teflon-coated magnetic stirring bar. Ensure that all of the material is washed out of the bomb. Place the bottle on a magnetic stirrer and stir the contents to assist the dissolution of any precipitated metal fluorides.[8] Transfer the sample to a 100 ml flask by means of a funnel, and wash the bottle and the funnel thoroughly with water. Adjust the volume to 100 ml in a volumetric flask and return the solution to the polypropylene bottle for storage.

PRECAUTIONS   The solution should not be kept in the glass volumetric flask for more than a few minutes, and the distilled water used for washing and dilution should be stored in polypropylene and not in glass reservoirs. The water should also have passed through a cation-exchange resin subsequent to distillation (Sec. 2.1.3) to remove traces of metals such as Si, K, and Na. Reagent blanks should be carried through the complete process. Gill and Kronberg have reported that standard SiO_2 solutions prepared from a slightly modified version of this digestion procedure are stable for 6 months or more.[9]

8.2.2 Procedure for Lithium Metaborate Fusion

The method described here is used extensively in the BCM laboratories as a sample dissolution step prior to the AAS measurement of major, minor, and some trace constituents. It is adequate for most nonsulfide materials and can be applied to sulfides subsequent to roasting them, although the roasting step may not be necessary if graphite crucibles are used. The reader should consult the paper by Bennett and Oliver[4] for a description of the application of lithium borate fluxes to a wide range of minerals, oxides, and ceramics (see also Sec. 4.3.6). The resulting solutions are stable for several months with no indication of silicate polymerization and precipitation.

Apparatus

Nonwetting platinum crucibles with lids, muffle furnace, 200 ml polypropylene beakers, Teflon-coated stirring bars, magnetic stirrer, Vycor™ tray, enamel tray, funnels, 200 ml graduated flasks.

Reagents

Lithium metaborate, anhydrous ($LiBO_2$, G. Frederick Smith Chemical Co.); hydrofluoric acid, 48%, BDH Chemicals Ltd., "Aristar" grade or equivalent; boric acid, reagent grade; nitric acid, reagent grade; cesium chloride solution—dissolve 25.34 g CsCl, BDH Chemicals Ltd., "Analar" grade or its equivalent, in 1000 ml of distilled, deionized water; water, distilled, deionized, and stored in well-cleaned polyethylene containers.

Procedure

Weigh accurately about 0.2 g of each sample into platinum crucibles, each containing 1.000 g of $LiBO_2$, mix the sample and the flux, place the covered crucibles in a Vycor™ tray, and slide the tray into a muffle furnace preheated to 1050°C. After 40 min, remove the tray and swirl each crucible individually. Put the tray back into the furnace and continue the fusion for 10 min more. Remove the tray from the furnace, allow the crucibles to cool, and place them in a shallow enamel basin containing 10% HNO_3 for 15 min to remove foreign matter from the bottoms of the crucibles. Immerse each crucible and lid in a 150 ml polyethylene beaker containing 100 ml of 4% v/v HNO_3. Add a Teflon-coated magnetic stirring bar to each crucible, cover the beakers, and stir the contents of the crucible and beaker on a magnetic stirrer for 2 hr or until all the melt has dissolved.

Add 3 ml of 48% HF and stir for 10 min, add 20 ml of H_3BO_3 solution (50 g/l) and stir for 10 min more. Remove the lids and crucibles, washing each thoroughly into the solution, and transfer the solutions quantitatively to 200 ml volumetric flasks. Add 10 ml of CsCl solution to each, dilute the solutions to 200 ml, and transfer them to polyethylene bottles for storage.

PRECAUTIONS   Reagent blanks should be carried through the complete procedure. The fusion time may have to be extended for an occasional sample. The stirring time must be sufficiently long to ensure complete dissolution of the melt. The solutions should not be in contact with the glass for more than 10 min, and the distilled, deionized water used should be stored in plastic containers.

8.3   DETERMINATION OF THE MAJOR AND MINOR ELEMENTS

The term "major and minor elements" refers here to Si, Al, Fe (total), Mg, Ca, Na, K, Ti, Mn, Ba, and Sr. Other elements of those listed in Sec. 1.3, such as P, Cl, V, Ni, Co, Cr, and Zr, will be covered in the next chapter on trace elements or have been already covered in Chapters 5 and 6. The solutions obtained by either of the dissolution techniques described in Sec. 8.2 can be used directly for the AAS determination of all of these elements in the concentration ranges they normally exhibit. The methods given here, however, will assume a lithium metaborate fusion, and the standards suitable for use are those described in Secs. 7.3.1 and 7.3.2.

The following description of the AAS measurement procedure is applicable to all of the elements listed above and will not be repeated for each element. The reader should refer to the description of the procedure and to Table 8.1.

Most of the interferences mentioned in the discussion of each element do not occur at a statistically significant level in the solutions from the fluoborate fusion techniques as used in the BCM and other laboratories. The exception is Ti, for which there is evidence that some small interference is present in the BCM method.

8.3.1.   AAS Measurement

Procedure

The following method is based on the sample decomposition scheme and the synthetic standards described in Secs. 8.2.2 and 7.3.1, respec-

TABLE 8.1   Instrumental Conditions Used for the Determination of Major Elements by AAS.

Analyte	Wave-length (nm)	Slit Width (nm)	Lamp Current (mA)	Flame Oxidant	Burner Angle (deg)	Expan-sion	References
Si	251.6	0.7	12	N_2O	0	2	3, 9–14
Al	309.3	0.7	13	N_2O	0	4	15–17
Fe	248.8	0.2	14	N_2O	0	3	18–22
Mg	285.2	0.7	14	N_2O	90	2	23–29
Ca	422.7	0.7	12	N_2O	90	1.5	28–32
Na	589.0	0.7	10	Air	45	1.4	5, 28, 33–36
K	766.5	2	10	Air	45	3	5, 28, 33–38
Ti	365.4	0.7	14	N_2O	0	14	6, 21, 28, 36–45
Mn	279.5	0.7	14	N_2O	0	8	28, 36, 37, 45–49
Ba	553.6	0.2	14	N_2O	0	20	27, 28, 50–56
Sr	460.7	0.2	10	N_2O	0	5	27, 28, 57–59

1. The base aspirant in all cases is 4% HNO_3 (v/v).
2. The instrument used to establish these conditions was a Perkin Elmer 107 in which the slit width has only three adjustments. This parameter, as well as the optical alignment, lamp current, burner height, and flame stoichiometry, will have to be optimized for each instrument and for each set of analyses.
3. A 3 sec integration interval was used in all cases. A uniform waiting period between the beginning of the aspiration of a solution and initiation of the integration sequence is mandatory (6 sec in the case of this instrument).
4. C_2H_2 is used as the fuel in all cases.

tively, and is the method used in the BCM laboratory. However, each laboratory will have to adjust the given instrument settings to those that will yield optimum results with its own equipment.

The instrument is turned on, the lamp current is adjusted, the flame is ignited and switched to $N_2O–C_2H_2$, and the instrument is allowed to stabilize for 20–30 min (see Table 8.1 for instrumental conditions). The instrumental parameters (fuel-to-oxidant-flow ratio, burner height, optics alignment, etc., see Note 2, Table 8.1) are optimized (see Sec. 7.6), and the measurement of the samples and standards is commenced. There are, however, several precautions which must be observed to ensure that satisfactory results are obtained. The time between the beginning of as-piration of the solution and the initiation of the measurement sequence must be uniform for all samples and standards, and it must be of sufficient duration that an equilibrium between the solution and the flame will be attained. A time of 6 sec was found to be sufficient in the case of the BCM, but it will have to be determined for each instrument. It has been found that a background reading using the base aspirant (Note 1, Table 8.1) is required after each sample and each standard reading and that this background must be subtracted from the preceding measurement.

Mr. P. Ralph of the BCM has developed a computer program which permits the running of samples in batches of up to 50 along with 12 synthetic standards. The samples and standards are run in a set sequence, replicate measurements are made, and the time of each measurement is recorded so that a time-based drift correction can be applied to the data. A small microprocessor with a timer is used to capture the data (see Fig. 2.5) on punched tape, which is then read into a terminal for treatment of the data by a large computer. This has led to the attainment of a relative standard deviation of about 1% (often less) on a routine basis.

The program developed for this purpose incorporates a time-based drift correction of the data derived from the time differences of the replicate measurements made (and lists the order of the polynomial required to fit it), an analysis of flame noise (initial and final) and photometric variance, the analytical calibration and an analysis of the variance about the best fitted "line". The relative standard deviation of the results for a standard sample replicated three times in each set of samples and the extent of agreement with the accepted value for that standard are calculated, as well as the relative standard deviation for each element as calculated from the replicate measurements made and the detection limit for each element. This information, as well as the analytical data for CO_2, S, P_2O_5, H_2O, and FeO and the total of the analysis, are produced as a printout, which is then checked for errors, the agreement between the hidden duplicates is assessed, and then the printout is forwarded to the geologist who submitted the samples. The program is a complex one and requires a large computer with adequate word length, memory, and precision to perform the calculations involved.

If no such program or computer facility is available, an alternative method is to make a preliminary analysis in order to be able to match standards with the samples. Each sample is bracketed with the two standards closest to it in concentration, and measurements are made on at least two other standards as well. The measurements are repeated three or four times to yield an acceptable average. Background corrections must be applied to each measurement, a calibration curve plotted, and the concentration of the unknown constituent then determined. A similar method has been described by Burdo and Wise[10] who report an absolute accuracy of 0.2–0.3%.

8.3.2 Silicon

The precision and accuracy of the determination of silicon (it is customarily reported as silica, SiO_2) in rocks and minerals by AAS is still the subject of some disagreement among analysts.[9-11] It is evident, however, from results appearing in the literature, that AAS is capable of producing results for Si over the wide concentration range encountered by most rock and mineral analysts (see also Sec. 1.4.1, Table 1.2) which equal in accuracy and precision those obtained by the classical gravimetric methods.

There are problems, of course, but it is possible to resolve them satisfactorily. The polymerization of SiO_2 in both standard and sample solutions has been overcome at the BCM by introducing HF and H_3BO_3 to complex the Si as a fluosilicate (see also Supp. Ref. 1). Burdo and Wise have described a technique to prevent silica polymerization utilizing a silicomolybdate complex.[10]

Interferences from several acids and elements in the determination of Si have been reported.[12, 13] Musil and Nehasilova[12] observed that Ni and mineral acids have a depressive effect, that Na, K, Al, Fe, Ti, Mn, V, Mg, and Cs enhance the Si signal, and that Cu and Zn give a negligible interference. Urbain and Carret have discussed the spectral source of the V enhancement of the Si signal in some detail[14] (Sec. 7.5.1). The interference due to Na is reported to reach a plateau at a concentration of about 6000 ppm.[3] It is thus apparent that the matrix of the standard must match that of the sample as closely as possible in order to account for possible interferences and to obtain an accurate calibration. The effect of these interferences has been rendered negligible under the conditions used in the BCM which, for the analysis of silicates, include the use of a fluoborate medium, suppression of ionization interferences with Cs, and the use of matching standards.

8.3.3 Aluminum

While the $N_2O-C_2H_2$ flame must be used to achieve acceptable sensitivity for the determination of Al by AAS, the technique is capable of producing very satisfactory analytical results. Examples of interference have been given by Van Loon[15] and by Cobb and Harrison[16] who found that this could be eliminated by the addition of a "releasing" agent (La^{3+}). Guest and MacPherson[17] observed that the addition of an easily ionized salt (in their case 6000 ppm Na^+) eliminated the interference effects due to K, Fe, Ti, Ca, and Mg observed in solutions containing no added Na salt. The addition of Na did not eliminate the effects they observed due to Si (suppression) and Mn (enhancement).

8.3.4   Iron

An air–C_2H_2 flame gives better sensitivity for iron than does a N_2O–C_2H_2 flame, but because iron is a major constituent of most silicates, this is not so important. The strength of the iron signal is not affected significantly by flame stoichiometry nor by the region in the flame through which the beam passes, except insofar as these parameters influence the magnitude of the depressant interference due to Cu, Co, and Ni.[18] Al and Si also depress the iron signal[19] under certain conditions. None of these interferences has been observed in the method used in the BCM for the analysis of silicate rocks. This indicates that, under the conditions described, the magnitude of the interferences is smaller than the variance resulting from instrumental and other factors (less than about 1% relative standard deviation).

Korkisch et al.[20] have reported the determination of micrograms per gram levels of Fe (as well as V and Cu) by AAS in the eluate from a Dowex 1-X8 ion–exchange preconcentration step. The rock sample was decomposed with a mixture of $HClO_4$ and HF. Heinrichs and Lange[21] carried out the microdetermination of Fe, Al, Mn, and Ti by flameless AAS subsequent to an $HClO_4$–HF decomposition of the silicate sample. This latter technique is potentially very useful in analyses for which only a small quantity of sample is available as, for example, in the study of a new mineral.

The general conditions for the determination of iron are discussed by Rains[22] and in a report by the Analytical Methods Committee of the Chemical Society (Supp. Ref 2).

8.3.5   Magnesium

There have been relatively few problems encountered in the application of AAS to the determination of Mg. Flames utilizing both air and N_2O as the oxidant have been used, and the detection limits are such that no preconcentration steps are required for the Mg concentrations commonly encountered in rocks and other geological samples. The air–C_2H_2 flame is capable of greater sensitivity, however, and is to be preferred for low concentrations of Mg.

There have been some interferences reported in the AAS determination of Mg. Panday and Ganguly[23] reported spectral interference in the Mg 285.213 nm line by the 285.214 nm line of Tb. This admittedly unusual spectral interference can be avoided by utilizing the alternative (but less sensitive) 202.582 nm Mg line. Komarck et al.[24] have discussed the influence of several ions, including binary and ternary systems, on the AAS determination of Mg with an air–C_2H_2 flame. At a concentration of 1000 ppm the SiO_3^{2-}, Be^{2+}, Zr^{4+}, Ti^{4+}, and F^- were found to decrease considerably the absorbance of a 1 ppm solution of Mg (ranging in magnitude from 32% for SiO_3^{2-} to 92% for F^-). PO_4^{3-}, Zn^{2+}, Ca^{2+}, and many other anions and cations are reported not to affect the absorbance by Mg at these concentrations, but they are found to influence the effect of an interfering ion. For example, 1000 ppm of Al^{3+} decreased the absorbance of a 1 ppm solution of Mg by 59%. In the presence of 200 ppm of each of PO_4^{3-} and Zn^{2+}, however, this same amount of Al decreased the signal by 67%, whereas in the presence of 200 ppm of Ca^{2+} there was 100% enhancement of the Mg absorbance. The addition of the organic chelating agents 8-hydroxyquinoline and 5-sulfosalicylic acid overcame the effects of the binary and ternary interfering ion systems.

Chen and Winefordner[25] investigated the interference of trace levels (from 0.3 to 30 ppm) of Al^{3+} on Mg and the influence that flame height, acetylene flow, and an electric field had on the effect of this interference in an air–C_2H_2 flame. They concluded that a slightly fuel-rich flame and the use of the higher parts of the flame for measurement will reduce the magnitude of the interference. The addition of HCl also reduced the extent of chemical interference. The interference of SO_4^{2-} in the determination of Mg at high analyte concentrations (up to 15 µg/ml) has been described.[26] The addition of an excess of a suitable releasing agent such as La or Sr salts overcomes the depressant effect of the sulfate on the absorbance of the Mg.

Abbey et al. have used both air and N_2O as the support gases[27] depending on the Mg concentration level. French and Adams[28] have described a microchemical technique for the determination of Mg in minerals by AAS using an air–C_2H_2 flame. A general discussion of the AAS determination of Mg is given by David.[29]

8.3.6 Calcium

The determination of calcium by AAS before the use of N_2O as a support gas was often unsatisfactory because of strong interferences from such common ions as SO_4^{2-}, PO_4^{3-}, and Al^{3+}, although the addition of a releasing agent minimized the interference effect. In an N_2O–C_2H_2 flame precautions must be taken to ensure that potential ionization interferences are recognized and overcome.

Magill and Svehla[30] have investigated the influence of 53 ions on the determination of Ca by AAS. They have concluded that, in an $N_2O-C_2H_2$ flame, elements with low ionization potentials are likely to be a source of interference, whereas in an air–C_2H_2 flame the interference is caused by the formation of refractory compounds by cations and anions. They also report that the addition of an easily ionized buffer cation such as K^+ overcame the interference in an $N_2O-C_2H_2$ flame, and the combination of the use of the standard addition technique and Sr as a releasing agent gave acceptable results in an air–C_2H_2 flame. The range of ions studied included most of the common metals, and anions such as MoO_4^{2-}, NO_3^-, SO_4^{2-}, SiO_3^{2-}, and $(COO)_2^{2-}$.

An earlier study[31] of the determination of Ca by AAS using an $N_2O-C_2H_2$ flame came to the same conclusion. That is, the addition of K^+ or some other buffer ion, with 5-sulfosalicylic acid as a protector, overcame most of the interferences. Some ternary ion systems (e.g., $Cu^{2+}/Al^{3+}/PO_4^{3-}$) continued to exhibit an effect on the absorption signal of the Ca.

The effect of the PO_4^{3-} ion on the Ca signal in flame spectrometry has been the subject of many investigations. The interpretation of the mechanism of the observed interference ranges from the suppression of Ca ionization by P-containing species to a "lateral diffusion interference effect." This latter interpretation is based on an observation that mineral acids in the aspirated solutions can affect the horizontal distribution of free atoms in the flame.[32]

In summary, most analysts use an N_2O supported flame with an "ionization buffer" such as K, Sr, or some other easily ionized metal,[27] but an air–C_2H_2 flame with a releasing agent added to the solution has also been used. A technique has been described for the determination of Ca using an $N_2O-C_2H_2$ flame in the microanalysis of minerals.[28]

8.3.7   Sodium

The use of flame ES for the determination of Na (and K) dates back to the early 1930s, and a large volume of literature concerning both this and the subsequent flame photometric method has evolved. The application of the AAS method to the determination of Na and K has been equally successful. Flame emission is, however, capable of yielding lower detection limits and many workers still prefer this method.

The major interferences in the determination of Na by AAS and other flame techniques have been discussed by Ure and Mitchell.[33] Interference may be encountered from the absorption in the 600 nm region by CaOH radicals at high Ca concentrations (Ref. 34, pp. 679–683). One means of overcoming this problem is to use reference standards having Ca concentrations similar to those known to be in the samples. Alternatively, if the Na concentration is at a level usually found in silicate rocks, the less sensitive analytical line at 330.3 nm can be used to avoid the CaOH interference.

In the case of the doublet 588.99 and 589.59 nm Na resonance lines, the former gives higher sensitivy (0.011 and 0.004 μg/ml/1% absorption, respectively),[33] and it should be isolated for use if that is required. A narrow bandwidth must be used with Ne and Ar-filled HCL sources to avoid spectral interferences from these filler gases.[35] The HCL should be used at a low current to avoid self-absorption and line broadening (Ref. 34, pp. 679–683).

The interferences from other alkali metals can be overcome by the use of an ionization buffer of sufficiently high concentration to render insignificant the effects of changes in concentration between samples and between samples and standards.

Acetylene contaminated with phosphine can introduce an interference from the phosphine band occurring between 530 and 690 nm. Because the more volatile phosphine is preferentially vaporized,[33] the magnitude of the interference will decrease as the tank of acetylene is depleted.

Several authors have described AAS methods for determining Na in rocks and minerals, but in general the prescribed conditions do not differ greatly.[11, 12, 36] Most often a lean C_2H_2–air flame is used with the burner set at an angle, and an ionization buffer is usually required (K and Cs are commonly used).

8.3.8   Potassium

The development of flame techniques for the determination of K is much the same as that described for Na. There are few spectral interferences reported for K.[33] Abbey et al. report that interference from a second-order Ar line limited the absorption sensitivity for K in an old Ar-filled HCL.[27]

Suppression of the K absorption signal by high concentrations of mineral acids has been reported, but this can be corrected for by using the same acid concentration in both the standard and the sample (Ref. 34, p. 662).

The addition of Na or Cs as an ionization suppressant is a necessary precaution to overcome the appreciable ionization of K at low concentrations in the air–C_2H_2 flame, although Boar and Ingram[36] do not include it in the method they described for analyzing coal ash and silicate rocks. A study of the interference of Na in the determination of K showed that interference occurred until the Na–K ratio exceeded 1000.[37] The interference is probably due to line overlap and thus is not helped by the use of an ionization suppressant.

In a report on an intelaboratory study of the determination of K (for age-dating purposes) in rocks and minerals, Rice[38] concludes that "a between-laboratory variation in flame spectrometric potassium results of less than 0.5% relative standard deviation is attainable." The methods used included both flame emission and absorption techniques. The addition of Cs as an ionization buffer ensured the elimination of any interference by Na in the AAS method.

8.3.9   Titanium

There is very little spectral interference encountered in the determination of Ti when either the 364.27 nm or the slightly more sensitive 365.35 nm line is used, nor does background interference from the $N_2O-C_2H_2$ flame constitute a problem. Because of the formation of the very stable TiO, a high temperature reducing flame is required to provide an appreciable fraction of the atoms in the ground state for the absorption signal. Even then the sensitivity of the AAS method is not as good for Ti as it is for most metals, and hence for samples having a low Ti content problems with the precision of measurement may be experienced.

There has been a significant amount of work done on the many interelement interferences which can affect the determination of Ti by AAS (Supp. Ref. 3). Most of the interferences have been interpreted as being related to condensed-phase or solute-vaporization effects.[39, 40] Panday[41] described a study of the interferences encountered in the determination of Ti in which the effect of members of the various periodic subgroups was observed. For the alkali and alkaline earth metals the elements having the smallest ionic radii (Li and Be, respectively) had the largest enhancement effect. Most other elements similarly enhanced the Ti absorption (B is an exception) with Al, usually present as a major constituent, having an appreciable effect even at a concentration of only 10 ppm and lower. All of the major interferences appear to be compensated for under the conditions described in Sec. 7.1 for the preparation of rock and mineral samples for analysis by AAS. This has also been found by other analysts using similar or generally similar conditions,[6, 27, 36, 42] although the BCM has seen evidence of some small but persistent interference.

As was mentioned above, the formation of TiO dictates the use of a hot reducing flame such as that produced by an $N_2O-C_2H_2$ mixture. Kirkbright et al.[43] have reported that sheathing the flame with Ar or N_2 to isolate it from atmospheric O_2 results in a significant increase in sensitivity because of improved conditions for maintaining a concentration of free atoms in the atom cell.

Heinrich and Lange[21] have reported the application of flameless AAS to the microdetermination of Ti, with Al, Fe, and Mn, in silicate and carbonate rocks following an $HF-HClO_4$ decomposition step. Attempts to determine Ti by indirect AAS methods in order to overcome interference problems or because of low Ti concentrations have been reviewed by Kirkbright and Johnson.[44]

Van Loon and Parissis[45] carried out a separation of Si from the sample solution before the determination of Ti. They further increased the sensitivity by adding $AlCl_3$ (750–1000 ppm). The separation of Si is not a common practice,[27, 36, 42] however, and under the solute conditions obtained by the sample decomposition procedures described earlier (Sec. 8.2) silica should not present a problem.

8.3.10 Manganese

The determination of Mn by AAS is relatively problem-free. An air–C_2H_2 flame is most often preferred, but N_2O works well as a support gas, and there are very few interferences in the AAS determination of Mn (Refs. 27, 36, 45, and Supp. Ref. 4). The suppression of the signal by Si can be eliminated by the addition of Ca (Ref. 34, pp. 632–637). Barnett investigated the effect of five acids in the AAS determination of Mn, Cu, Cr, and Ni.[46] He concluded that, for Mn, a single slot burner and an air–C_2H_2 flame gave the least interference from acids and that under these conditions no appreciable interference is experienced with $HClO_4$, H_2SO_4, H_3PO_4, HNO_3, and HCl in concentrations up to 2% (v/v). Above this level the standards and the samples should have similar acid concentrations. Na is reported to have some effect on the absorption signal from a 0.5 ppm Mn solution for a Na–Mn ratio of 20:1, but the addition of an equal amount of Na to the standards eliminated this problem.[37] It is reported that a higher optimum measuring height above the burner top is necessary for more concentrated Mn solutions.[47]

The determination of trace concentrations of Mn in silicate and carbonate materials has been reported,[48] and Willis has described a method for the aspiration of a water slurry of very fine (<44μm) rock powder directly into a flame for crude geochemical measurements.[49]

8.3.11 Barium

The determination of Ba by AAS, like that of Ca, requires an $N_2O-C_2H_2$ flame in order to obtain a sufficiently high free-atom concentration[27] and to avoid the numerous interferences otherwise encountered. Magill and Svehla[50] investigated the effect of 50 different ions on the AAS determination of Ba and found that the use of an $N_2O-C_2H_2$ flame eliminates virtually all interferences and that the sensitivity obtained was much higher than that obtained with an air–C_2H_2 flame. Maruta et al.[51] have reported interference with the Ba absorption signal in an N_2O–C_2H_2 flame from alkali and alkaline earth elements, the magnitude of the interference being dependent on burner height and on the choice of analytical line. They report an increased interference effect with increasing atomic number within a periodic group. The effect of matrix variation on the flameless AAS determination of trace amounts of Ba in silicates is discussed by Cioni et al.[52] who conclude that matrix effects can only be avoided by the use of ion-exchange techniques. This technique is also described by Frache and Mazzucotelli.[53] Cioni et al. also suggest the use of an ion-exchange separation prior to the regular AAS determination of Ba in rocks. Little difficulty was experienced in the determination of Ba in carbonate rocks by flameless AAS.[54]

Abbey et al.[27] determined Ba in silicate rocks following an $LiBO_2$ fusion—HF–H_3BO_3 dissolution procedure. Na rather than K was added as an ionization buffer to avoid the precipitation of KBF_4, and no chemical interference effects were observed. Warren and Carter determined trace amounts of Ba in silicate rocks by AAS following decomposition of the sample in a Teflon bomb.[55] They report no interference other than that from boric acid which was found to be constant over the range from 0.4 to 1.0%.

It is found at the BCM that the effective detection limit for Ba in rocks using an $LiBO_2$ fusion method is about 50 ppm and that there is significant uncertainty in the accuracy of the determination below about 150 ppm. At levels above this, however, the results are acceptable and no interference is found. It is imperative that care be taken in adjusting the instrument and optimizing the measurement conditions.

A comparative study of the methods for the determination of Ba and Sr in rocks was reported by Ingamells et al.[56] who do not favor the use of AAS.

8.3.12   Strontium

Magill and Svehla[57] report that it is possible to determine Sr by AAS using an air–C_2H_2 flame if La is used as a releasing agent to eliminate most interference effects. They also found no interference effects when N_2O is used as a support gas, provided that an ionization buffer is added. Carter et al. state that the addition of La as a releasing agent eliminated the observed depressant effect of Al and enhancement effect of Ca and Mg on the absorption signal using an N_2O–C_2H_2 flame.[58] There are no spectral interferences reported for the 460.73 nm absorption line of Sr.[29]

Abbey[27] and Medlin et al.[59] have reported the determination of Sr in silicate rocks following an $LiBO_2$ fusion but with different dissolution procedures.

References

1. L. R. P. Butler, in J. A. Dean and T. C. Rains, Eds., *Flame Emission and Atomic Absorption Spectrometry*, Vol. 3, Dekker, New York, 1975.

2. F. Bea Barredo and L. Polo Diez, *Talanta*, **23**, (1976).

3. R. J. Guest and D. R. MacPherson, *Anal. Chim. Acta*, **71**, 233 (1974).

4. H. Bennett and G. J. Oliver, *Analyst*, **101**, 803 (1976).

5. M. Cremer and J. Schlocker, *Am. Mineral.*, **61**, 318 (1976).

6. B. Bernas, *Anal. Chem.*, **40**, 1682 (1968).

7. D. E. Buckley and R. E. Cranston, *Chem. Geol.*, **7**, 273 (1971).

8. F. J. Langmyhr and K. Kringstad, *Anal. Chim. Acta*, **35**, 131 (1966).

9. R. C. O. Gill and B. I. Kronberg, *At. Absorp. Newsl.*, **14**, 157 (1975).

10. R. A. Burdo and W. M. Wise, *Anal. Chem.*, **47**, 2360 (1975).

11. H. Bennett, *Analyst*, **102**, 153 (1977).

12. J. Musil and M. Nehasilova, *Talanta*, **23**, 729 (1976).

13. O. H. Kriege, J. Y. Marks, and G. G. Welcher, in J. A. Dean and T. C. Rains, Eds., *Flame Emission and Atomic Absorption Spectrometry*, Vol. 3, Dekker, New York, 1975.

14. H. Urbain and G. Carret, *Analusis*, **3**, 110 (1975).

15. J. C. Van Loon, *At. Absorp. Newsl.* **7**, 3 (1968).

16. W. D. Cobb and T. S. Harrison, *Analyst*, **96**, 764 (1971).

17. R. J. Guest and D. R. MacPherson, *Anal. Chim. Acta*, **78**, 299 (1975).

18. J. M. Ottaway, D. T. Coker, W. B. Rowston and D. R. Bhattarai, *Analyst*, **95**, 567 (1970).

19. A. P. Ferris, W. B. Jepson, and R. C. Shapland, *Analyst*, **95**, 574 (1970).

20. J. Korkisch, I. Steffan, and H. Gross, *Mikrochim. Acta*, **I**, 263 (1976).

21. H. Heinrichs and J. Lange, *Z. Anal. Chem.*, **265**, 256 (1973).

22. T. C. Rains, in J. A. Dean and T. C. Rains, Eds., *Flame Emission and Atomic Absorption Spectrometry*, Vol. 3, Dekker, New York, 1975.

23. V. K. Panday and A. K. Ganguly, *Spectrosc. Lett.*, **9**, 73 (1976).

24. J. Komarek, J. Jambor, and L. Sommer, *Z. Anal. Chem.*, **262**, 91 (1972).

25. C. T. Chen and J. D. Winefordner, *Can. J. Spectrosc.*, **20**, 87 (1975).

26. M. S. Cresser and D. A. MacLeod, *Analyst*, **101**, 86 (1976).

27. S. Abbey, N. J. Lee, and J. L. Bouvier, Geological Survey of Canada, Paper 74–19, 1974.

28. W. J. French and S. J. Adams, *Analyst*, **99**, 551 (1974).

29. D. J. David, in J. A. Dean and T. C. Rains, Ed., *Flame Emission and Atomic Absorption Spectrometry*, Vol. 3, Dekker, New York, 1975.

30. W. A. Magill and G. Svehla, *Z. Anal. Chem.*, **268**, 177 (1974).

31. J. Komarek, J. Jambor, and L. Sommer, *Z. Anal. Chem.*, **262**, 94 (1972).

32. A. C. West, V. A. Fassel, and R. N. Kniseley, *Anal. Chem.*, **45**, 242 (1973).

33. A. M. Ure and R. L. Mitchell, in J. A. Dean and T. C. Rains, Eds., *Flame Emission and Atomic Absorption Spectrometry*, Vol. 3, Dekker, New York, 1975.

34. G. F. Kirkbright and M. Sargeant, *Atomic Absorption and Fluorescence Spectroscopy*, Academic, New York, 1974.

35. D. C. Manning, J. Trent, S. Sprague, and W. Slavin, *At. Absorp. Newsl.* **4**, 255 (1965).

36. P. L. Boar and L. K. Ingram, *Analyst*, **95**, 124 (1970).

37. J. Ramirez-Munoz, *Anal. Chem.*, **42**, 517 (1970).

38. T. D. Rice, *Talanta*, **23**, 359 (1976).

39. O. W. Kriege, J. Y. Marks, and G. G. Welcher, in J. A. Dean and T. C. Rains, Eds., *Flame Emission and Atomic Absorption Spectrometry*, Vol. 3, Dekker, New York, 1975.

40. J. Y. Marks and G. G. Welcher, *Anal. Chem.*, **42**, 1033 (1970).

41. V. K. Panday, *Anal. Chim. Acta,* **57**, 31 (1971).

42. F. J. Langmyhr and P. E. Paus, *Anal. Chim. Acta*, **43**, 397 (1968).

43. G. F. Kirkbright, M. Sargeant, and T. S. West, *Talanta*, **16**, 1467 (1969).

44. G. F. Kirkbright and H. N. Johnson, *Talanta*, **20**, 433 (1973).

45. J. L. Van Loon and C. M. Parissis, *Analyst*, **94**, 1057 (1969).

46. W. B. Barnett, *Anal. Chem.*, **44**, 695 (1972).

47. M. de Waele and W. Harjadi, *Anal. Chim. Acta*, **45**, 21 (1969).

48. C. W. Fuller and J. Whitehead, *Anal. Chim. Acta*, **68**, 407 (1974).

49. J. B. Willis, *Anal. Chem.*, **47**, 1752 (1975).

50. W. A. Magill and G. Svehla, *Z. Anal. Chem.*, **268**, 180 (1974).

51. T. Maruta, T. Takeuchi, and M. Suzuki, *Anal. Chim. Acta*, **58**, 452 (1972).

52. R. Cioni, A. Mazzucotelli, and G. Ottonello, *Anal. Chim. Acta*, **82**, 415 (1976).

53. R. Frache and A. Mazzucotelli, *Talanta*, **23**, 389 (1976).

54. R. C. Hutton, J. M. Ottaway, T. C. Rains, and M. S. Epstein, *Analyst*, **102**, 429 (1977).

55. J. Warren and D. Carter, *Can. J. Spectrosc.*, **20**, 1 (1975).

56. C. O. Ingamells, N. H. Suhr, F. C. Tan, and D. H. Anderson, *Anal. Chim. Acta*, **53**, 345 (1971).

57. W. A. Magill and G. Svehla, *Z. Anal. Chem.*, **270**, 177 (1974).

58. D. Carter, J. G. T. Regan and J. Warren, *Analyst*, **100**, 721 (1975).

59. J. H. Medlin, N. H. Suhr, and J. B. Bodkin, *At. Absorp. Newsl.* **8**, 25 (1969).

Supplementary References

1. F. B. Barredo and L. P. Diez, *Talanta,* **37,** 69 (1980).
2. Analytical Methods Committee, *Analyst,* **103,** 643 (1978).
3. J. J. La Brecque, *Appl. Spectrosc.,* **53,** 389 (1979).
4. R. F. Sanzolone and T. T. Chao, *Talanta,* **25,** 287 (1978).

CHAPTER 9

THE DETERMINATION OF TRACE CONSTITUENTS IN ROCKS AND MINERALS BY ATOMIC ABSORPTION SPECTROSCOPY

9.1 GENERAL

At one time or another, an analyst is likely to be requested to determine most of the elements in the periodic table, with the exception of most of the lanthanides, the actinides other than U and Th, the noble gases, and a few other elements. The ones selected for inclusion in this chapter are those that are most likely to be requested of an analyst dealing with rocks and minerals, and which usually occur in minor or trace concentrations. They are also, of course, those that can be determined by AAS techniques.

The reason for their determination may vary from a need to complete a detailed determination of major and minor constituents on a single sample, to that of providing trace element data for a large number of varied samples collected during geochemical exploration. While the trace elements may be the same for these two extremes, the way in which the determinations are made for each application can be very different.

The determination of trace constituents presents particular problems to the analyst, who must be concerned with the possibility of even minor contamination from reagents and other sources, with detection limits, preconcentration steps, and other factors besides the usual ones of sample decomposition and dissolution, instrument stability, statistical evaluation of analytical data, and the like. The determination of constituents at the parts per million and, especially, at the parts per billion level will often push the capability of the analyst and the analytical equipment to their limit. The problems likely to be encountered in this type of an analysis should never be underestimated.

Preconcentration techniques were discussed in Sec. 7.2, the preparation of standard solutions in Sec. 7.3, and instrumentation in Secs. 7.4, 7.5, and 7.6. Sample decomposition is discussed in Chapter 4, with two methods given in detail in Sec. 8.2. Other decomposition methods used in trace analysis will be described in the next section.

9.2 SAMPLE DECOMPOSITION

The problem of sample decomposition has been discussed in general terms in Chapter 4, and methods for the decomposition of rock and min-

eral samples by HF in a Teflon bomb and by fusion with lithium metaborate flux were described in Sec. 8.2. The Teflon bomb and lithium
metaborate fusion methods have been used for decomposition prior to
trace element determination but are not of general application. The methods described in this section will be those most commonly used in the
analysis of rock and mineral samples for trace constituents, together with
some routine methods used for the analysis of samples for geochemical
exploration studies.

9.2.1   Cold Extraction Method for Soils and Silts

The cold extraction (or "buffer" extraction) method takes many forms,
the intent being to extract metallic ions loosely adsorbed on sediments
and soils following their leaching from oxidized deposits by natural processes. The technique finds wide application to samples from tropical and
subtropical environments but much less for those from areas, such as
most of Canada, where little oxidation of deposits has occurred in the
relatively short postglacial era.[1, 2] Cold extraction procedures used in
geochemical laboratories commonly involve treatment with EDTA (1–2%
aqueous solution) and ammonium citrate or hydroxylamine hydrochloride
solutions. Dilute nitric and hydrochloric acids have also been used, as
have acetic acid and some other organic acids, either alone or in combination with each other.

There is no widely accepted convention for the technique of cold extraction. It generally involves the mixing of a weighed amount of stream
or lake sediment, or soil sample, with a portion of the extractant solution
and letting it sit for a period ranging from a few minutes (with shaking)
to overnight before determining the concentration of the element(s) of
interest by colorimetric, AAS, or some other technique. The main concern
is that all samples must be treated in exactly the same manner so that a
statistically valid compilation of the resulting data can be interpreted to
identify geochemical anomalies. The particular extractant and treatment
to be used may often be suggested by the geologist or geochemist sub-
mitting the samples, in consultation with the analyst, because the nature
of the samples and the sample environment and the specific analytes
required will dictate the type and strength of extractant needed and the
extraction time. It is thus not feasible to give a specific procedure here.

9.2.2   Hot Extraction Method for Soils and Silts

The term "hot extraction" refers to an attack on the sample by hot
acid solution (100°C or greater), either a single acid or a mixture of them.
The most common method of attack is done with a mixture of HNO_3 and

$HClO_4$ in a 3:1 ratio. Perchloric acid is used both because of its strong oxidizing capability and because it is then possible to carry out the digestion more efficiently at temperatures up to about 180°C. HNO_3 is added to preoxidize any organic matter present and thus eliminate the possibility of an explosive reaction with the $HClO_4$ (Sec. 4.2.4). There are some minerals, such as chromite, pyroxene, magnetite, and molybdenite, which are not attacked completely, or even at all, by this acid mixture, and in many cases the completeness of the attack is not always reproducible.[3] Reference should be made to the monograph on sample decomposition by Doležal et al.[4] The following is an example of the use of the HNO_3–$HClO_4$ mixture. Transfer a weighed (1 g) sample into a calibrated test tube, add 6 ml of the 3:1 HNO_3–$HClO_4$ acid mixture, and digest the contents (in a sand bath or in aluminum blocks drilled to accept the test tubes) for 1 hr at <200°C. If frothing becomes a problem, the addition of a drop of kerosene to the mixture before heating will often prevent it. Cool, dilute to a predetermined volume, shake or otherwise mix the contents, and either allow the solids to settle or centrifuge the mixture before aspirating the supernatant solution into the flame of the AA unit.

A volume of 15 ml is usually sufficient but may be greater depending on the concentration of the elements of interest and other factors; a tenfold dilution or more may be necessary for some measurements. The sample is allowed to settle with or without centrifuging (preferably it should be decanted) to prevent blocking of the aspirator orifice.

Caution. Hot $HClO_4$, that is, approaching fuming temperature, will react explosively with organic matter, and it is imperative that sufficient HNO_3 be present to oxidize any organic matter before the $HClO_4$ reaches fuming temperature during digestion. More HNO_3 may have to be added in exceptional circumstances.

9.2.3 Mixed Acid Attack with HF for Rocks and Minerals

A general discussion of the efficacy of HF in the decomposition of rocks and minerals is given in Sec. 4.2.1 and in Ref. 4. The procedure described below is used to decompose rock and mineral samples prior to the determination of many of the trace elements commonly requested of the BCM laboratory.

Weigh accurately 1 g of sample into a 100 ml Teflon beaker, add 6 ml of aqua regia, 2 ml of $HClO_4$, and 6 ml of HF. Cover the beaker with a polyethylene lid and place it on a sand bath held at 200°C for 1 hr. Remove the lid (wash the condensate on the lid into the beaker), and evaporate the contents to light $HClO_4$ fumes. Add another 6 ml of HF and evaporate to dense $HClO_4$ fumes, but not to dryness. Add distilled, deionized water, 5 ml of HCl, warm the contents to ensure complete solution, and transfer it to a graduated flask or other vessel of appropriate volume (usually not less than 25 ml), and dilute to the correct volume prior to determining the analytes of interest by means of AAS.

Variations on this digestion technique include larger samples (for which an additional HF treatment is required as well as larger volumes of HF for each digestion), fusion of any residue remaining from the acid attack (e.g., chromite and other refractory minerals), and concentration of the analyte into a smaller volume by ion exchange, solvent extraction, co-precipitation, or some other technique (see Sec. 7.2).

PRECAUTIONS   Of the several precautions to be observed, an obvious one is to keep the sand bath temperature less than 190–200°C in order not to deform the Teflon beakers. The lids should not be left on the beakers for an extended period of time to avoid the formation of a layer of silica gel. The sample should not be allowed to go to complete dryness in order to prevent loss of Pb.

9.3   CONTAMINATION, BLANKS, AND THE DETECTION LIMIT

9.3.1   Contamination

Perhaps the most important precautions to be taken in trace analysis are those that will guard against contamination of the sample or sample solution. This is a very serious problem, especially when samples of many different types and element concentrations are being analyzed in the same laboratory (highly mineralized ore samples and lake sediment samples, for example). A thorough discussion of the control of contamination in trace element determination is given by Zief and Mitchell, and it is recommended reading for the analyst involved in these determinations.[5] Of interest also are reviews by Tölg[6] and by Zief and Horvath[7] of problems encountered in trace analysis.

Some commonsense recommendations are worth mentioning here. Those samples that have high concentrations of elements occurring at trace levels in other samples must be kept separate from the latter at all stages of the sample preparation and decomposition procedures. These include the use of separate comminution equipment (in a different location if possible, and equipped with a dust control device), separate glassware and other laboratory utensils, and, in the case of ultratrace work (parts per billion levels), a separate laboratory having a specially filtered air supply and a positive air pressure relative to other rooms.

The specially filtered air supply and positive pressure are intended to eliminate airborne contaminants which can be a source of serious problems. These airborne particles commonly contain metallic elements which may originate from corroded plumbing and electrical fixtures and laboratory equipment, from painted surfaces, and from dust brought in from the outside. Laboratory fixtures should, if possible, be made of an inert material or, failing this, be coated with an epoxy paint. Laboratory equipment should be covered with a protective plastic jacket which is removed only when that particular instrument or piece of equipment is in use.[5] The lead content of automobile exhausts is a serious source of airborne outside contamination which, unless guarded against, will cause high blanks and irreproducible results in the determination of Pb by flameless AAS.

Because reagents are a ubiquitous source of contamination, the use of special grades and/or purification steps are common requirements in trace element analysis. Dabeka et al. have described the procedures for subboiling distillation and storage of high-purity acids and water.[8] They also report their experience that, while no significant inorganic contamination resulted from the use of polypropylene or TPX beakers, those made of Teflon–FEP were found to be a source of Cr, Fe, Mn, and Ni contamination. Si and Zr were introduced into the solution from Vycor™ beakers, Ti and Al from those of linear polyethylene, but no observable contamination (as measured by spark source mass spectrometry and carbon rod AAS) resulted from the use of transparent quartz (99.9%) beakers. Polypropylene utensils were found to become discolored and to contaminate the high-purity acids with organic material under long-term storage conditions. Moody and Lindstrom[9] have discussed in detail the preparation and cleaning of containers for trace analysis.

9.3.2 The Blank

The determination of the blank is of utmost importance in trace analysis because, in most cases, detection limits are not determined by the sensitivity of the instrumental or other measuring technique involved but by the variability of the blank.[5] Thus enough blanks (Tölg recommends 20)[6] must be run with every batch of trace analyses to enable the analyst to obtain a reliable value for the standard deviation of the blank. The minimizing of the variability of the blank can be accomplished only through adequate safeguards against contamination. The blank value should preferably be lower than the corresponding value for the analyte by a factor of 10, but in rare instances, where the blank determination is very precise, a blank value higher than that of the analyte can be subtracted to give a reliable result.[5]

In obtaining a value for the blank in the determination of trace elements in geological materials, the matrix of the solution used for the blank must match that of the samples as closely as possible and should be carried through the whole procedure. Because a true blank is difficult to obtain in many instances (e.g., the matching of a sediment matrix), the analyst should ensure that he has available to him a variety of reference materials with well-established trace element values. This will allow him to report his values with some confidence that the acid concentrations, instrumental settings, and other variables have been controlled to a sufficient extent.

9.3.3   The Detection Limit

The detection limit has been defined by Winefordner as "the smallest concentration C_L that can be detected with reasonable certainty,"[10] and he describes it with the relationship:

$$C_L = \frac{3s_b}{S}$$

where s_b is the standard deviation of the blank and S is the sensitivity or slope of the analytical calibration curve. Skogerboe gives a similar defintion of C_L except that $3s_b$ is replaced by the t statistic for the number of measurements made and the confidence level required.[11] In practice the value of $3s_b$ usually corresponds to approximately a 90% confidence level, and analysts sometimes use $6s_b$ if a higher confidence level is required. The value of s_b should be determined from at least 20 measurements as both Tölg and Winefordner have pointed out, and it will include contributions from contamination and instrument noise as well as manipulative factors.[12] This emphasizes again the previous contention that the blank is of critical importance to successful trace element determination.

9.4   DETERMINATION OF TRACE ELEMENTS USING AAS

The successful measurement of trace element concentrations using AAS techniques generally taxes the capability of both the analyst and the instrument. It requires that the instrumental conditions be carefully optimized (Sec. 8.3.1), that the instrument be allowed to stabilize before measurements are made, and that carefully prepared standards be measured frequently. The analyst has to be aware of such factors as possible interferences, preconcentration techniques, the capabilities and limitations of the instrument, and other matters discussed in Chapter 7.

All of these requirements, as well as those involving possible contamination of the sample and the determination of blanks discussed in Sec. 9.3, will have to be considered by the analyst before the optimum detection limit can be achieved, the precision of the measurement made acceptable, and the accuracy adequate for the purpose at hand.

General Procedure

The determination of most elements in silicate materials can be done by digesting the sample as described in Sec. 9.2.3, optimizing the instrumental parameters (Sec. 8.3.1), and making the AAS measurement using the conditions in Table 9.1 as a guide. It must be remembered that each time a new set of measurements is to be made, the instrumental conditions

TABLE 9.1   Selected Conditions for the Determination of Trace Constituents by AAS.

Analyte	Wavelength (nm)	Slit Width[a] (nm)	Flame Oxidant	References
Sb[b]	231.15; 217.58	0.16	—	13, 17
As[b]	197.20	0.32	—	13, 14, 18–21
Bi[b]	306.77	0.32	—	4, 14–16, 24–28
Cd	228.80; 326.11	0.32	Air	14, 15, 24, 29–31
Cr	357.87	0.16	N_2O	4, 15, 16, 30, 32–42
Co	240.72; 304.4	0.08	Air	15, 16, 36, 37, 43–48
Cu	324.75; 327.40	0.32	Air	15, 16, 36, 37, 42, 43, 45, 49–54
Au	242.80; 267.60	0.32	Air	15, 37, 55–70
Pb	217.00; 283.31	0.32	Air	16, 27, 30, 36, 46, 49, 53, 71–75
Hg[c]	253.65	0.32	—	76–82
Mo	313.26	0.32	N_2O	15, 16, 30, 37, 43, 46, 49, 83–91
Ni	232.0	0.04	Air	16, 30, 36, 37, 42, 43, 48, 50, 53, 92–95
Pt	265.9	0.16	Air	58, 61, 62, 96–109
Se[b]	196.03	0.16	—	14, 16, 19, 26, 37, 45, 110–116
Ag	328.07; 338.29	0.32	Air	15, 16, 37, 106, 117–126
Te[b]	214.27	0.16	—	14, 37, 111, 114, 127–131
Sn[b]	225.48	0.16	—	4, 14, 16, 26, 111, 132–136
W	255.13; 400.88	0.32	N_2O	15, 85, 137–143
Zn	213.86; 307.59	0.32	Air	15, 30, 46, 49, 52, 53, 93, 144, 145

[a] Specified for IL 351. Operator should optimize for own instrumentation.
[b] Hydride generation and heated cell atomization.
[c] Cold vapor technique.

must be optimized (optical alignment, aspiration rate, flame stoichiometry, lamp current, burner height, and other parameters must all be considered.) The standards must match the samples as closely as possible in matrix composition (major constituents, acid concentrations, and analyte concentration), and sufficient blanks must be run so that a reliable estimate of the detection limit can be made. See Appendix 3 for suggested sources for standard stock solutions.

The foregoing general procedures (Sec. 9.2 et seq.) will serve as a reference in the following element by element discussion unless a specific procedure is given.

9.4.1 Antimony

Antimony can be determined by ordinary flame AAS techniques, by carbon rod or some other thermal volatilization method, and also by hydride generation combined with flameless AA. The first technique is better than the other two for samples containing Sb as a major constituent, or even as a minor constituent. Because the possible detection limit is lower in the other two methods, however, they are a better choice for the determination of Sb when it is present at concentrations <100 μg/g. The thermal volatilization technique using a solid rock or mineral sample may be subject to severe matrix interference which necessitates the extraction of the Sb into a nonaqueous solvent. The hydride generation technique (using a borohydride to generate SbH_3) is capable of detecting a few nanograms with a minimum of sample treatment, and it produces precise results with little opportunity for contamination.[13] One of the advantages of the technique is that it can be used for many other elements (As, Bi, Ge, Pb, Sn, Se, and Te) as well (Ref. 14; Supp. Ref. 1). It is this hydride generation procedure that is given in detail in the following.

For the determination of Sb by AAS, the 206.83 and 217.58 nm lines are more sensitive than the 231.15 nm line but suffer from more noise than the latter, especially the 206.83 nm line (Ref. 15, pp. 192–196). The use of the 231.15 nm line avoids the interference by Cu and Pb observed when the 217.58 nm line is used (Ref. 16, pp. 548–550). The spectral bandwidth must be set to exclude the nonresonance atomic line at 217.9 nm and an ionic line at 217.0 nm which could be a source of nonlinearity in the calibration.[17] The Sb signal can be affected by the acid concentration, and it is necessary that the acid strengths of the samples and standards be matched when flame techniques are used.

Welsch and Chao[18] have described a novel method for the determination of trace levels of Sb in geological materials. It involves heating the sample mixed with NH_4I (predried at 105°C overnight) at 350°C for

10 min and extracting the SbI_3 thus generated into 10% HCl and then into trioctylphosphine oxide-methyl isobutyl ketone (TOPO-MIBK). The amount of Sb is measured by aspirating the TOPO-MIBK layer into the air–C_2H_2 flame of an AAS unit. They report a working range of 1–40 ppm Sb in the sample; up to 2000 ppm of Cu, Pb, Zn, Sn, As, or Hg caused no interference in the measurements.

Reagents

A stock solution of Sb (1000 µg/ml) is prepared by dissolving Fisher certified antimony potassium tartrate in 6 M HCl. Solutions prepared by diluting the stock solution should be freshly made. See also Appendix 3.

A 2% w/v aqueous solution of reagent grade sodium borohydride should be freshly made for each set of measurements.

HCl and HNO_3 should be of reagent grade.

Nitrogen is required as a flow gas.

Apparatus

Micropipettes capable of dispensing 500 µl of solution accurately and reproducibly; a hydride generation system and atom cell as shown in Fig. 9.1 with sufficient hydride generation cells to last for the entire run; an atomic absorption spectrometer equipped with a strip chart recorder (or with a peak area integration facility) and, as an option, a power source for an electrodeless discharge lamp and an Sb electrodeless discharge lamp; test tubes calibrated to 10 ml.

Procedure

Add 10 ml of freshly prepared aqua regia to 0.1–0.5 g of sample weighed into a test tube calibrated to 10 ml. Allow the samples to sit at room temperature for 1.5 hr and in a water bath at 90°C for 2 hr. Mix the samples two or three times during this digestion period (a vortex mixer works well). Cool the solutions and dilute them to the 10 ml calibration mark, mix, and allow solids to settle.

Prepare a set of standards containing from 0 to 50 ng/ml Sb in 1.5 M HCl using the 1000 µg/ml stock solution.

Turn on the instrument, the burner, and the gas flow and adjust each to its approximate settings.

Dilute a 1 ml (1000 µg) aliquot from each sample to 10 ml using 1.5 M HCl. Add 2 ml of the sodium borohydride solution to a set of hydride generation cells, attach the first one to the atomization cell as shown in Fig. 9.1 and inject into it a 1 µl aliquot from a 10 ng/ml standard. Optimize the instrumental settings and the gas flow (about 2.0 l/min) with a series

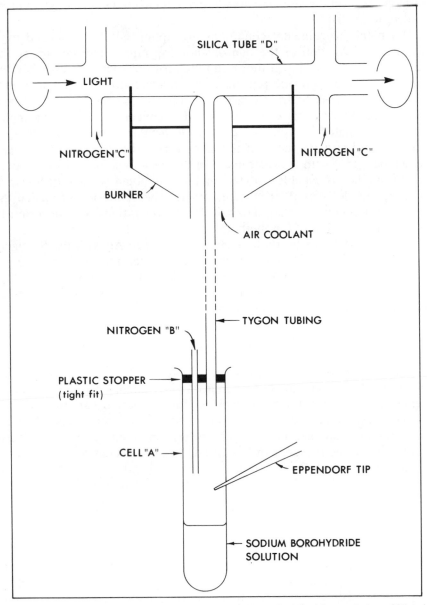

Figure 9.1. Hydride generator and atomizing cell. Reproduced with permission of Elsevier Scientific Publishing Company and the author (Ref. 13).

of aliquots from the 10 ng/ml standard, each injected into a fresh hydride generation cell.

When the instrumental conditions are optimized, as evidenced by a maximum deflection of the pen on the strip chart recorder (or by the maximum peak area integration), run the series of standards and then the samples interspersed with repeats of the standards by injecting microliter aliquots into fresh hydride generation cells containing 2 ml of 2% m/v sodium borohydride for each injection. The concentration in the samples can be calculated by comparing the peak heights obtained on the strip chart recording with those of the standards or by comparing the peak areas measured with the instrument in the peak area integration mode.

This method is capable of detection limits of about 80 ng/g for Sb when a 0.5 g sample is used. It is necessary for the operator to practice the injection procedure so that variability of that operation does not seriously affect the precision of the method.

Some interference by up to 5000 ppb of Ni and Ag with the Sb signal in the 0–50 ng/ml calibration range used has been reported, but the addition of sufficient EDTA to give a solution strength of 0.01 M EDTA decreased this interference significantly.[13] There was no interference observed for similar concentrations of Fe, Co, or Cu.

9.4.2   Arsenic

For arsenic, as was found for Sb, the preferred method for the determination of trace amounts of As in geological samples is that of hydride generation (as arsine). Several workers have reported variations of this method. Terashima[19] and Moruta and Sudoh[20] have both described essentially the same method in which arsine, generated by the use of powdered zinc in the presence of $SnCl_2$, HCl, and KI, is carried by a stream of Ar into an Ar–H_2 flame and the As measured with a conventional HCL at 193.7 or 197.2 nm. Terashima[19] reports a working range of 0.04–4000 μg/g in the original sample. Another variation, described by Aslin[13] and by Thompson and Thomerson,[14] involves generating the arsine with sodium borohydride and sweeping it onto a heated quartz tube where the As is released and measured in a "flameless" condition. Smith et al.[21] used a similar method with the difference that the arsine is generated in a disposable syringe and injected into the atomizer cell. They also report that, for siliceous materials, a fusion of the sample at 550°C for 15 min in a Ni crucible with a flux of KOH and MgO is necessary for a total extraction of the As. Aslin[13] reports a satisfactory extraction of As with an aqua regia attack for several geological standard reference materials.

Bajo (Supp. Ref. 2) has warned about the possibility of loss of As(III) by volatilization during the HF decomposition of silicates.

The determination of As using flame techniques suffers from the fact that the three absorption lines exhibited by As are all below 200 nm, and sensitivity is decreased because of absorption of the radiation by the optics, air, and flame gases. The 193.70 nm line is preferred because it is more sensitive than the 197.20 nm line, and less subject to absorption by flame gases than the 188.99 nm line (Ref. 16, pp. 550–552). Abercrombie and Massotti report that an Ar (entrained air)–H_2 flame transmits about 85% of the radiation at 193.70 nm, whereas an air–acetylene flame will transmit only about 38% (Ref. 15, pp. 205–209). Kirkbright et al. have described a burner designed to produce a nitrogen-separated air–acetylene flame which lowers the flame background and noise levels significantly.[22]

When enough As is present that detection limits are not of concern, a lean oxidizing air–acetylene flame can be used with the measurement made at 193.70 nm. Interfering species under these conditions include Al, Cr, Mg, Mn, Mo, Ni, Sn, Ti, and nitrate ions when present at a concentration ratio of 10:1 with respect to As (Ref. 15, p. 207). Kirkbright et al. encountered interference from Ca, V, Fe, Cr, and Cu as well as Mg, Al, Ni, and Mo when present at a concentration 100 times greater than that of As.[22]

The procedure described for the determination of As is the same as that for Sb (Sec. 9.4.1), and it is based on work by Aslin[13] and by Thompson and Thomerson.[14] Wauchope[23] has extended a version of it to the determination of As in soils, water, and plants. Refinements to the procedure, including automation, have been introduced by several workers (Supp. Refs. 1, 3–6).

Reagents

A 1000 µg/ml stock solution of As(V) is prepared by dissolving Na_3AsO_4 in dilute NaOH and acidifying with HCl. Dilutions of this stock solution should be freshly made when required. See also Appendix 3.

A 2% w/v aqueous solution of reagent grade $NaBH_4$ should be freshly made for each set of measurements.

HCl and HNO_3 should be of reagent grade.

Nitrogen is required as a flow gas.

Apparatus

Micropipettes, hydride generation system and cells, 10 ml test tubes, and AAS instrument, as described in Sec. 9.4.1.

Procedure

The procedure to be followed is similar to that given in Sec. 9.4.1, but using the As(V) calibration solutions.

This method is capable of detection limits of about 0.16 μg/g (ppm) for As when a 0.5 g sample is used. It is necessary for the operator to practice the injection procedure so that variability of that operation does not seriously affect the precision of the method. Problems have been encountered with NaBH$_4$ as a source of As contamination.[13, 14]

There has been some interference reported from up to 5000 ppb of Ni in the calibration range used (up to 50 ng/ml of As), but the addition of EDTA to give a solution strength of 0.01 M EDTA decreased this interference significantly.[13] There was no interference observed for similar levels of Fe, Co, Cu, or Ag.

9.4.3   Bismuth

Comparatively little is found in the literature concerning the determination of Bi in geological materials. Doležal et al. report that most minerals containing Bi dissolve readily in HNO$_3$, aqua regia, or HNO$_3$–H$_2$SO$_4$ (Ref. 4, pp. 48, 56). Heinrichs and Lange[24] discuss the determination of trace levels of Bi with flameless AAS techniques. Reichel and Bleakley[25] concentrated trace amounts of Bi (as well as Se, Te, As, Sb, Fe, Pb, and Sn) from blister copper by coprecipitation with La(OH)$_3$. The elements were then determined by AAS. Jones[26] evaluated the application of this technique to the determination of Bi (and many of the other elements listed above) in sulfide concentrates and found the technique to be successful. The extraction of Bi from solutions of geological materials for subsequent measurement by AAS has been described by Viets (Supp. Ref. 7) and Donaldson (Supp. Ref. 8).

In flame AAS, few significant chemical interferences have been reported in the determination of Bi. The 306.77 nm line is subject to interference from the OH band head, but the use of a fuel-rich air–C$_2$H$_2$ flame overcomes most of that problem. The linear range for the 306.77 nm line is reported to extend from 1 to 400 μg/ml according to Baker et al. (Ref. 15, pp. 196–197). They also point out that the Bi line at 223.06 nm should be used only if the analyst is sure that there is no Cu in the HCL, because of Cu lines in that region. Kirkbright and Sargent, on the other hand, advocate the use of the 223.06 nm line rather than the 306.77 line because of the noise introduced by the OH band interference at the latter wavelength (Ref. 16, pp. 558–559).

Bi can be determined by passing BiH_3, generated in a hydride generation cell, into a heated atomization cell for a "flameless" AAS measurement, as reported by Pollock and West[27] and by Thompson and Thomerson.[14] The former used Mg and $TiCl_3$ in HCl to generate the hydride, which was then measured in an H_2–air-entrained flame. The latter employed $NaBH_4$ for hydride generation and a quartz cell heated in an air–C_2H_2 flame as the atomization chamber. Refinements of this procedure, including automation, have been reported (Supp. Refs. 1, 9). The detection limits reported for this method are on the order of a few nanograms per milliliter. The method described here is similar to that described for As and Sb, and it is based on work done by Hall of the GSC.[28]

Reagents

A 1000 μg/ml stock solution of Bi is prepared by dissolving Bi metal in 10 M HNO_3, boiling gently to dispel the brown fumes, cooling, and diluting to volume. Dilutions of this stock solution should be freshly made when required.

Other reagents as for Sec. 9.4.2, but using the Bi calibration solutions.

Apparatus

As described in Sec. 9.4.2.

Procedure

As described in Sec. 9.4.2. This method is capable of detection limits of about 80 ng/g for Bi when a 0.5 g sample is used. It is necessary for the operator to practice the injection procedure so that variability of that operation does not seriously affect the precision of the method.

9.4.4   Cadmium

The sensitivity of the AAS technique is very high for Cd and both flame[29, 30] and flameless[24, 31] methods for the analysis of rocks and minerals have been reported in the literature (Supp. Ref. 7). The most sensitive line is at 228.80 nm; a less sensitive one at 326.11 nm can be used with higher Cd concentrations. Dean and Hanamura report a sensitivity of 0.03 μg/ml and a working range of 0.5–5 μg/ml for the former and a sensitivity of 20 μg/ml and a working range of 400–4000 μg/ml for the latter (Ref. 15, pp. 76–79). The Cd HCL must be set at a low current to preclude self-absorption. Silicon is reported to depress the Cd signal to some extent, but there are very few other chemical interferences observed in the air–C_2H_2 flame most commonly used in the AAS determination of Cd (Ref. 14, pp. 562–565).

Sighinolfi and Santos[31] have reported a 40 ppb detection limit for Cd in silicate materials when an HF–$HClO_4$ digestion and an Amberlite LA-2-xylene–ethylenediamine extraction were combined with a graphite furnace AAS measurement. The sensitivity of the method was limited by Cd contamination of the reagents.

Robert and Mallett[29] fused samples of Zn, Pb, and Cu sulfide concentrates in Na_2O_2 and followed with a water leach and acidification with HNO_3. They measured Cd and several other metals with flame AAS and reported a 50 ppm detection limit for Cd.

Agemian and Chau[30] digested finely pulverized sediment samples in a Teflon-lined bomb with a mixture of HF, HNO_3, and $HClO_4$ acids. The solution was stabilized with boric acid, and Cd and 19 other elements were measured by flame AAS.

Procedure

See the general procedure given in Sec. 9.4.

9.4.5   Chromium

One of the most serious problems in the determination of Cr is that of ensuring complete sample dissolution. Much has been written about the refractory nature of such Cr minerals as chrome spinel, chromite, and others, most of which present a dissolution problem to the analyst.

Rodgers has published a review of several techniques for decomposing chrome spinel, and he recommends the preparation of a peroxide frit of the mineral at 480°C for 7 min as an effective method.[32] Pyroxenes require separate treatment. He also states that chrome-rich spinels require both long fusion times and multiple fusions in order to be completely dissociated in a sodium carbonate–borax flux. Cresser and Hargitt have described a double fusion method which consists of fusing the sample first with $NaHSO_4$ and then with Na_2CO_3 in a Pt crucible.[33] Ferruginous samples are leached with acid prior to the fusions, and any residue from the fusion must be treated with HF. There is some loss of Pt from the crucible in this fusion, and it should be used sparingly. Bennett and Oliver have reported that chrome ores and refractories can be decomposed by a fusion mixture consisting of 1 part of sample–10 parts of $LiBO_2$–12.5 parts of $Li_2B_4O_7$ (see Table 4.1).[34]

Mixed acid attacks do not effect a complete dissolution of most refractory Cr minerals and, in addition, fuming with $HClO_4$ can result in low Cr results because of the volatilization of chromyl chloride (CrO_2Cl_2). Phosphoric acid will decompose chrome spinels slowly and the 72.6%

azeotrope of $HClO_4$ will dissolve chromites after 6–8 hr of boiling[32] with, however, the danger of the loss of chromyl chloride (Ref. 35, pp. 315–316). The combination of a mixed acid attack using a Teflon bomb to obtain a more effective dissolution of silicate rocks and sediments has been reported.[30, 36]

Doležal et al. have discussed the decomposition of Cr minerals using a variety of methods, and reference should be made to their work if problems are encountered.[4]

The determination of Cr in silicate materials by AAS has been successful using both flameless and flame techniques. A review of flameless AAS summarizes the analytical conditions used by analysts for determining Cr in a variety of matrices.[37] Several interferences have been reported in the determination of Cr by flame AAS (Supp. Ref. 10). A number of reports have been published on the effect of the valence state of Cr on the absorption signal under different flame conditions. Green[38] concluded that the maximum absorption for Cr is obtained by converting all of it to the Cr(III) oxidation state (with H_2O_2 in acidic medium) and using an $N_2O-C_2H_2$ flame for the measurement. Kraft et al.[39] found that there is a difference between signals from equal concentrations of Cr(III) and Cr(IV) in both air- and N_2O-supported flames, although the effect is less significant in the latter flame. They state that the addition of K (2 g/l as KCl) and the use of an $N_2O-C_2H_2$ flame eliminates the differences. It is reported elsewhere that no difference is observed between solutions of the two oxidation states if an oxidizing air–C_2H_2 flame is used.[40] Cresser and Hargitt[41] have related the pH dependence of the absorption signal of Cr(VI) in an air–C_2H_2 flame to the chromate–dichromate equilibrium. No such pH effect was seen when an $N_2O-C_2H_2$ flame was used.

Barnett[42] found that the interferences from five mineral acids with the Cr signal were similar in both air–C_2H_2 and $N_2O-C_2H_2$ flames.

Kriege et al. (Ref. 15, pp. 164–168) have reviewed observed interelement interferences and find that those due to Co, Fe, Mn, and Ti persist even when an $N_2O-C_2H_2$ flame is used. A lean flame eliminates the interference due to Mn and Ti. The addition of K or Al (as chlorides) suppresses the interference due to Mn and Ti but enhances that of Fe. Kirkbright and Sargent (Ref. 16, pp. 576–580) also recommend the use of a lean air–C_2H_2 flame to minimize interferences due to Fe, Ni, and others. They also report that chemical interferences are less pronounced in an $N_2O-C_2H_2$ flame.

Procedure

The fusion should be performed as outlined in Sec. 8.2.2 with the exception that a longer fusion time may be required for some highly refractory minerals. The detection limit obtained with the dilution factor specified in Sec. 8.2.2 (0.2 g of sample diluted to 200 ml) is about 10 μg/g. This can be improved by using a lower dilution factor, up to a limit of approximately 0.2 g of sample in 50 ml.

The standards used should be multi-element solutions as described in Sec. 7.3.1. A guide to the AAS measurement conditions is given in Table 9.1, but precautions should be taken to avoid or minimize the problems discussed above.

9.4.6 Cobalt

Both flame and flameless AAS techniques are reported to have been applied successfully to the AAS determination of Co in rocks and minerals. A review of general conditions for the flameless AAS determination of Co is given by Fuller.[37] Schweizer has described the use of a graphite furnace for the determination of Co and five other metals in carbonate rocks.[43] Riandy et al.[44] give the optimum conditions for the determination of Co and several other metals in granite and basaltic rocks by flameless AAS, and Ohta and Suzuki[45] describe the flameless determination of Co and Cu in rocks. Cruz and Van Loon[46] evaluated nonflame techniques for Co and several other metals, and they have concluded that flameless techniques do not offer much of an advantage over flame techniques when "heavy inorganic matrices" are encountered.

Co is usually determined using a lean air–C_2H_2 flame with a narrow spectral bandwidth. The latter precaution is necessary because of the presence of nonabsorbing lines close to the Co 240.72 nm line used for the analysis. Without a narrow bandwidth both the linear range and the sensitivity will suffer (Ref. 15, pp. 226–232; Ref. 16, pp. 581–584). Some interferences have been reported for the determination of Co by flame AAS. Govett and Whitehead[47] report errors in the AAS determination of Co, Pb, Zn, and Ni in the presence of Fe, Al, Ca, Mg, Na, and K. The interference was reported to be mainly an enhancement due to background absorption for concentrations up to 1 ppm of the analyte and 4000 ppm total cations. At a higher concentration of the analyte or of major elements however, suppression of the signal is observed. Thus as the authors report, a generally applied background correction can lead to negative errors. Warren and Carter[36] have also reported the necessity of using background correction for Co and Ni. They worked at a 1000 ppm total cation concentration level and did not observe the suppression effect reported by Govett. Cresser and MacLeod[48] reported that, for 50 μg/ml Co, Ni, or Mg solutions prepared as sulfates and as chlorides, the sulfate solutions gave a depressed response relative to the chloride solutions. Most chemical interferences can be overcome by the use of an N_2O–C_2H_2

flame, but its use for Co reduces the sensitivity by a factor of about 2.5 over the use of an air–C_2H_2 flame (Ref. 15, pp. 226–232).

Procedure

See the general procedure given in Sec. 9.4.

9.4.7   Copper

Numerous flameless AAS methods have been reported for the determination of Cu in geological materials.[37, 43, 45] Owens and Gladney[49] applied them subsequent to a $LiBO_2$ fusion and HNO_3 dissolution of coal fly ash, and Fuller and Whitehead[50] have described a mixed acid attack on silicate glasses prior to flameless measurement techniques. Copper is usually abundant enough, however, and AAS is sufficiently sensitive in the case of Cu, that recourse to flameless techniques and the concomitant problems[46] is not a frequent necessity.

There are few interferences in the determination of Cu by AAS using an air–C_2H_2 flame (Ref. 15, pp. 66–70; Ref. 16, pp. 585–590). Barnett[42] concluded that, in an air–C_2H_2 flame, HCl and HNO_3 gave the least interference of the five mineral acids which he studied ($HClO_4$, H_2SO_4, and H_3PO_4 were the others). He also found that the burner design was a factor and that the acid concentrations of the standards and samples should be matched for optimum results. Fujiwara et al.[51] reported that the interference of mineral acids on the response of Cu in an AAS determination was a function of both flame stoichiometry and measuring height in the flame. They also determined the degree of interference in a fuel-rich flame to be in the order HNO_3, $HClO_4$ > HCl > HBr > HI.

Most Cu minerals do not present a dissolution problem and many procedures have been reported in the literature. The addition of peroxide to a mineral acid attack on rocks[52] and on soils and sediments[53] has been described, but the more conventional methods of sample digestion are to be recommended.[36] The analysis of different minerals of copper (determining "oxide" copper in the presence of chalcopyrite, for example) has been discussed by Young (Ref. 54, pp. 32–38).

Procedure

See the general procedure given in Sec. 9.4.

9.4.8   Gold

Because the 242.80 and 267.60 nm resonance Au lines are not overly sensitive in AAS (0.3 and 0.6 μg/ml, respectively) (Ref. 15, pp. 73–74) and because low levels of Au are of economic interest, many schemes have been used for the preconcentration of gold prior to the AAS measurement of it. A predominant preconcentration scheme involves solvent extraction into methylisobutylketone (MIBK), dibutyl sulfide, butyl acetate, or one of many others (see Sec. 7.2.1). Das and Bhattacharyya have reviewed the separation of Au by solvent extraction.[55] Another widely used preconcentration technique is to collect the Au into a Pb[56] or a Sn[57, 58] button or into a Ag prill[59] using fire assaying (see Sec. 7.2.3). Trace quantities of Au may also be concentrated by chromatographic separation, and Sukiman has reported using the resin Amberlite XAD-7 for this purpose.[60]

Sen Gupta[61] has reviewed the methods for the determination of the precious metals as have Beamish and Van Loon in a well referenced and useful book on the subject.[62]

Both flame and flameless techniques have been used successfully for the determination of Au. The introduction of the MIBK or some other extract directly into the thermal atomizer for the flameless AAS determination of Au is not unusual.[37, 63–65] Riandy et al. have described optimum flameless conditions for the determination of Au.[44]

The determination of Au in an air–C_2H_2 flame is subject to some interference from Pt and Pd, but not from most of the more common elements (Ref. 16, pp. 598–600). Elliott and Stever have, however, reported interference from Al, Fe, and Ca.[66] Dixon et al. found that SO_4^{2-} in concentrations greater than 0.3% interferes with the AAS determination of the noble metals,[67] and suppression of the Au signal by H_2SO_4 has been reported by others as well (Ref. 16 above). Suppression has also been attributed to the presence of CN^-.[68] Special precautions were found to be necessary to ensure complete sample decomposition when Cu sulfide ores were being subjected to an acid attack.[69] It was found that it was necessary to roast the sample at 600°C for 2 hr (with a preroasting at 480°C for 2 hr if arsenopyrite was present), to follow this with a 2-hr digestion with 50% v/v HCl at 90°C and then to add HNO_3 to complete the attack with aqua regia.

Procedure

Two procedures will be described for the determination of Au. The first covers the analysis of an Ag prill obtained by fire assay techniques[70] (method A, see Sec. 7.2.3 and 12.3) and the second involves an acid attack on the sample followed by an MIBK extraction (method B). In both cases the analyst must be aware of the possibility of subsampling problems resulting from the presence of "free" gold in the sample. Subsample size should be as large as possible to minimize this problem (see Sec. 3.2.2).

a. Method A

Reagents

Standard fire assay reagents as described by Haffty et al.[70]

1000 ppm standard Au solution. Prepared by dissolving the element in aqua regia and adjusting to 20% v/v of the acid. Store in a dark plastic bottle to protect it from light. Dilute solutions made from this should be adjusted to 20% v/v aqua regia also and made up fresh each day.

High purity HCl and HNO_3. Some batches of reagent grade acids have enough Au in them to affect the detection limits, and high purity acids are recommended.

Apparatus

Standard fire assay furnace, pots, cupels, and so on;[70] volumetric flasks (25 ml); atomic absorption spectrometer and Au HCL.

Procedure

Place each Ag prill (obtained by a standard fire assay technique) in a 25 ml volumetric flask. Ensure that enough fire assay blanks are available to determine the Au content contributed by the litharge and other reagents. The sample size used for obtaining the Ag prill should be as large as is convenient (30–40 g can usually be treated with a standard fire assay charge). Add 2 ml of dilute HNO_3 (1 : 7) to each flask, warm them to effect a parting, and dissolve the gold by adding 1.25 ml of concentrated HNO_3 and 3.75 ml of concentrated HCl. Dilute the solutions in the flasks to the correct volume and measure them immediately using a lean air–C_2H_2 flame. Background readings should be made between readings. The standards (0, 1, 2, 3, 4, and 5 ml of 25 ppm Au solution in 25 ml flasks) should have appropriate amounts of concentrated HNO_3 and HCl added to provide a 20% aqua regia concentration when diluted to the correct volume. This method is capable of detection limits of 10 ppb of Au in the sample, provided that the blank is low enough and a sufficiently large sample is available.

b. Method B

This is the method described by Rubeska et al.[65]

Reagents

Stock Au solution.   Dissolve 0.1 g Au in 10 ml aqua regia, evaporate to near dryness, dissolve in 2 N HCl, and dilute to 100 ml with 2 N HCl. Store in the dark.

0.2 M dibutyl sulfide (DBS).   Dissolve 14.6 g of DBS in 500 ml of toluene. Store in a dark bottle which is kept tightly closed.

Analytical reagent grade HF, $HClO_4$, HNO_3, and HCl.

Reagent grade NH_4NO_3.

Apparatus

Porcelain crucibles (to hold 35 g); 150 ml Teflon beakers, 400 ml glass beakers; 250 ml separatory funnel; test tubes; centrifuge, Bunsen burner, muffle furnace, water bath, and sand bath; atomic absorption spectrometer and Au HCL.

Procedure

Mix 25 g of pulverized sample in a porcelain crucible with 10 g of powdered NH_4NO_3 and sinter this by heating it slowly over a Bunsen burner for 15 min and then in a muffle furnace at 650°C for 30 min. Transfer the sintered material to a Teflon beaker, break it up with a plastic stirring rod, add 30 ml of HF and 20 ml of $HClO_4$, and heat at less than 200°C on a sand bath until dense $HClO_4$ fumes are evolved, add another 30 ml of HF, and again heat to $HClO_4$ fumes. Transfer the residue to a 400 ml glass beaker, add 3 g of NaCl, 60 ml of concentrated HCl, and let the solution stand for 20 min. Add 20 ml of concentrated HNO_3, boil the contents for 30 min, and evaporate the solution on a water bath to obtain wet salts. Dissolve these in a few milliliters of 2 N HCl with heating and, while hot, transfer the contents to a test tube and centrifuge them to separate the undissolved residue.

Transfer the solution to a 250 ml separatory funnel, add 5 ml of 0.2 M DBS in toluene, and shake for 5 min. Separate the organic phase and aspirate it into the flame of an atomic absorption spectrometer that has been optimized as described previously. Aspirate a 5% KCN solution, with a few drops of H_2O_2 added, to prevent a slight memory effect between readings. *Note.*   Proper venting of the AAS unit is mandatory when aspirating KCN solutions.

Extract standards of suitable Au concentration (depending on the Au concentration in the samples) in a similar manner and measure them at the same time as the unknowns. The DBS gives a small background signal which must be taken into account.

This method is capable of detection limits of 20 ppb when a 25 g sample is used. Iron and most other metals are not extracted with the gold (Pd is coextracted quantitatively), and thus backwashing (of uncertain efficacy in any case) is not necessary as it is with MIBK. In addition, the low solubility of DBS in HCl precludes the necessity of having to pretreat the HCl as is necessary with MIBK. Strong and Murray-Smith[69] have described a method using MIBK for high sulfide ores, but it could be readily adapted to regular samples.

9.4.9   Lead

The use of AAS for the determination of Pb in rocks and minerals is widespread and presents few problems (Supp. Ref. 7). Chemical interferences are limited to the precipitation of Pb by certain anions, but the addition of EDTA as a complexing agent overcomes most of these interferences. Both the 217.00 and the 283.31 nm Pb absorption lines have been used. The former is more sensitive but is subject to background interference from the air–C_2H_2 flame and the latter is more commonly used (Ref. 16, pp. 616–620). In addition, if Cu is present in the cathode (as in the case of a multielement lamp for Cu as well as Pb), the Cu resonance line at 216.51 nm can give rise to a spectral interference with the 217.00 nm Pb line, and the 283.31 nm line should be utilized instead.

Flameless AAS techniques have been used to determine Pb in geological samples which have been decomposed by mixed acid attack in a Teflon bomb,[30, 36, 71] by HNO_3 attack for carbonate rocks,[72] and by a $LiBO_2$ fusion,[49] while others have simply mixed the powdered sample with graphite and volatilized the Pb directly without a specific dissolution step.[73] The conclusions of Cruz and Van Loon[46] should be kept in mind, however, when a flameless technique is being considered (see sec. 9.4.6).

The concentration of Pb by means of ion-exchange resins[74] and by solvent extraction (Ref. 53; Supp. Ref. 11) prior to measurement by AAS has been reported, as has the determination of Pb generated as a hydride and atomized in a heated cell (Ref. 27, Supp. Refs. 1, 12). Govindaraju et al. have described the determination of Pb in rocks by dipping an iron screw rod wetted with ethyl alcohol into the powdered sample and then introducing the coated screw rod into an air–C_2H_2 flame and measuring the peak produced on a strip chart recorder. The method is reported to be capable of producing a relative standard deviation of 15% at a level of 10 ppm.[75]

The AAS determination of Pb using an air–C_2H_2 flame subsequent to mixed acid digestion of a geological sample is the most common tech-

nique, as is indicated in the review Korkisch and Cross gave as an introduction to their paper.[74] The digestion can be carried out either at elevated
temperatures and pressures in a Teflon bomb, or in open Teflon beakers,
and the results obtained for Pb using standard AAS measurements can
be reproducible and accurate. Note that there is a chance of loss of Pb
by volatilization if the mixed acid digestion mixture is taken to dryness.

Procedure

See the general procedure given in Sec. 9.4.

9.4.10   Mercury

The cold vapor-flameless AAS method is the most common method for
the determination of Hg. It has the important advantages of being a simple,
rapid, and reliable method capable of very low detection limits. Ure[76] has
reviewed this method, and Chilov[77] has produced a more general review
of the determination of Hg.

Koirtyohann and Khalil have investigated the variables in the determination of Hg by the cold vapor AAS technique.[78] They have concluded
the H_2SO_4 concentration and temperature have a significant effect on the
partition function (which is about 0.4) of Hg when equilibrated between
a gaseous and an aqueous phase. The addition of oxidizing agents can
prevent low Hg values due to the premature reduction of Hg by components in plastic laboratory equipment but the presence of more than
1 ppm of Au, Ag, or Pt will result in a low Hg value.

Variations on the technique have included volatilization of the Hg from
the sample in an induction furnace;[79] absorbing the Hg (released by heating the sample to 750°C) in Ag mesh which is heated to 380°C to desorb
the Hg and release it to the measuring cell;[80] absorbing the Hg on Au foil
which is heated to 750°C to release the Hg to the measuring cell.[81]

The method described here will be based on the well-tested reduction–aeration procedure reported by Hatch and Ott.[82]

Reagents

10% w/v $SnCl_2$ in concentrated HCl.
$MgClO_4$ (the unhydrated form).
HCl, H_2SO_4, HNO_3, analytical reagent grade (the Mallinckrodt HNO_3
was found to be low in Hg).

Standard Hg solution (1000 μg/ml).   dissolve 1.354 g $HgCl_2$ in 1 *M*
H_2SO_4 and dilute to 1000 ml with 1 *M* H_2SO_4. Prepare 1 μg/ml solution
fresh each day.

Apparatus

150 ml Erlenmeyer flask, water bath, and an AAS unit with Hg HCL and strip chart recorder. Cold vapor apparatus consisting of a 150 ml Erlenmeyer flask with a two-hole rubber stopper; a drying tube filled with $MgClO_4$, with glass wool packing at either end; a 10 or 15 cm absorption cell with quartz windows; and a cylinder of N_2 or a circulating pump.

Procedure

Weigh 0.500 g of sample into a 150 ml Erlenmeyer flask, add 20 ml concentrated HNO_3, 1 ml concentrated HCl, mix, and allow to stand for 10 min. Place the flask on a hot water bath held at 90°C for 1.5 hr, swirling the flask occasionally and then remove, allow to cool, and add about 20 ml of water.

Allow the HCL and instrument to warm up prior to the analysis, renew the $MgClO_4$ in the drying tube if there is any indication that it is damp, establish the N_2 flow, and align the absorption cell in the optical path of the AAS.

Make up standard solutions containing 0, 5, 10, 25, and 50 ng of Hg, and having the same acid concentration as the samples, in 150 ml Erlenmeyer flasks just prior to the addition of the $SnCl_2$ solution.

The measurement sequence described below should be performed for each standard and sample sequentially, with identifying marks made on the resulting peak on the chart of the strip chart recorder.

Add 10 ml of $SnCl_2$ solution to the Erlenmeyer flask and immediately connect it to the evolution train for which the N_2 flow has been established. The Hg vapor released from the acid solution by the $SnCl_2$ is swept from the flask through the $MgClO_4$ column and through the absorption cell before being vented. The absorption of the signal is recorded on the strip chart recorder, and the peak heights of the samples are compared with those of the standards for calibration purposes. Blanks and standards should be run frequently (after every tenth sample measurement).

This method is capable of quickly detecting <5 ng/g (ppb) of Hg in the original sample with good precision and accuracy.

9.4.11   Molybdenum

The sensitivity of the atomic absorption signal for Mo is relatively weak, and the concentration of the element is usually so low in most geological materials that either a large sample or a preconcentration step (usually solvent extraction) is necessary for the determination of Mo by

AAS. Hutchison has extracted the benzoin α-oxime complex into chloroform after an $HF-HClO_4-HNO_3$ acid attack.[83] Kim et al. have studied the extraction of the thiocyanate complex into MIBK[84] and also into long-chain alkylamines,[85] following a mixed acid attack in PTFE beakers similar to that described in the general procedure. Rao reported the extraction of phosphomolybdate into MIBK subsequent to a $LiBO_2$ fusion.[86] Another group has reported a preconcentration of Mo by means of an anion-exchange resin following either a mixed acid attack or an alkaline fusion (Na_2CO_3) of the geological sample.[87]

Agemian and Chau measured Mo directly in an $N_2O-C_2H_2$ flame following digestion of the sample in a Teflon bomb,[30] and Sutcliffe also used an $N_2O-C_2H_2$ flame directly subsequent to an $H_2SO_4-HNO_3$ acid attack.[88]

The determination of Mo using flameless AAS, although not common, has been reported following both a $LiBO_2$ fusion of fly ash[49] and a mixed acid $(HF-HClO_4)$ digestion of carbonate rocks.[43] One disadvantage of employing flameless AAS in the determination of Mo, aside from the general matrix problems mentioned by Cruz and Van Loon,[46] is that in order to be confident that all of the Mo has been removed from the atomizer, long atomization times and high temperatures are required (Ref. 37, p. 75).

The determination of Mo by flame AAS requires the use of reducing fuel-rich conditions for both the air-C_2H_2 and the $N_2O-C_2H_2$ flames (Ref. 16, pp. 643–646). In the latter case the fuel-to-oxidant ratio should be adjusted so that the flame is not quite luminous. Sturgeon and Chakrabarti have reported that the free-atom population density is confined to a small area. They also concluded that the sensitivity is affected by flame stoichiometry as well as the chemical nature of the organometallic compound being aspirated and the complex–solvent combination used for the extraction.[89]

There are numerous elements, anions, and acids which interfere with the determination of Mo in both air- and N_2O-supported flames. Dilli et al.[90] have reported that dilute HNO_3 and, to a lesser extent, dilute HCl provide favorable conditions but that H_2SO_4 (severe depression) and H_3PO_4 (small enhancement) should be avoided. They also found that the air-C_2H_2 flame is capable of greater precision than is the $N_2O-C_2H_2$ flame. Sutcliffe, however, reports that the $N_2O-C_2H_2$ flame is more sensitive and that HNO_3 does interfere, as do $HClO_4$, H_2SO_4, and H_3PO_4, whereas HCl does not to any significant extent.[88] Sutcliffe also found that the addition of Al was only partially effective in removing the interference due to these acids, whereas it was a good means of removing that due to some of the more common elements found in geological materials (not including Si, which he never investigated). The Al also enhanced the Mo signal very significantly. Ramarkrishna et al. listed a long series of anions and cations which interfere with the Mo absorption in an $N_2O-C_2H_2$ flame.[91] They also report the use of Al to help overcome many interference problems.

Spectral interferences include that from V at both the 390.30 and the 313.26 nm line and from Cr at the former (Ref. 15, pp. 171–175).

The method described here is taken from that described by Kim et al.[84]

Reagents

1000 ppm Mo stock solution (see Appendix 3).

10% w/v $SnCl_2$ solution. Warm 20 g of analytical grade reagent ($SnCl_2\cdot2H_2O$) and a small piece of analytical grade Sn metal in 20 ml of 10 M HCl. When the solution is clear, cool and dilute it to 200 ml with distilled water, filter if necessary, and add a piece of Sn metal.

10% w/v KSCN solution.

MIBK, redistilled with the 116–118°C fraction collected for use.

Analytical grade reagents. KSCN, ascorbic acid, HCl, HF, $HClO_4$, NaF.

Apparatus

100 ml Teflon beakers, sand bath, 100 ml separatory funnel, 10 ml volumetric flask; atomic absorption spectrometer, $N_2O-C_2H_2$ flame.

Procedure

Weigh 1 g of pulverized sample into a Teflon beaker, moisten it with distilled water, and add 4 ml of 70% $HClO_4$ and 10 ml of 40% HF. Evaporate the solution to $HClO_4$ fumes at 200°C (or slightly less) on a sand bath. Add another 10 ml of HF and evaporate to dryness. Wash down the walls of the beaker with 1 M HCl, add 20 ml extra, cover, and boil gently to dissolve the residue. Cool the solution and transfer it to a 100 ml separatory funnel, using 1 M HCl to wash the beaker, and cover until the total volume is 40–50 ml. Add 1 ml of 10% w/v KSCN solution and enough ascorbic acid to bleach the red ferric thiocyanate color. If a white suspension appears (due to Ti), add 0.5 g of NaF and shake the solution until it clears. Add 2 ml of 10% $SnCl_2$ solution, 5 ml of MIBK, and shake vigorously for 2 min. Allow the phases to separate, drain the aqueous layer into a second separatory funnel and the organic phase into a 10 ml volumetric flask. Wash the first funnel with 4.5 ml of MIBK, add this to the second funnel, and shake for 1 min. When the phases have separated, discard the aqueous layer and add the MIBK to that in the 10 ml volumetric flask. Wash the second funnel with sufficient MIBK to fill the flask to the correct volume and measure the absorbance, using MIBK as a blank.

The instrument conditions must be adjusted very carefully to optimize the absorption signal,[89] using the information given in Table 9.1 as a guide.

9.4.12 Nickel

The general conditions for the determination of Ni by flameless AAS are given by Fuller (Ref. 37, pp. 75–76). The same author has also described the flameless AAS determination of Ni in lead-based[92] and other glasses.[50] Schweizer[43] has reported that matrix interferences are observed in the determination of Ni in carbonate and silicate rocks when a graphite furnace is used for the electrothermal atomization of the sample subsequent to an HF–HClO$_4$ acid digestion. He concluded, however, that the technique was "an accurate and rapid method," citing a coefficient of variation of 3.9% obtained for Ni at a concentration less than 5 μg/ml.

Willis[93] has reported the determination of Ni and several other elements by aspirating a water slurry of finely pulverized (<44 μm particle size) rock samples directly into a flame. The accuracy is limited by the different atomization efficiencies obtained for different rock types. Where an error factor of about 2 is acceptable, as in some geochemical survey applications, the technique could prove to be useful.

The choice of absorption line for the determination of Ni in geological materials using flame AAS techniques has presented a few problems. The nonabsorbing line at 231.98 nm causes curvature of the calibration because it is not resolved from the 232.00 nm absorption line in most monochromaters. It is thus necessary to go to a less sensitive line (of which there are several) (Ref. 16, p. 648) when linearity over a greater concentration range is required. The 341.48 nm line is frequently the one chosen.

Few interferences are encountered in the determination of Ni by flame AAS techniques. Sundberg[94] has reported interferences from Zn, Fe(III), Cu, Co, Mn(II), and Cr(III) in both oxidizing and reducing air–C$_2$H$_2$ flames. The interferences can be overcome by a careful adjustment of the observation height. Cresser and MacLeod[48] have observed an interference (suppression) at the 50 μg/ml Ni level from a high concentration of SO$_4^{2-}$ ions, introduced when sulfate rather than chloride salts were used to make up standards. The interference was found to be dependent on the stoichiometry of the flame, with increased suppression in a fuel-rich flame. The interference was reported to be overcome by diluting the solution and either adding a releasing agent or taking measurements in the upper part of a fuel-lean flame. Barnett[42] found little interference from acid concentrations $\leqslant 8\%$ (v/v) in the AAS determination of Ni.

Many digestion techniques have been suggested for the determination of Ni in geological materials, ranging from a mixed acid attack (HNO_3–HF–$HClO_4$) on sediment and rock samples in a digestion bomb[30, 36] to an HNO_3–H_2O_2 attack for soils and sediments.[53] Hannaker and Hughes[95] describe the dissolution of rock samples using HF, HNO_3 and either $HClO_4$ or H_2SO_4 in a Teflon crucible. They then separate and concentrate many trace elements by solvent extraction for subsequent measurement by flame AAS techniques. Ni is extracted as an EDTA complex into MIBK along with Cu, Co, Cr, Zn, Ag, and others.

Procedure

See the general procedure given in Sec. 9.4. One precaution which should be observed is that some Ni minerals require extended fuming with $HClO_4$ to effect a complete dissolution.

9.4.13   Platinum

As is the case with Au (Sec. 9.4.8), most analytical methods incorporating AAS to determine Pt include a preconcentration step. The fire assay concentration of Pt into Sn,[58] NiS,[96] and Pb[97] buttons as well as Ag prills,[98, 99] has been reported, with subsequent measurement of the Pt concentration by both flame and flameless AAS. Coombes and Chow (Supp. Ref. 13) compared measurement of the Pt concentrated in an Ag prill by AAS, ES, and XRF and concluded that the AAS method was preferable. Solvent extraction preconcentration has been used by Simonsen[100] and by Swider[101] subsequent to the acid decomposition of the host matrices. Stanton and Ramankutti[102] have described a sample attack using a Br_2–HBr solution followed by an MIBK separation of Au and Te, a reduction of the Pt with Sn^{2+}, and subsequent Pt extraction into MIBK. The ketone extract is aspirated directly into an air–C_2H_2 flame for measurement of the Pt concentration with a detection limit of 0.1 ppm.

The direct measurement of Pt and other precious metals (Au, Ag, and Pd) using flameless AAS following an aqua regia and HF attack has been described by Fryer and Kerrich.[103]

Sen Gupta[61] and Beamish and Van Loon[62] have reviewed numerous methods for the determination of Pt, and the latter authors have included some subjective evaluation of various methods to the advantage of the reader. Adriaenssens and Knoop[104] have studied the optimal conditions for flameless AAS Pt determinations, as has Everett,[105] and optimum conditions for using an air–C_2H_2 flame have been reported by Heinemann.[106] Eckelmans et al.[107] have reported that the mutual interference shown by Au, Pt, and Pd in an air–C_2H_2 flame during an AAS measurement can be eliminated by the use of the dithizonates of the metals. Nicolas and Jones[108] have reported that the addition of dithiocarbamate overcomes the same mutual interferences. Both sets of authors ascribe the elimination of the mutual interferences to the formation of noble metal complexes. Pitts et al.[109] have reported that the use of an N_2O–C_2H_2 flame gives lower sensitivity and that all interferences are eliminated, including those from high concentrations of other precious metals.

Procedure

The procedure given here is that described by Stanton and Ramankutti.[102] This method has been chosen because neither fire assay nor flameless AAS capabilities are required and thus the method can be used in most laboratories. If either or both of these two capabilities are available, the analyst may prefer to use the methods referred to earlier in the general discussion.

Reagents

HBr (49% w/w).

Bromine solution.   Mix 20 ml of Br_2 with 1 liter of HBr.
4-methylpentan-2-ketone (MIBK).
HCl (36% w/w).

SnCl₂ solution.   Dissolve 500 g $SnCl_2 \cdot 2H_2O$ in HCl and dilute to 1 liter with HCl.

Standard Pt solution.   (1) 100 μg Pt/ml: dissolve 50 mg of Pt sponge in 10 ml of bromine solution and dilute to 500 ml with HBr; (2) 5 μg Pt/ml: dilute 5 ml of 100 μg/ml solution to 100 ml with 2 *M* HBr (make up fresh each day).

Method

Early in the day weigh 25 g of sample into a 400 ml beaker, add 100 ml of bromine solution, mix well, swirl occasionally during the day, and allow to stand overnight. Boil to expel the excess Br_2, cool, dilute to 250 ml with water, and allow the undissolved material to settle. Take 100 ml of the clear solution and extract with 15 ml and then with 10 ml portions of MIBK, shaking the solution for 2 min each time. The organic phase may be analyzed for Au or Te, but if this is not required, discard it. Add 25 ml of $SnCl_2$ solution to the aqueous phase, and extract with 15 ml and then with 10 ml portions of MIBK. Wash the combined organic phases twice by extraction with 25 ml volumes of 2 M HBr, dilute to 25 ml with more MIBK, and aspirate the organic extract into an air–C_2H_2 flame. Carry out extractions of Pt standards, in the range of 0–10 μg, from 2 M HBr solutions in a similar manner for preparation of a calibration curve. The optimization of the instrumental conditions should be carried out as described earlier (Sec. 8.3.1).

This method is capable of a detection limit of .1 ppm, but the use of 200 ml of the acid solution instead of 100 ml, and extracting with 10 ml of ketone instead of 25 ml, can further lower this limit.

9.4.14   Selenium

As was found for Sb and As, the preferred method for the determination of trace amounts of Se in geological samples is that of hydride generation combined with an AAS measurement. Thompson and Thomerson[14] have reported a detection limit of 0.0018 μg/ml using the technique of a heated tube. Hermann[110] reports an absolute detection limit of 0.03 μg and Smith[111] one of 0.05 μg when the hydride was swept into an Ar–H_2 flame. Under the latter conditions Smith found that Ag, Cu, Ni, Pd, Pt, Rh, Ru, and Sn give severe suppression (50%) and Au, As, Cd, Co, Fe, Ge, Pb, Sb, and Zn more moderate suppression (between 10 and 50%). It was found that Se(VI) gave a very weak signal when compared to Se(IV), and the latter oxidation state is the one from which the hydride should be generated.[19] Pierce and Brown[112] reported on an automated method for determining Se by using a Technicon Analyser™ system to introduce the reagents to the sample. They also studied potential interferences from a number of cations and anions as a function of the order of reagent addition. When $NaBH_4$ was added before HCl they found, for a 1 μg/l Se concentration, 100% suppression of the signal in the presence of less than 1 mg/l of Cd^{2+}, Co^{2+}, and Sr^{2+}, and the same degree of suppression in the presence of 3.3 mg/l of Cd^{2+}, Co^{2+}, Cu^{2+}, Ag^+, Sr^{2+}, Sn^{2+}, V^{2+}, MnO_4^-, VO_3^-, $S_2O_8^{2-}$, and NO_3^-. At higher concentrations of Pb^{2+}, Ni^{2+}, and Zn^{2+}, complete suppression was also found. Virtually all of these interferences were overcome when HCl was added prior to the $NaBH_4$, with the exception of some residual suppression from the anions at concentrations of 3.3 mg/l and higher.

The determination of Se using ordinary flame techniques suffers from the fact that the most sensitive line is below 200 nm (at 196.03 nm). This results in decreased sensitivity because of poor spectrometer optical properties in some instances but, more seriously, from background absorbance in the air–C_2H_2 flame. The more transparent but cooler inert gas–H_2 flames result in serious interelement interferences (Ref. 16, pp. 672–674).

Kirkbright and Ranson[113] have observed that a low flame background is observed in an $N_2O-C_2H_2$ flame at the 193.7 nm As and 196.03 nm Se resonance lines and that shielding the flame with N_2 results in complete transparency. The dectection limit for Se was not improved over that obtained with an air–C_2H_2 flame, however, because of its low sensitivity in an $N_2O-C_2H_2$ flame.

Severne and Brooks[114] have reported obtaining a detection limit of 0.1 μg/g for Se and Te in geological materials by using a mineral acid digestion, coprecipitation of the Se and Te on an As carrier (with hypophosphorus acid), dissolution of the precipitate in dilute HNO_3 and the AAS determination of Se and Te in an N_2-H_2 flame. Golembeski[115] used an HNO_3–HF digestion for geological materials and extracted the Se from an HCl–HBr solution into phenol dissolved in benzene. The benzene phase was then evaporated. A water leach of the residue was evaporated in a Ta boat, and the Se concentration was measured by heating the boat in the air–C_2H_2 flame of an AAS unit. He reports results at concentrations as low as 0.058 μg/g with an uncertainty of ± 0.001 μg/g.

Jones[26] has reported the determination of Se in sulfide concentrates by standard flame techniques with a detection limit of 2.5 μg/g.

A detection limit of 1.4×10^{-10} g of Se was reported by Ohta and Suzuki[45] for the determination using electrothermal atomization techniques. Fuller (Ref. 37, p. 79) lists conditions and interferences for this method which has been only infrequently applied to the determination of Se in geological materials.

Procedure

The procedure and the reagents are very much as described in Sec. 9.4.1 with the exception of the standard Se(IV) solution. A selenious acid solution containing 1000 μg Se/ml is prepared as follows (Ref. 116, pp. 377–378). Dissolve 250 mg Se in concentrated HBr–Br solution (1 ml liquid Br + 25 ml concentrated HBr); nearly neutralize the excess Br with SO_2 (from a standard gas cylinder of Se-free SO_2) while shaking vigorously, and complete the neutralization by adding dropwise a slight excess of 5% w/v phenol solution; dilute to 250 ml, and ensure that all SO_2 has been neutralized, or Se will be precipitated by reduction from the H_2SeO_3 solution. Dilute solutions should be made up fresh on a daily basis.

9.4.15   Silver

The use of AAS for the determination of Ag has shown the method to be very sensitive and subject to few interferences (Ref. 16, pp. 676–679).

The 328.07 nm absorption line is the most sensitive for Ag, with a detection limit of about 0.005 μg/ml in an air–C_2H_2 flame. However, this line requires a narrow bandpass (about 0.3 nm) to separate the 327.40 nm line of Cu which is usually present in an Ag HCL (Ref. 15, pp. 70–72). The 338.29 nm line, although less sensitive by a factor of about 2, exhibits a linear calibration over a larger concentration range than does the line at 328.07 nm.

Both flameless and flame techniques have been employed in the determination of Ag by AAS in sulfides, rocks, and other geological matrices. Langmyhr et al.[117] have described the application of both to sulfide ores with an emphasis on the direct atomization from the solid state in a graphite furnace. They found that the precision obtained by the direct atomization compared favorably with that obtained by an acid digestion followed by flameless and flame AAS measurements. In another paper, Langmyhr et al.[118] have reported on the flameless AAS measurement of Ag in silicate rocks both by direct atomization from the solid state and by atomization from a solution obtained by a mixed acid digestion of the samples. Fuller has reported a detection limit of 0.1 picogram (pg) using a graphite furnace for the AAS determination of Ag in an aqueous solution (Ref. 37, p. 80). Bogdanova and Dobretsova[119] have found that Ag can accumulate in the graphite of an atomizer cell during the flameless determination of Ag in aqueous solutions or in benzene and toluene extractions. Ag can also be retained on glassware and silica micropipettes. Washing of the atomizer with an aqueous NH_3 solution or with a solution of dithizone in benzene will overcome the problem.

Adriaenssens and Verbeek[120] have determined Ag in cupellation prills obtained by fire assay techniques using an air–C_2H_2 flame and a KCN–K-EDTA solution subsequent to an acid digestion of the prill. A detection limit of about 50 ppb can be obtained using this approach, and the precision is reported to be about 2%. Ag can also be concentrated by extraction into MIBK[121] and other nonaqueous solvents prior to aspiration into a flame (Supp. Ref. 7). Pierce et al.[122] have described the use of a Technicon Instrument Corporation Auto Analyser™ to automate the extraction and AAS measurement of Ag.

Because the AAS method for the determination of Ag is a very sensitive one, it is usually unnecessary to include a preconcentration step. Direct aspiration of an aqueous solution obtained by acid digestion of the sample into an air–C_2H_2 flame is the common method for determining Ag in samples of geological materials. The acid decomposition procedure differs for different matrices. For example, Ng[123] used EDTA and thiosulfate to keep Pb and Ag in solution, respectively, following digestion of the sample with acid when determining Ag in sulfide minerals. Royal and Mallett[124] have described an $HNO_3-HCl-Br$ method for the decomposition of sulfides and concentrates containing less than 100 ppm of Ag, and another method, involving a Na_2O_2 fusion followed by an HNO_3 leach, for those containing higher concentrations of Ag. Walton[125] employed an $HF-HCl-HNO_3-HClO_4$ acid decomposition procedure followed by the addition of 5 ml of HCl and 30 ml of concentrated ammonia solution per 100 ml of solution to overcome the interference from residual silicate ions. All three of the above methods used an air–C_2H_2 flame for the AAS measurement of Ag. Heinemann[106] has discussed the optimization of conditions for the AAS determination of Ag. Minkkinen[126] suggests the use of background correction to overcome problems encountered when samples high in Ca are analyzed.

Procedure

The procedure described below has been used in the BCM laboratory for the determination of low concentrations of Ag in rock and mineral samples. It has an effective detection limit of about 0.15 μg/g of sample.

Reagents

Standard solutions.   Dissolve 0.2000 g of Ag metal in dilute HNO_3 and dilute to 500 ml in a volumetric flask to give a 400 ppm stock solution. Dilute 10 ml of this stock solution to 1000 ml just before use to obtain a solution containing 4 ppm Ag. Add 0, 5, 10, 15, 20, and 25 ml aliquots of this solution to 100 ml volumetric flasks, each containing 6 ml $HClO_4$, 15 ml HCl, and 2.5 ml HNO_3 and dilute to volume.

Acids.   Analytical reagent grade HCl, HNO_3, HF, and $HClO_4$.

Ag metal.   High purity (99.999%) Ag.
Distilled, deionized water.

Apparatus

Volumetric flasks of 100, 500, and 1000 ml capacity; Ag and D_2 HCLs; AA spectrometer, with background correction facility if possible; analytical balance; 100 ml Teflon beakers and lids.

Method

Weigh 5 g of sample into a Teflon beaker, add 6 ml of $HClO_4$ and 5 ml of HF, and evaporate to $HClO_4$ fumes on a sand bath at 200°C or less. Repeat the evaporation with two more additions of 5 ml of HF. Cool, add 2.5 ml of HNO_3 and 15 ml of HCl, dilute to 50 ml, cover, and warm to dissolve the salts. Cool, transfer the solution to a 100 ml flask, and dilute to 100 ml. Optimize the AAS measurement conditions (see Sec. 8.3.1 and

Table 9.1) and measure the Ag content of the standards and the sample using a D_2 lamp for background correction.

9.4.16 Tellurium

The AAS determination of Te in geological materials has involved many different digestion and extraction techniques, the latter being required because of the small concentrations of Te usually found and the relatively poor sensitivity of AAS for Te. Watterson and Neuerburg[127] digested -80 mesh rock samples in HNO_3 and extracted the Te into HBr. They then coprecipitated the Te and As and subsequently extracted it into MIBK and measured it by AAS with an air–C_2H_2 flame. They report a detection limit of 5 ppb. For sulfide concentrates in which the Te content may be appreciably higher, Nicolas[128] used a Na_2O_2–Na_2CO_3 fusion to decompose the sample. A three-phase liquid–liquid extraction of Te [as Te(IV)] into 5% diantipyrinylmethane in a $3:7$ benzene–$CHCl_3$ solution was performed prior to determining Te by AAS down to 10 ppm. Nazarenko et al.[129] describe the use of a mixed acid digestion followed by a double extraction. The first extraction into MIBK, with Te present as Te(VI), removed Fe(III) and Cr(VI). The Te was then reduced to Te(IV), extracted into MIBK from a 5 M HCl medium, and the solvent phase was aspirated into an air–C_2H_2 flame. A detection limit as low as 0.06 ppm was reported for silicate rocks, Ni–Cu ores, pyrite, and other kinds of samples. Severne and Brooks[114] report a detection limit of 0.1 ppm when a geological sample is digested in HNO_3–HF, the Te coprecipitated with As from a 6 M HCl leach, the precipitate dissolved in concentrated HNO_3, and the solution diluted and aspirated into an H_2 flame.

Beaty[130] has described the determination of Te in rocks using a graphite furnace atomizer following an HF decomposition and an extraction with MIBK. A detection limit of 70 pg (70×10^{-12} g) of Te is reported. Fuller (Ref. 37, p. 81) describes the general conditions for the AAS determination of Te using electrothermal atomization with a detection limit of 50 pg.

For the general determination of Te in geological samples, the most convenient method is hydride generation of TeH_2 by $NaBH_4$ and subsequent measurement by AAS in a heated measuring cell. Thompson and Thomerson[14] and others[131] have described the method and Smith[111] has discussed possible interferences (see Supp. Refs. 14 and 15).

Procedure

The procedure, reagents, and equipment are very much as described in Sec. 9.4.1 with the exception of the standard Te solutions. A 100 μg Te/ml solution is prepared by dissolving 0.100 g Te in 10 ml of warm HCl with dropwise addition of HNO_3, and then diluting to 1000 ml. Solutions containing 100 ng Te/ml or less should be prepared just before use by the appropriate dilution of the stock solution. The AAS settings are optimized according to Table 9.1, and the detection limit obtainable with this method is 5 ng Te/g.

9.4.17   Tin

The determination of Sn in geological samples by flame AAS techniques is not very common. The air–H_2 flame is the most sensitive but is subject to chemical interferences because of its low temperature (Ref. 16, pp. 697–699). Guimont et al.[132] have applied it successfully to the determination of 1–50 ppm of Sn in sediments by heating the sample with NH_4I to separate the Sn from possible interferences. The SnI_4 sublimate is dissolved in a mixture of acetic acid–1% HCl (1 : 4), filtered, and aspirated into the flame. Welsch and Chao[133] describe a modification of this method in which the SnI_4 is dissolved in 5% HCl, extracted into trioctylphosphine oxide (TOPO)–MIBK, and aspirated into an N_2O–C_2H_2 flame. The detection limit is reported to be 2 ppm Sn. Mensik and Seidemann[134] have used the same method as Welsch and Chao for Sn concentrations from 50 ppm to 1%. For samples containing 0.5–75% Sn, a solution of Fe Cl_2 is added to the sublimate, the pH is adjusted to 7.2–7.6 with NH_4OH, the solution centrifuged and the supernatant liquid discarded, the precipitate dissolved in HCl, and the resulting solution aspirated into an air–H_2 flame. Jones[26] has described the determination of Sn in sulfide concentrates based on a Na_2O_2 fusion of the sample. Moldan et al.[135] have given a good account of the various aspects of the determination of Sn by the flame AAS technique, including flame conditions, interference, and other parameters.

The development of the hydride generation method has enabled the analyst to determine Sn at very low levels with much greater ease and speed than was previously possible. Vijan and Chan[136] have reported a semiautomated method for this technique. Thompson and Thomerson[14] have described the technique in detail, and Smith[111] has discussed the interferences likely to be encountered (see also Supp. Refs. 1, 16).

Procedure

The procedure is very much the same as that described in Sec. 9.4.1 with the exception of the standard Sn solution. This can be prepared by dissolving 0.100 g of Sn in 15 ml of warm HCl and diluting to 1 liter to give a solution containing 100 µg Sn/ml. More dilute solutions should be made up from this just before use. If the Sn is present as cassiterite, this mineral will have to be decomposed by a NaOH fusion prior to hydride generation.

9.4.18 Tungsten

The determination of W by AAS suffers from a lack of sensitivity, and as a result preconcentration techniques are required for the determination of trace level concentrations. A fuel-rich $N_2O-C_2H_2$ flame is preferred, with the 255.13 nm absorption line being the most sensitive (Ref. 15, pp. 174–175). Edgar[137] reports that the addition of Na_2SO_4 to a 100 μg/ml solution of W containing 2000 μg/ml of Mo, Mn, Cr, Cu, Mg, K, Al, or Ni removed all interference from the latter. These elements enhanced the W signal by up to 62% when no Na_2SO_4 was present and a reducing $N_2O-C_2H_2$ flame was used.

Keller and Parsons[138] determined W in silicate ores by fusing the sample with $LiBO_2$ and leaching the melt with hot dilute HCl. The solution was made basic with NH_4OH to dissolve the precipitated tungsten, centrifuged, and the W concentration measured at the 400.88 nm absorption line using an $N_2O-C_2H_2$ flame. Rao[139] also used a $LiBO_2$ fusion to get tungsten ore into solution, but the melt was leached with dilute H_3PO_4. This resulted in a phosphotungstate complex which was extracted into di-isobutyl ketone (DIBK) containing Aliquat 336. The extracted W was measured at the 400.88 nm line using an $N_2O-C_2H_2$ flame. The method is applicable to ores containing more than 0.5% W.

Quin and Brooks[140] used an $HF-HNO_3$ mixture (1:1) in polypropylene beakers to decompose W ores. The residue was leached in a KOH solution which was centrifuged and then aspirated into an $N_2O-C_2H_2$ flame. Tindall[141] used a low temperature HCl digestion (60–70°C) followed by treatment of the residue with $NH_4F \cdot HF$, the addition of Superfloc to help settle the precipitate, and then aspiration of the supernatant solution into an $N_2O-C_2H_2$ flame. The method is applicable to samples with W concentrations varying from 0.05 to 72% WO_3.

Kim et al.[85] investigated the application of long-chain alkylamines to the preconcentration of Mo, W, and Re prior to AAS measurement and extended it to the determination of trace levels of W in geological samples.[142] Topping[143] has published a thorough review of analytical methods for W.

Procedure

The procedure described below is based on that of Kim et al.[142] and is capable of a detection limit of 3 ppm in the sample.

Reagents

Standard W solution.   Dissolve 1.7941 g of $Na_2WO_4\cdot2H_2O$ in water and dilute to 1 liter to obtain a stock solution containing 1000 μg W/l.

SnCl₂ solution (20% w/v).   Dissolve 40 g $SnCl_2\cdot2H_2O$ in 40 ml of 10 *M* HCl, containing a small piece of analytical reagent grade Sn metal, with warming until the solution is clear. Dilute 200 ml with water and add another piece of Sn metal.

NaSCN, 50% w/v.

Alamine 336 in $CHCl_3$, 1% v/v.

MIBK. Redistill and use the fraction boiling at 116–118°C. All other reagents should be of analytical reagent grade.

Apparatus

Teflon beakers (100 ml); Pyrex beakers (25 and 50 ml); separating funnels (100 ml); weighing bottles with lids (25 ml); steam bath; sand bath; volumetric flask (25 ml); atomic absorption spectrometer with a W HCL and N_2O–C_2H_2 flame.

Method

Weigh 1 g of sample into a 100 ml Teflon beaker, moisten with water, add 4 ml of 70% $HClO_4$ and 10 ml of 40% HF. Evaporate the solution on a sand bath at 200°C, or slightly less, to dense white $HClO_4$ fumes, add another 10 ml of 40% HF and repeat the evaporation. Wash down the sides of the beaker with about 25 ml 1 *M* HCl, cover the beaker, and boil the contents gently to dissolve the residue, cool, and transfer the solution to a 50 ml glass beaker. Add a piece of granular Sn metal (about 1 g), 1 ml of 20% stannous chloride solution, and boil for 20 min. Cool the solution rapidly (in ice water) and add 2 ml of 50% NaSCN solution and allow to stand for 20 min. Transfer to a 100 ml separatory funnel, rinse beaker and remaining Sn metal with cold water, and dilute to a final volume of 50 ml. Add 10 ml of 1% Alamine 336 in chloroform, shake vigorously for 2 min, and drain the organic phase into a 25 ml beaker (or a 25 ml weighing bottle for low W samples). Wash the aqueous phase with a second 10 ml portion of Alamine 336–$CHCl_3$ solution and warm the combined extracts on a water bath to evaporate the $CHCl_3$. If the sample is thought to have a low W concentration, the organic phase should be evaporated instead in a 25 ml weighing bottle, 1–5 ml of MIBK added (1 ml for 20 μg of W and 5 ml for 25–100 μg of W), the weighing bottle capped, the solution warmed to dissolve the residue, and the absorbance of this solution measured against an MIBK blank to obtain the

W concentration. If the W concentration is expected to be high (>100 μg of W present), the organic phase should be collected in a 25 ml beaker, the residue remaining from the evaporation of the $CHCl_3$ warmed with a few milliliters of MIBK, and the solution transferred to a 25 ml volumetric flask. The beaker is rinsed with several washings of MIBK and the solution diluted to volume. The AAS measurement is made using the information in Table 9.1 as a guide, with the conditions optimized as described in Sec. 8.3.1.

9.4.19 Zinc

The determination of Zn by AAS is subject to few interferences, with the depression of the signal by Si being the only one likely to be encountered in geological samples. An oxidizing air–C_2H_2 flame is usually used, with the measurements made at the most sensitive line, 213.86 nm, for samples of low concentration and at the 307.59 nm line for those with higher concentrations (>150 μg/ml) (Ref. 15, pp. 74–76). Because Zn is relatively abundant in most geological samples, and the sensitivity of AAS is adequate, a preconcentration step is usually not necessary; Armannsson[144] has described the use of a dithizone extraction procedure for determining Zn in rocks and sediments (see also Supp. Ref. 7).

There have been applications of flameless techniques to the determination of Zn in geological samples. Owens and Gladney[49] injected a dilute HNO_3 solution, obtained by leaching fly ash fused with $LiBO_2$, into a graphite furnace and found an uncertainty in measurement of about 10% at the 200 ppm Zn level. Langmyhr et al.[118] have determined Zn in silicate rock samples by a direct atomization from the solid state in a graphite furnace. They compared the results with those obtained for the same samples using an acid digestion and flame atomization. As might be expected, the standard deviation of the former method was substantially higher in two of the three standard samples analyzed by both methods but not to the extent that the direct atomization method is discredited. As Cruz and Van Loon[46] have concluded, the flameless technique very seldom has an advantage over the use of a flame in mineral matrices.

Few problems are encountered in dissolving Zn minerals in preparation for subsequent measurement by flame AAS. Krishnamurty et al.[53] have described the use of an HNO_3–H_2O_2 digestion at 100°C for soils and sediments; Wall[145] used a Na_2O_2 fusion followed by a water leach, and Agemian and Chau[30] an HNO_3–$HClO_4$–HF digestion in a Parr acid digestion bomb for lake sediments. Sanzolone and Chao[52] digested the sample with HF–HCl–H_2O_2 in a Teflon beaker, and Willis[93] aspirated a slurry of the finely pulverized sample directly into the flame. This latter method gives results only good to $\pm 50\%$ of the concentration, but it is rapid and could find applications in geochemical survey programs.

Procedure

See the general procedure given in Sec. 9.4.

References

1. P. M. Bradshaw, D. R. Clews, and J. L. Walker, *Mining in Can.*, 25–33. (Aug. 1970).

2. A. A. Levinson, *Introduction to Exploration Geochemistry*, Applied Publ. Ltd., Calgary, Alta., 1974, pp. 247–249.

3. J. R. Foster, *Can. Inst. Min. Metal. Bull.*, **85** (1973).

4. J. Doležal, P. Povondra, and Z. Sulcek, *Decomposition Techniques in Inorganic Analysis*, American Elsevier, New York, 1968.

5. M. Zief and J. W. Mitchell, *Contamination Control in Trace Element Analysis*, Wiley, New York, 1976.

6. G. Tölg, *Talanta*, **19,** 1489 (1972).

7. M. Zief and J. Horvath, *Laboratory Equipment Digest*, 47–64 (Oct. 1976).

8. R. W. Dabeka, A. Mykytiuk, S. S. Berman, and D. S. Russell, *Anal. Chem.*, **48,** 1203 (1976).

9. J. R. Moody and R. M. Lindstrom, *Anal. Chem.*, **49,** 2264 (1977).

10. J. D. Winefordner, Ed., *Trace Analysis*, Wiley, New York, 1976.

11. R. K. Skogerboe, in J. A. Dean and T. C. Rains, Eds., *Flame Emission and Atomic Absorption Spectrometry*, Vol. 1, Dekker, New York, 1969.

12. R. R. Liddell, *Anal. Chem.*, **48,** 1931 (1976).

13. G. E. M. Aslin, *J. Geochem. Explor.*, **6,** 321 (1976).

14. K. C. Thompson and D. R. Thomerson, *Analyst*, **99,** 595 (1974).

15. J. A. Dean and T. C. Rains, Eds., *Flame Emission and Atomic Absorption Spectrometry*, Vol. 3, Dekker, New York, 1975.

16. G. F. Kirkbright and M. Sargent, *Atomic Absorption and Fluorescence Spectroscopy*, Academic, New York, 1974.

17. L. De Galan and G. F. Samaey, *Spectrochim. Acta*, **24B,** 679 (1969).

18. E. P. Welsch and T. T. Chao, *Anal. Chim. Acta*, **76,** 65 (1975).

19. S. Terashima, *Anal. Chim. Acta*, **86,** 43 (1976).

20. T. Maruta and G. Sudoh, *Anal. Chim. Acta*, **77,** 37 (1975).

21. R. G. Smith, J. C. Van Loon, J. R. Knechtel, J. L. Fraser, A. E. Pitts and A. E. Hodges, *Anal. Chim. Acta*, **93,** 61 (1977).

22. G. F. Kirkbright, M. Sargent, and T. S. West, *At. Absorpt. Newsl.*, **8,** 34 (1969).

23. R. D. Wauchope, *At. Absorpt. Newsl.*, **15,** 64 (1976).

24. H. Heinrichs and J. Lange, *Z. Anal. Chem.*, **265,** 256 (1973).

25. W. Reichel and B. G. Bleakley, *Anal. Chem.*, **46,** 59 (1974).

26. E. A. Jones, Natl. Inst. Metall. Rep. 1787, 1976.

27. E. N. Pollock and S. J. West, *At. Absorpt. Newsl.*, **12,** 6 (1973).

28. G. Aslin Hall, private communication, 1978; H. F. Schaffer, private communication, 1978.

29. R. V. D. Robert and R. C. Mallett, Natl. Inst. Metall. Rep. 1746, 1975.

30. H. Agemian and A. S. Y. Chau, *Anal. Chim. Acta*, **80,** 61 (1975).

31. G. P. Sighinolfi and A. M. Santos, *Mikrochim. Acta*, **1976 I,** 477 (1976).

32. K. A. Rodgers, *Mineral. Mag.*, **38,** 882 (1972).

33. M. S. Cresser and R. Hargitt, *Anal. Chim. Acta*, **82,** 203 (1976).

34. H. Bennett and G. J. Oliver, *Analyst*, **101,** 803 (1976).

35. J. A. Maxwell, *Rock and Mineral Analysis*, Interscience, New York, 1968.

36. J. Warren and D. Carter, *Can. J. Spectrosc.*, **20,** 1 (1975).

37. C. W. Fuller, *Electrothermal Atomization for Atomic Absorption Spectrometry*, The Chemical Society, London, 1977.

38. H. C. Green, *Analyst*, **100,** 640 (1975).

39. G. Kraft, D. Lindenberg, and H. Beck, *Z. Anal. Chem.*, **282,** 119 (1976).

40. M. E. Britske and A. N. Savel'eva, *Zh. Anal. Khim.*, **31,** 2042 (1976).

41. M. S. Cresser and R. Hargitt, *Talanta*, **23,** 153 (1976).

42. W. B. Barnett, *Anal. Chem.*, **44,** 695 (1972).

43. V. B. Schweizer, *At. Absorpt. Newsl.*, **14,** 137 (1975).

44. C. Riandy, P. Linhares, and M. Pinta, *Analusis*, **3,** 303 (1975).

45. K. Ohta and M. Suzuki, *Talanta*, **22,** 465 (1975).

46. R. B. Cruz and J. C. Van Loon, *Anal. Chim. Acta*, **72,** 231 (1974).

47. G. J. S. Govett and R. E. Whitehead, *J. Geochem. Explor.*, **2,** 121 (1973).

48. M. S. Cresser and D. A. MacLeod, *Analyst*, **101,** B6 (1976).

49. J. W. Owens and E. S. Gladney, *At. Absorpt. Newsl.*, **15,** 95 (1976).

50. C. W. Fuller and J. Whitehead, *Anal. Chim. Acta*, **68,** 407 (1974).

51. K. Fujiwara, H. Haraguchi, and K. Fuwa, *Anal Chem.*, **47,** 1670 (1975).

52. R. F. Sanzolone and T. T. Chao, *Anal. Chim. Acta*, **86,** 163 (1976).

53. K. V. Krishnamurty, E. Shpirt, and M. M. Reddy, *At. Absorpt. Newsl.*, **15,** 68 (1976).

54. R. S. Young, *Chemical Phase Analysis*, Charles Griffin and Co., London, 1974.

55. N. R. Das and S. N. Bhattacharyya, *Talanta*, **23,** 535 (1976).

56. P. E. Moloughney, *Talanta*, **24,** 135 (1977).

57. G. H. Faye and P. E. Moloughney, *Talanta*, **19,** 269 (1972).

58. P. E. Moloughney and G. H. Faye, *Talanta*, **23**, 377 (1976).

59. J. C. Van Loon, *Z. Anal. Chem.*, **246**, 122 (1969).

60. S. Sukiman, *Anal. Chim. Acta*, **84**, 419 (1976).

61. J. G. Sen Gupta, *Miner. Sci. Eng.*, **5**, 207 (1973).

62. F. E. Beamish and J. C. Van Loon, *Recent Advances in the Analytical Chemistry of the Noble Metals*, Pergamon, New York, 1972.

63. R. Machiroux and Doan Thi Kim Anh, *Anal. Chim. Acta*, **86**, 35 (1976).

64. G. P. Sighinolfi and A. M. Santos, *Mikrochim. Acta*, **1976 II**, 33 (1976).

65. I. Rubeska, J. Korecková, and D. Weiss, *At. Absorpt. Newsl.*, **16**, 1 (1977).

66. E. V. Elliott and K. R. Stever, *At. Absorpt. Newsl.*, **12**, 60 (1973).

67. K. Dixon, D. J. Nicolas, R. V. D. Robert, and E. van Wyck, Natl. Inst. Metall. Rep. 1739, 1975.

68. E. Adriaenssens and F. Verbeek, *At. Absorpt. Newsl.*, **12**, 57 (1973).

69. B. Strong and R. Murray-Smith, *Talanta*, **24**, 1253 (1974).

70. J. Haffty, L. B. Riley, and W. D. Goss, *U. S. Geol. Surv. Bull.*, **1445** (1977).

71. C. Block, *Anal. Chim. Acta*, **80**, 369 (1975).

72. W. C. Campbell and J. M. Ottaway, *Talanta*, **22**, 729 (1975).

73. D. D. Siemer and Horng-Yih Wei, *Anal. Chem.*, **50**, 147 (1978).

74. J. Korkisch and H. Cross, *Talanta*, **21**, 1025 (1974).

75. K. Govindaraju, G. Mevelle, and C. Chouard, *Anal. Chem.*, **46**, 1672 (1974).

76. A. M. Ure, *Anal. Chim. Acta*, **76**, 1 (1975).

77. S. Chilov, *Talanta*, **22**, 205 (1975).

78. S. R. Koirtyohann and M. Khalil, *Anal. Chem.*, **48**, 136 (1976).

79. Y. Kuwae and T. Hasegawa, *Anal. Chim. Acta*, **84**, 185 (1976).

80. G. L. Corte and L. Dubois, *Mikrochim. Acta*, **1975 I**, 69 (1975).

81. R. A. Nicholson, *Analyst*, **102**, 399 (1977).

82. W. R. Hatch and W. L. Ott, *Anal. Chem.*, **40**, 2085 (1968).

83. D. Hutchison, *Analyst*, **97**, 118 (1972).

84. C. H. Kim, C. M. Owens, and L. E. Smythe, *Talanta*, **21**, 445 (1974).

85. C. H. Kim, P. W. Alexander, and L. E. Smythe, *Talanta*, **22**, 739 (1975).

86. P. D. Rao, *At. Absorp. Newsl.* **10**, 118 (1971).

87. J. Korkisch and H. Gross, *Talanta*, **20**, 1153 (1973).

88. P. Sutcliffe, *Analyst*, **101**, 949 (1976).

89. R. E. Sturgeon and C. L. Chakrabarti, *Anal. Chem.*, **48**, 677 (1976).

90. S. Dilli, K. M. Gawne, and G. W. Ocago, *Anal. Chim. Acta*, **69**, 287 (1974).

91. T. V. Ramarkrishna, P. W. West, and J. W. Robinson, *Anal. Chim. Acta*, **44**, 437 (1969).

92. C. W. Fuller, *At. Absorpt. Newsl.* **14**, 73 (1975).

93. J. B. Willis, *Anal. Chem.*, **47**, 1752 (1975).

94. L. L. Sundberg, *Anal. Chem.*, **45**, 1460 (1973).

95. P. Hannaker and T. C. Hughes, *Anal. Chem.*, **49**, 1485 (1977).

96. K. Dixon, E. A. Jones, S. Rasmussen, and R. V. D. Robert, Natl. Inst. Metall. Rep. 1714, 1975; through *Anal. Abstr.*, No. 2B245, **31**, (1976).

97. A. Diamantatos, *Anal. Chim. Acta,* **94**, 49 (1977).

98. R. J. Coombes, A. Chow, and R. Wageman, *Talanta*, **24**, 421 (1977).

99. A. Tello and N. Sepúlveda, *At. Absorpt. Newsl.* **16**, 67 (1977).

100. A. Simonsen, *Anal. Chim. Acta*, **49**, 368 (1970).

101. R. T. Swider, *At. Absorpt. Newsl.* **7**, 111 (1968).

102. R. E. Stanton and S. Ramankutti, *J. Geochem. Explor.*, **7**, 73 (1977).

103. B. J. Fryer and R. Kerrich, *At. Absorpt. Newsl.*, **17**, 4 (1978).

104. E. Adriaenssens and P. Knoop, *Anal. Chim. Acta*, **68**, 37 (1973).

105. G. L. Everett, *Analyst*, **101**, 348 (1976).

106. W. Heinemann, *Z. Anal. Chem.*, **280**, 127 (1976).

107. V. Eckelmans, E. Graauwmans, and S. DeJaegere, *Talanta*, **21**, 715 (1974).

108. D. J. Nicolas and E. A. Jones, Natl. Inst. Metall. Rep. 1826, 1976; through *Anal. Abstr.*, No. 6B164, **33** (1977).

109. A. E. Pitts, J. C. Van Loon, and F. E. Beamish, *Anal. Chim. Acta*, **50**, 195 (1970).

110. R. Hermann, *At. Absorpt. Newsl.*, **16**, 44 (1977).

111. A. E. Smith, *Analyst*, **100**, 300 (1975).

112. F. D. Pierce and H. R. Brown, *Anal. Chem.*, **48**, 693 (1976).

113. G. F. Kirkbright and L. Ranson, *Anal. Chem.*, **43**, 1238 (1971).

114. B. C. Severne and R. R. Brooks, *Talanta*, **19**, 1467 (1972).

115. T. Golembeski, *Talanta*, **22**, 547 (1975).

116. W. Horwitz, Ed., *Official Methods of Analysis of the Association of Official Agricultural Chemists*, 10th ed., Association of Official Agricultural Chemists, Washington, DC, 1965.

117. F. J. Langmyhr, R. Solberg, and L. T. Wold, *Anal. Chim. Acta*, **69**, 267 (1974).

118. F. J. Langmyhr, J. R. Stubergh, Y. Thomassen, J. E. Hanssen, and J. Doležal, *Anal. Chim. Acta*, **71**, 35 (1974).

119. V. I. Bogdanova and I. L. Dobretsova, *Zh. Anal. Khim.*, **32**, 1717 (1977); through *Anal. Abstr.*, No. 4B40, **34** (1978).

120. E. Adriaenssens and F. Verbeek, *At. Absorpt. Newsl.*, **13**, 41 (1974).

121. A. N. Chowdhury, A. K. Das, and T. N. Das, *Z. Anal. Chem.*, **269**, 284 (1974).

122. F. D. Pierce, M. J. Gortatowski, H. D. Mecham, and R. S. Fraser, *Anal. Chem.*, **47**, 1132 (1975).

123. W. K. Ng, *Anal. Chim. Acta*, **63**, 469 (1973).

124. S. J. Royal and R. C. Mallett, Natl. Inst. Metall. Rep. 1797, 1976; through *Anal. Abstr.*, No. 1B34, **33** (1977).

125. G. Walton, *Analyst*, **98**, 335 (1973).

126. P. Minkkinen, *At. Absorpt. Newsl.* **14**, 71 (1975).

127. J. R. Watterson and G. J. Neuerburg, *J. Res. U. S. Geol. Surv.*, **3**, 191 (1975).

128. D. J. Nicolas, Natl. Inst. Metall. Rep. 1794, 1976; through *Anal. Abstr.*, No. 1B99, **33** (1977).

129. I. I. Nazarenko, G. E. Kalenchuk, I. V. Kislova, *Zh. Anal. Khim.*, **31**, 498 (1976); through *Anal. Abstr.*, No. 5B151, **31** (1976).

130. R. D. Beaty, *At. Absorpt. Newsl.*, **13**, 38 (1974).

131. L. P. Greenland and E. Y. Campbell, *Anal. Chim. Acta*, **87**, 323 (1976).

132. J. Guimont, A. Bouchard, and M. Pichette, *Talanta*, **23**, 62 (1976).

133. E. P. Welsch and T. T. Chao, *Anal. Chim. Acta*, **82**, 337 (1976).

134. J. D. Mensik and H. J. Seidemann, Jr., *At. Absorpt. Newsl.*, **13**, 8 (1974).

135. B. Moldan, I. Rubeška, M. Mikšovsky, and M. Huka, *Anal. Chim. Acta*, **52**, 91 (1970).

136. P. N. Vijan and C. Y. Chan, *Anal. Chem.*, **48**, 1788 (1976).

137. R. M. Edgar, *Anal. Chem.*, **48**, 1653 (1976).

138. E. Keller and M. L. Parsons, *At. Absorpt. Newsl.* **9**, 92 (1970).

139. P. D. Rao, *At. Absorpt. Newsl.*, **9**, 131 (1970).

140. B. F. Quin and R. R. Brooks, *Anal. Chim. Acta*, **65**, 206 (1973).

141. F. M. Tindall, *At. Absorpt. Newsl.* **16**, 37 (1977).

142. C. H. Kim, P. W. Alexander, and L. E. Smythe, *Talanta*, **23**, 573 (1976).

143. J. J. Topping, *Talanta*, **25**, 61 (1978).

144. H. Armannsson, *Anal. Chim. Acta*, **88**, 89 (1977).

145. G. Wall, Natl. Inst. Metall. Rep. 1798, 1976.

Supplementary References

1. J. F. Chapman and L. S. Dale, *Anal. Chim. Acta*, **111**, 137 (1979).

2. S. Bajo, *Anal. Chem.*, **50**, 649 (1978).

3. I. Rubeška and V. Hlavinkova, *At. Absorpt. Newsl.* **18**, 5 (1979).

4. D. E. Fleming and G. A. Taylor, *Analyst*, **103**, 101 (1978).

5. J. R. Knechtel and J. L. Fraser, *Analyst*, **103**, 104 (1978).

6. G. F. Kirkbright and M. Toddia, *Anal. Chim. Acta*, **100**, 145 (1978).

7. J. G. Viets *Anal. Chem.*, **50**, 1097 (1978).

8. E. M. Donaldson, *Talanta*, **26**, 1119 (1979).

9. C. Y. Chan, M. W. A. Baig, and A. E. Pitts. *Anal. Chim. Acta,* **111,** 169 (1979).

10. M. Ihnat, *Can. J. Spectrosc.,* **23,**112 (1978).

11. P. J. Aruscavage and E. Y. Campbell, *Talanta,* **26,** 1052 (1979).

12. K. Jin, M. Tago, H. Yoshida, and S. Hikime, *Bunseki Kagaku,* **27,** 759 (1978); through *Anal. Abstr.,* No. 2B125, **37** (1979).

13. R. J. Coombes and A. Chow, *Talanta,* **26,** 991 (1979).

14. G. P. Sighinolfi, A. M. Santos, and G. Martinelli, *Talanta,* **26,** 143 (1979).

15. T. T. Chao, R. F. Sanzolone, and A. E. Hubert. *Anal. Chim. Acta,* **96,** 251 (1978).

16. V. F. Hodge, S. L. Seidel, and E. D. Goldberg. *Anal. Chem.,* **51,** 1256 (1979).

PART IV

Methods for Determination of Major, Minor, and Trace Elements in Rocks and Minerals Using X-Ray Fluorescence Spectroscopy

CHAPTER 10

X-Ray Fluorescence Spectroscopy—General Considerations

CHAPTER 11

The Determination of Major, Minor, and Trace Constituents in Rocks and Minerals by X-Ray Fluorescence Spectroscopy

X-RAY FLUORESCENCE SPECTROSCOPY—GENERAL CONSIDERATIONS

In keeping with the approach used in the chapters on AAS, no attempt will be made to discuss the theoretical aspects of XRF. For this the reader should refer to basic texts on the subject (Refs. 1–7, Supp. Ref. 1). As well, little attention will be given to the (increasingly frequent) application of energy dispersive XRF techniques to the analysis of geological and related materials (Supp. Refs. 2–4). This latter omission is not as serious as it may first appear, because much of the following discussion on topics such as sample preparation, preconcentration, and calibration is applicable to both wavelength and energy dispersive techniques.

10.1 SAMPLE PREPARATION

Samples of geological materials can be exposed to an X-ray beam in several forms. These include loose powders, pressed pellets, fused discs, and thin films such as those from a coprecipitation preconcentration step. The selection of the appropriate medium must be based on such parameters as the nature of the analytical problem, the amount of sample available, the element(s) to be determined, the detection limit required, the time available for the analysis, and the degree of reliability required of the data. So that the discussion of these different media will be of most use to the analyst faced with a variety of analytical jobs, each medium will be considered in terms of its advantages and disadvantages, as well as of the method of its preparation.

10.1.1 Loose Powders

Loose powders may be exposed to the primary X-ray beam in different ways, depending on whether the geometry of the instrument is inverted, upright, or inclined (see Ref. 7, pp. 739–741). However, regardless of the type of presentation used, the analysis of loose powders is subject to many errors. One of the more serious of these is due to segregation on the basis of particle size and/or density during the filling of the sample cell. This problem can be overcome to a certain extent by ensuring that the samples and standards have the same particle size distribution or are all ground to the same particle size. Both of these conditions are difficult to satisfy, and the problem of segregation is a serious drawback to this form of sample. Another problem associated with the analysis of loose powders is that of particle size effects in nonhomogeneous samples (Ref. 1, pp. 114–120). A general rule of thumb is that the largest particle size should not be larger than the critical thickness of the sample for the lowest energy (longest wavelength) analyte line to be used. In most instances it is sufficient to grind samples to less than 325 mesh (45 μm particle size), but if very light elements such as Na and Mg are to be determined, problems can still arise. If loose powders are to be used, care must be taken to ensure that the samples are packed to give a reproducible packing density and that a flat surface is exposed to the primary X-ray beam. This latter requirement is most easily satisfied in spectrometers that have the X-ray tube situated below the sample, because the sample holder will usually have a thin (6 mμ or less) sheet of Mylar surface against which the sample can be packed. Care must be taken, however, to avoid open air pockets and to ensure uniform packing density.

In geological applications the use of loose powders is most commonly confined to the analysis of samples collected for geochemical exploration work where rapid results are important, a relative accuracy and precision of 10–15% is adequate, and heavier metals (with relatively short wavelength analyte lines) such as Ba and Rb are of interest.[8] Levinson and dePablo have described the application of this type of sample preparation to geochemical samples to be analyzed for elements heavier than iron.[9] Berri et al. have published a thorough discussion of particle size effects in energy-dispersive XRF.[10]

10.1.2   Briquettes

Powders can be analyzed in the form of briquettes or pressed pellets. The briquette has many advantages over a loose powder. There is much less risk of contaminating the inside of the instrument with dust, a more reproducible sample preparation is possible, internal standards can be added as desired, and a finished surface is available for exposure also to a primary X-ray beam (as opposed to the secondary emission used in XRF analysis) enabling elements such as Ti to be determined readily. However, the precautions mentioned in the previous section, that is, ensuring that the particle size is smaller than the critical thickness of the sample for the longest wavelength analyte line to be used, must be observed in this method as well. A disadvantage is that the frequent and/or prolonged exposure of such briquettes to the primary X-ray beam leads to a gradual deterioration of their surface and their use as standards or monitors is thus limited.

Briquettes can be made with or without binder or backing. These latter can take many forms. Aluminum or plastic cups can be placed in the die, the sample mixture added, the plunger inserted, and the pressure applied. The result is a disc with only one surface available for analysis. It should be noted that when boric acid is used in conjunction with a vacuum system, the boric acid will eventually coat the internal parts of the instrument.

Fabbi has described both a die and a procedure designed to produce a briquette having a hard plastic-like casing around the sample mixture.[11] A casting sleeve is placed inside a die sleeve and enough sample powder to yield a disc of critical thickness is poured in and compacted with a casting piston. The casting sleeve and piston are removed and 20–30 ml of methyl cellulose are poured evenly over the sample, the die piston is inserted, and the contents are pressed at 2100 kg/cm² (30,000 psi) to form a briquette with a protective methyl cellulose surface. A smooth surface for measurements is obtained by placing a cleaned glass flashlight lens in the die before the addition of the sample powder.

The most popular method of forming briquettes involves the mixing of the finely pulverized sample with up to 15% by weight of a binder such as polyvinyl alcohol, cellulose nitrate, boric acid, or a commercial substance such as Somar-mix (Somar Laboratories, Inc.) or Liquid Binder (Chemplex Industries, Inc.) in a mixer mill of some sort and pressing the mixture at a minimum pressure of 1400 kg/cm² (20,000 psi) in a cylindrical die. This forms a firm briquette without backing. If a thermosetting binder is used (methyl methacrylate, for example) or one with a low melting point (such as phenol formaldehyde), the briquette can be placed on a glass plate and heated in a drying oven at a temperature which will cause the binder either to set or to melt, depending on the kind used. The cooled briquette is much more robust and less susceptible to crumbling and breaking than the first type, and does much to overcome problems associated with dust infiltrating the instrument and the contamination of one sample by another.

It is necessary to ensure that both samples and standards have the same packing density, and thus a standardized method of sample preparation is required. This can be achieved by adhering to a fixed grinding time, sample weight, binder concentration, briquetting pressure, and pressing time. Each laboratory must establish its own operating conditions, and the choice will be determined by the types of samples to be analyzed and the equipment available.

Many workers have used briquettes for the determination of a variety of elements over a large concentration range. Leake et al. report on the analysis of 1550 rock powders for 38 major and trace elements. They fabricated the briquettes by mixing 6 g of -250 mesh (<63 μm) sample with 1 g of phenol formaldelyde and pressing the mixture at 4200 kg/cm^2 (60,000 psi) for 5 min. The briquettes were then placed on a glass plate and heated in a drying oven at 110°C for 30 min.[12]

Goodman has described the rapid determination of Sn in soils and stream sediments by XRF using briquettes fabricated with Somar binder. He stresses the necessity of grinding to <53 μm to avoid particle size effects. He also points out that there is a minimum sample thickness which must be maintained because of the penetrating Sn K$_\alpha$ X-rays. In his case, 6 g of sample ground with a 0.5 g Somar tablet and pressed in a briquette 23 mm in diameter was sufficient.[13] Table 10.1 (from Ref. 13) shows the variation in the apparent concentration of tin as a function of sample weight.

Giauque et al. pressed 2 g of finely ground sample in a 1.27 cm diameter lucite ring at 500 kg/cm^2 using Scotch tape as a bottom to the ring. They determined 26 trace and 2 major elements using an energy-dispersive XRF system.[14] The same workers also report the fabrication of thin briquettes using sulfur both as a binder and as a means of overcoming the matrix absorption effect of the sample. They mixed 100 mg of finely ground sample with 400 mg of sulfur powder, pressed it at 1000 kg/cm^2

TABLE 10.1   The Variation in Tin as a Function of Sample Weight.[a]

Sample Weight (grams)	Sample A (ppm)	Sample B (ppm)
1	96	13
2	165	15
4	238	17
6	248	20
8	250[b]	19[b]
12	247	19

Reproduced with permission of *Economic Geology*.[13]

[a] All analyses performed using a 23 mm diameter Siemens sintered coal sample holder.

[b] Correct value as determined on replicate samples.

in a 2.54 cm diameter die, and determined 40 elements in geochemical samples and coal fly ash using an energy dispersive system.[15]

When using pressed powder briquette, or another form of sample preparation, it is necessary to consider the possible buildup of S on the surface after extended exposure of the sample to the vacuum system of the spectrometer, as when one of the samples is a frequently used standard or a monitor.[17] Kanaris-Sotiriou and Brown have reported this problem when analyzing soils.[16] They found that a thin film of polycarbonate or Mylar over the surface of the disc minimized the problem. The use of a vacuum pump oil that is low in S will also help.

10.1.3 Fused Discs

The most common method of preparing rock samples for subsequent analysis by XRF is the fusion of the powdered sample with a borate salt (first proposed by Claisse[18]) followed by casting of the melt in a mold to form a disc for presentation to the spectrometer.

There are several advantages in favor of this approach (Ref. 7, pp. 751–762). The sample is rendered homogeneous, particle size problems are eliminated, synthetic standards can more readily be manufactured, a heavy absorber or an internal standard can be added easily, and matrix differences between samples are minimized. The sample and standard discs are durable, easily stored, and can be reused many times without deterioration from exposure to the X-ray beam. Le Maitre and Haukka have reported that a $Li_2B_4O_7$ disc exposed to Cr radiation (40 kV, 40 mA) showed evidence of deterioration (reduction in count rates for the major elements) only after 50 hr of irradiation.[19]

The major disadvantages resulting from the fusion of the sample to form a disc are the dilution factor which can affect the determination of trace constituents, reduction of the fluorescence intensity from elements of low atomic number because of dilution and absorption, and the destruction of the sample (Ref. 5, pp. 33-11, 33-12).

There are numerous variations of the basic technique, with each laboratory developing its own set of specific conditions. The selection of parameters such as choice of flux (see Table 4.1), flux-to-sample ratio, sample size, fusion time and temperature, disc size, casting technique and storage is based on the equipment available and the analysis to be performed. The choice of specific lithium borate fluxes was discussed in Sec. 4.3.6, but the analyst should also refer to the paper by Bennett and Oliver.[20] West et al. chose $Na_2B_4O_7$ with $NaNO_3$ as flux because of the higher solubility of samples containing high metal oxide concentrations.[21]

There are numerous published methods of XRF analysis using fused discs manufactured in different ways. Norrish et al. have described in detail the casting of a glass disc in a brass ring,[22, 23] and Hebert and Street similarly used nickel-plated copper rings.[24] In both cases the disc was pressed with a plunger against a carbon surface held at about 250°C and the disc then cooled gradually to prevent shattering. Harvey et al.[25] used the same flux described by Norrish and Hutton[22] ($Li_2B_4O_7$ 47.03%; Li_2CO_3 36.63%; La_2O_3 16.34%), but they poured the melt into a heated duralumin mounting platten and pressed it with a heated plunger assembly to form the bead. The bead and platten were cooled by placing them in a heated ceramic cooling block (complete with cover) held at 200°C on a hot plate. After 10 min the cooling block was moved off the hot plate and allowed to cool to room temperature, at which time the bead came free of the platten. For the determination of sulfur, $NaNO_3$ was added to the flux (13.33 g/kg of flux).

A more convenient method of sample preparation involves the use of a nonwetting Pt–Au alloy mold (95% Pt, 5% Au). The fused melt is poured into the heated mold and cooled. No two laboratories adopt the same technique, but all start with a heated mold (some at 200°C, some higher) and cool the poured melt slowly to room temperature. Claisse has developed an apparatus which stirs and heats six crucibles at one time (see Fig. 2.1). When the fusion is complete, the fused material is poured into six molds by turning a lever.[26] An automatic model is also available. Other mechanized or automated means of producing fused glass discs have also been described.[27–29]

The flux-to-sample ratio can go as high as 9:1, and this high dilution raises problems when trace constituents are to be determined. Haukka and Thomas[30] report on a flux-to-sample ratio of 2:1 using $LiBO_2$ as the flux. They give the compositional limitations for the preparation of low-dilution glass discs, and it is evident that the technique is applicable to most samples encountered by the rock and mineral analyst. The sample is cast in a manner similar to that described by Norrish and Hutton,[22] but they give additional comments concerning the disc preparation and the precautions to be observed.[31]

The most convenient means of manufacturing the fused disc to be used for future analysis is to pour the fused melt into a heated mold made of the 95:5 Pt–Au alloy mentioned earlier. On cooling, the glass disc separates readily from the mold. These molds can be purchased from Johnson, Matthey and Mallory or Engelhard Industries, or they can be fabricated in a machine shop as described by van Willigan et al.[32]

Procedure

The following method is used routinely at the GSC laboratories[33] for the determination of major, minor, and some trace elements in the same disc. Weigh 5 g (± 0.002 g) of $Li_2B_4O_7$ and 0.3 g (± 0.002 g) of LiF into a Pt–Au crucible, make a well in the flux, introduce into it 0.2–0.5 g of NH_4NO_3 (depending on the S content of the sample) and 1.0 g (± 0.2 mg) of sample, and mix carefully. Clamp the crucible in the Claisse Fluxer[26] and position the Pt–Au molds above the crucibles. Ignite the burners and preheat the crucibles for 6 min without agitation, then fuse for 6 min with the automated stirring motion; continue to heat for 1 min more without agitation, pour the contents of the crucibles into the molds using the lever on the apparatus, and allow the discs to cool. Carefully invert the molds over asbestos pads and catch the discs as they drop out. Place discs in small labeled envelopes.

Notes. The mold and crucibles must be positioned carefully according to the manufacturer's directions to ensure that all of the melt will be poured into the mold when the crucible is tipped. The Pt–Au crucibles are readily prepared for reuse by treatment with hot $1:4$ HNO_3.

The casting surface of the mold must have a mirror finish for best results and should be polished as often as required. This can be done using Al_2O_3 polishing powder (1 μm size, Carver Metallurgical Ltd. Micropolish). The mold is clamped under a drill press having a polishing rod fitted into the chuck. The rod has a piece of self-adhesive lapidary material fastened on the bottom surface of a cylinder of the same diameter as the disc portion of the mold (32 mm). A small amount of Micropolish is placed in the mold, and the polishing rod is lowered onto the mold to begin the polishing.

If there is a problem with the disc shattering on cooling (because of the presence in the sample of 400 ppm or more of Cu, for example) the addition of NH_4Br will help to rectify this.

10.1.4 Thin Films

The use of a thin film is often a convenient technique for analyzing a small amount of sample (such as a small mineral grain) or for determining trace constituents concentrated by solvent extraction, coprecipitation, ion exchange, or some other means. The use of a thin film has the advantage that, as long as absorption of radiation by the sample is negligible, the intensity of the secondary fluorescence will be linearly proportional to the mass per unit area of the sample in the thin film (Ref. 5, pp. 36-2–36-5, Ref. 7, pp. 622–623). The sample thickness which will permit the maintenance of the condition of negligible absorption varies with the wavelength of the analyte, but in most instances it falls between 0.03 and 6 μm (Ref. 4, p. 136), with the longer wavelength lines requiring the smaller thickness.

There are several ways to prepare thin film for XRF analysis. The simplest is to spread the powdered sample as a thin layer on scotch tape or between two sheets of Mylar film stretched over a glass cell and held in place with O-rings. The sample may be put into solution and the solution evaporated on a Mylar film or on a piece of filter paper. Knapp et al[34] describe the eluting of sodium diethyldithiocarbamate-chelated metals from a Chromosorb W-DMCS column using chloroform. The elutant is caught in a special Teflon container with polished surfaces to avoid wetting. A disc of filter paper placed at the bottom of the container absorbs the elutant and, following evaporation of the chloroform, the paper is glued to a Teflon disc for XRF measurement. They found that duplicate samples could be obtained by this technique. Luke[35] and Mitchell et al.[36] have described the collection of small precipitates on Millipore paper discs for subsequent measurement of trace elements by XRF.

The important requirements in the preparation of the various types of thin film samples are that the sample thickness be uniform, that the sample exhibit negligible absorption of the incident and fluorescent X-rays, and that the sample preparation be reproducible. This latter consideration is critical in order to achieve reliable calibration graphs.

10.2   PRECONCENTRATION TECHNIQUES

The determination of trace element constituents by XRF subsequent to a preconcentration step of one type or another is a procedure in common use. The preconcentration step may involve ion exchange, coprecipitation, or chelation, among other techniques. In rock and mineral analysis these three are used frequently and have found some application to XRF analysis.

10.2.1   Ion Exchange

Ion-exchange techniques have been used for preconcentration of trace constituents in geological materials for subsequent analysis by AAS (see Sec. 7.2.2), and many of these techniques can be easily adapted for XRF measurements. Kashuba and Hines,[37] Leyden et al.,[38] and Blount et al.[39] have all described the preparation of pressed pellets or discs from ion-exchange resins following the concentration procedure. The technique is straightforward and involves drying the fine resin (usually <100 mesh), mixing it with a binder such as cellulose powder, and pressing it in a die. Blount et al.[39] spread the dried resin in a heated die and added enough melted paraffin to cover the resin. A die piston with 0.05 mm clearance is inserted and pressure applied with a C clamp to force the excess paraffin out of the die around the loosely fitting die piston. The die is then cooled in an ice bath to solidify the paraffin, and the pellet is removed. The resin used was Chelex 100 (Dowex A-1), which is highly selective for divalent and polyvalent metal ions and effects a good separation of transition metals from alkali metal solutions.

Other workers have reported the use of ion-exchange impregnated paper or membrane (Ref. 7, p. 789, and Ref. 40). Once the paper has been in contact with the solution long enough to extract the ions of interest, it is dried and placed in a suitable sample cell for XRF measurement. By means of an anion-exchange impregnated paper the determination of sulfate is also possible.[41]

In general, ion-exchange techniques can be used to concentrate a group of ions or to separate out a specific ion for measurement. There is much good information in the literature on both partition coefficients and on the adsorption and elution conditions necessary to obtain the desired results.[42-47] The application of these techniques to the determination of trace constituents in rocks and minerals in XRF may present an occasional modification problem, but their use can lower the effective detection limit of the XRF method to a significant degree.[48] The bienniel *Reviews* issue of *Analytical Chemistry* usually has numerous references to the application of ion-exchange separations to geological materials and should be consulted.

10.2.2 Coprecipitation

Preconcentration by coprecipitation has several advantages. The effective detection limit of XRF is improved substantially, large enough subsamples can be taken to ensure that subsampling errors frequently encountered when determining trace constituents are minimized (Sec. 3.2.1), and the method lends itself to thin film techniques (Sec. 10.1.4) so that matrix effects are eliminated. In addition, standardization is greatly simplified.

Date[49] has described the preparation of trace element reference materials for geological applications using a coprecipitation gel technique. The standards were intended for use in ES, but they can be used for XRF or other instrumental measurement techniques with little or no modification of the preparation methods.

Luke[35] has published an extensive discussion of coprecipitation techniques and conditions for preconcentration of trace elements prior to the XRF measurement step. His discussion included precipitants and coprecipitants, interferences, filtration, preparation of calibration graphs, and XRF measurement for a wide range of elements. Püschel[50] and Ackermann et al.[51] have also described the XRF determination of trace elements after concentration by a coprecipitation technique. Detection limits of 0.1–0.5 µg of analyte in 50 ml of solution have been reported.

Mitchell et al.[36] and Kessler et al.[52] have described micro-coprecipitation concentration techniques for subsequent XRF measurement of trace constituents. The former report detection limits in the order of 10 or 20 ppb of contaminants such as Cu in ultrapure reagents; Kessler et al. were able to measure 10–100 ng of Cu, Mn, and other metals. Both groups used special filtration equipment designed to concentrate the precipitate in a 1–2 mm diameter area on the filter disc. The measurement step requires the use of specialized X-ray milliprobe equipment (see also Ref. 7, pp. 791–794).

10.2.3   Chelation

Chelation and solvent extraction techniques have been used extensively for the preconcentration of trace elements from geological material for measurements by AAS and other techniques, but this method has not found much application to XRF measurement. The technique used by Knapp et al.[34] has been described previously (Sec. 10.1.4). Leyden and Luttrell[53] preconcentrated metals by means of chelating groups immobilized by silylation and used the silica gel substrate as the matrix for the XRF measurement.

10.3   STANDARDS

Analysis by XRF is a comparative method, and it follows that the use of good and reliable standards is a requirement for a successful analysis. The standards chosen for a specific analysis should match the unknown sample as closely as possible in matrix composition, particle size, physical shape, and texture, although matrix corrections can make this less critical. In addition, the concentration of analyte in the unknown should be bracketed by its concentrations in the standards.

These requirements are not trivial, and it frequently happens that the application of XRF to a specific problem is precluded because of the lack of a suitable standard. Standards for use in the XRF analysis of rocks and minerals may be natural, synthetic, or a combination of both.

10.3.1   Natural Standards

Natural standards may be primary or secondary. A primary natural standard is a sample of rock, mineral, or other geological material that has been analyzed exhaustively, often in several different laboratories, by more than one method. There are numerous primary standards available, most of which will be found listed in Appendix 1. Unfortunately there are three major weaknesses associated with these standards. As Abbey[54, 55] has pointed out, significant discrepancies have been found in the values reported by different laboratories for specific constituents in the same reference material. He also points out that there has been little or no coordination among world agencies of the various requirements for establishing a primary standard. A third weakness is that there are several elements for which few if any of these standard materials are certified. In some instances, the standard materials may not be homogeneous for all of the elements listed.[56, 57]

One precaution to be observed in the use of standard reference materials has been emphasized by Engels and Ingamells[56] when they stated that ". . . use of a geostandard as a reference in procedures for which it is not intended or certified may lead to erroneous results and to enormous waste of time and effort." They also discuss several factors for consideration during the preparation of geological standard reference materials (geostandards).

Secondary natural standards are those created by an organization for its internal use only. They are usually standardized against primary natural standards using XRF or other analytical methods. If the secondary natural standard is carefully prepared (i.e., adequately homogenized, reduced to a sufficiently small particle size) and the analytical work properly done, the secondary standard can successfully replace the comparatively valuable and rare primary standard. This consideration is not as important in XRF as in other methods, however, since the standard pellets or fused glass discs can be reused over a long period of time once they have been prepared, and the consumption of primary standard material is thus much less than that for other more destructive methods. The secondary standards can serve also to fill in gaps in matrix types or in concentration ranges not covered by primary standards, and thus their preparation is usually found to be necessary.

10.3.2 Synthetic Standards

Synthetic standards are those made from laboratory chemicals rather than from naturally occurring substances. They have the advantage over natural materials that they can be manufactured to match most geological matrices and to cover the concentration range required. The most satisfactory synthetic standards for XRF work are fused discs. The fusion helps to ensure homogeneity, the disc is durable, and the fluxes can be chosen to solubilize a wide range of source materials for the manufacture of multielement standards.

In the preparation of a set of major and trace level multielement standards for the BCM laboratory, V. Vilkos experienced many basic difficulties either not mentioned or not easily found in the literature. Initially four basic lithium borate glasses, each containing six trace elements of interest in specified concentrations, were prepared. Other glasses contained most of the major elements in varying concentrations. Both sets of glasses were analyzed by AAS and ES to confirm the concentrations. Small, carefully weighed portions of these master glasses were then fused together with added flux and cast in a mold to form the desired standard discs for measurement. The discs had concentrations of 24 trace elements, each varying from nil to 500, 1000, 1500, or 2000 ppm, depending on the element and its anticipated upper concentration limit in the majority of samples submitted for analysis. The discs were not intended to be used as standards for major or minor elements, but most of the major elements found in common rocks were included. All of these standards contained 15% La_2O_3 as a heavy absorber, but current matrix correction techniques make this precaution unnecessary for any future set of similar standards.

As in any project of this nature, many problems were encountered and a few of them are listed for information:

The melts of the master glasses tended to creep over the edge of the crucible because of long fusion times required for the complete dissolution of some salts, except when a nonwetting Au–Pt (5:95) crucible was used.

Spectrographically pure grades of reagents are mandatory, and their purity should be confirmed by ES or other methods before use.

Many salts are difficult to weigh because of their hygroscopic nature (e.g., Al_2O_3, La_2O_3, K salts).

While the tetraborate salt must be used to dissolve Al_2O_3, the metaborate must be used for SiO_2 (see Table 4.1).

The addition of NH_4Br (0.1 g for 12 g of melt) assists greatly when pouring the melt and also decreases the frequency of disc failure on cooling. If NH_4Br is used, the fusion melt should be poured into the mold in the furnace (a hot procedure) in order to prevent the formation of bubbles in the disc.

Some elements are volatilized in part or completely, with As and Tl being the most notable.[58]

Silver salts were reduced or decomposed and the Ag was alloyed with the crucible metal.

Some dried powders (e.g., TiO_2) were very difficult to weigh because of an electrostatic effect.

A very stable balance table and a balance capable of weighing to ± 0.01 mg are required.

Meyers et al. have also described the preparation of glass reference standards for trace element analysis.[58] Of the five standards prepared, one was a blank and the four others each contained 46 elements at concentration levels of 500, 50, 5, and 0.5 ppm respectively. They report the loss of Re, Tl, and Se through volatilization, but As was retained.

10.3.3   Natural/Synthetic Standards

Natural/synthetic standards are formed when selected elements are added in one form or another to geological materials. The addition can be made by pipetting prepared solutions onto a powdered rock matrix or by adding the elements as salts. Leake et al. have described the latter method.[12] In either case, care must be taken to mix thoroughly the base matrix and the added material if the reference material is to be in the form of a pressed pellet. Fusion of the mixture to form a glass disc will assure a homogeneous reference material for subsequent calibration purposes, but any weight loss during the fusion must be accounted for when calculating the concentration of the added element(s).

10.4   INSTRUMENTATION

As has been previous practice, no attempt will be made here to discuss the theoretical aspects of XRF instrumentation. The discussion will be concerned instead with the selection of different instrumental parameters to optimize the conditions for the general analytical procedures likely to be encountered in the analysis or rocks and minerals.

10.4.1   Type of Instrument

The most important decision that must be made in deciding upon XRF instrumentation is in the choice between a wavelength dispersive (WD) and an energy dispersive (ED) system. The latter first became commercially available in the early 1970s and is a relatively young technique when compared with the 20 or so years that wavelength units had been marketed prior to the introduction of ED systems. Nevertheless, as a result of significant improvements that have been made, the ED system is much superior to the WD system for certain applications.

The advantages and disadvantages of ED with respect to WD have been discussed by Jenkins[59] and he summarized them as follows:

Advantages of ED versus WD

1. Operationally faster than the single channel WD system
2. Less costly than the multichannel WD system
3. Measures all elements
4. Display of data improves ease of interpretation;
5. Fewer geometric constraints

Disdvantages of ED versus WD

1. Input count rate limited to 5×10^4 counts/sec
2. Resolution not as good by 1 or 2 orders of magnitude.
3. Requires use of computer for quantitative analysis calculations
4. Occurrence of misleading artifacts in acquired spectra
5. Requires a supply of liquid nitrogen

In the application of XRF to the analysis of rocks and minerals, particularly for the determination of trace and minor constituents, the foregoing disadvantages of the limited input count rate and the poorer resolution militate strongly against the choice of an ED system because, when measuring analytes of low concentrations, the X-rays from all of the constituent elements determine the count rate rather than only those from the analyte present in low concentration. Thus the detection limit of the ED system is not as low as that of the WD system (generally by a factor of from 2 to 10). In addition, because most geological samples are characterized by a complex compositional matrix, there are many instances of spectral line overlap in which the 150 eV resolution of the ED system (sometimes greater at the counting rates required for some analyses) is simply not adequate to resolve them, even with the help of deconvolution techniques. In these instances recourse must be had either to a WD system which offers a resolution better by at least an order of magnitude or to some other analytical technique.

Wavelength dispersive XRF systems come in two modes, single channel and multichannel. In the former, each element and background measurement must be made sequentially, whereas in the latter mode it is possible to make multiple measurements simultaneously. As a consequence of this, the multichannel instrument is capable of much faster analytical times for multielement analyses, and it has found extensive application in cement, metallurgical, and other industries where a fixed suite of elements is to be determined. In a laboratory required to do a wide variety of analyses, however, the versatility of the sequential single channel instrument would be a considerable advantage over the limited 20–30 element capability of most simultaneous units.

10.4.2 Sample Excitation

The selection of sample excitation parameters involves a choice of X-ray tube targets and the excitation voltage. There are numerous targets available, the most common of which are Cr, Cu, Mo, Rh, Ag, W, Pt, and Au, but almost any metal except the very volatile ones can be used. In rock and mineral analysis it is advisable to have available at least one Cr target tube and one heavy element target tube (W or Au, for example). The Cr tube will be used for the determination of light elements such as Si and Al. The second tube will be used for the determination of heavier elements such as Zr and Mo. Dual target tubes are available from certain manufacturers.

A general set of rules for the selection of a tube target and the excitation voltage has been given by Adler (Ref. 1, p. 35) which is still valid, and is reproduced here.*

1. The most effective radiation for exciting an element has a wavelength just shorter than the absorption edge (about 0.2 Å shorter) which leads to the emission of a particular spectral series. For example, the Cu K_α line at 1.54 Å is very effective in exciting the Fe K spectrum; the Fe K edge occurs at 1.7 Å.

2. Because the choice of targets is limited, it is necessary mainly to depend on the continuous spectrum for excitation; thus it is essential to choose an X-ray tube with a target having the highest atomic number. We have already shown that the intensity of the continuum increases with atomic number.

3. As a corollary to rule 2, it is generally necessary to apply a maximum voltage to the X-ray tube, to ensure that the largest possible number of unknown elements in the specimen will be excited to fluorescence. It will also ensure the greatest possible intensity of fluorescence of each element.

4. The X-ray tube is operated at lower voltages only in the special case when selective excitation is desired. For example, if it is desired to determine Hf in the presence of Zr, by operating the tube at 17 kV, the Hf L spectrum ($V_{min} = 11$ kV) will be excited whereas the Zr K spectrum ($V_{min} = 18$ kV) will not be. This is of course accomplished at the cost of severe attenuation of the Hf L spectrum.

5. On occasion, it may be necessary to change X-ray tubes when, for example, a characteristic line of the target interferes with the element line. Such interferences can occur when the two lines have the same wavelength

* With the permission of American Elsevier Publishing Company, Amsterdam and the author.

and also when a higher order reflection of the target line coincides with or is near the line sought. This type of interference can occur commonly enough to warrant the purchase of two X-ray tubes with different targets.

The power supply for the X-ray tube should be capable of delivering both a variable voltage and a variable current, that is, 10 to at least 50 kV, 5–50 mA, with an accuracy of ±0.05%. This means that it must supply a minimum of 2.5 kW (50 mA × 50 kV) of power; some units supply more by increasing the voltage to as much as 100 kV. There is some limited use of the higher kilovolt setting to excite the shorter lines of some heavier elements. Because most commercial instruments meet all of these requirements, however, they are seldom a problem to the analyst.

A thorough discussion of all aspects of excitation can be found in Ref. 7, Chapter 4, and less detailed discussions in Ref. 1, Chapter 3, in Ref. 4, Chapter 1, and in Ref. 6, Chapter 1, among others.

10.4.3   Dispersion of Fluorescent X-Rays

a.   *Collimators*

In current instruments there are multiple slit (Soller) devices placed between the specimen and the analyzing crystal (the primary, source, or divergent collimator) and/or between the analyzing crystal and the detector (the secondary, detector, or receiving collimator). They are made of a series of regularly spaced metal foils or plates placed edgewise to the X-ray beam so that a more or less parallel beam arrives at the crystal or detector because of the exclusion of most of the randomly scattered X-rays from the beam path.

Increasing the collimation (that is, by narrowing the slit width or lengthening the foil so that the X-ray beam passes through a longer collimator) has the effect both of increasing the resolution of the instrument (because of narrower peaks and less "tailing") and of decreasing the intensity of the analyte signal. The necessity for a trade-off between resolution and intensity can thus arise here as in any other instrumental technique. Most instruments enable the analyst to select either a fine or a coarse primary collimator. Fortunately, at wavelengths greater than about 0.3 nm (where fluorescent intensity is low), coarse collimation is adequate for first-order K and L lines, obviating the need for a trade-off in these cases.

b.   *Crystals*

The dispersion of X-rays in WD instruments is effected by means of crystals. Each crystal has its own interplanar spacing and other properties that must be considered in choosing a particular crystal for a specific job. A crystal with an interplanar spacing d cannot diffract wavelengths that are greater than $2d$ in length. Thus crystals with large interplanar d spacing are necessary for the diffraction of X-rays from the light elements such as Mg and Na. These same crystals cannot be used, however, for the diffraction of shorter wavelengths because of the higher background and lower resolution that will result. It is necessary to use a crystal having a smaller d spacing.

It is also necessary to consider the thermal expansion of the crystal which will alter its d spacing in proportion to the temperature change. For high precision work it is necessary to keep the crystal chamber at a constant temperature ($\pm 0.5°C$), and most instruments have this capability. The extent of the thermal expansion that is possible is demonstrated graphically by Jenkins and De Vries (Ref. 4, p. 38), whereby crystals exhibit an increasing thermal expansion in the following order: topaz, quartz (101), ADP, EDDT, gypsum, LiF (200), LiF (220), and, highest of all to a large extent, PE.

When the analyzing crystal is bombarded with X-rays, it will emit secondary X-rays from its constituent atoms. This crystal fluorescence is usually a problem only during the measurement of wavelengths greater than about 0.35 nm (3.5 Å), that is, for elements lighter than K or Ca, and can usually be overcome by pulse height selection techniques.

When using crystals of low symmetry (e.g., topaz or EDDT), the analyst must be on the lookout for possible abnormal reflections. These can usually be recognized because they are much broader than normal reflections (Ref. 4, p. 40). One such abnormal reflection observed in the BCM laboratory and reported by Post and Jenkins[60] is the "parasitic" one from the 110 plane in a LiF 220 crystal. The d spacing of the LiF 110 planes is the same as that in the NaCl crystal used in XRF measurement, and the abnormal reflections can be identified by comparison with the NaCl wavelength data found in the wavelength tables.[61] The problem should not occur if the LiF crystal is cut only after careful orientation of the crystal boule.

The integral reflection coefficient is a measure of the diffraction efficiency of a crystal that varies over the wavelength range diffracted by the crystal. Gilfrich et al. have divised a means of measuring the integral reflection coefficient and have reported their results for several commonly used crystals.[62]

The search for new crystals and improved crystal technology has resulted in occasional introduction of new crystal substances. An example is in the acid phthalate series[63] where ammonium, sodium, and potassium (KAP) crystals were introduced first, followed by the rubidium (RbAP) and then the thallium (TlAP) crystals. Wybenga has evaluated the use of the TlAP crystal in XRF compared to that of KAP, RbAP, and PE (pentaerythritol) crystals.[64] He concludes that the TlAP crystal is superior to KAP and RbAP crystals for the light elements (F to Si) and compares favorably with PE for Si and Al. Fregerslev reports that the RbAP crystal deteriorates with time.[65]

A list showing the more common crystals that are available and the elements for which they are used is given in Table 10.2

Other sections of LiF crystals have been used to obtain better resolution (but at the expense of intensity, as is usual) with LiF (420) having a $2d$ spacing of 0.1802 nm and LiF (422) having one of 0.1645 nm. At very short wavelengths, however, the geometric factor (the percentage of the X-ray beam intercepted by the analyzing crystal) ensures that there is very little loss of intensity when the (420) crystal is used.[66]

10.4.4 X-Ray Detectors

This section will deal only with gas flow proportional counters and scintillation counters, the two types most commonly used in commercially available WD-XRF units, and the solid state detectors used in ED-XRF systems will not be discussed. A more detailed coverage of the detectors used in WD systems will be found in Refs. 4–7, and of solid state detectors in Ref. 6, 7, and 67.

TABLE 10.2 Analyzing Crystals Commonly Available for XRF Spectrometers.

Crystal	Designation	$2d$ (nm)	Element Range: K_α Lines	Element Range: L_α Lines
Aluminum fluorsilicate (hydrated)	Topaz	0.2712	W to Cr	U to Ba
Lithium fluoride (220)	LiF (220)	0.2848	W to Cr	U to Ba
Lithium fluoride (200)	LiF (200)	0.4028	W to K	U to Cd
Pentaerythritol	PE	0.8742	K to Al	Sn to Rb
Ethylenediamine ditartrate	EDDT	0.8808	K to Al	Sn to Rb
Ammonium dihydrogen phosphate	ADP	1.065	Al to Na	
Calcium sulfate dihydrate	Gypsum	1.519	Al to Na	
Thallium acid phthalate	TlAP	2.590	Si to F	
Potassium acid phthalate	KAP	2.663	Si to F	
Rubidium acid phthalate	RbAP	2.612	Si to F	

a. Gas Flow Proportional Counter

The "gas" of the gas flow proportional counter is usually Ar/10% CH_4 (P. 10), a gas mixture which can diffuse through the very thin entrance window required for the measurement of long-wavelength X-rays. The spectrometer is operated in either the vacuum or the flush mode for these measurements, in order to prevent a gas buildup inside the spectrometer. The counter is operated at a voltage which results in an amplification of the signal that it receives , and it has two advantages. This first is that the mean pulse amplitude output of the counter is directly proportional to the energy of the incident X-ray photon initiating the signal. This means that a method of energy discrimination (pulse height selection) can be used to eliminate undesirable signals. The second advantage is that this counter has a very low dead time, on the order of 0.5 μsec. Thus even at count rates of 10^5 counts/sec, losses due to dead time are less than 20%.

Because gas flow proportional counters are designed for the detection of low energy X-rays, the selection of a window material is critical. If one is measuring the K lines of Mg, Na, or F, for example, the use of a Mylar film 4–6 μm thick as the window of the counter will result in the absorption of most of the X-rays, and in this case Mylar of 1 μm thickness, or stretched polypropylene, must be used. A thorough discussion of window materials is given in Ref. 7, pp. 221–226, but a useful guide to the thickness of window material to use with different wavelengths is given by Jenkins and De Vreis (Ref. 4, p. 60) as follows:

For λ of 0.1–0.8 nm, use 4 or 6 μm Mylar or 4 μm polycarbonate.

For λ of 0.6–2 nm, use 1 μm stretched polypropylene.

For λ longer than 2 nm, use collodion supported on a high transmission grid.

Gas flow proportional counters are commonly used for the measurement of emissions from the K lines of elements lighter than As and scintillation counters for those from elements heavier than Fe, with some overlap. This is demonstrated in Fig. 10.1, where the count rates of the two counters are compared. For comparison with these the count rates from a flow counter and a scintillation counter mounted in series are also included.

b. Scintillation Counters

As can be seen in Fig. 10.1, the scintillation counter is adequate only for wavelengths shorter than about 0.18 nm. The typical scintillation

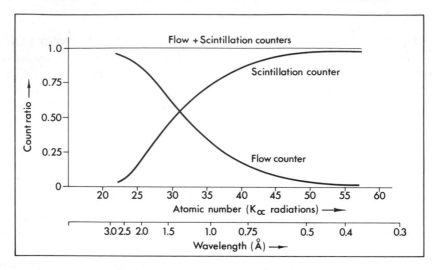

Figure 10.1.   Response versus wavelength for flow and scintillation counters. Reproduced with permission of N. V. Philips Gloeilampenfabrieken (Ref. 4).

counter used in XRF work consists of a NaI (Tl) phosphor crystal mounted on a photomultiplier tube. The inherent noise of the photomultiplier tube, combined with the absorption of long wavelength X-rays by the casing of the crystal, are the reasons for the low energy X-ray limitation of this detector.

The output of the scintillation counter is proportional to the energy of the incident X-ray photon, and thus pulse height selection can be used to clean up this output as well as that of the gas flow proportional counter. A second advantage shared by the two, and mentioned previously, is that the dead time of both the proportional and the scintillation counters is low (0.5 and 0.2 μsec, respectively). However, because the scintillation counter requires more energy to produce a signal and the efficiency of the photomultiplier tube is poor, the degree of resolution of this counter is not as good as that of the gas flow proportional counter (51% at Fe K_α versus 15%, respectively (Ref. 4, p. 65)).

10.4.5   Pulse Height Selection

The signals generated by the two detector systems discussed previously can vary over a wide spectrum of energies because of escape peaks, second, third, or higher order lines, and other features. It is necessary that only a narrow range of energy be selected from this total signal for counting purposes in the scaler-counting circuit, and this is achieved by means of pulse height selection.

The selection is done by setting a pulse height discriminator to pass only signals of voltages higher than a certain base level, 5 V, for example. This effectively eliminates detector and amplifier noise as well as some low energy background signals and improves the signal-to-noise ratio. A voltage "window," for example, 12 V, is then set so that no pulse having a voltage greater than the sum of the base voltage and the window width (5 V + 12 V = 17 V in this case) will be passed to the detector. In this way a pulse height selection of signals from 5 to 17 V is established (see Fig. 10.2).

In order for pulse height selection to be effective it must be set with fairly tight windows (narrow energy distributions), and often this means cutting the tails off a Gaussian-shaped peak. Any factor which affects the shape of this peak (i.e., a distortion) or the voltage of its maximum, as in a peak shift, will necessarily affect the fraction of the signal that will be passed through the window. As might be expected, this will give rise to serious errors in a quantitative analysis. Jensen and De Vreis have discussed many of the problems associated with pulse height selection (Ref. 4, pp. 75–88), and the analyst should be aware of them.

Peak distortion can result from an additional peak(s) occurring at the same wavelength as that of the peak being measured (escape peaks, for example) or from those occurring at other wavelengths (e.g., higher order lines from other elements present as major constituents). Peak shift can

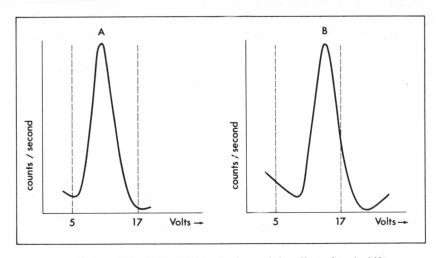

Figure 10.2.   Pulse height selection and the effect of peak shifts.

result from a change in the voltage applied to the detector, if the detector is subjected to high X-ray intensity, and from a change in the diameter of the anode in the gas flow counter or in the pressure and composition of the gas. A dirty anode wire can also cause a peak shift as well as a significant loss of detector resolution.

10.4.6   Corrections for Matrix Effects

Because rock and mineral analysts are presented with samples of widely varying composition, they must concern themselves with the effects that the different sample compositions will have on their quantitative measurements. This is because absorption and enhancement effects due to varying elemental concentrations disrupt the linear relationship between concentration and fluorescent X-ray intensity.

One means of overcoming the matrix differences between samples, and between samples and standards, is to add a heavy metal absorber to all of them (sometimes called "high-absorber dilution"). This involves adding sufficient La_2O_3, $BaSO_4$, H_2WO_4, or some other heavy metal salt to both samples and standards in order to raise the mass absorption coefficient of each to a level such that element concentration changes between the samples will have no significant effect on the concentration–intensity relationship.

A second technique is known as "low-absorber dilution." In this case the sample is diluted with a low absorption matrix to the extent that attenuation of the fluorescent X-rays of the analyte by interfering elements of the sample matrix is insignificant. A disadvantage is that the dilution factor may be so large that the intensity of the analyte line is insufficient for the determination.

A third method of sample treatment which avoids matrix effects is to prepare the specimen as a thin film. This is accomplished by collecting a coprecipitated analyte on some form of filter, or by concentrating it in an ion-exchange paper, for example. In any case, if the layer is sufficiently thin that primary and secondary X-ray lines are not significantly absorbed, then each atom interacts with the primary beam and fluoresces independently of others, and the analyte line intensity is a linear function of the analyte concentration (Sec. 10.1.4). Govindaraju and Montanari have reported the application of this to the determination of six major elements in rock samples.[68]

Each of the above methods has drawbacks which the analyst would like to avoid if possible. One approach that is widely used is that of mathematical matrix corrections based on theoretical and/or empirical correction coefficients (Supp. Refs. 5, 6). The coefficients are a measure of the interelemental influences which give rise to the matrix effects. Recent reviews by Jenkins[69] and MacDonald[70] have summarized the development and use of this matrix correction technique, and most XRF texts discuss it in detail (Ref. 4, pp. 134–143, Ref. 6, pp. 386–406, and Ref. 7, Chap. 15). The literature is replete with articles describing various applications of the technique, but most are based on empirical correction coefficients or "concentration" coefficients. These are coefficients which, when multiplied by the concentration of the interfering element, are used to give a correction for the fluorescent X-ray intensity of the analyte line. Some of the more commonly used empirical equations as listed by Jenkins[69] are as follows:

Linear model $\qquad \dfrac{W_i}{R_i} = K_i$

Lachance–Traill $\qquad \dfrac{W_i}{R_i} = K_i + \sum \alpha_{ij} \cdot W_j$

Rasberry–Heinrich $\qquad \dfrac{W_i}{R_i} = K_i + \sum \alpha_{ij} \cdot W_j + \sum \beta_{ik} \cdot \dfrac{W_k}{1 + W_i}$

Claisse–Quintin $\qquad \dfrac{W_i}{R_i} = K_i + \sum \alpha_{ij} \cdot W_j + \sum \gamma_{ij} \cdot W_j^2$

where W_i is the concentration of element i, R_i is the characteristic line intensity for element i, and K_i is the slope of the analyte calibration curve. The α_{ij} term represents the correction for the absorbing effect that element j has on element i, and the β_{ij} of the Rasberry–Heinrich model is a correction for enhancement. The higher order terms in the Claisse–Quintin equation are invoked to account for "crossed effects" such as those attributable to a third element.

The influence coefficients are determined empirically by carefully measuring the appropriate fluorescent X-ray intensities in a set of well-analyzed standards. The analyst can then use multiple regression techniques, graphical techniques, or calculations employing a physical constants-based algorithm to obtain the coefficients[71–73] (see also Supp. Refs. 5, 6). Haukka and Thomas have described a similar technique.[30] As Jenkins pointed out, this empirical means of determining the coefficients can be susceptible to several sources or error, six of them being:

1. Failure to adequately separate instrumental and matrix dependent effects

2. Poor judgement on the part of the analyst as to whether or not a correction term should really be included

3.   Poor technique on the part of the analyst in the determination of the coefficients

4.   Poor quality and/or range of calibration standards

5.   Inadequacy of the regression analysis program used in the determination of the coefficients

6.   Application of the technique in cases where the specimens under analysis are insufficiently homogeneous

Because the scattered background intensity at any particular wavelength is linearly related to the reciprocal of the sample mass absorption coefficient,[74] it is possible to use a measurement of the Compton-scattered radiation for the matrix correction for a series of elements.[14, 15, 75–79] This means that it is not necessary to measure the major element concentrations in order to make matrix corrections for the determination of minor or trace elements. The only step that is necessary is the measurement of the Compton-scattered background for each unknown sample and a calculation of the matrix correction at each analyte wavelength to be measured. This technique overcomes several of the complications listed above, but the quoted relative accuracy of 2–5% (Ref. 76) is not as good as that obtained by other methods. It is adequate, however, for most geochemical and virtually all trace element determinations, and the application of this technique in the BCM laboratory has resulted in a relative accuracy for Sr as low as 0.5%. An added advantage of this method is that the same background measurement is used for the background correction for most elements to be measured in the sample.[76] This saves a lot of measurement time when a series of elements is to be determined in a large number of samples.

De Jongh[80] and Tertian[81, 82] have described a "differential" coefficient method which was extended by Claisse et al.[83, 84] The concept is to determine coefficients for smaller ranges of concentration so that the corrections made for small concentration and matrix effect differences are more accurate than when more general coefficients are used.

Bougault et al.[85] have shown that the general relationship between net peak intensity and analyte concentration can be expressed in a form that includes terms for instrumental interferences. Using this expression to correct for these interferences, they have succeeded in enhancing the application of XRF to the determination of trace elements in rocks. Corrections for interferences equivalent to a maximum of 200 ppm of the analyte resulted in no apparent loss of accuracy.

References

1.   I. Adler, *X-Ray Emission Spectrography in Geology*, Elsevier, New York, 1966.

2. L. S. Birks, *X-Ray Spectrochemical Analysis*, Interscience, New York, 1959.

3. J. G. Brown, *X-Rays and Their Applications*, Plenum, New York, 1966.

4. R. Jenkins and J. L. De Vries, *Practical X-Ray Spectrometry*, 2nd ed., Philips Technical Library, Springer-Verlag, New York, 1969.

5. E. F. Kaelble, Ed., *Handbook of X-Rays*, McGraw-Hill, New York, 1967.

6. H. A. Liebhafsky, H. G. Pfeiffer, E. H. Winslow, and P. D. Zemany, *X-Rays, Electrons, and Analytical Chemistry*, Wiley-Interscience, New York, 1972.

7. E. P. Bertin, *Principles and Practice of X-Ray Spectrometric Analysis*, 2nd ed., Plenum, New York, 1975.

8. I. B. Brenner, L. Argov, and H. Eldad, *Appl. Spectrosc.*, **29**, 423 (1975).

9. A. A. Levinson and L. de Pablo, *J. Geochem. Explor.*, **4**, 399 (1975).

10. P. F. Berri, T. Furuta, and J. R. Rhodes, *Adv. in X-Ray Anal.*, **12**, 612 (1968).

11. B. P. Fabbi, *X-Ray Spectrom.*, **1**, 39 (1972).

12. B. E. Leake, G. L. Hendry, A. Kemp, A. G. Plant, P. K. Harrey, J. R. Wilson, J. S. Coats, J. W. Aucott, T. Lünel, and R. J. Howarth, *Chem. Geol.*, **5**, 7 (1969/1970).

13. R. J. Goodman, *Econ. Geol.*, **68**, 275 (1973).

14. R. D. Giauque, R. B. Garrett, and L. Y. Goda, *Anal. Chem.*, **49**, 62 (1977).

15. R. D. Giauque, R. B. Garrett, and L. Y. Goda, *Anal. Chem.*, **49**, 1012 (1977).

16. R. Kanaris-Sotiriou and G. Brown. *Analyst*, **94**, 780 (1969).

17. G. R. Lachance, Geological Survey of Canada, private communication.

18. F. Claisse, Quebec Department of Natural Resources, P.R. 327, 1956.

19. R. W. Le Maitre and M. T. Haukka, *Geochem. Cosmochim. Acta*, **37**, 708 (1973).

20. H. Bennett and G. J. Oliver, *Analyst*, **101**, 803 (1976).

21. N. G. West, G. L. Hendry, and N. T. Bailey, *X-Ray Spectrom.*, **3**, 78 (1974).

22. K. Norrish and J. T. Hutton, *Geochim. Cosmochim. Acta*, **33**, 431 (1969).

23. K. Norrish and B. W. Campbell, in J. Zussman, ed., *Physical Methods in Determinative Mineralogy*, 2nd ed., Academic, New York, 1977.

24. A. J. Hebert and K. Street, Jr., *Anal. Chem.*, **46**, 203 (1974).

25. P. K. Harvey, D. M. Taylor, R. D. Hendry, and F. Bancroft, *X-Ray Spectrom.*, **2**, 33 (1973).

26. R. Le Houillier and S. Turmel, *Anal. Chem.*, **46**, 734 (1974).

27. G. Wronka and W. Becker, *Arch. Eisenhuettenwes.*, **42**, 645 (1971); through *Anal. Abstr.*, **22**, No. 4641 (1972).

28. A. Wittmann, J. Chmeleff, and H. Herrmann, *X-Ray Spectrom.*, **3**, 137 (1974).

29. A. A. Tunney and H. Hughes, Br. Steel Corp. Open Rep. GS/TECH/243/ 1/72/C; through *Anal. Abstr.*, **25**, No. 1362 (1973).

30. M. T. Haukka and I. L. Thomas, *X-Ray Spectrom.*, **6**, 204 (1977).

31. I. L. Thomas and M. T. Haukka, *Chem. Geol.*, **21**, 39 (1978).

32. J. H. H. G. van Willigan, H. Kruidhof, and E. A. M. F. Dahmen, *Talanta*, **18**, 450 (1971).

33. J. L. Bouvier and G. R. Lachance, private communication.

34. G. Knapp, B. Schreiber, and R. W. Frei. *Anal. Chim. Acta*, **77**, 293 (1975).

35. C. L. Luke, *Anal. Chim. Acta*, **41**, 237 (1968).

36. J. W. Mitchell, C. L. Luke, and W. R. Northover, *Anal. Chem.*, **45**, 1503 (1973).

37. A. T. Kashuba and C. R. Hines, *Anal. Chem.*, **43**, 1758 (1971).

38. D. E. Leyden, R. E. Channell, and C. W. Blount, *Anal. Chem.*, **44**, 607 (1972).

39. C. W. Blount, W. R. Morgan, and D. E. Leyden, *Anal. Chim. Acta*, **53**, 466 (1971).

40. R. E. Van Grieken, C. M. Bresseleers, and B. M. Vanderborght, *Anal. Chem.*, **49**, 1326 (1977).

41. B. Schreiber and P. A. Pello, *Anal. Chem.*, **51**, 783 (1979).

42. F. W. E. Strelow, A. H. Victor, C. R. van Zyl, and C. Eloff, *Anal. Chem.*, **43**, 870 (1971).

43. F. W. E. Strelow, C. J. Liebenberg, and A. H. Victor, *Anal. Chem.*, **46**, 1409 (1974).

44. F. W. E. Strelow *Anal. Chem.*, **32**, 1185 (1960).

45. J. Korkisch and S. S. Ahluwalia, *Talanta*, **14**, 155 (1967).

46. F. W. E. Strelow and H. Sondrop, *Talanta*, **19**, 1113 (1972).

47. R. Cesareo, S. Sciuti, and G. E. Gigante, *Int. J. Appl. Radiat. Isotopes*, **27**, 58 (1976).

48. C. W. Blount, D. E. Leyden, T. L. Thomas, and S. M. Guill, *Anal. Chem.*, **45**, 1045 (1973).

49. A. R. Date, *Analyst*, **103**, 84 (1978).

50. R. Püschel, *Talanta*, **16**, 351 (1969).

51. G. Ackermann, R. P. Koch, H. Ehrhardt, and G. Sanner, *Talanta*, **19**, 293 (1972).

52. J. E. Kessler, S. M. Vincent, and J. E. Riley, Jr., *Talanta*, **26**, 21 (1979).

53. D. E. Leyden and G. H. Luttrell, *Anal. Chem.*, **47**, 1612 (1975).

54. S. Abbey, Geological Survey of Canada, Paper 77-34, 1977.

55. S. Abbey, *Geostand. Newsl.* **1**, 39 (1977).

56. J. C. Engels and C. O. Ingamells, *Geostand. Newsl.*, **1**, 51 (1977).

57. C. O. Ingamells and P. Switzer, *Talanta*, **20**, 547 (1973).

58. A. T. Meyers, R. G. Havens, and W. W. Niles, in E. L. Grove, Ed. *Glass*

Reference Standards for Trace Element Analysis of Geologic Materials in Developments in Applied Spectroscopy, Vol. 8, Plenum, New York, 1970.

59. R. Jenkins, *Norelco Rep.*, **25**, (3), 12 (1978).

60. B. Post and R. Jenkins, *X-Ray Spectrom.*, **1**, 161 (1972).

61. E. W. White and G. G. Johnson, Jr., *X-Ray and Absorption Wavelengths and Two-Theta Tables* 2nd ed., American Society for Testing Materials, ASTM Data Series DS 37A, 1970.

62. J. V. Gilfrich, D. B. Brown, and P. G. Burkhalter, *Appl. Spectrosc.*, **29**, 322 (1975).

63. R. Vié le Sage and B. G. Grubis, *X-Ray Spectrom.*, **2**, 189 (1973).

64. F. T. Wybenga, *X-Ray Spectrom.*, **7**, 33 (1978).

65. S. Fregerslev, *X-Ray Spectrom.*, **6**, 89 (1977).

66. R. Jenkins, *Phillips Analytical Equipment Bull.*, 7000.38.1900.11 (July 1971).

67. R. Woldseth, *X-Ray Energy Spectrometry*, Kevex Corporation, Burlingame, CA., 1973.

68. K. Govindaraju and R. Montanari, *X-Ray Spectrom.*, **7**, 148 (1978).

69. R. Jenkins, *Adv. in X-Ray Anal.*, **22**, 281 (1979).

70. G. L. Macdonald, *CRC Critical Reviews in Analytical Chemistry*, pp. 301–303, Jan. 1975.

71. G. R. Lachance and R. J. Traill, *Can. J. Spectrosc.*, **11**, 43 (1966).

72. S. D. Rasberry and K. F. J. Heinrich, *Anal. Chem.*, **46**, 81 (1974).

73. M. Kirchmayer and B. Dziunikowski, *X-Ray Spectrom.*, **7**, 164 (1978).

74. F. Claisse and M. Quintin, *Proc. 13th Colloq. Spectrosc. Int.*, (Ottawa, 1967), Adam Hilger Ltd., London, 1968.

75. R. C. Reynolds, *Am. Mineral.*, **52**, 1493 (1967).

76. C. E. Feather and J. P. Willis, *X-Ray Spectrom.*, **5**, 41 (1976).

77. F. Bazan and N. A. Bonner, *Adv. in X-Ray Anal.*, **19**, 381 (1976).

78. L. Leoni and M. Daitta, *X-Ray Spectrom.*, **6**, 181 (1977).

79. R. Vié le Sage, J. P. Quisefit, R. Dejean de la Bâtie, and J. Faucherre, *X-Ray Spectrom.*, **8**, 121 (1979).

80. W. K. de Jongh, *X-Ray Spectrom.*, **2**, 151 (1973).

81. R. Tertian, *X-Ray Spectrom.*, **4**, 52 (1975).

82. R. Tertian, *X-Ray Spectrom.*, **8**, 117 (1979).

83. F. Claisse and T. P. Thinh, *Anal. Chem.*, **51**, 954 (1979).

84. G. Fréchette, J. C. Hébert, T. P. Thinh, R. Rousseau, and F. Claisse, *Anal. Chem.*, **51**, 957 (1979).

85. H. Bougault, P. Cambon, and H. Toulhoat, *X-Ray Spectrom.*, **6**, 66 (1977).

Supplementary References

1. G. J. Oliver, British Ceramic Research Assoc., Spec. Publ. 98, 75, 1979.

2. W. C. Campbell, *Analyst*, **104**, 177 (1979).

3. D. Newell, British Ceramic Research Assoc., Spec. Publ. 98, 123, 1979.

4. K. Matsumoto and K. Fuwa, *Anal. Chem.*, **51**, 2355 (1979).

5. G. R. Lachance, *X-Ray Spectrom.*, **9**, 195 (1980).

6. G. R. Lachance and R. M. Rousseau, Geological Survey Canada, Paper 80-1A, 390, 1980.

CHAPTER 11

THE DETERMINATION OF MAJOR, MINOR, AND TRACE CONSTITUENTS IN ROCKS AND MINERALS BY X-RAY FLUORESCENCE SPECTROSCOPY

11.1 GENERAL

The determination of an element in a geological sample using XRF techniques involves several stages prior to the measurement step. These include sample preparation, preconcentration (in the case of certain trace determinations), standard preparation or selection, and the selection of instrumental parameters. All of these have been discussed in some detail in Chapter 10, and it will be assumed that the reader has referred to the appropriate sections of that chapter in subsequent discussions. Each element will be discussed (from the point of view of WD-XRF) with respect to possible interferences or special problems, with appropriate references.

There are several generalities that can be used as guides in the selection of measurement conditions. Those applicable to the selection of crystal, X-ray path (medium), detector, and collimator have been put in graphic form by various authors and equipment manufacturers for easy reference. Figures 11.1 and 11.2, taken from the operating manual provided by Philips Electronics Ltd. to accompany its PW 1450 X-ray spectrometer, will serve as guides in the selection of instrumental parameters.

Figure 11.3 has been adapted from a General Electric Company wall chart and shows the relative excitation efficiency of various X-ray tubes as plotted against analyte wavelengths. This can be used as a guide in selecting an X-ray tube with the best target element for the specific needs of the individual laboratory.

11.2 CONSIDERATIONS FOR INDIVIDUAL ELEMENTS

The elements will be discussed in the same order in which they are found in Table 11.1, alphabetically according to their names.

11.2.1 Aluminum ($K_{\alpha_{1,2}}$ 0.8339 nm, 1.486 keV)[3, 4, 23, 25, 28, 33, 34, 40]

Because Al is so light in weight and its K_α X-ray is of such a long wavelength, it is difficult to obtain acceptable quantitative results on

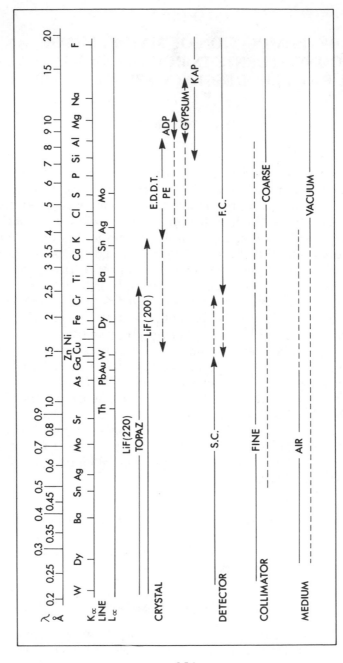

Figure 11.1. Choice of conditions for XRF. Reproduced with permission of N. V. Philips Gloeilampenfabrieken.

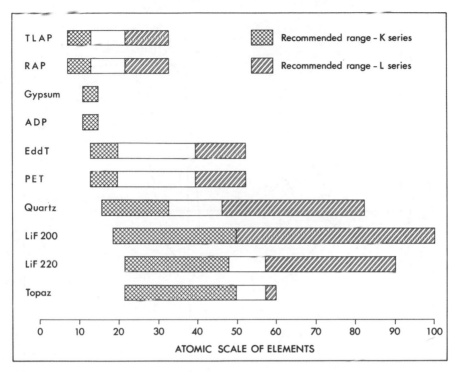

Figure 11.2. Crystal selection for element determination by XRF. Reproduced with the permission of N. V. Philips Gloeilampenfabrieken.

samples that have not been fused to provide a glass disc for measurement purposes. With the use of a PE or KAP crystal (see Table 10.2), resolution is good enough that there should be no interelement interferences in most rock or mineral samples encountered, although the Br L_α line at 0.8375 nm may cause a problem if the Br concentration is much greater than that of Al. The excitation of Al K_α by the absorption of Si K_α radiation should be kept in mind during the calibration steps.[40]

11.2.2 Antimony ($K_{\alpha_{1,2}}$ 0.0472 nm, 26.27 keV; L_{α_1} 0.3439 nm, 3.60 keV)[5, 35]

It is not likely that there will be interelement interferences when the Sb $K_{\alpha_{1,2}}$ line of 0.0472 nm is used. The K 0.3454 nm $K_{\beta_{1,3}}$ line may interfere with the use of the Sb L_{α_1} line at 0.3439 nm if trace amounts of the latter are in the presence of high K concentrations. The use of pressed powder

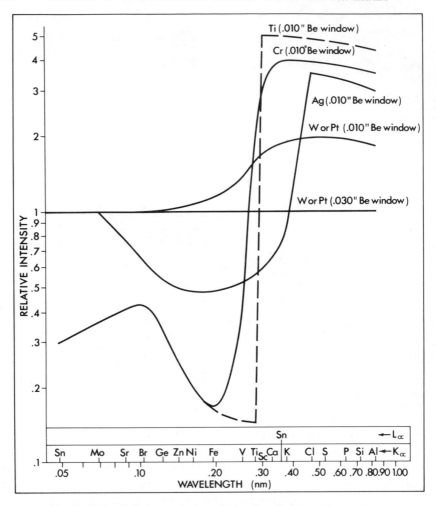

Figure 11.3.   Relative excitation efficiencies of various X-ray tubes.

samples has been successful with the $K_{\alpha_{1,2}}$ line, but it is not likely that it would be with the L_{α_1} line.

11.2.3   Arsenic ($K_{\alpha_{1,2}}$ 0.1177 nm, 10.53 keV)[5, 15, 21, 35, 40]

The most obvious, and the most serious, interelement interference is that from the Pb L_{α_1} radiation at 0.1175 nm. Because of this the use of the As $K_{\alpha_{1,2}}$ line in the presence of substantially higher concentrations of Pb is precluded. There are several reports of the successful use of

TABLE 11.1 Analytical Conditions for the Analysis of Selected Elements by XRF.

Element	Analyte Line	λ (nm)	keV	Possible Interferences	Tube	Crystal	Detector	References
Aluminum	Kα	0.8339	1.486	Br Lα	Cr	PE/KAP/RbAP	Flow	3, 4, 23, 25, 28, 33, 34, 40
Antimony	Kα	0.0472	26.27		W/Au	LiF 200/220	Scin.	5, 35
	Lα	0.3439	3.60	K Kβ	Cr	LiF 200	Flow	
Arsenic	Kα	0.1177	10.53	Pb Lα	Mo/Rh/Ag	LiF 200/220	Flow & Scin.	5, 15, 21, 35, 40
Barium	Kα	0.0387	32.06	I Kβ	W/Au	LiF 200/220	Scin.	5, 16, 17, 31, 40
	Lα	0.2776	4.46	Ti Kα; Sc Kβ	Cr	LiF 200	Flow	
Bismuth	Lα	0.1144	10.84	As Kα; Pb Lα	Mo/Rh/Ag	LiF 200/220	Flow & Scin.	30, 36
Bromine	Kα	0.1041	11.91	Hg Lα	Mo/Rh/Ag	LiF 200/220	Scin.	5
Cadmium	Kα	0.0536	23.11	Rh target tube	W/Au	LiF 200/220	Scin.	5, 15
Calcium	Kα	0.3359	3.69		Cr	LiF 200/220	Flow	3, 4, 23, 28, 33, 40
Cerium	Kα	0.0359	34.57	Cs Kβ; W/Au target tubes	Pt	LiF200	Scin.	5, 40
	Lα	0.2562	4.84		Cr	LiF 200/220	Flow	
Cesium	Kα	0.0402	30.85	Te Kβ	W/Au	LiF 200/220	Scin.	5, 40
	Lα	0.2892	4.29		Cr	LiF 200	Flow	
Chlorine	Kα	0.4729	2.62	Mo target tube	Cr	LiF 200/	Flow	
Chromium	Kα	0.2291	5.41	V Kβ	Mo/Rh/Ag	PE/EDDT	Flow	9, 25, 39, 40
						LiF 200	Flow	1, 5, 15, 20, 23, 25, 28, 31, 40
Cobalt	Kα	0.1790	6.92	Fe Kβ	Mo/Rh/Ag	LiF 200/220	Flow	5, 15, 25, 30, 40
Copper	Kα	0.1542	8.04	Fe K absorption edge	Mo/Rh/Ag	LiF 200/220	Flow	2, 5, 15, 25, 35, 40
Iron	Kα	0.1937	6.40	—	Mo/Rh/Ag	LiF 200/220	Flow	1, 3–5, 20, 23, 28, 33, 34, 40

TABLE 11.1 (Continued)

Element	Analyte Line	λ (nm)	keV	Possible Interferences	Tube	Crystal	Detector	References
Lead	L_α	0.1175	10.55	As K_α	Mo/Rh/Ag	LiF 200/220	Flow & Scin.	2, 5, 15, 25, 31, 35, 40
Magnesium	K_α	0.9890	1.25	—	Cr	KAP/RbAP	Flow	3, 4, 23, 25, 28, 33, 34, 40
Manganese	K_α	0.2103	5.89	Cr K_β	Mo/Rh/Ag	LiF 200	Flow	3–5, 15, 23, 28, 33, 34, 40
Molybdenum	K_α	0.0711	17.44	Zr K_β; U L_β	Rh/Ag	LiF 200/220	Scin.	5, 25, 26, 35
Nickel	K_α	0.1659	7.47	Co K_β; Fe K absorption edge	Mo/Rh/Ag	LiF 200/220	Flow	2, 5, 10, 15, 30, 34, 40
Niobium	K_α	0.0748	16.58	Y K_β	Rh/Ag	LiF 200/220	Scin.	5, 18, 22, 29, 40
Phosphorus	K_α	0.6158	2.01	—	Cr	PE/EDDT	Flow	3, 4, 23, 28, 32–34, 40
Potassium	K_α	0.3742	3.31	—	Cr	PE/EDDT	Flow	3, 4, 23, 28, 32–34, 40
Rubidium	K_α	0.0927	13.37	Br K_β	Mo/Rh/Ag	LiF 200/220	Scin.	2, 5, 6, 12, 15, 29, 31, 34, 40
Selenium	K_α	0.1106	11.21	W/Au target tubes	Mo/Rh/Ag	LiF 200/220	Flow & Scin.	5
Silicon	K_α	0.7126	1.74	—	Cr	PE/KAP/RbAP	Flow	3, 4, 23, 28, 33, 34, 40
Sodium	K_α	1.1910	1.04	—	Cr	KAB/RbAP	Flow	3, 4, 23, 25, 28, 40
Strontium	K_α	0.0877	14.14	—	Mo/Rh/Ag	LiF 200/220	Flow & Scin.	2, 5, 6, 10, 12, 15, 29, 31, 40
Sulfur	K_α	0.5373	2.307	—	Cr	PE/EDDT	Flow	9, 28, 32, 38, 40

Element	Line				Target	Crystal	Detector	References
Tantalum	L_α	0.1522	8.14	—	Mo/Rh/Ag	LiF 200/220	Flow	5, 22
Tellurium	K_α	0.0453	27.38	—	W/Au	LiF 200/220	Scin.	5
	L_α	0.3289	3.77	In K_β	Cr	LiF 200	Flow	
Thorium	L_α	0.0956	12.97	Bi L_β	Mo/Rh/Ag	LiF 200/220	Flow & Scin.	5, 7, 40
Tin	K_α	0.0492	25.19	Ag K_β; Cd K_β	W/Au	LiF 200/220	Scin.	5, 35, 40
	L_α	0.3600	3.44	—	Cr	LiF 200	Flow	
Titanium	K_α	0.2750	4.51	Sc K_β; Ba L_α	Cr	LiF 200	Flow	1, 3–5, 8, 19, 20, 23, 28, 33, 34, 40
Tungsten	L_α	0.1476	8.40	Yb L_β	Mo/Rh/Ag	LiF 200/220	Flow & Scin.	1, 5, 14, 20, 25–27
Uranium	L_α	0.0911	13.61	Rb K_α; Br K_β; Bi K_β; Te K_β	Mo/Rh/Ag	LiF 200/220	Flow & Scin.	5, 7, 13, 35, 37, 40
Vanadium	K_α	0.2505	4.95	Ti K_β	Cr	LiF 200	Flow	5, 11, 34, 40
Yttrium	K_α	0.0830	14.93	Rb K_β; Pb L_β	Mo/Rh/Ag	LiF 200/220	Scin.	5, 18, 29, 35
Zinc	K_α	0.1436	8.63	Fe K absorption edge	Mo/Rh/Ag	LiF 200/220	Flow & Scin.	2, 5, 15, 31, 34, 40
Zirconium	K_α	0.0787	15.74	Sr K_β; Th L_β	Rh/Ag	LiF 200/220	Scin.	2, 10, 24, 29, 31, 40

pressed powder discs for measurement. The quantitative retention of As during fusion for the preparation of a glass disc is dubious, however, and caution is advised.

11.2.4   Barium ($K_{\alpha_{1,2}}$ 0.0387 nm, 32.06 keV; L_{α_1} 0.2776 nm, 4.46 keV)[5, 16, 17, 31, 40]

The $K_{\beta_{1,3}}$ lines of I at 0.0385 nm could be a source of interference in the use of the Ba $K_{\alpha_{1,2}}$ line. The Ti K_α line at 0.2750 nm and the Sc $K_{\beta_{1,3}}$ line at 0.2780 nm lie very close to the Ba L_{α_1} line, and a correction for the Ti present is usually necessary. Both fused discs and pressed powder pellets have been used for the determination of Ba, and if the pressed powder form is used, the particle size must be smaller than 250 mesh. If the K_α line is used, the critical thickness of the pellet or disc is of concern, and unless enough sample is available to form a sufficiently thick sample, the L_α line should be used instead.

11.2.5   Bismuth (L_{α_1} 0.1144 nm, 10.84 keV)[30, 36]

There are few references to the determination of Bi in geological materials by XRF, chiefly because its generally low level (<1 ppm) requires a preconcentration step. There are no likely interferences, however, and pressed powder pellets have been used successfully. Blount et al. report a detection limit of 0.2 ppm for Bi using an ion-exchange resin preconcentration.[30] Stanton also preconcentrates the Bi prior to the XRF measurement.[36]

11.2.6   Bromine ($K_{\alpha_{1,2}}$ 0.1041 nm, 11.91 keV)[5]

Br is very seldom determined by XRF, but if it is, a potentially serious interference from the 0.1040 nm Hg L_{β_2} line, if Hg is present at a concentration of twice or more that of Br, is a possible problem.

11.2.7   Cadmium ($K_{\alpha_{1,2}}$ 0.0536 nm, 23.11 keV)[5, 15]

Trace Cd determinations should not be attempted with a Rh tube because of the 0.0535 nm K_{β_2} line, but other than this there are no likely interferences. Because of its generally low abundance (<1 ppm), preconcentration is usually required before Cd can be determined by XRF in most geological materials not associated with Zn mineralization.

11.2.8 Calcium ($K_{\alpha_{1,2}}$ 0.3359 nm, 3.69 keV)[3, 4, 23, 28, 33, 40]

There are no interferences likely to be encountered in the XRF determination of Ca, and while it has been determined in pressed powder pellets, most workers prefer to measure it in fused discs in order to avoid particle size effects.

11.2.9 Cerium ($K_{\alpha_{1,2}}$ 0.0359 nm, 34.57 keV; L_{α_1} 0.2562 nm, 4.84 keV)[5, 40]

Both W and Au X-ray tubes may cause difficulties with trace Ce determinations using the $K_{\alpha_{1,2}}$ line because of second order lines. In addition, the Cs $K_{\beta_{1,3}}$ lines at 0.0354 and 0.0355 nm could create problems. There is very little geochemical interest in Ce, reflected in the lack of references to be found on its determination by XRF.

11.2.10 Cesium ($K_{\alpha_{1,2}}$ 0.0402 nm, 30.85 keV; L_{α_1} 0.2892 nm, 4.29 keV)[5, 40]

The Te $K_{\beta_{1,3}}$ lines at 0.0400 and 0.0411 nm may be a source of interference with the $K_{\alpha_{1,2}}$ Cs line, but it is not a likely one. Few references are found on the determination of Cs by XRF, but it has been measured successfully with pressed powder discs.

11.2.11 Chlorine ($K_{\alpha_{1,2}}$ 0.4729 nm, 2.62 keV)[9, 25, 39, 40]

Because of the low energy of the K_α Cl line (2.62 keV) it is unlikely that the measurement of pressed powder discs will give acceptable results unless the powder is very finely ground. Some workers have reported their successful use when all material passed through a 250 mesh screen in one case[40] and a 400 mesh screen in another.[39] The latter reported a detection limit of 10 ± 7 ppm. There is a possible interference with the determination of trace levels of Cl when a Mo X-ray tube is used because of the 0.4725 nm L_{α_1} Mo line.

11.2.12 Chromium ($K_{\alpha_{1,2}}$ 0.2291 nm, 5.41 keV)[1, 5, 15, 20, 23, 25, 28, 31, 40]

The V $K_{\beta_{1,3}}$ line (0.2284 nm) will interfere with the Cr $K_{\alpha_{1,2}}$ line, and has been measured as the equivalent of 6.7 ppm of Cr for each 100 ppm of V present.[40] Thus the V concentration must be measured in order to correct the Cr value. In addition, some W X-ray tubes contain Cr, and provision must be made to correct for this. Leake et al. did so by "regularly obtaining the ratio of the background on one side of the Cr peak to the Cr K_α peak in distilled water. This factor is then multiplied by the measured background in each sample and the resulting value substracted from the peak".[40]

11.2.13   Cobalt ($K_{\alpha_{1,2}}$ 0.1790 nm, 6.92 keV)[5, 15, 25, 30, 40]

The proximity of the Fe K_{β_1} line (0.1791 nm) to the Co $K_{\alpha_{1,2}}$ line, combined with the usually major and highly variable Fe concentrations found in most geological materials, result in a complicated interference effect with the measurement of Co. It is possible to obtain sufficient resolution to separate them by using a long collimator,[40] but the reduced count results in a poorer detection limit than might otherwise be expected.

11.2.14   Copper ($K_{\alpha_{1,2}}$ 0.1542 nm, 8.04 keV)[2, 5, 15, 25, 35, 40]

There are no likely interferences to be encountered other than that from the Fe K absorption edge and a gradual migration to the surface of Cu from the substrate on which the target element (Cr, W, Au, etc.) is plated to form the X-ray tube target. This can result in a tube-derived Cu signal in older tubes which must be accounted for (see Cr). Finely divided powders can be pressed into pellets for the measurement of Cu but, again, fused glass discs are preferable.

11.2.15   Iron ($K_{\alpha_{1,2}}$ 0.1937 nm, 6.40 keV)[1, 3–5, 20, 23, 28, 33, 34, 40]

There are few problems associated with the determination of Fe by XRF other than those associated with particle size effects (in the case of pressed powder pellets), matrix differences, and the selection of appropriate background measurement locations and instrumental parameters. Most geological samples have Fe as a major constituent, and minor interferences are not likely to have a significant effect on the Fe measurement.

11.2.16   Lead (L_{α_1} 0.1175 nm, 10.55 keV)[2,5,15,25,31,35,40]

The $K_{\alpha_{1,2}}$ line of As at 0.1177 nm cannot be resolved from the Pb L_{α_1} line, and the interference requires a correction. It is not possible to measure small concentrations of one in the presence of large concentrations of the other using these lines. The measurement of the $L_{\beta_{1,2}}$ line of Pb at 0.0983 nm is one alternative, but the 0.0956 nm L_{α_1} line of Th is a possible source of interference.

11.2.17 Magnesium ($K_{\alpha_{1,2}}$ 0.9890 nm, 1.25 keV) [3, 4, 23, 25, 28, 33, 34, 40]

Because of the low energy of the Mg K_α line, the use of a fused glass disc is much preferable to a pressed powder pellet, even with a very finely ground sample. There have been reports of the measurement of Mg in pressed powder discs, but variable mineralogical matrices and other factors result in errors. There are no serious interferences likely from adjacent lines, but third order Ca K_α reflections have been reported to require two background measurements for adequate corrections.[40]

11.2.18 Manganese ($K_{\alpha_{1,2}}$ 0.2103 nm, 5.89 keV) [3–5, 15, 23, 28, 33, 34, 40]

There is a possibility of some interference in the measurement of trace amounts of Mn from the $K_{\beta_{1,3}}$ line of Cr at 0.2085 nm if Cr is present as a major constituent, but there are no other likely interferences. Again, fused glass discs are the preferred sample form, but pressed powder pellets have been measured successfully.

11.2.19 Molybdenum ($K_{\alpha_{1,2}}$ 0.0711 nm, 17.44 keV) [5, 25, 26, 35]

The most serious interference in the determination of traces of Mo is the Zr $K_{\beta_{1,3}}$ line at 0.0702 nm. The line cannot be resolved, and thus Zr must be measured (Zr $K_{\alpha_{1,2}}$) to obtain a correction for its contribution to the Mo $K_{\alpha_{1,2}}$ signal. Another possible but unlikely interference is that from the 0.0710 nm L_{β_3} and the 0.0720 nm L_{β_1} lines of U if it is present in significant concentrations. Mo is measured successfully in either pressed powder pellets or fused glass discs.

11.2.20 Nickel ($K_{\alpha_{1,2}}$ 0.1659 nm, 7.47 keV) [2, 5, 10, 15, 30, 34, 40]

The Co $K_{\beta_{1,3}}$ line at 0.1621 nm is a possible interference, and the Fe K absorption edge at 0.1744 nm could affect the calibration slope for Ni. In addition, there is at least one instance reported of the need to correct for a Ni impurity in a W target tube. The correction was made by "obtaining the ratio of counts on W_{LL} to Ni K_α in distilled water, cellulose, and pure silica."[40]

11.2.21 Niobium ($K_{\alpha_{1,2}}$ 0.0748 nm, 16.58 keV) [5, 18, 22, 29, 40]

Nb can be determined readily using either pressed powder pellets or fused glass discs, but the Y $K_{\beta_{1,3}}$ line at 0.0741 nm requires a measurement of Y to make a correction in the Nb K_α intensity measurement. The second order La $K_{\alpha_{1,2}}$ line at 0.0744 nm is a possible problem if La is added as a heavy absorber, as is the U K_{β_2} line at 0.0755 nm if U is present to any significant extent.[22]

11.2.22   Phosphorus ($K_{\alpha_{1,2}}$ 0.6158 nm, 2.01 keV)[3, 4, 23, 28, 32–34, 40]

The high labor costs involved in the determination of P by most other techniques makes the application of XRF to this measurement very desirable, and numerous papers have been published on the subject. Both pressed powder pellets and fused glass discs have been measured, the latter giving more reliable results because of an elimination of particle size effects on the low energy P K_α line. There are no likely interferences that will not be eliminated by selection of the correct crystal and other instrumental parameters. Tanaka et al. have reported a detection limit of about 1 ppm P using pressed pellets.[32]

11.2.23   Potassium ($K_{\alpha_{1,2}}$ 0.3742 nm, 3.31 keV)[3, 4, 23, 28, 32–34, 40]

There are no likely interferences in the determination of K by XRF, and both pressed powder pellets and fused glass discs have been measured, the latter with more precise results. Because K lies close to the overlap in range between crystals (LiF 200 is good from W to K and PE and EDDT from K to Al; see Table 10.2), the selection of the preferred crystal is not obvious as in some other instances. The PE or EDDT crystal is usually selected over the LiF 200 because the 2θ angle is only about 50° in the two former crystals rather than the much higher angle of 136.69° in the latter.

11.2.24   Rubidium ($K_{\alpha_{1,2}}$ 0.0927 nm, 13.37 keV)[2, 5, 6, 12, 15, 29, 31, 34, 40]

In most instances Rb is one of the elements for which the XRF technique gives its most precise results (see also Sr). There are few interferences, the Br $K_{\beta_{1,3}}$ line at 0.0933 nm being an exception. The BCM laboratory has determined that 80 ppm Br are equivalent to 1 ppm of Rb. In addition, the Au L_{α_1} line at 0.0926 nm could be a problem if a Au target X-ray tube is used. The most efficient tube for exciting the Rb $K_{\alpha_{1,2}}$ line is one with a Mo target, but the BCM laboratory has obtained a detection limit of 10 ppm with the use of a Au target tube.

11.2.25   Selenium ($K_{\alpha_{1,2}}$ 0.1106 nm, 11.21 keV)[5]

Because of its generally very low concentration in geological samples, and the availability in most laboratories of the sensitive hydride generation/AAS technique for its determination, Se is seldom determined by XRF. There are, however, few interferences with the latter method, the W L_{α_1} line being one exception when a W target X-ray tube is used. A preconcentration technique is usually required.

11.2.26 Silicon ($K_{\alpha_{1,2}}$ 0.7126 nm, 1.74 keV)[3, 4, 23, 28, 33, 34, 40]

XRF is one of the most satisfactory methods for determining Si in geological materials, with the best results being obtained with fused glass discs. No likely interferences will be encountered if the appropriate instrumental parameters are used.

11.2.27 Sodium ($K_{\alpha_{1,2}}$ 1.1910 nm, 1.04 keV)[3, 4, 23, 25, 28, 40]

Na is the lightest element commonly determined on a routine basis by XRF techniques, although F can be determined under ideal circumstances and if present at a sufficiently high concentration. Although Na results have been reported for pressed powder pellets, the results are likely to be much inferior to those obtained using fused glass discs. There are no likely interferences in the determination of Na by XRF, but the instrument itself must be in good repair with clean crystals, a good vacuum, and a clean flow proportional counterwire.

11.2.28 Strontium ($K_{\alpha_{1,2}}$ 0.0877 nm, 14.14 keV)[2, 5, 6, 10, 12, 15, 29, 31, 40]

As in the case of Rb, Sr is an ideal element for determination by XRF techniques, and the application of XRF Rb/Sr determinations for geochronological work is widespread. There are no likely interferences to be encountered, and with a Mo target X-ray tube, detection limits for both Rb and Sr approach 1 or 2 ppm with very good precision even at that level.

11.2.29 Sulfur ($K_{\alpha_{1,2}}$ 0.5373 nm, 2.307 keV)[9, 28, 32, 38, 40]

S is conveniently determined by XRF on fused glass discs to which a nitrate salt has been added as part of the flux to prevent the loss of S during fusion. The determination of S in pressed powder pellets has been reported, but both particle size and matrix effects introduce uncertainties. There exists the possibility of a gradual buildup of S from the oil of the vacuum pump on frequently used discs, unless a sulfur-free oil is used. There are no likely interferences in the determination of S by XRF. Terashima has reported a detection limit of 10 ppm even when pressed pellets are used[9] and Tanaka et al. a limit of about 1 ppm.[32]

11.2.30   Tantalum (L_{α_1} 0.1522 nm, 8.14 keV)[5, 22]

There are few interferences in the determination of Ta by XRF, but its low natural occurrence (~2 ppm) and a general lack of geochemical interest in it combine to result in few references to its determination in geological materials.

11.2.31   Tellurium ($K_{\alpha_{1,2}}$ 0.0453 nm, 27.38 keV; L_{α_1} 0.3289 nm, 3.77 keV)[5]

As in the case of Se, the generally low level of Te found in most geological samples, and the ready access to a more sensitive and rapid hydride generation/AAS method, result in few references to the determination of Te by XRF. There is a possible interference from the 0.0455 nm $K_{\beta_{1,3}}$ line of In, but there is none for the Te L_{α_1} line.

11.2.32   Thorium (L_{α_1} 0.0956 nm, 12.97 keV)[5, 7, 40]

The determination of Th by XRF is much faster than that by chemical methods, is just as reliable, and is capable of a lower detection limit (4 ppm in the BCM laboratory) than is achieved readily by gamma-ray spectrometry (Sec. 6.2.2). Neutron activation analysis is a competitive method, however, if it is available. The 0.0955 nm L_{β_2} Bi line is a possible interference to be encountered in the XRF determination of Th. Pressed powder pellets are adequate if they are properly prepared.

11.2.33   Tin ($K_{\alpha_{1,2}}$ 0.0492 nm, 25.19 keV; L_{α_1} 0.3600 nm, 3.44 keV)[5, 35, 40]

XRF is a convenient method for the determination of Sn present in silts and other geological materials at the 2–3 ppm level without the need for preconcentration. There is the possibility of an interference from the Ag $K_{\beta_{1,3}}$ line at 0.0497 nm and a more remote one from the 0.0476 nm Cd $K_{\beta_{1,3}}$ line. There are no likely interferences when the Sn L_{α_1} line is used.

11.2.34   Titanium ($K_{\alpha_{1,2}}$ 0.2750 nm, 4.51 keV)[1, 3–5, 8, 19, 20, 23, 28, 33, 34, 40]

The 0.2780 nm $K_{\beta_{1,3}}$ line of Sc is a possible interference, but at the much higher concentration of Ti generally encountered it is not a serious concern. There have been numerous references to the determination of Ti by XRF, and with adequate sample preparation and with properly selected instrumental parameters, no problems are encountered.

11.2.35   Tungsten (L_{α_1} 0.1476 nm, 8.40 keV) [1, 5, 14, 20, 25–27]

Because of the labor intensive aspect of other techniques for the determination of W (colorimetric, for example), and because AAS does not provide adequate sensitivity, the determination of W by XRF is inherently attractive. There are problems, however, such as the possible interference from the 0.1476 L_{β_1} line of Yb (2 ppm Yb ≡ 1 ppm W), and the 20% difference in W L_{α_1} line intensity per unit concentration encountered when the sample matrix changes from scheelite to wolframite. Carr-Brion and Payne attempted to minimize this by using the 0.0211 nm W $K_{\alpha_{1,2}}$ line, but the poor signal-to-background ratio resulted in unsatisfactory detection limits.[27] For the determination of W in silt samples (>80 mesh) using pressed discs, the BCM laboratory obtained a detection limit of 3–4 ppm. The fact that different W minerals give different measurement intensities is a persistent problem, one best eliminated by the use of fused glass discs.

11.2.36   Uranium (L_{α_1} 0.0911 nm, 13.61 keV) [5, 7, 13, 35, 37, 40]

The determination of U by XRF is not uncommon, although other rapid instrumental methods are available, NAA being the most noteworthy of these. The 0.0927 nm $K_{\alpha_{1,2}}$ line of Rb can interfere in the determination of low concentrations of U if a few hundred parts per million of Rb are present, and its intensity must be measured in order that a correction can be made. The most efficient tube for exciting U L_α lines is one with a Mo target, as is the case for Rb and Sr. The Br $K_{\beta_{1,3}}$ line at 0.0933 nm is a possible interference, as are the $K_{\beta_{1,3}}$ lines of Bi at 0.0952 and 0.0938 and the 0.0906 nm Te $K_{\alpha_{1,2}}$ line. In the absence of these interfering elements, however, the BCM has achieved a detection limit of 3 ppm in silt samples using an Au target tube.

11.2.37   Vanadium ($K_{\alpha_{1,2}}$ 0.2505 nm, 4.95 keV) [5, 11, 34, 40]

Although V appears to be of little geochemical interest, its determination by XRF has been reported by numerous workers. There is an interference from the $K_{\beta_{1,3}}$ line of Ti at 0.2514 nm so that it is necessary to measure Ti in order to make a correction, but even with this a precise measurement for V is difficult to obtain.

11.2.38   Yttrium ($K_{\alpha_{1,2}}$ 0.0830 nm, 14.93 keV) [5, 18, 29, 35]

Yttrium is readily determined by XRF, and the BCM laboratory has achieved a dectection limit of about 0.5 ppm using a tube with an Au target. There is interference from the 0.0829 nm Rb $K_{\beta_{1,3}}$ line, which is sufficiently close that Rb must be measured in order to make a correction for its contribution to the Y K_α line. There is also the chance of interference from the Pb L_{γ_1} line at 0.0840 nm, but the magnitude of the interference is quite small, 3000 ppm Pb equivalent to 1 ppm Y, as calculated by P. Ralph of the BCM laboratory.

11.2.39   Zinc ($K_{\alpha_{1,2}}$ 0.1436 nm, 8.63 keV)[2, 5, 15, 31, 34, 40]

There are no likely interferences in the determination of Zn by XRF other than a possible one from the Fe K absorption edge. Both pressed powder pellets and fused discs have been used for measurement. Fabbi et al.[34] have reported a detection limit of 7 ppm for Zn when measuring pressed powder pellets, but the sample must be finely ground to pass a 250 mesh screen to avoid particle size effects.

11.2.40   Zirconium ($K_{\alpha_{1,2}}$ 0.0787 nm, 15.74 keV)[2, 10, 24, 29, 31, 40]

Because of the refractory nature of the common Zr minerals and the lack of sensitivity of AAS for the element, XRF is a preferred method for the determination of Zr. There is interference from the 0.0783 nm $K_{\beta_{1,3}}$ line of Sr which must be corrected for by measuring the Sr concentration. There is also a possible minor interference from the L_{β_2} line of Th at 0.0794 nm, the magnitude of which the BCM laboratory has estimated as 460 ppm Th being the equivalent of 1 ppm Zr. Both pressed powder pellets and fused disc samples have been used successfully with a detection limit of about 5 ppm for pressed pellets reported by one group of workers.[40]

References

1.   C. E. Austen *Anal. Abstr.*, 2B22, **35** (1978).
2.   K. H. Lieser, E. Breitwieser, P. Burpa, M. Röber, and R. Spatz, *Mikrochim. Acta*, **1978 I** 363.
3.   I. L. Thomas and M. T. Haukka, *Chem. Geol.*, **21**, 39 (1978).
4.   M. T. Haukka and I. L. Thomas, *X-Ray Spectrom.*, **6**, 204 (1977).
5.   R. D. Giaque, R. B. Garrett, and L. Y. Goda, *Anal. Chem.*, **49**, 1012 (1977).
6.   E. A. T. Verdurmen, *X-Ray Spectrom.*, **6**, 117 (1977).
7.   G. W. James, *Anal. Chem.*, **49**, 967 (1977).
8.   A. E. Hubert and T. T. Chao, *Anal. Chim. Acta*, **92**, 197 (1977).
9.   S. Terashima, *Chishitsu Chosasko Geppo*, **27**, 185 (1976); through *Anal. Abstr.* 2B57, **34** (1978).
10.   H. Bougault, P. Cambon, and H. Toulhoat, *X-Ray Spectrom.*, **6**, 66 (1977).
11.   C. Riddle and T. E. Smith, *X-Ray Spectrom.*, **6**, 18 (1977).

12. A. Turek, C. Riddle, and T. E. Smith, Can. J. Spectrosc., **22**, 20 (1977).

13. J. W. Rowson and S. A. Hontzeas, Can. J. Spectrosc., **22**, 24 (1977).

14. G. Domel and A. M. E. Balaes, Natl. Inst. Metall. Rep. 1817, 1976.

15. R. Lichtfuss and G. Bruemmer, Chem. Geol., **21**, 51 (1978).

16. E. Murad, Spectrochim. Acta, **30B**, 433 (1975).

17. E. Murad, Spectrochim. Acta, **30B**, 10 (1975).

18. L. Leoni and M. Saitta, X-Ray Spectrom., **5**, 29 (1976).

19. A. E. Hubert and T. T. Chao, Anal. Chim. Acta, **92**, 197 (1977).

20. C. E. Austin, National Institute for Metallurgy, Republic of South Africa, Rep. 1883, 1977.

21. H. Tanaka, U. Moriguchi, T. Yamamoto, and G. Hashizume, Jpn. Analyst, **21**, 57 (1972); through Anal. Abstr., No. 100, 26 (1974).

22. G. Domel and B. G. Russell, Natl. Inst. Metall. Rep. 1514 1973.

23. C. E. Austin and B. G. Russell, Natl. Inst. Metall. Rep. 1599, 1974.

24. E. Donderer, Neues Jb. Miner., ML. (5), 183 (1974); through Anal. Abstr., 6B76, **29** (1975).

25. D. A. Pantony and P. W. Hurley, Analyst, **97**, 497 (1972).

26. W. T. Elwell and D. F. Wood, Analytical Chemistry of Molybdenum and Tungsten, Pergamon, New York, 1971.

27. K. G. Carr-Brion and K. W. Payne, Analyst, **93**, 441 (1968).

28. C. Palme and E. Jagoutz, Anal. Chem., **49**, 717 (1977).

29. E. Jagoutz and C. Palme, Anal. Chem., **50**, 1555 (1978).

30. C. W. Blount, D. E. Leyden, T. L. Thomas, and S. N. Guill, Anal. Chem., **45**, 1045 (1973).

31. E. Murad, Anal. Chim. Acta, **67**, 37 (1973).

32. H. Tanaka, T. Ozaki, Y. Moriguchi, K. Hiroyuki, and G. Hashizume, Jpn. Analyst, **23**, 333 (1974); through Anal. Abstract., No. 1H30, **29** (1975).

33. P. K. Harvey, D. M. Taylor, R. D. Hendry and F. Bancroft, X-Ray Spectrom., **2**, 33 (1973).

34. B. P. Fabbi and F. L. Espos, U. S. Geological Survey, Prof. Paper 800B, B147–B150, 1972.

35. V. Machacek, Cas. Miner. Geol., **17**, 171 (1972); through Anal. Abstr., No. 2827, **24** (1973).

36. R. E. Stanton, Proc. Australasion Inst. Min. Metall., **240**, 113 (1971); through Anal. Abstr., No. 4207, **23**(1972).

37. N. H. Clark and J. G. Pyke Anal. Chim. Acta, **58**, 234 (1972).

38. P. Richter, Z. Anal. Chem., **258**, 287 (1972).

39. B. P. Fabbi and F. L. Espos, Appl. Spectrosc., **26**, 293 (1972).

40. B. L. Leake, G. L. Hendry, A. Kemp, A. G. Plant, P. K. Harrey, J. R. Wilson, J. S. Coats, J. W. Aucott, T. Lünel, and R. J. Howarth, Chem. Geol., **5**, 7 (1969/70).

PART V

Selected Other Methods of Analysis

CHAPTER 12

Other Analytical Methods

OTHER ANALYTICAL METHODS

12.1 OPTICAL EMISSION SPECTROSCOPY

One of the earliest applications of the spectroscope, following its invention in 1859 by Kirchhoff and Bunsen, was in the qualitative analysis of geological materials. This association has continued to this day, albeit with certain rises and falls in methodological popularity as other methods superseded it in analytical interest. Indeed, there has been of late a notable resurgence in interest in the use of ES as an analytical tool, and, because of its recognized capacity for multielement determinations, it has regained some of the ground previously lost to such arrivals on the analytical scene as AAS and XRF.[1-3]

It is virtually impossible, and certainly unwise, to state that a particular ES procedure is the best one. There are as many of these as there are spectroscopists; each analytical approach will be governed by a unique set of circumstances that will determine the procedure to be followed, and each spectroscopist will solve the problem in his or her own way. All that can be done here is to describe some equipment and methods, to draw attention to advantages and limitations, and to give some examples of application.

A number of sources of detailed information on the basic theory of ES are readily available, and reference should be made to these as required,[4-7] as well as to some that deal specifically with the application of ES to the analysis of rocks, ores, minerals, ceramics, and soils.[8-14] Comprehensive reviews of the latest developments in this field are found in the *Fundamental* and in the *Applications Reviews* issues published every two years by *Analytical Chemistry*[15, 16] and in the *Annual Reports on Analytical Atomic Spectroscopy* published by The Chemical Society.[17]

12.1.1 Applications

The first widespread use of ES in geological work was that of Goldschmidt's school in the 1930s (Ref. 8, p. 88), and the technique has since contributed much to the generation of data for studies in mineralogy, petrology, geochemistry, geochronology, and meteoritics. The *dc arc* can excite about 70 elements, and it is the most popular form of excitation; the *ac spark*, which provides the highest energy of excitation, is also used but usually with solutions or under special conditions.[11] Many laboratories have found ES to be a useful tool for quick qualitative and semiquantitative analyses of unknown samples, especially those available in limited amounts. It is also used for the routine monitoring of samples for unexpected but significant element concentrations, although ED-XRF, the electron microprobe, and the scanning electron microscope find favor also. Its use as a quantitative tool is, however, not as widespread. The procedures for developing, fixing, and reading the photographic record are time-consuming, and very close attention to operating details is necessary to ensure satisfactory reproducibility. There is the problem of sample inhomogeneity when only a few milligrams of material are actually burned in the energy source, thus necessitating close control over the quality of the sample powder.[6] A notable advantage of ES is that no preliminary separation of elements prior to arcing is usually required, but geological spectra do tend to be complex, and the spectrometer used must be capable of providing adequate dispersion (0.25–0.70 nm) of the spectral lines. The development of multichannel instruments using photoelectric detectors, now frequently coupled to a minicomputer for operating control, data processing, and evaluation, has made quantitative analysis by ES much less labor-intensive.

The most common geological application of dc arc ES has been the semiquantitative or quantitative determination of minor or trace element concentrations, with only occasional use for the determination of major concentrations, but the recent development of new excitation sources may change this. Laqua[18] has listed 13 fundamental properties of ES as a general method, including such attributes as its multielement analysis facility, coverage of a wide concentration range, small consumption of sample, and high selectivity, all of which have application to the analysis of geological samples; he considers that the analysis of solid samples is a field of application where ES often offers unique solutions to complicated analytical problems. Anyone who has been faced with what are termed "not ordinary" samples will be in full agreement.

12.1.2   Instrumentation

There are essentially two basic choices open to the analyst who is considering the addition of ES to laboratory operations. A *spectrograph* employs a photographic plate or film as detector and provides maximum versatility of use. A *multichannel spectrometer* is a spectrograph in which the photographic detector is replaced by several exit slit-photoelectric detectors located at appropriate points on the focal curve of the instrument so as to monitor the characteristic emission of specific elements. Obviously the latter instrument will be faster and more reproducible than the photographic recording instrument, particularly when a computer is built in or interfaced with it to control operation, process the data, and print out the analytical results. As usual, however, there are limitations resulting from the lack of flexibility of the system. There are physical limits to the number of detectors that can be mounted, and it is not easy to make changes in the detector array once it is established, thus limiting the number of elements that can be determined. Significant differences between the compositional matrices of the sample and standards, such as the presence of an unsuspected element, may result in large spectral interference and resulting errors.

The simple choice outlined in the preceding paragraph is, of course, made considerably more complex by the new developments in available detectors, including cathode ray tube displays, photo-diode arrays, and the use of both phototransistors and photoresistors. The traditional excitation sources are now being challenged by the plasma torch, considered to be chiefly responsible for the rekindling of interest in the use of ES. The review by Veillon (Ref. 7, Chapter 6) of available instrumentation and new developments is a useful guide to the analyst who must decide the best choice of instrumentation for his or her particular set of parameters. These will include such considerations as the number of samples to be handled, the elements to be determined and the number of elements per sample, the detection limits required and the desired accuracy and precision, the skill of the operator, and the cost per sample.

12.1.3 Methods

Although it was stated at the beginning that the very nature of ES analysis precludes the setting forth of a "best" method, it will be useful to indicate just what is involved in spectrographic analysis by describing three methods developed by W. H. Champ and currently in use at the GSC. It must be stated immediately that a detailed description of these methods, with all the ancillary requirements, precautions, and procedural steps, would occupy the rest of this chapter, and it is thus necessary to go the other extreme, that of presenting only the bare outline of the methods.

a. *Photographic Recording Method*

In the Analytical Chemistry Section of the GSC it is current practice to use the ES method almost exclusively for the determination of trace level concentrations of elements. It is necessary, of course, to adapt the operating procedure to suit the compositional nature of the sample, which may be a silicate, sulfide, carbonate, or highly ferruginous, to mention a few possibilities. This can be done by developing a procedure that is closely tailored to one compositional type, but this approach is productive only when a large and continuing flow of such samples is assured. The other approach is to develop a general procedure which is applicable to a range of compositions and to accept limitations in detection limits for some elements as a result. Even then, the compositional range is not open-ended and an ES laboratory dealing with a variety of samples must have a number of basic procedures on tap.

Quantitative

This is a general method for silicate rocks, utilizing a 3.4 m Wadsworth photographic plate recording instrument, having a reciprocal linear dispersion of 0.52 nm in the first order over a wavelength range of 222.5–490.0 nm. It provides a reasonably rapid analysis over a wide concentration range and without being overly restrictive as to the sample type to which it applies. Results are reported, for the most part, on the basis of a *single* exposure and are expected to be within ± 15% of the reported value.

The sample, ground to 200 mesh or finer, is mixed in a fixed proportion with high purity graphite and an alkali salt buffer containing Li and K as control element and fluxing agents, and Eu and Pd as internal standards. A portion of the mixture is packed into the crater of a deep-bore high purity graphite electrode of small diameter which is made the anode in an argon–oxygen jet-controlled dc arc, and is burned under controlled conditions using a tapered graphite rod as cathode. The spectra are recorded on two 10 × 25 cm photographic plates which are processed, and the transmittances of selected analytical and internal (or external) standard spectral line pairs are measured on a microphotometer. The values obtained are converted to element concentrations either through manual graphical calculation or by means of an off-line minicomputer. Table 12.1 lists the elements and their concentration ranges determined by this method.

Semiquantitative

There has been frequent reference, in previous chapters, to the variable composition of geological samples. The photographic recording method just described, and the direct reading one to follow, are "general" methods which allow *reasonable* latitude in compositional variability. This semiquantitative method is intended to provide compositional data for a

TABLE 12.1 Elements and Concentration Ranges Determined by a Photographic Recording Quantitative Analysis Method.

Element	Range (ppm)	(%)	Element	Range (ppm)	(%)
Si	100	2.0	La	30	2.0
Al	500	2.0	Mo	10	2.0
Fe	30	2.0	Nb	50	2.0
Ca	200	2.0	Nd	200	2.0
Mg	200	2.0	Ni	7	2.0
Ti	5	2.0	Pb	200	2.0
P	2000	2.0	Sc	3	0.30
Mn	10	2.0	Sn	50	2.0
Ag	1	0.10	Sr	5	2.0
As	1000	2.0	Th	150	2.0
B	20	1.0	U	200	2.0
Ba	5	2.0	V	10	1.0
Be	1	0.50	W	200	2.0
Ce	150	2.0	Y	3	2.0
Co	7	2.0	Yb	1.5	0.30
Cr	3	1.5	Zn	150	2.0
Cu	2	2.0	Zr	10	1.0

wide range of sample types, excluding only those containing much organic material, and metals and alloys.

The method is based upon the complete volatilization of a fixed weight of sample plus buffer mixture in an argon–oxygen (70–30%) jet-controlled dc arc, and using the 3.4 m Wadsworth photographic recording instrument. Up to 55 elements are determined *semiquantitatively* by comparison with external reference spectra of Fe oxide and selected reference samples. When the type of sample is known in advance, or the information is obtained subsequently, it is possible to achieve near-quantitative results by the judicious selection of reference spectra. Because the sample is more diluted than that for the other methods, in order to ensure generality of application and minimal matrix effect, the method is less sensitive for trace elements.

The sample must be ground to at least − 150 mesh. A weighed portion (10 mg) is mixed with 100 mg buffer (Li_2CO_3, LiF, K_2SO_4, and high purity graphite), from which a weighed charge (35 mg) is taken for loading into a high purity graphite preformed electrode which is made the anode in the dc arc, with a high purity graphite tapered rod as cathode. The spectra are recorded on photographic plates as in the previous method, and selected spectral lines are compared with reference spectra using a comparator-microphotometer. See Table 12.2 for the elements and range of concentrations determined.

b. *Direct Reading Method*

This is largely an adaptation of the photographic recording general quantitative method described in Sec. 12.1.3a for the elements listed in

TABLE 12.2   Elements and Concentration Ranges Determined by a Photographic Recording Semiquantitative Analysis Method.

Element	Range (ppm)	Range (%)	Element	Range (ppm)	Range (%)
Si	100	a	In	50	5.0
Al	500	a	La	50	10
Fe	100	a	Mo	70	a
Ca	100	a	Nb	50	10
Na	2000	a	Nd	200	7.0
Mg	10	a	Ni	15	a
Ti	30	a	Pb	300	a
P	3000	a	Pt	100	1.0
Mn	15	a	Sb	1000	a
Ag	5	0.3	Sc	7	0.5
As	5000	a	Sn	100	a
Au	100	5.0	Sr	2	3.0
B	50	1.5	Ta	3000	10
Ba	10	a	Te	5000	a
Be	1.5	a	Th	300	a
Bi	150	a	Tl	500	a
Cd	500	a	U	500	a
Ce	300	10	V	20	a
Co	15	10	W	500	a
Cr	10	a	Y	20	a
Cu	10	a	Yb	5	0.5
Ga	100	3.0	Zn	1000	a
Ge	150	3.0	Zr	50	a
Hf	200	3.0			

a Upper limit of determination is 30%.

TABLE 12.3   Elements and Concentration Ranges Determined by Direct Reading Quantitative Analysis Methods.

Element	Range (ppm)	(%)	Element	Range (ppm)	(%)
Si	500	2.0	Cr	5	7.0
Al	300	2.0	Cu	7	12.5
Fe	300	2.0	La	100	2.0
Ca	20	2.0	Mo	50	1.0
Na	50	1.0	Ni	10	2.0
Mg	500	2.0	Pb	700	10.0
Ti	100	2.0	Sb	500	4.0
Mn	50	1.5	Sn	200	3.0
Ag	5	0.45	Sr	10	7.0
As	2000	12.5	V	20	1.25
B	50	2.0	Y	40	1.5
Ba	5	4.0	Yb	4	0.22
Be	3	0.35	Zn	200	7.0
Ce	200	3.5	Zr	20	1.75
Co	10	1.0			

Table 12.3, again with the expected concentration range to be covered. The instrument used is a 1.5 m Jarrel–Ash direct reading optical spectrometer having 41 readout channels. The method is intended to provide a rapid "routine" analysis for the selected elements, again without severe restrictions on sample composition and with results generally comparable in both accuracy and precision to those of the photographic method (i.e., within ± 15% of the value reported). Because the system is much less flexible than the photographic method, there are limitations on its applicability to samples of unknown composition. It does have the advantage of being a faster method and is an invaluable tool in providing abundant trace element data on large numbers of samples having a generally similar composition.

The preparation of the electrode charge is the same as that used for the quantitative photographic method, except that the Eu and Pd are omitted. The volatilization of the mixture is carried out as for the latter, that is, for a fixed time period (90 sec), but the output of the photomultiplier tubes which measure the intensities of individual spectrum lines are fed to an on-line minicomputer which automatically applies corrections for background and interelement effects and converts the corrected intensity measurements to analytical values using calibration curves based on synthetic and international reference materials. A high speed printer is used to produce the results in a format suitable for use by the submitter.

12.1.4   The Plasma Excitation Torch

Brief mention has already been made of this excitation source which is now the subject of much study. Although the initial work dates from 1959, the first commercial plasma excitation sources for use with direct reading ES instruments were not introduced until the early 1970s. The availability of these, together with enthusiastic (and frequently overoptimistic) claims for the source as a panacea for all ES ills, has resulted in its application to the analysis of a wide variety of materials, one of the first of which was the determination of major, minor, and trace elements in geological materials. Many of the more extravagant claims are being modified in light of current investigations, but there is little doubt now that ES has received a much-needed shot in the arm to make it more competitive with XRF and AAS.

It is first necessary to distinguish between the three types of plasma sources now available, differing in the number of electrodes which are necessary to sustain the plasma:

1. Dc argon plasma jet (2- or 3-electrode)[5, 19, 20, 25]
2. Microwave induced plasma cavities (single electrode)[21–23, 25]
3. Inductively coupled plasma torches (electrodeless)[24–28]

In most geochemical laboratories, and in many others, the preference appears to favor the inductively coupled plasma (ICP) torch, and the majority of published papers emphasize the use of this type of excitation. *The ICP Information Newsletter*, now several years old, is a prime source for current information on the subject and includes a detailed bibliography covering the period of 1959–1977.[29] Only the ICP excitation source will be discussed further.

A plasma is a gas in which a significant fraction of the atoms or molecules is ionized and is thus able to interact, or *couple*, with a magnetic field that is oscillating at about 27 MHz. To form and sustain a stable plasma at the open end of an assembly of quartz tubes requires a pattern of two or three argon flows. The torch usually consists of three concentric tubes (of quartz glass, except for the inner tube which may be either quartz or borosilicate glass) surrounded by a Cu induction coil connected to a high frequency (RF) current generator. The sample, most commonly as an aerosol generated from a solution of the sample by either a pneumatic or an ultrasonic nebulization technique, is injected into the plasma by an Ar gas flow. The temperature attained by the plasma is very high, about 10,000°K, in the region of maximum eddy current flow, and sample particles are heated to 7000–8000°K. The use of an Ar–N$_2$ ICP has been investigated in order to reduce the consumption of costly Ar, but it was found that the Ar ICP yielded significantly better detection limits for the elements studied.[30]

Solid samples, in finely powdered form, can be excited by ICP-ES, but in its present state of development ICP should be viewed primarily as an excitation source for liquid samples (and for gaseous samples, of course, such as hydrides), and the current literature supports this view. What, then, are the positive and negative features of the technique, assuming the use of liquid or gaseous samples (either natural or prepared) in terms of its application to the analysis of rocks and minerals:

Positive

1. There is excellent sensitivity for most elements (low detection limits); the very high temperature dissociates molecular species, including refractory compounds, and thus eliminates chemical interference.

2. The signal response is linear over a wide concentration range (up to 5 decades) from the detection limit, because the low background and the small subsample size cause negligible self-absorption. It is thus possible to determine elements present at major, minor, and even some trace concentration levels in a single solution without the need for further dilution.

3. The use of solutions ensures homogeneity of the subsample introduced into the plasma and permits considerable flexibility in the choice of sample weight and physical nature of the original sample. It also facilitates easy adjustment of concentration levels, the matching of solution parameters such as acid concentration, and the preparation of appropriate standards.

4. The technique emphasizes the multielement capability of ES. It is possible to determine 40 or more elements simultaneously with a polychromator, or up to 20 elements sequentially with a monochromator, and this lends itself to the automation of sample handling, and the use of computerized data reduction, recording, and reporting.

5. ICP does not require electrodes and thus realignment problems are not encountered.

Negative

1. Because of the high temperature involved, the heavier elements produce complex spectra that will likely cause serious line interference, and it will be necessary to use spectral apparatus having a high resolving power to minimize this. Recourse may also be necessary to dilution and/

or to preliminary separations. Boumans[31] has described a project designed to produce a new tabulation of spectral interferences for ICP spectra which will enable the user, either manually or by means of a computer program, to select a set of the most sensitive lines that are interference-free under the conditions dictated by the spectral apparatus used, and the analytes.

2.   The presence of >1% dissolved solids is undesirable because it may cause clogging of the nebulizer system. This may be overcome by dilution, but this in turn may reduce the concentration level of some trace elements below the limit of detection (putting the sample into solution in the first place may do this also). Aspiration systems and torches capable of accommodating solutions having higher concentrations of dissolved solids have been developed, but clogging remains a significant limitation.

3.   The more complex and time-consuming sample preparation required to produce solutions offers greater opportunity for contamination at the trace element concentration level, and more rigorous (and costly) control of the procedure may be required.

4.   Volatile elements may be lost during the dissolution of the sample.

5.   Emission by major constituents present in the solution (>1000 ppm) may cause "stray light" which will affect the background level in samples but not in standards, unless the latter are matched to the sample, a time-consuming and not always feasible step.

6.   Recent studies have shown that the plasma plume is not spatially homogeneous in terms of temperature or excitation energy, and thus the optimum height above the torch orifice for observation of the emission will differ for each analytical line. Significant changes in matrix composition may cause temperature variations to occur, and the optimum position could vary from sample to sample, making difficult the determination of corrections.

The relative performance of ICP-ES and AAS has been studied by Ediger and Wilson.[32] They found that the ICP detection limits for Ba, B, Cd, Mn, P, Ti, W, U, V, and Zr were for the most part as good or better than those for flame AAS, but less than those obtainable with AAS using a graphite furnace. Chemical and ionization interferences are much less for ICP, but background and spectral interferences were, as one might expect, more prevalent with ICP. The precision attainable with AAS is considered to be superior, and the use of AAS is more practical for the routine determination of 5–10 elements in a large number of samples. ICP is capable of true simultaneous multielement determinations, with the right equipment, but the relative costs of equipment are an important consideration. The AAS nebulizers, which require a higher gas flow, are less prone to clogging than the ICP nebulizers, and can thus tolerate much higher concentrations of dissolved solids. To conclude, each technique has its own particular advantages over the other, and a selection should be based on a consideration of all of these.

Berman and McLaren[33] have reviewed the application of ICP to geochemical analysis and consider its sensitivity to be intermediate between that of flame and graphite furnace AAS when the more common pneumatic nebulization of the sample is used. The use of ultrasonic nebulization results in an improvement of at least one magnitude in ICP detection limits, but requires desolvation of the sample aerosol.

It has been mentioned previously that the multielement capability of ES may be utilized in either a simultaneous or a sequential mode. Direct reading systems for both ES and ICP-ES suffer from the fact that only a limited number of wavelength positions are available for analyte and background measurements, and the analyst is likely to become "locked in" to his or her initial choice of instrumental array. With such variable sample matrices as those encountered in the analysis of geological samples, the limited number of wavelength sites from which to choose in order to overcome a particular interference presents a distinct problem because of the difficulty usually encountered in modifying the instrumental configuration. The best of both worlds may possibly be attained as a result of the development of ICP-ES instrumentation primarily for operation in the sequential mode, but which can be programmed to measure selectively any series of wavelengths in a sequential manner. Although the speed of operation is less than that attainable in the simultaneous mode, there is a greater degree of flexibility that is worth consideration.

The use of a selective extraction technique (19% Aliquat 336-MIBK) to eliminate interferences from major elements in a variety of geological materials and increase the sensitivity of such trace elements as Ag, Au, Bi, Cd, Cu, Pb, and Zn is reported by Motooka et al.[34] The determination of the rare earth elements is of much interest to petrologists and geochemists, and Brokaert et al.[35] have studied the application of ICP to their determination in rocks and in such minerals as monazite and bastnaesite. They found it to be very satisfactory, with both photographic and photoelectric recording instruments, for rare earths present at very different concentration levels in the samples (e.g., Eu ranging from approximately 5 ppm in a carbonatite quartz rock to about 500 ppm in monazite). Minor and trace elements in silicate rocks have been determined by Uchida et al. following acid dissolution of the sample in a sealed Teflon vessel (16 hr at room temperature for 0.5 g) and removal of Si by subsequent volatilization to avoid the interference effect of varying solution viscosity. Good results were obtained for Ba, Cr, Cu, Li, Sc, Sr, V, and Zr, with those for Co and Ni being less satisfactory.[36] The use of a hydride preconcentration of As (see Sec. 9.4.2) and direct nebulization into an ICP has been reported,[37] and this technique should be applicable to the determination of Sb, Bi, Ge, Te, and other hydride-forming elements as well.

The above are but a few examples of the application of ICP-ES to the analysis of rocks and minerals. The current state of the art is readily determined from the comprehensive reviews[15-17] referred to at the beginning of this section.

12.2   NEUTRON ACTIVATION ANALYSIS

Neutron activation analysis (NAA) is a well-established technique for the analysis of geological materials, and the principle and applications have been described in several texts and review papers,[38-43] notably the papers written for a short course on NAA in the geosciences.[44] In his overview of NAA in geochemistry, Haskin[45] has outlined the advantages and disadvantages of the technique and provided historical perspective to the progress that has occurred since the discovery of the neutron in 1932 and the enunciation of the principles of NAA by von Hevesy and Levi in 1936. As noted by Haskin, most of the advances during the past decade have been based on major improvements in technology (detectors, multichannel analyzers, and computers), and most have been made by researchers developing better methods to obtain data, as opposed merely to demonstrating a new capability.

The important role of NAA in studying small and precious geological samples such as meteorites and lunar samples has been discussed by Haskin[45] and Frey;[46] indeed, Smales[47] found that NAA was the most common technique used in determining the elemental composition of lunar rocks. Although NAA has some application in major element analysis of rocks and minerals, especially in the determination of the principal constituents oxygen and silicon, some major elements, such as magnesium and calcium, are not readily determined by *instrumental* NAA, and thus NAA is not competitive with XRF as an analytical technique for major element analysis. Rather it should be regarded as complementary to XRF because of its application to the determination of trace elements. A useful aspect of this application is the flexibility to use *instrumental* NAA for samples with relatively high trace element contents and to utilize *radiochemical* NAA for samples with relatively low trace element contents.[46]

12.2.1 Major and Minor Element Determination

An interesting example of the determination by NAA of major and minor elements in small, valuable samples, using a 14 MeV neutron generator and an isotopic neutron source (^{252}Cf), is found in the analysis of lunar samples as described by Janghorbani et al.[48] The results yield a satisfactory total, and there is very good precision (e.g., 98.6 ± 0.6%, 99.6 ± 0.7%), but these are not sufficient in themselves to prove the quality of these determinations. Morgan and Ehmann[49] have provided evidence that the method is capable of producing accurate values with a relative standard deviation of <1%. A comparison of the accepted value for Si in AGV-1, 27.58%, and that for BCR-1, 25.47%, with the NAA values of 27.71 and 25.69%, respectively, shows that the accuracy of the technique is as acceptable as its precision.[50] It is pertinent to note that the method is capable of the direct determination of oxygen, an element that is very difficult, if not impossible, to determine in geological materials by other means.

12.2.2 Trace and Ultratrace Element Determinations

There are two distinct routes employed in the determination of trace and ultratrace constituents by NAA (Ref. 51, pp. 270–294). The first involves a series of chemical processes to isolate the element or group of elements of interest; these are almost always carried out following the exposure of the sample to the neutron flux. This technique has been used for many years and has been reviewed extensively.[52, 53] The separation techniques include classical chemical group precipitations, coprecipitation, and ion exchange, among others. An example of this approach is shown in Table 12.4, reproduced from Ref. 51, p. 283. The method is capable of achieving very low detection limits but is labor intensive, open to manipulative errors, and is most suited for those nuclides with fairly long half-lives.

The second approach to trace NAA work requires the introduction of the irradiated sample into a counting device capable of resolving the radiation on the basis of its energy (similar to ED-XRF detectors). The development of very sensitive Ge(Li) gamma-ray spectrometers in the last decade and the interfacing of them to computers for spectral stripping and other data manipulation techniques has enabled analysts to avoid the longer chemical separation procedures without sacrificing sensitivity. Reeves and Brooks have given a list of the detection limits attainable using this technique (Ref. 51, pp. 287–289). Similar lists are also given for both thermal and fast-neutron activation (Ref. 54, pp. 284–326).

TABLE 12.4 Multielement Analysis with Chemical Separations and γ Spectrometry

Elements	Sample	Notes
Lanthanides	Rocks, minerals	Alkali fusion, group separation; lanthanides separated as hydroxides for Ge(Li) γ spectrometry
Al, As, Au, Ba, Ca, Cd, Co, Cr, Cs, Cu, Fe, Ga, Hf, Hg, In, K, Mo, Na, Ni, Np, Pa, Rb, Re, Sb, Sc, Sr, Ta, W, Zn, Zr, and 12 lanthanides	Rocks	Eight elements determined instrumentally before separation; others by acid dissolution and separation into six groups of 2–20 elements for Ge(Li) γ spectrometry
Ag, As, Au, Ba, Br, Cl, Co, Cr, Cs, Cu, Fe, Ga, Hf, In, K, Ni, Mn, Rb, Sb, Sc, Se, Sr, Ta, Zn, and 12 lanthanides	Lunar fines, basalt	Alkali fusion; separation into 12 groups; Ge(Li) and NaI(Tl) γ spectrometry
As, Au, Co, Cs, Ga, Ge, Hg, Mo, Os, Re, Sb, Sc, Se, Tl, Zn,	Rocks, chondrites	Separation into groups for Ge(Li) γ spectrometry, using solvent extraction, ion exchange, distillation
Au, Cu, Ga, La, Mn, Sb	Glass	Acid dissolution, Ge(Li) γ spectrometry after separation of Na, Ta on HAP
Cr, Cs, P, Rb	Silicates	Alkali fusion; hydroxides precipitated; Rb, Cs, Cr by Ge(Li) γ spectrometry on filtrate
Lanthanides (12)	Lunar samples	Separated as hydroxides, converted into oxalates for Ge(Li) γ spectrometry
Ba, Cs, Rb, Sr, and 8 lanthanides	Ultramafic rocks	Alkali fusion; Rb, Cs precipitated as tetraphenylborates; ion exchange and solvent extraction gives remaining elements in three fractions
Au, Ir, Os, Pd, Pt, Ru	Geological materials	Alkali fusion; noble metals extracted on to chelating resin for Ge(Li) γ spectrometry
Ag, Au, Bi, Br, Cd, Co, Cs, Ga, Ge, In, Ir, Ni, Rb, Re, Sb, Se, Te, Tl, U, Zn	Meteorites, lunar and terrestrial samples	Alkali fusion; extensive chemical separations for NaI(Tl) γ counting; some β and X-ray counting also

Reproduced with permission of John Wiley & Sons, Ltd. (Ref. 51, p. 283).

Several contributors to Ref. 44 have discussed the relative merits of the instrumental and radiochemical routes to NAA. A tabulation of the analytical and instrumental problems associated with the determination of individual elements by instrumental NAA is given in that publication (Ref. 44, p. 120). Although Haskin[45] has suggested that the need for the radiochemical approach may decline as further technological refinements occur in instrumentation, the radiochemical route is still a necessary method for a large group of elements (Ref. 44, p. 136).

The success of the delayed neutron counting method for the determination of uranium has been described by numerous authors.[55–58] Bowie et al.[55] refer to three factors that make delayed neutron counting preferable for the determination of U: (1) *high specificity*, there being only slight interference from Th; (2) *high sensitivity*, capable of detecting 0.03 μg U with a flux of 5×10^{12} n/cm^2-sec; and (3), *high precision*, with a relative standard deviation of 1% at the 50 μg level largely independent of matrix. In fact, the slight Th interference mentioned previously can be eliminated via measurement of, and correction for, the Th content using a fast neutron flux.

Cumming[56] has used the same procedure to determine Th. The sensitivity for Th is significantly less than that for U, but Cumming was able to determine Th at the parts per billion level. Although the precision was not as good as that achieved for U, Cumming states that a similar precision for Th can be obtained with a reasonable number of repeat irradiations, if desired.

12.2.3 Sources of Error

As in the case for any other method of analysis, NAA is subject to some errors. Among the more common sources of error are:

1. Differences in counting geometry between samples and standards
2. Differences between the flux experienced by the samples and the standards (flux inhomogeneity)
3. Matrix differences between the standards and samples so that the attenuation of flux is different for each
4. Spectral overlap in the gamma-ray spectrum
5. Errors in manipulation if chemical separations are performed

An evaluation of error in the application of NAA to the analysis of rocks has been presented by Rosenberg[59] who investigated it with respect to the determination of the lanthanide elements in the U.S. Geological Survey standard rocks W-1 and G-1. He has estimated the average precision found to be 11.6%, which includes 5.2% due to flux inhomogeneity,

2.5% due to counting geometry differences, and 2.5% due to matrix effects.

12.2.4   Advantages

The advantages of NAA include its very low ultimate sensitivity for most elements, the possibility of the nondestructive analysis of very small samples, a minimum of sample preparation (and hence of contamination), and, with the 14 MeV method, the possibility of doing major elements nondestructively, including oxygen.[60] In the case of the delayed neutron techniques for the determination of U, the system can be highly automated and results can be generated very quickly.

12.2.5   Disadvantages

The major disadvantages of NAA is the need to have a neutron source, either in the form of a highly radioactive species such as ^{252}Cf or from very costly sources such as nuclear reactors, cyclotrons, or the like. In addition, the samples are often radioactive for several months after irradiation, and provision for safe storage and disposal must be made.

12.3   FIRE ASSAYING

12.3.1   Applications

Fire assaying is widely used for the determination of Au, Ag, and the Pt metals in all types of samples ranging from exploration silts, drill cores, and chip samples through to the finished bars poured in the refinery. It is also used as a preconcentration mechanism to concentrate the noble metals from large samples (in excess of 30 g if necessary) into either a Cu, Pb, or Sn button, a Cu–Ni–Fe alloy, NiS, or an Ag prill (pellet) for subsequent analysis by chemical, AAS, ES, or NAA techniques.

The sample is mixed with specific fluxes in a crucible and the mixture fused. The fluxes are chosen to provide a slag phase which dissolves most of the base metals and other gangue materials and, usually, a molten Pb phase which acts as a collector for the noble metals. The Pb button is separated from the slag, and the Pb is oxidized by heating in a cupel in a stream of air to leave an Au–Ag prill which will also contain other precious metals. The Au, Ag, and other metals can then be determined in the prill by any of several methods.

12.3.2 General

Fire assaying has been used since ancient times for the isolation of Au and Ag, and there are biblical references to crucible, fining pot (cupel), furnace, and refining of silver and "trying" gold by fire in several different places.[61] Wertime has pointed out that the oldest evidence of silver refining is cupel buttons found at Mahamatlar and dated to be from the late third millenium.[62]

The surprisingly high quality of silver refining and of silver assaying in the Middle Ages is pointed out by Watson in his discussion on Au and Ag trial plates deposited in the Pyx Stronghold of the Royal Mint.[63] The earliest silver plate (from 1279 A.D.) was made to a prescribed standard of 11 ounces, 2 pennyweights (dwt) per troy pound, or 925 parts per 1000. It was found to have an actual composition of 921 parts per 1000 of Ag, 61.9 of Cu, 8.1 of Pb, and 3.3 of Au. The first Au plate, from 200 yr later (1478 A.D.), was made to a prescribed standard of 23 carats, $3\frac{1}{2}$ grains (994.8 parts per 1000), and was found to contain 993.5 parts of Au, 5.15 of Ag, and 1.35 of Cu and other base metals.

The determination of Au, Ag, and other precious metals by fire assaying techniques is still common today, although very few texts have been published on the subject since about 1945. Two of the standard references available are those by Bugbee[64] and by Shepard and Dietrich.[65] A comparison of these earlier books with the work of Haffty et al.,[61] or with the methods used in any operating fire assay laboratory, will show that no significant changes in the fire assay techniques have occurred, with the exception of the introduction of collection schemes other than the classical Pb one because Os, Ir, and Ru are not alloyed with the Pb button. Some equipment has been improved, however, of which electronic balances and electrical furnaces are the most obvious examples. Other methods of analysis for the determination of the precious metals are, of course, available.[66-68] The discussion following will be based on the more recent publication by Haffty et al.[61] and will be restricted to the common Pb collection scheme. For those interested in the other collection techniques, reference should be made to the work of Beamish and Van Loon.[66-68]

12.3.3 Fluxes

The seven fluxing materials commonly used in fire assaying are:

Na_2CO_3.   A powerful basic flux which forms alkali sulfides, silicates, and aluminates in the fused mixture. It can be considered to be an oxidizing and desulfurizing agent.

PbO (litharge).   A less basic flux than Na_2CO_3, it also acts as an oxidizing and desulfurizing agent. When PbO is reduced in the fluid fusion, the resulting Pb acts as a collector for the precious metals.

SiO₂ (silica).   A strongly acidic flux which forms silicates with the metal oxides. It is added if the ore is low in SiO_2 to give a more fluid melt and to protect the pot from attack by PbO.

$Na_2B_4O_7$ (borax or sodium tetraborate).   Also a strongly acidic flux, capable of dissolving (or fluxing) virtually all metal oxides. It also lowers the fusion temperature of all slags.

CaF_2.   Increases the fluidity of the melt and is used when the Al content of the sample exceeds 1%.

Flour.   Provides a source of C to act as a reducing agent for the reduction of PbO to the necessary Pb metal.

KNO_3 (niter).   Decomposes in the fusion to provide a source of oxygen to oxidize the sulfides in highly reduced sulfide ores.

A guide to the fluxing of various ores likely to be encountered is given in Table 12.5 This has been drawn directly from a more detailed tabulation to be found in Ref. 61 and should not be used to the exclusion of this latter, nor of the use of Refs. 64 and 65, or any other basic text on fire assaying.

One of the first steps in fire assaying is to determine the reducing potential (RP) of the sample by means of a preliminary fusion with a flux as shown in Table 12.5. For example, if 3 g of the sample are fused with a given flux and a Pb button weighing 15 g is obtained, the RP of the ore is $15 \div 3 = 5$. The RP values for several common minerals and reagents are listed in Table 12.6 (taken from Ref. 61), as well as the oxidizing potentials (OP) for several species (an OP of 4.2 for KNO_3 means that 1 g of KNO_3 will oxidize 4.2 g of Pb).

As an example of the use of these values, consider a 15 g sample of pyrite (RP = 11) which is to be analyzed for Au and Ag. The amount of KNO_3 required to ensure that a 30 g Pb button will result is calculated as follows:

Reducing effect of sample	$15 \times 11 = 165$ g Pb
Weight of Pb button required	30 g
Ore equivalent to be oxidized	$165 - 30 = 135$ g
Amount of KNO_3 required	$135 \div 4.2 = 32.1$ g

The 15 g sample used above is close to the $\frac{1}{2}$ assay ton weight used in numerous laboratories on the North American continent, and no changes in the charges given in Table 12.5 are necessary if this weight is used.

TABLE 12.5 Guide to the Fluxing of Selected Ores (Adapted from Haffty et al., Ref. 61[a]).

Sample Type	Sample Weight	Charge (g)						
		Na_2CO_3	PbO (Litharge)	SiO_2 (Silica)	$Na_2B_4O_7$ (Borax)	CaF_2	Flour	KNO_3 (Niter)
Preliminary fusion for RP	3	10	46	3	1	0	0	0
Basalt	15	30	35	4–8	35	1	3.2	0
Copper sulfides	15	20	100	10	5	1	0	7.5
Dolomite	15	30	34	6	35	0	3.4	0
Fluorite rock	15	25	40	15	8	0	3.0	0
Galena (RP = 3.41)	15	20	50	5	3	0	0	5.3
Hematite	15	25	60	12–15	7	1	3.8	0
Kaolin rock[b]	15	20	50	12	10	5	2.8	0
Magnetite	15	25	50	15	8	4	4.0	0
Manganese ores	15	30	35	6–12	35	1	3.2	0
Pyrite (RP = 10.56)	15	30	60	12	10	0	0	30
Pyrrhotite (RP = 5.89)	15	35	70	10	10	0	0	15
Quartz	15	20	40	0	3	0	3.0	0
Rhyolite	15	20	50	1	3	1	2.8	0
Black sand	15	30	40	20	8	5	2.5–3.8	0
Sphalerite (RP = 7.00)	15	30	90	12	10	0	0	18
Sulfides (massive) (RP = 8.07) twice for 15 g	7.5	35	70	12	10	0	0	7.0
Tuff	15	20	50	1	3	1	2.8–3.0	0

[a] With permission of the authors and the U.S. Geological Survey.
[b] Also add 10 g K_2CO_3.

TABLE 12.6 The Reducing and Oxidizing Potentials of Selected Minerals and Reagents (from Haffty et al., Ref. 61[a]).

	Mineral or Reagent	RP	OP
	Flour	10–11	—
FeS_2	Pyrite	11	—
PbS	Galena	3.4	—
Cu_2S	Chalcocite	5	—
$FeAsS$	Arsenopyrite	7	—
Sb_2S_3	Stibnite	7	
$CuFeS_2$	Chalcopyrite	8	
ZnS	Sphalerite	8	
FeS	Pyrrhotite	9	
Fe	Metallic iron	4–6	
C	Carbon	18–25	
Fe_2O_3	Hematite	—	1.3
MnO_2	Pyrolusite	—	2.4
Fe_3O_4	Magnetite	—	0.9
KNO_3	Niter	—	4.2
	Magnetite-ilmenite	—	0.4–0.6

[a] With permission of the authors and the U.S. Geological Survey.

The assay ton weight (29.166 g) is such that if a 1 assay ton sample is taken, the weight of the resulting Ag prill in milligrams is equivalent to the number of troy ounces of Ag in 1 ton of 2000 pounds avoirdupois of the original material from which the sample was taken. The result is reported in troy ounces per ton. In countries which use the metric system, the result is reported in parts per million (1 oz/ton = 34.286 ppm).

12.3.4 Fusion

The fusion of the mixture of sample and fluxes in the crucible is carried out in an electric furnace at a temperature of about 900°C for 15–20 min, then at 1000°C for another 15–20 min. If a significant amount of KNO_3 has been added or if the sample is high in iron oxide, there is a chance that the fusion may boil over, and a close watch should be kept. If it does start to froth, the furnace door should be opened to cool the melt slightly.

If chromite ores are to be assayed, the sample should be fused in five 3 g lots, in separate pots, each with a charge of 30 g Na_2CO_3, 35 g PbO, 10 g SiO_2, 30 g $Na_2B_4O_7$, 1 g CaF, and 3.5 g flour. The five Pb buttons can then be combined in the scorification step (see Sec. 12.3.5) prior to cupellation.

When the fusion is complete, the crucibles are removed from the furnace and the melts poured into conical iron moulds in which the heavier molten Pb sinks to the bottom of the cone beneath the slag. When the mould has cooled sufficiently for the melt to have solidified, the Pb button is separated from any slag adhering to it by pounding the button into a cubic shape. It is now ready for scorification (used if the button is too large, two or more are to be combined, or when samples high in Cu or Ni have been fused and it is necessary to purge the Pb button of them) or cupellation.

12.3.5 Scorification

Scorification is an oxidizing fusion designed to eliminate base metals from the Pb button and/or to reduce the amount of elemental Pb in the button to the desired weight of about 30 g. It is carried out with a mixture of $Na_2B_4O_7$ and SiO_2 as fluxes in a shallow dish (scorifier). The scorification proceeds at a temperature of 950°C with the furnace door open to admit air for the oxidation of the Pb to PbO. The latter is dissolved in the slag together with base metals such as Cu and Ni which may have been present in the original button.

12.3.6 Cupellation

Cupellation is the process used since antiquity for the separation of Au and Ag from Pb. The process consists of rapidly oxidizing the Pb to PbO in a cupel at a temperature of about 840°C in a furnace with the door left open to supply air for the oxidation. The Pb buttons are placed in preheated cupels which absorb most of the PbO formed during the cupellation process and the Ag, Au, and other precious metals are left as a prill at the center of the cupel.

There are, of course, numerous precautions to be taken. When significant amounts of Cu, Ni, and other transition metals are present in the Pb button, precious metals may be lost to the cupel, and an intermediate scorification is necessary. If the temperature of the cupellation is too low, the PbO may form a crust over the button which is then said to be "frozen." Alternatively, if the cupellation temperature is too high, Ag may be volatilized. The maintenance of the correct temperature for cupellation requires practice and experience, especially if samples with high Ag content are to be analyzed.

12.3.7   Parting

Once the Au–Ag prill has been weighed, the next problem is to separate the Ag from the Au so that the amount of Au and, by difference, the amount of Ag can be calculated. This "parting" process is accomplished by treating the weighed prill with a dilute (1:9) HNO_3 solution which will dissolve out the Ag present, leaving the Au behind. The resultant Au prill is rinsed to rid it of $AgNO_3$, dried, annealed at 800°C for 5 min, cooled, and weighed.

It is necessary to ensure that the Ag:Au ratio is greater than 5:1. If this is not the case, the Ag will not be parted from the Au, and the sample must be *inquarted*, that is, a weighed amount of Ag must be added to the sample, preferably at the fusion step.

12.3.8   Advantages

The most obvious advantage of fire assaying is that it uses a much larger subsample than do most other techniques (15–30 g or more), and this is important in reducing the subsampling errors encountered in many spotty gold ores and other noble metal samples. The technique is also an efficient means of separating the noble metals from gangue minerals and concentrating them either in the base metal button or in the precious metal prill for subsequent determination by chemical or instrumental methods. This is used in the determination of Os and Ru, for example, in which the Pb button is decomposed for further analytical treatment because of the likelihood of loss of these volatile elements during cupellation.

12.3.9   Disadvantages

The major disadvantages of fire assaying are that it is labor intensive, uses relatively expensive reagents and supplies, generates a lot of heat, and requires a significant amount of training of new employees before they are able to produce satisfactory results on a routine basis. In spite of these disadvantages, it is a technique that has proven its value over thousands of years and it is still widely used.

12.4   ELECTRON MICROPROBE ANALYSIS*

12.4.1   General

The first edition of this book did not include an account of electron microprobe analysis, but in view of the widespread use of the method

---

* This section was written by Dr. A. G. Plant, Head of the Mineralogy Section, Geological Survey of Canada.

and in recognition of the fact that most mineral analyses published in the geological literature are now obtained by this means, such an omission is no longer justified. The instrument resulted from the combination of X-ray spectroscopy with the techniques of electron optics, and the practical development of electron microprobe analysis was undertaken in 1948 as a Ph. D. project by R. Castaing at the University of Paris.[69] The name derives from the essential feature whereby a fine electron beam, directed at the point to be analyzed, generates X-rays characteristic of the elements present in the sample. The wavelength (or energy) of the lines in the X-ray spectrum identifies the elements present and provides a qualitative analysis. A quantitative analysis may be obtained by comparing the intensities of the X-ray lines from the sample with those from standards of known composition and applying instrumental and matrix corrections (see Sec. 12.4.3). The principles of electron microprobe analysis, including electron optics, X-ray generation, and X-ray spectroscopy, have been described in several textbooks and many review papers,[70-78] and reference to these will lead the interested reader to a more complete understanding of the subject, as well as to the related technique of scanning and transmission electron microscopy.

The first commercial microprobe appeared in 1958, and it has since established itself as an indispensable tool in mineralogical studies notwithstanding the fact that valency states cannot be distinguished quantitatively and that some constituents such as H_2O and CO_2 cannot be determined. The introduction and availability of the instrument has probably had as great an effect on geological science as the petrological microscope did in the nineteenth century.

An early review of the applications of the electron microprobe to mineralogy was that of Agrell and Long[79] who reported data on three investigations: (1) exsolution in pyroxenes, (2) zoned spinel, and (3) nickel content of kamacite and taenite. These examples demonstrated 20 years ago the fundamental attribute of the instrument, whereby a quantitative mineral analysis may be obtained *in situ* from an area as small as 1 μm in diameter. In addition, compositional data may be obtained from line traverses across minerals and mineral assemblages, and also by scanning the electron beam over a chosen area and displaying the X-ray intensity of a selected element by modulation of the z-axis of an oscilloscope. In general, these data may be obtained nondestructively, and the applications of these techniques have revolutionized our knowledge of the chemistry of minerals and of the subtleties of Nature.

Before the advent of the microprobe, chemical data could only be obtained for minerals which could be prepared as concentrates suitable for chemical analysis. In effect this limited our knowledge to minerals that occurred in sufficient quantities for, and were amenable to, concentration and separation. Aside from their time-consuming preparation, these separates suffered from two serious disadvantages: a chemical analysis of such a mineral separate represents the *average* composition of a large number of mineral grains and suppresses completely the *inter* and *intra* grain variation; also, a bulk mineral separate will invariably contain contaminants, either as minute inclusions within the grains or as discrete grains not rejected by the separation process. Frequently these contaminant grains affect the trace and minor element content of the analysis, and are a source of error in comparative studies between chemical analyses and microprobe analyses. These disadvantages are overcome with the electron microprobe, and the ability to obtain *in situ* mineral compositional data forms the basis of all applications to mineralogy.[80]

12.4.2 Instrumentation

In its simplest form, the electron microprobe consists of an electron-optical system which focuses an electron beam onto the surface of a sample, an optical microscope which allows the area of interest to be selected, a stage on which the sample and standards are mounted and which permits a selected point to be positioned in the electron beam, and X-ray spectrometers for the measurement of the X-ray spectra generated by the electron beam. The references[70–78] cited earlier discuss the principles involved in the design and operation of all of these components of the electron microprobe.

In common with other analytical techniques, electron microprobe analysis has enjoyed the benefits made possible through major advances in technology, and during the past decade the widespread use of the minicomputer and the introduction of the energy dispersive spectrometer have been significant. Rucklidge[81] has discussed various aspects of the use of a minicomputer both for operation and automation of the microprobe itself and for the processing of the analytical data. The interest in instrument automation has been directed particularly at the operation of wavelength-dispersive spectrometers so that they could be programmed to scan through a selected wavelength range or sequentially to preselected wavelengths, and proceed to measure the peak and background intensities on both the sample(s) and the standard(s) through the use of a computer-controlled stage. The development of sophisticated automated systems coincided with the introduction of the energy-dispersive spectrometer with which the complete X-ray spectrum from a sample, both the characteristic lines and the continuum, is recorded simultaneously, although in most systems elements of an atomic number less than 11 are excluded. The various components in an energy-dispersive spectrometer system, including the lithium-drifted silicon detector, and their principles of operation have been described in several texts,[70, 77, 82] and these authors have discussed the advantages and disadvantages of the system. Much effort has been expended in recent years, both by manufacturers and by individual analysts, to the development of computer programs for the conversion of energy-dispersive spectra to quantitative analyses. Although the many advantages of the energy-dispersive spectrometer are counterbalanced to a degree by its inherently poorer resolution compared to that of the wavelength spectrometer, it has been demonstrated that an energy-dispersive spectrometer can be used to obtain quantitative mineral and rock analyses with an accuracy and precision comparable to those obtained with a wavelength-dispersive spectrometer.[82–84]

12.4.3 Quantitative Analysis

In addition to the optimization of instrumental parameters, the care and thought given to a number of other factors are of importance in obtaining accurate quantitative analyses.[70, 71, 76, 77] These include (1) sample preparation (reliable quantitative data can only be obtained when specimens and standards have a flat, well-polished surface free from relief); (2) surface coating (electrically insulating specimens, such as most minerals, require a conducting film, usually of vacuum-evaporated carbon, and the careful control of film thickness has been discussed by Sweatman and Long[76]); and (3) choice of standards. As with all comparative instrumental techniques, electron microprobe analysis requires the use of well-characterized standards that ideally possess certain properties.[75, 76, 85] The following quotation from *Short Course in Microbeam Techniques* (Ref. 75, p. 179) summarizes the situation:

> Acquisition of a good collection of standards is one of the most pressing and difficult tasks facing an analyst who is attempting to establish a new laboratory; it is an on-going task for analysts in well established laboratories. Ideally, a standard should be close in composition to the unknown, of precisely known composition, absolutely homogeneous, stable beneath an intense electron beam, and not subject to atmospheric alteration. The material should be available in coarse particles that are easily mounted and polished. Another useful attribute is cathodoluminescence. Normally it is impossible to obtain standards which have all these properties and compromises must be made.

Castaing[69] showed that to a first approximation the intensity of a characteristic X-ray line is directly proportional to the weight concentration of the element from which it originates. However, this relationship depends in a complicated manner on the composition of the sample, and matrix corrections are used to convert sample intensity-to-standard intensity ratios into concentrations.[70, 72, 86] The measured intensities must first be corrected for characteristics of the counting system as well as for background, the chief source of which is the X-ray continuum.[70] These corrected intensities are then converted to weight concentrations by applying matrix corrections which take account of three effects known as the atomic number (Z), absorption (A), and fluorescence (F) effects. This procedure, known as the ZAF method, is described in considerable detail by Reed[70] who provides a critical assessment of the equations currently in use. Because of the complexity of the equations, use of a computer program is mandatory.[86, 87] Using these procedures, it is possible to obtain analyses with an accuracy of at least ± 2 relative percent.[70] Limits of detection depend upon a number of factors, and Reed[70] suggests that it is typically 100 ppm. In most routine analyses it is probably higher than this, though in favorable circumstances and with a counting time of 1000 sec the ultimate limit of detection using a wavelength-dispersive spectrometer is 10 ppm.[70] Detection limits are considerably higher in the case of the energy-dispersive spectrometer.[83]

As the ZAF correction method is mathematically complex and computer data reduction was much less generally available in the early 1960s, simpler approaches known as the empirical or α-factor method have been developed to correct for matrix effects.[88-92] The correction coefficients are determined empirically or by calculation from first principles and apply only to the instrumental conditions (accelerating voltage and X-ray take-off angle) for which they were determined. The method has enjoyed considerable success in the analysis of geological samples, particularly silicate minerals, and most of the thousands of mineral analyses obtained from the lunar samples were calculated by this method. It has also been applied to the analysis of heavy element matrices.[93]

Although the electron microprobe is universally used for the analysis of minerals, it may also be used for the analysis of rocks. Several methods have been suggested for the preparation of representative samples, such as pressed powder, flux fusion, and direct fusion. Each of these methods has particular advantages and disadvantages, and the subject has been summarized by Schimann and Smith.[84] They describe the use of an optical furnace to produce direct-fusion glasses from a wide range of rock compositions. The use of a larger diameter electron beam, generally known as a defocused or broad beam, in combination with an energy-dispersive spectrometer is desirable for the analysis of prepared rock samples or of fine-grained matrices in polished thin sections.[94, 95]

References

1. P. W. J. M. Boumans, *Philips Tec. Rev.*, **34**, 305 (1974).

2. I. Rubeška, *Analusis*, **4**, 314 (1976).

3. V. A. Fassel, *Anal. Chem.*, **52**, 340A (1980).

4. B. F. Scribner and M. Margoshes, "Emission Spectroscopy," in I. M. Kolthoff and P. J. Elving, Eds., *A Treatise on Analytical Chemistry*, Vol. 6, Part 1, Wiley-Interscience, New York, 1965, p. 3347.

5. M. Slavin, *Emission Spectrochemical Analysis*, Vol. 36, Chemical Analysis Series, P. J. Elving and I. M. Kolthoff, Eds., Wiley-Interscience, New York, 1971.

6. A. L. Gray, "Optical Emission Spectroscopy," in T. Mulvey and R. K. Webster, Eds., *Modern Physical Techniques in Material Technology*, Oxford University Press, London, 1974, p. 232.

7. J. D. Winefordner, *Trace Analysis*, Vol. 46, Chemical Analysis Series, P. J. Elving, J. D. Winefordner, and I. M. Kolthoff, Eds. Wiley-Interscience, New York, 1976.

8. L. H. Ahrens and S. R. Taylor, *Spectrochemical Analysis*, 2nd ed., Addison-Wesley, Reading, MA, 1961.

9. P. R. Barnett and E. C. Mallory, Jr., *Techniques of Water Resources Investigations of the United States Geological Survey*, Book 5, 1971, Chap. A2.

10. K. Govindaraju, *Analusis*, **2**, 367 (1973).

11. M. H. Timperley, *Spectrochim. Acta*, **29B**, 95 (1974).

12. H. Bennett, in A. W. Nicol, Ed., *Physicochemical Methods of Mineral Analysis*, Plenum, New York/London, 1975; Chap. 12.

13. R. E. Stanton, *Analytical Methods for Use in Geochemical Exploration*, Wiley, New York, 1976, Chap. 7.

14. R. D. Reeves and R. R. Brooks, *Trace Element Analysis of Geological Materials*, Vol. 51, Chemical Analysis Series, P. J. Elving, J. D. Winefordner, and I. M. Kolthoff, Eds., Wiley-Interscience, New York, 1978.

15. W. J. Boyko, P. N. Keliher, and J. M. Malloy, *Anal. Chem.*, **52**, No. 5, 53R (1980).

16. J. I. Dinnin, *Anal. Chem.*, **51**, No. 5, 144R (1979).

17. J. B. Dawson and B. L. Sharp, Eds., *Annual Reports on Analytical Atomic Spectroscopy*, Vol. 8, The Chemical Society, London, 1979.

18. K. Laqua, *Pure Appl. Chem.*, **49**, 1595 (1977).

19. E. W. Golightly and J. L. Harris, *Appl. Spectrosc.*, **29**, 233 (1975).

20. D. C. Bankston, S. E. Humphris, and G. Thompson, *Anal. Chem.*, **51**, 1218 (1979).

21. K. Govindaraju, G. Mevelle, and C. Chouard, *Anal. Chem.*, **48**, 1325 (1976).

22. J.-O. Burman, B. Boström, and K. Boström, *Geol. Fören. Forh.*, **99**, Part 2, 102 (1977).

23. J.-O. Burman and K. Boström, *Anal. Chem.*, **51**, 516 (1979).

24. V. A. Fassel, *Pure Appl. Chem.*, **49**, 1533 (1977).

25. P. Tschöpel, "Plasma Excitation in Spectrochemical Analysis," in G. Svehla, Ed., *Comprehensive Analytical Chemistry*, Vol. IX, Elsevier, Amsterdam, 1979, Chap. 3, pp. 173–293.

26. J. A. Corbett, Min. Research Laboratory, CSIRO (Australia), Invest. Rep. 121, Feb. 1977.

27. R. M. Barnes, Ed., *Applications of Inductively Coupled Plasma to Emission Spectroscopy* (1977 Eastern Analytical Symp.), Franklin Institute Press, Philadelphia, PA, 1978.

28. P. W. J. M. Boumans, *ICP Inf. Newsl.*, **5**, 181 (1979); from *Optica Pura y Aplicada*, **11**, 143 (1978).

29. *ICP Inf. Newsl.* **4**, No. 1, (1978); R. M. Barnes, Ed., Department of Chemistry University of Massachusetts, Amherst, MA 01003.

30. A. Montoser and J. Mortazavi, *Anal. Chem.*, **52**, 255 (1980).

31. P. W. J. M. Boumans, *Spectrochim. Acta*, **35B**, 57 (1980).

32. R. D. Ediger and D. L. Wilson, *Ato. Absorp. Newsl.*, **18**, 41 (1979).

33. S. S. Berman and J. W. McLaren, "Application of the Inductively Coupled Plasma to Geochemical Analysis," Preprint, *Proc. Geoanalysis 1978* (Symp. on the Analysis of Geological Materials, Ottawa, Ont., May 1978).

34. J. M. Motooka, E. L. Mosier, S. J. Sutley, and J. G. Viets, *Appl. Spectrosc.*, **33**, 456 (1979).

35. J. A. C. Brokaert, F. Leis, and K. Laqua, *Spectrochim. Acta*, **34B**, 73 (1979).

36. H. Uchida, T. Uchida, and C. Iida, *Anal. Chim. Acta,* **116**, 433 (1980).

37. R. C. Fry, M. B. Denton, D. L. Windsor, and S. J. Northway, *Appl. Spectrosc.*, **33**, 399 (1979).

38. G. E. Gordon, K. Randle, G. G. Goles, J. B. Corliss, M. H. Beeson, and S. S. Oxley, *Geochim. Cosmochim. Acta*, **32**, 369 (1968).

39. A. O. Brunfelt and E. Steinnes, *Anal. Chim. Acta*, **48**, 13 (1969).

40. J. Hertogen and R. Gijbels, *Anal. Chim. Acta*, **56**, 61 (1971).

41. A. O. Brunfelt and E. Steinnes, Eds., *Activation Analysis in Geochemistry and Cosmochemistry*, (Proc. of the NATO Advanced Study Institute, Kjeller, Norway, Sept. 1970), Universitetsforlaget, Oslo, 1971.

42. D. DeSoete, R. Gijbels, and J. Hoste, *Neutron Activation Analysis*, Wiley-Interscience, London, 1972.

43. J. W. Jacobs, R. L. Korotev, D. P. Blanchard, and L. A. Haskin, *J. Radioanal. Chem.*, **40**, 93 (1977).

44. G. K. Muecke, Ed., *Neutron Activation Analysis in the Geosciences*, Mineralogical Association of Canada, Spec. Publ., May 1980.

45. L. A. Haskin, "An Overview of Neutron Activation Analysis in Geochemistry," in G. K. Muecke, Ed., *Neutron Activation Analysis in the Geosciences*, Mineralogical Association of Canada, Spec. Publ., 1980, Chap. 1.

46. F. A. Frey, "Applications of Neutron Activation Analysis in Mineralogy and Petrology," in G. K. Muecke, Ed., *Neutron Activation Analysis in the Geosciences*, Mineralogical Association of Canada, Spec. Publ., 1980, Chap. 7.

47. A. A. Smales, "The Place of Activation Analysis in Geochemistry and Cosmochemistry," in A. O. Brunfelt and E. Steinnes, Eds., *Activation Analysis in Geochemistry and Cosmochemistry*, Universitetsforlaget, Oslo, 1971, p. 17.

48. M. Janghorbani, D. E. Gillu, and W. D. Ehmann, "Application of 14 MeV and Cf-252 Neutron Sources to Instrumental Neutron Activation Analysis of Lunar Samples, "in *Analytical Methods Developed for Application to Lunar Sample Analyses*, American Society for Testing and Materials, ASTM, STP 539, 1973, pp. 128–139.

49. J. W. Morgan and W. D. Ehmann, *Anal. Chim. Acta*, **49**, 287 (1970).

50. F. J. Flanagan, *Geochim. Cosmochim. Acta*, **33**, 81 (1969).

51. R. V. Reeves and R. R. Brooks, *Trace Element Analysis of Geological Materials*, Wiley, New York, 1978.

52. F. E. Beamish, K. S. Chung, and A. Chow, *Talanta*, **14**, 1 (1967).

53. R. H. Gijbels, *Pure Appl. Chem.*, **49**, 1555 (1977).

54. M. L. Parsons, in J. D. Winefordner, Ed., *Trace Analysis, Spectroscopic Methods for Elements*, Wiley, New York, 1976.

55. S. Bowie, M. Davis, and D. Ostle, *Uranium Prospecting Handbook*, Institute of Mineralogy and Metallurgy, London, 1972, p. 95.

56. G. L. Cumming, *Chem. Geol.*, **13**, 257 (1974).

57. S. Amiel, *Anal. Chem.*, **34**,1683 (1962).

58. N. H. Gale, *Radioactive Dating and Methods of Low Level Counting*, Atomic Energy Agency Vienna, 1967, pp. 431–452.

59. R. J. Rosenberg, *Accuracy in Trace Analysis: Sampling, Sample Handling and Analysis* (Proc. 7th IMR Symp.), National Bureau of Standards, Spec. Publ. 422, Vol. II, 1241, 1976.

60. A. Volborth, *Elemental Analysis in Geochemistry, A. Major Elements*, Elsevier, Amsterdam, 1969, Chap. 13.

61. J. Haffty, L. B. Rile, and W. D. Gross, *U.S. Geol. Surv. Bull.*, **1445** (1977).

62. T. A. Wertime, *Science*, **182**, 875 (1973).

63. J. H. Watson, "A Description of the Ancient Gold and Silver Trial Plates Deposited in the Pyx Stronghold of the Royal Mint," HMSO, London, 1962.

64. E. E. Bugbee, *A Textbook of Fire Assaying*, 3rd. ed., Wiley, New York, 1940.

65. O. C. Shepard and W. F. Dietrich, *Fire Assaying*, McGraw-Hill, New York, 1940.

66. F. E. Beamish, *The Analytical Chemistry of the Noble Metals*, Pergamon, New York, 1966.

67. F. E. Beamish and J. C. Van Loon, *Recent Advances in the Analytical Chemistry of the Noble Metals*, Pergamon, New York, 1972.

68. F. E. Beamish and J. C. Van Loon, *Analysis of Noble Metals, Overview and Selected Methods*, Academic, New York, 1977.

69. R. Castaing, Ph.D Thesis, University of Paris, 1951.

70. S. J. B. Reed, *Electron Microprobe Analysis*, Cambridge University Press, Cambridge, 1975.

71. J. V. P. Long, in J. Zussman, Ed., *Physical Methods in Determinative Mineralogy*, 2nd. ed. Academic, New York, 1977.

72. J. I. Goldstein and H. Yakowitz, Eds., *Practical Scanning Electron Microscopy*, Plenum, New York, 1975.

73. C. A. Andersen, Ed., *Microprobe Analysis*, Wiley, New York, 1973.

74. T. D. McKinley, K. F. J. Heinrich, and D. B. Wittry, Eds., *The Electron Microprobe*, Wiley, New York, 1966.

75. D. G. W. Smith, Ed., *Short Course in Microbeam Techniques*, Mineralogical Association of Canada, Toronto, 1976.

76. T. R. Sweatman and J. V. P. Long, *J. Petrol.*, **10**, 332 (1969).

77. D. R. Beaman and J. A. Isasi, *Electron Beam Microanalysis*, American Society for Testing and Materials, Spec. Tech. Publ. 506, 1972.

78. D. G. W. Smith and J. C. Rucklidge, *Adv. In Geophys.*, **16,** 57 (1973).

79. S. O. Agrell and J. V. P. Long, in A. Engström, V. Cosslett, and H. Pattee, Eds., *X-Ray Microscopy and X-Ray Microanalysis*, Elsevier, Amsterdam, 1960.

80. A. G. Plant, in D. G. W. Smith, Ed., *Short Course in Microbeam Techniques*, Mineralogical Association of Canada, Toronto, 1976.

81. J. C. Rucklidge, in D. G. W. Smith, Ed., *Short Course in Microbeam Techniques*, Mineralogical Association of Canada, Toronto, 1976.

82. D. G. W. Smith, Ed., *Short Course in Microbeam Techniques*, Mineralogical Association of Canada, Toronto, 1976.

83. A. C. Dunham and F. C. F. Wilkinson, *X-Ray Spectrom.*, **7, 50** (1978).

84. K. Schimann and D. G. W. Smith, *Can. Mineral.*, **18,** 131 (1980).

85. E. Jarosewich, J. A. Nelen, and J. A. Norberg, *Smithsonian Contrib. to the Earth Sci.*, **22,** 68 (1979).

86. G. Springer, in D. G. W. Smith, Ed., *Short Course in Microbeam Techniques*, Mineralogical Association of Canada, Toronto, 1976.

87. D. R. Beaman and J. A. Isasi, *Anal. Chem.*, **42,** 1540 (1970).

88. T. O. Ziebold and R. E. Ogilvie, *Anal. Chem.*, **36,** 322 (1964).

89. G. R. Lachance and R. J. Traill, *Can. Spectrosc.*, **11**, 43 (1966).

90. R. J. Traill and G. R. Lachance, *Can. Spectrosc.*, **11**, 63 (1966).

91. A. E. Bence and A. L. Albee, *J. Geol.*, **76**, 382 (1968).

92. A. L. Albee and L. Ray, *Anal. Chem.*, **42**, 1408 (1970).

93. G. J. Pringle, Geological Survey of Canada, Paper 79-1C, 111, 1979.

94. A. A. Chodos, A. L. Albee, and J. E. Quick, in R. E. Ogilvie and D. B. Wittry, Eds., *Abstracts of the 8th Int. Conf. on X-Ray Optics and Microanalysis*, (Boston, 1977).

95. R. A. F. Grieve and A. G. Plant, *Geochim. Cosmochim. Acta*, Suppl. 4, 667 (1973).

APPENDIX 1

STUDIES IN "STANDARD SAMPLES" FOR USE IN THE GENERAL ANALYSIS OF SILICATE ROCKS AND MINERALS. PART 6: 1979 EDITION OF "USABLE VALUES" (GSC PAPER 80-14)

Abstract

Reviews already published on the state of "standard samples" of silicate rocks and minerals, as well as of some samples of other materials that can be used as reference standards for the general analysis of silicate rocks and minerals, have been updated. Usable values of varying degrees of reliability are suggested for major, minor and trace constituents of nearly one hundred different samples. Presentation of the information has been improved with a view to making it more convenient for readers' use.

Résumé

On a mis à date les études sommaires déja publiées sur l'état d'échantillons de roches dits "standards" et de minéraux silicatés, ainsi de quelques échantillons d'autres matériaux dont on peut se servir comme standards de référence pour l'analyse génerale des roches et des minéraux silicatés. On suggère des valeurs utilisables à divers niveaux de certitude pour des composants majeurs, mineurs et en traces dans presque cent échantillons particuliers. La présentation des renseignements a été améliorée afin de la faire plus convenable pour l'usage des lecteurs.

INTRODUCTION

Increasing use of analytical instrumentation has created a demand for "standard samples" of rocks and minerals for use in calibrations. As a result, geological institutions in a dozen different countries have produced a variety of such materials. Other materials with comparable compositions (glasses, refractories, slags, etc.) have also found use as calibration standards in rock and mineral analysis. In testing

NATURE OF THE RAW DATA

Among the many collaborative analytical programs for proposed reference samples of rocks that have been conducted over the last 30 years, one feature in common has been the wide dispersal of results reported for essentially all constituents. Various possible explanations for this state of affairs have been proposed. Flanagan (1976a) has referred to the debate between those who hold that sample

408

new analytical methods and in the comparison of analytical results emanating from different laboratories, such reference materials can be an invaluable tool in the hands of the analyst. (The term "reference materials" is preferred to "standard samples" because the latter name suggests a greater degree of reliability than can realistically be attributed to the derived compositional values.)

Although the originators of rock reference samples have collaborated in providing analytical data for one another's samples, all have gone their separate ways in the selection and preparation of their materials, in their methods for assuring homogeneity and in the manipulation of the highly incoherent results reported by participating laboratories. Recent years have seen a growing trend toward exchange of information, with a view to bringing order out of this chaotic situation. Sessions devoted to the subject of reference samples of geological materials have formed part of several international conferences. An international journal, Geostandards Newsletter (K. Govindaraju, editor, c/o CRPG, 54500 Vandoeuvre-Nancy, France) has been established, as has also an International Study Group on Reference Materials (T.W. Steele, chairman, c/o National Institute for Metallurgy, Randburg 2125, South Africa).

This paper presents background information on a large number of geological reference materials and usable numerical values for the concentrations of many constituents in them. Also included is a discussion of possible reasons for the discrepancies in collaborative analytical data and of various schemes that have been proposed for resolving those discrepancies.

inhomogeneity is the main culprit and those who consider interlaboratory variability to be more significant. Other ideas put forward concern the use of inadequate subsampling procedures and the absence of information on statistical parameters of analytical methods used.

In an effort designed to compare the possible effects of inhomogeneity with those of interlaboratory factors, Flanagan (personal communication, 1979) has requested that all analyses of three new samples (USGS-BIR-1, -DNC-1 and -W-2) be performed in randomized sequence according to a specified experimental design – i.e. two subsamples from each of three randomly selected bottles. Results were then to be used in analysis-of-variance computations. Incomplete data on the three samples mentioned and on eight earlier ones (Flanagan, 1976b) suggest that inhomogeneity is rarely present to any significant extent, at least among the samples examined.

Ridley et al. (1976) reported experimental data emphasizing the effect of particle size on homogeneity. The validity of their conclusions, however, was limited because the work was done on a single sample in only one laboratory. Ingamells and Switzer (1973) studied the effects of particle size, mineralogical composition, and subsample weight on the scatter of results, concluding that a minimum sample weight ("the sampling constant") could be computed for the determination of each constituent in a particular sample within specified confidence limits. Jaffrezic (1976) demonstrated a linear correlation between particle size and variance when plotted on log-log paper. Maessen et al. (1976) suggested that evaluation of results would be facilitated if collaborating analysts were to provide sufficient statistical information on their analytical methods.

All of the foregoing proposals have merit. Unfortunately, their application in practice is rather difficult. Derivation of sampling constants for each constituent of one sample – let alone a set of two, three or six – would entail a great deal of additional work. Fortunately, most analysts know intuitively what constitutes a safe sample size, as revealed by the low incidence of evidence of inhomogeneity effects in those programs where analysis-of-variance computations have been used. More information on statistical parameters of analytical methods would of course be useful, but again hardly practical. Many contributors to collaborative programs, as well as compilers of the data, have been lax in the information provided on "method used". Too often they have merely listed the technique used in final measurement (XRF, AAS, etc.) in cases where the earlier steps in the procedure may have had a greater bearing on overall reliability than did the measurement technique.

Some programs have endeavored to overcome the homogeneity problem by preparing fused glass samples of controlled composition (e.g. de la Roche and Govindaraju, 1973b). The dispersion of analytical data on such glasses was found to be no better than on the more heterogeneous rock samples, thus pointing once more to the greater importance of interlaboratory factors. In programs involving two or more samples – generally of considerably different compositions – bias in results from a particular laboratory has generally been evident on all of the samples. Thus programs based on a single sample material may well fail to reveal useful information.

Imbalance of available compositional data for the various constituents of reference rocks has reflected the availability and extent of application of suitable analytical methods. Thus comparatively little is known about the concentrations of such "difficult" trace elements as Ag, Au, Bi, Br, Cd, Ge, I, the platinum group, some of the less abundant rare earths, Ta, Th, W, etc. Even with certain more

the sheer bulk of work required. Few rock programs, therefore, have attempted to organize controlled replicate schemes; those who have done so have found that many contributors fail to live up to the requirements. Inevitably, analytical results on rock samples have displayed greater dispersion than has been the case with ores.

Many procedures have been proposed for deriving "provisional", "preliminary", "tentative", "magnitude", "average", "preferred", "best", "proposed", "recommended", "certified", "attested" or "guaranteed" values, in most cases without definition of the adjective used. Many of the procedures have involved the mean of values remaining after rejection of those lying beyond one, two or three standard deviations from the mean of all the original data. Details of the system used by each originating agency are given below in the notes on the samples they have issued.

A common limitation of most of the above schemes is a failure to account for skewed distributions, a common occurrence with rocks, particularly for some trace elements. Christie and Alfsen (1977) have attempted to overcome the effects of skew by computing a "mode", using a data-transformation calculation, to produce a "Gamma Central Value". Ellis et al. (1977) introduced a "Dominant Cluster Mode" to overcome the effect of skewness. Abbey et al. (1979) proposed a "Moving Histogram Mode" and a "Moving Histogram Transformation Mode" as simplified approximations of the Dominant Cluster and Gamma modes, but also presented evidence to question the value of any "mode" approach to the problem.

Because of the many uncertainties in the final values assigned to the concentrations of constituents in rock reference samples, the term "usable values" has been proposed. Essentially, its meaning is that the value in question may be used with caution, always in combination with values for the same element in other samples and preferably with some understanding of its limitations.

410

abundant constituents, the fact that they cannot ordinarily be determined by optical or X-ray spectroscopy has meant that comparatively fewer values are available for carbon dioxide, water, and ferrous iron. With carbon dioxide, many contributors have failed to report on whether their results represent total carbon, as determined by combustion methods, or carbonate carbon, as evolved by acid treatment.

DERIVATION OF "USABLE VALUES"

Given a highly incoherent set of results for the determination of each constituent of a proposed reference sample, the originator is faced with the difficult problem of estimating the "true" concentration. In the analysis of ores, where a relatively small number of constituents are of interest, where collaborative analysis is done according to a specified replicate sampling design, and where the collaborating analysts are specialists in the handling of similar materials, a systematic and relatively rigorous statistical treatment of the raw data is frequently applied. Thus Sutarno and Faye (1975) have used a scheme in which a few outlying results are rejected according to established statistical rules, and within laboratory and interlaboratory means and variances are computed. From these, it is possible to deduce a "certification factor", indicating whether a particular value has been sufficiently well established to justify certification, to deduce a reliable recommended value, and to assign confidence limits.

With rocks, matters are complicated by the wide variation in composition between samples, by the many constituents (whose contents may range from over 50 per cent to less than 1 ppm) that are of interest, by the varied backgrounds of the participating laboratories, and hence by

THE "SELECT LABORATORIES" METHOD

Assuming that the originators of a particular sample have more information about it than has anyone else, the originators' own recommended values generally were used in compiling the Tables in this paper, except that all values were converted to the "dry basis" where suitable H_2O^- data were available. In the few cases where the originators' values gave rise to discrepancies, they were replaced by values derived by the "select laboratories" method outlined below. The same method was also used where the originators did not recommend usable values.

Although by no means conclusive, there is a large body of evidence to support the view that interlaboratory bias, or systematic error, is the major cause of incoherent results. It may then be argued that the work of some contributing laboratories is more reliable than that from others and therefore should lead to more dependable values. In an extreme example of that approach, Ingamells (1978) proposed that the evaluation of a reference sample be based on analyses done in only two reputable laboratories, each using mutually independent methods. Only where those two disagree would a third laboratory be called in to resolve the differences. Unfortunately, with rock samples, it would be very difficult to find two, let alone three laboratories that would be dependable for all of the many components of interest.

The select laboratories method has undergone numerous modifications as it has been applied over the years, but the principle remains the same: By some arbitrary set of parameters, all results for each constituent of each sample

411

are categorized as "good", "fair" and "poor". Each laboratory is given a "rating" on the basis of its relative number of good, fair and poor results. All laboratories with a rating exceeding a specified value are considered "select" and the results reported by them are used to arrive at a usable value.

Details are as follows:

1. All raw data are converted to the dry basis, where H_2O^- values are reported.

2. Where fewer than three results are available for a particular constituent of a particular sample, no value is assigned.

3. Where three or four results are available, the median is taken as usable value, but only if results are close to one another and three independent methods have been used. Such values are tabulated with question marks.

4. Where five to nine results are available, the median is used, with a question mark, regardless of the methods involved and how closely the results agree.

5. Where ten or more results are available, the mean, $\bar{x}$, and standard deviation, s_1, are calculated. All results lying below $x-s$ and above $x+s$ are classified as "poor".

6. After setting aside the poor results, the mean of the remainder, x_1, and standard deviation, s_1, are calculated. All results lying below $\bar{x}_1 - s_1$ and above $\bar{x}_1 + s_1$ are classified as "fair".

7. After setting aside the fair results, those remaining are classified as "good".

8. For each contributing laboratory, the total number of good (N_g), fair (N_f), and poor (N_p) results are determined.

9. Each laboratory is given a rating, R, where

$$R = \frac{N_g - N_p}{N_g + N_f + N_p} \times 100$$

VERIFICATION OF USABLE VALUES

No known test can prove the validity of a concentration value derived from a mass of incoherent data. The fact that different statistically-based selection procedures can produce somewhat different final values casts doubt on the validity of at least some of those procedures. Two rough criteria are used in the Tables in this paper as measures of the validity of listed values: the summation and the "compatibility of the iron oxides".

The summation criterion considers closeness of the total to 100 per cent as a measure of the quality of the listed values, provided all constituents present at a significant level have been included. Unfortunately, compensating errors can result in "good" summations. "Bad" summations can result from adding in certain trace elements that have already been accounted for as part of some major element, if the latter has been determined by a chemical method (e.g. Sr and Ca). Incorrect summations can also result from the reporting of some elements as oxides where they may occur in the metallic state.

The iron-oxide compatibility test can be applied only where usable values for ferrous, ferric and total iron are derived independently from their respective reported values. If the derivation procedure is sound, the derived value for total iron based on individual results reported as such (i.e. Fe_2O_3 TR) should be close to Fe_2O_3 TC (i.e. $1.1113 FeO + Fe_2O_3$), where FeO and Fe_2O_3 are the derived values for those constituents based on individual results reported for them. Because ferric iron is normally determined by difference in an ordinary rock analysis, some originators have derived values only for ferrous and total iron, then obtained a derived value for ferric by difference, instead of deriving a value for Fe_2O_3 from individually reported values, each of which was calculated by difference from individual results for FeO and Fe_2O_3 T. Such a procedure unfortunately renders this test meaningless.

PRESENTATION OF THE DATA

The Tables have been arranged somewhat differently from those in the preceding paper of this series (Abbey, 1977a) in the hope of making the information more useful to the reader. Table 1 lists the samples alphabetically according to sample type. In Table 2, the samples themselves are listed alphabetically and identified with sample type, source and country of origin, and they are referred to more detailed descriptive notes. Table 3 lists the "complete analyses" of all samples in roughly geographic sequence and in alphabetic order within each issuing institution. The order of presentation of major and minor constituents has been changed from that used previously (Abbey, 1977a), to approximate that used by most other publications, by listing titania between silica and alumina, and manganese between ferrous and magnesium oxides. For the first time in this series, this table also includes samples whose analyses are incomplete or for which firm usable values are not available, in order to give a general idea of overall composition. Table 4 lists essentially the same information as Table 3, but gives the samples in descending order of concentration for each major and minor constituent. In Table 5, the "trace elements" are presented in much the same manner, as ppm (μg/g), ppb (ng/g), ppt (pg/g) or ppq (fg/g), along with equivalent percentages of appropriate oxides, where they are high enough to affect the summation.

Most of the samples listed in the Tables are silicate rocks; a few are compositionally similar materials such as micas, feldspars and clays. Some additional samples have been arbitrarily selected for inclusion because they contain the same major and minor constituents as do silicate rocks, but in different proportions, and, therefore, may be useful as

10. All laboratories with ratings exceeding a specified level are considered "select". Originally, the specified level was 50, but changes introduced in the method since the last paper in this series (Abbey, 1977a) have resulted in a general reduction of rating levels, so the specified level has been reduced, generally to 40, occasionally to 30.

11. For each constituent of each sample, results reported by the select laboratories are examined. Any outlier among the select values that differs from its nearest neighbor by as much as or more than the latter differs from the opposite extreme is rejected.

12. If fewer than five select values are available, a subjective choice is made between their median and the median of all the original data, as usable value, but reported with a question mark.

13. Where five or more select values are available, both their mean and median are calculated. A subjective choice between the two is then made to establish an "unquestioned" usable value. Generally, the median is favoured because it is less affected by the distribution.

It should be noted that the question mark is used in several different circumstances to indicate varying degrees of uncertainty. How it has been applied where usable values are those recommended by the originators is described in the notes for each originating organization. It is inevitable, therefore, that it can have somewhat variable significance in going from one group of samples to another.

413

high or low points in calibrations or because of their trace-element contents. They should, however, be used with caution because their compositions may result in unexpected interferences, some may contain excessive organic matter, others may be difficult to fuse, etc. They include fly ash, bauxite, glass, iron ore, refractories, sand, etc. They have been included because their compositions appear to be close enough to those of silicate rocks to be of value. A number of others are available in the publications of the various issuing agencies.

CCRMP – CANADIAN CERTIFIED REFERENCE MATERIALS PROJECT

(Contact: Dr. H.F. Steger, Co-ordinator, CCRMP, c/o Canada Centre for Mineral and Energy Technology, 555 Booth St., Ottawa, Canada, K1A 0G1)

This project, operating originally under the then Canadian Association for Applied Spectroscopy, is now an activity of the Canada Centre for Mineral and Energy Technology, but also involves a number of other Canadian government bodies and some private industries. A catalog of all of their samples is available (Steger, in press).

Syenite sample SY-1 and sulphide ore sample SU-1 were issued some years ago, and a compilation of available analytical data was published by Sine et al. (1969). However, neither of those samples ever attained the status of a reliable reference material for rock analysis. SY-1 is now exhausted. SU-1 is now available as an ore standard, certified for a few selected base metals.

"Ultramafic rock" samples UM-1, UM-2 and UM-4 are available for use as standards for certain components by means of specific tests. They are not intended for use as reference samples in general rock analysis.

Table 1

Listing of Samples According to Type

Sample Type	Samples
Andesite	AGV-1, JA-1
Anorthosite	AN-G
Ash	NBS-1633
Basalt	BCR-1, BE-N, BHVO-1, BIR-1, BM, BR, JB-1, JB-2
Bauxite	BX-N, NBS-69a, NBS-69b, NBS-697
Biotite	Mica Fe
Calc-silicate	M-3
Clay	KK, NBS-97a, NBS-98a
Diabase	DNC-1, W-1, W-2
Diorite	DR-N
Dolerite	I-3
Dunite	DTS-1, NIM-D
Feldspar	BCS-375, BCS-376, FK-N, NBS-70a, NBS-99a
Gabbro	MRG-1, SGD-1A
Glass	NBS-91, VS-N
Glauconite	GL-O
Granite	G-2, GA, GH, GS-N, I-1, MA-N, NIM-G, SG-1A
Granodiorite	GSP-1, JG-1
Iron Ore	ES-681-1
Jasperoid	GXR-1
Kaolinite	KK
Kyanite	DT-N

Syenite samples SY-2 and SY-3, and gabbro MRG-1 became available some years ago as "uncertified standards". Results of a worldwide program of collaborative analysis are given in a "comprehensive report" (Abbey, in press). The values listed in the Tables were derived by the select laboratories method described above.

Of the many other materials available from this source, four soils and a blast-furnace slag have been included in this paper for possible use in silicate rock and mineral analysis.

The soils, SO-1, SO-2, SO-3 and SO-4, are described by Bowman et al. (1979a,b). Assigned values of varying degrees of confidence for many major, minor and trace elements were derived by the method of Sutarno and Faye (1975), although the authors admitted that "...some subjectivity was required in identifying outliers". They also stated that "...it is evident that the capability of the analyst is the most important factor in determining the reliability of results as both good and poor data were generated by all methods". Their listing of "methods" includes AA, XRF, ES, NAA, COLOR, TITR, GRAV, all of them merely measurement techniques, none of them sufficiently specific. For certain trace elements, the spread of results reported by contributing laboratories is as bad as or worse than what is generally observed with rock samples. There is some evidence that differing chemical pre-treatments were used by different laboratories in methods where solutions were required. The summations show a small but persistent negative bias, suggesting the possibility that the analyses are incomplete, or that the validity of the derivation procedure is open to question. The iron-oxide compatibility test cannot be applied because the presence of major quantities of organic matter precluded the determination of ferrous iron.

Larvikite	ASK-1
Latite	QLO-1
Lujavrite	NIM-L
Marl	MO8-1
Mica	Mica Fe, Mica Mg
Norite	NIM-N
Peridotite	PCC-1
Phlogopite	Mica Mg
Pyroxenite	NIM-P
Rhyolite	RGM-1
Sand	FK, NBS-81a, NBS-165a, SS
Schist	ASK-2, M-2, SDC-1
Sediment	GXR-3, MAG-1
Serpentine	SW, UB-N
Shale	SCo-1, SGR-1, TS
	BCS-267, BCS-313, NBS-81a,
	NBS-165a, SS
Silica	
Sillimanite	BCS-309
Slag	BCS-367, ES-878-1, SL-1
Slate	TB
Soil	GXR-2, GXR-5, GXR-6, SO-1,
	SO-2, SO-3, SO-4, SOIL-5
Syenite	NIM-S, NS-1, STM-1, SY-2, SY-3
Tonalite	T-1
Trap	ST-1A

Nevertheless, these soil samples may prove useful in rock analysis, particularly because of their usable values for certain trace elements, but caution should be exercised in view of the high organic contents and the other reservations outlined above. The values listed here with question marks are some of those given by Bowman et al. (1979a, b) as "uncertified, for information only".

Individual results for the blast furnace slag, SL-1, are much more coherent than those for the soil samples (Mason and Bowman, 1977), suggesting conditions similar to those that usually apply to ores. The use of the procedure of Sutarno and Faye (1975) therefore can not be questioned in this case. However, the summation is somewhat low, and if an oxygen-for-sulphur correction were applied (assuming all the sulphur to be present as sulphide), the summation would fall under 99 per cent, suggesting that not all constituents have been accounted for. Little information is available on trace elements. Values listed with question marks are those given by Mason and Bowman (1977) as "for information only".

This slag sample may be useful in rock analysis calibrations, particularly where high calcium and low silica points are required. It should be noted that the sample contains appreciable non-carbonate carbon (J. Kelly, Steel Co. of Canada, verbal communication).

USGS – UNITED STATES GEOLOGICAL SURVEY

(Contact: F.J. Flanagan, Liaison Officer, Geological Survey, U.S. Department of the Interior, Reston, Va. 22092, U.S.A.)

United States Geological Survey samples G-1 and W-1 are probably the best known reference samples of silicate rocks, early work on them having been published in 1951. The supply of G-1 has long been exhausted and, therefore, has not been included in the work of this series. W-1 was available

unfortunately none for the most unusual one of all, SGR-1, which apparently contains major amounts of carbonate and organic matter (petroleum?). The report contains a great mass of analytical data, but the latter is unfortunately grossly imbalanced, in terms of the amounts of work done on the various samples, for the various constituents and by various methods in many different laboratories. Values for many constituents of these last eight samples are listed in this report, but most of them must be regarded as only preliminary. However, the data in Flanagan (1976b) clearly indicate that the samples are sufficiently homogeneous for most practical purposes.

An unfortunate feature about the above eight samples is that Flanagan's (1976b) report included no data beyond 1972, and although many additional results have appeared in the literature, no comprehensive compilation nor any recommended values have been published. It is rather surprising that the one institution that has apparently produced more reference samples of rocks than any other organization has not provided more complete compositional information. Absence of recommended values, however tentative, severely limits the usefulness of those samples.

More recently, three additional samples have appeared: basalt BIR-1 and diabases DNC-1 and W-2. The last of these is an intended replacement for W-1, which was originally reported as exhausted by Flanagan (1973). No replacement for DTS-1, listed as exhausted at the same time, has appeared.

Another report (Myers et al., 1976) listed individual results and median values for certain trace elements and most major and minor constituents on four synthetic glasses of rock-like composition, GSB, GSC, GSD and GSE. These samples are intended for use only in U.S. Geological Survey laboratories and are not available for general distribution. Their compositions, therefore, are not given in this paper.

USGS-AEG – U.S. GEOLOGICAL SURVEY AND ASSOCIATION OF EXPLORATION GEOCHEMISTS

(Contact: U.S. Geological Survey, Federal Center, Denver, Colorado 80225, U.S.A.)

Six samples, GXR-1 to GXR-6 inclusive, have been prepared for use as reference materials in geochemical exploration. They represent a variety of compositions, some of which are close to those of silicate rocks. Gladney et al. (1979), having observed the wide scatter of results obtained in U.S.G.S. "in-house" and "round-robin" analyses, attempted to resolve the discrepancies by undertaking repeat analyses, using a variety of analytical techniques, although they relied heavily on neutron-activation methods for most trace elements. On the basis of their own results, Gladney et al. (1979) then listed recommended values for a variable number of constituents of all six samples.

Because of the inevitability of interlaboratory bias, it is difficult to accept the results from one laboratory as a firm basis for recommended values, particularly where only one analytical method was used. As a result, some of the values for the GXR samples tabulated in this report are shown with question marks.

These samples should prove useful in general rock analysis because of the many usable values on certain trace elements. However, it must be remembered that they were designed for use in geochemical exploration and that certain constituents in them (e.g. arsenic, copper, mercury, molybdenum, lead, antimony, selenium, tungsten) could cause problems in analytical procedures for certain other elements.

until relatively recently, and is included here to provide continuity with earlier reports in the series. Most of the listed values for W-1 in this report are based on those of Flanagan (1973).

A large compilation of data was published on six later samples, andesite AGV-1, basalt BCR-1, dunite DTS-1, granite G-2, granodiorite GSP-1, and peridotite PCC-1, (Flanagan, 1969), but no values were recommended. A later publication (Flanagan, 1976c) included additional results and some "recommended", "average" and "magnitude" values. Those values were the same as those listed earlier (along with similar values for many other samples) by Flanagan (1973).

Some contradictions arose between Flanagan's (1973) values and Abbey's (1972) values, resulting in a Critical Comment and Reply (Abbey, 1975b; Flanagan, 1975). In two recent papers in this series (Abbey, 1975a, 1977a), Flanagan's values were given precedence, except where there were apparent errors, omissions or other discrepancies. Subsequently, because of the appearance of Flanagan's (1976c) second compilation on the six samples, the Editor of Geostandards Newsletter suggested that the select laboratories method be applied to all the data now available. This was done and a new set of usable values published by Abbey (1978). The procedure used in that case differed in a few details from that outlined above, and the values listed in this report are those from Abbey (1978) with a few minor corrections.

In recent years, eight additional rock samples were prepared: basalt BHVO-1, marine sediment MAG-1, quartz latite QLO-1, rhyolite RGM-1, schist SDC-1, shales SCo-1 and SGR-1, and syenite STM-1. Flanagan (1976b) gave background information for seven of the samples, but

417

NBS – NATIONAL BUREAU OF STANDARDS (U.S.A.)

(Contact: Office of Standard Reference Materials, Room B311, Chemistry Building, National Bureau of Standards, Gaithersburg, Md. 20234, U.S.A.)

Of the many standard reference materials issued by this agency, only the potash feldspar 70a and the soda feldspar 99a fall within the composition range of silicate rocks. Meinke (1965a,b) gave their compositions as "provisional", but NBS Special Publication 260 presented the same data without qualification. Certificate values are given for most of the major and minor elements, but no information is available on trace elements.

In addition to the two feldspars, this compilation includes two clays, three bauxites, two sands, a glass and a fly ash from NBS. The additional samples were selected from the many others available from the same source because of their potential usefulness in the analysis of silicate rocks and minerals. They are certified mainly for major and minor components, except for the fly ash, which is certified for certain trace elements of particular interest in environmental studies. Additional analytical work on that sample, done in four different laboratories (Ondov et al., 1975), has served to confirm some "uncertified" values and has provided much useful data for a number of additional elements.

NBS also offers some "Trace Element Standards". Of those, feldspar 607 is certified only for rubidium and strontium. Glass samples 610 to 617 inclusive have been spiked with some 36 trace elements, but certified values have been established for only four to eight elements per sample. There is also some disadvantage in the fact that one sample

The analysts involved were evidently expert in feldspar analysis, and the data produced were therefore more coherent than those usually obtained with rock samples, where the need for results for many additional components necessitates the participation of many laboratories with variable backgrounds.

Several additional BCS samples have been included in this compilation because of their possible usefulness in silicate rock and mineral analysis. They include a sillimanite, an iron ore, a silica brick, a "high purity" silica, and a blast furnace slag. The slag sample suffers from the same summation problems as does CCRMP-SL-1. It will be noticed that the iron ore sample is now a "Eurostandard", ES-681-1. In the most recent paper in this series (Abbey, 1977a), an iron ore sample, BCS 302/1, was listed. Compositions of the two are so nearly the same that it could be concluded that one is a replacement for the other, or even that ES-681-1 represents a more thorough re-analysis of the material previously designated as BCS 302/1. It will be noticed that no summation for that sample is shown in Table 3. Even after correcting for sulphur (assuming it is all present as sulphide), the sum of the values in Table 3 is in excess of 101 per cent, suggesting that some of the assigned values are questionable. If all of the carbon is assumed to be in the non-carbonate form, the total is reduced to about 96 per cent. It may therefore be concluded that some carbonate carbon is present.

A major drawback in most BCS samples is the absence of information on trace elements, but a recent announcement (British Ceramic Research Association, 1979) suggested that steps are being taken to overcome the problem.

418

contains the maximum concentration of all the trace elements, whereas the others appear to be mere dilutions with the "pure" base materials. Further, the samples are available only as wafers, one or three millimetres thick. They therefore appear to be of little value for general rock analysis, except in special techniques which can use samples in that shape.

NBS has also issued a set of "mineral glasses for microanalysis", mainly for use in microbeam techniques. They are of little interest in general rock analysis.

There is a rapid turnover of some of the NBS materials, with the result that some of the samples listed here are no longer available. Similarly, new ones may become available before this report is published. NBS Publication 260 is revised frequently, listing only the compositions of available samples. Users of samples no longer available must therefore carefully guard their original certificates or maintain a continuing file of old NBS catalogues.

BCS – BRITISH CHEMICAL STANDARDS

(Contact: Bureau of Analysed Samples, Newham Hall, Newby, Middlesbrough, Teesside TS8 9EA, England)

As is the case with NBS, this agency offers a variety of reference samples of many different types, including some "Eurostandards", originating in several continental European countries. Details are given in their List 461, published in 1977, which also lists the compositions of all of their samples.

Two BCS samples, soda feldspar 375 and potash feldspar 376, fall within the composition range of silicate rocks. Their Certificates of Analysis list all analytical data reported by the collaborating laboratories. In this case, the number of components determined and the number of participating analysts are both small. The results are in excellent agreement with one another, so there need be no hesitation in accepting arithmetic means as usable values.

QMC – QUEEN MARY COLLEGE (U.K.)

(Contact: Dr. A.B. Poole, Department of Geology, Queen Mary College, University of London, Mile End Road, London E1 4NS, England)

This group produced four reference samples several years ago, but is apparently no longer involved with such materials. Available analytical data were listed in a "Third Report" (Poole, 1972), from which usable values have been derived by an earlier version of the select laboratories method.

The samples are aplitic granite I-1, dolerite I-3, pelitic schist M-2, and calcsilicate M-3. Relatively small quantities of these samples were prepared and it is not known whether they are still available.

The analytical data on these samples include very few results for H_2O^-. Analyses, which did not include that determination, were therefore taken as being on the dry basis. The resulting uncertainty would affect only those constituents present at relatively high levels. Question marks have therefore been used with all usable values exceeding 10 per cent. They have also been used where uncertainty exists for the more usual reasons.

ASK – ANALYTISK SPORELEMENT KOMITE (Scandinavia)

(Contact: Dr. Olav H.J. Christie, Mass Spectrometric Laboratory, University of Oslo, Box 1048, Oslo 3, Norway)

Two samples from this group, larvikite ASK-1 and schist ASK-2, fall within the composition range of silicate rocks. A third, ASK-3, is an iron sulphide, of more interest in ore analysis.

The three samples were analyzed for a selected number of trace elements in a small number of laboratories, all located in the "Nordic" countries. Recommended values, arrived at by the highly commendable procedure of a round-table discussion by the collaborating analysts, were published by Christie (1975), and those values are listed in this paper. "Information values" for some major and minor constituents are listed in Table 3 to give a general idea of composition, but not in Table 4 because they are not usable values. Usable values for certain trace elements are given in Table 5, all without question marks.

IRSID – INSTITUT DE RECHERCHES DE LA SIDÉRURGIE (France)

(Contact: G. Jecko, Station d'Essais, Maizières-lès-Metz (57), France)

This institute has produced many reference samples of value in metallurgical industries. Of those, only two have been selected for inclusion in this compilation because their compositions may prove useful in the analysis of silicate rocks and minerals. They are blast furnace slag ES-878-1 (a Eurostandard formerly listed as LOI-1) and ferriferrous marl MO8-1.

Most values given in Tables 3 and 4 are those listed as "most probable" on the certificates provided with the samples. Values shown with question marks were reported by the originators in a private communication; they did not appear on the certificates.

CRPG – CENTRE DE RECHERCHES PÉTROGRAPHIQUES ET GÉOCHIMIQUES (France)
ANRT – ASSOCIATION NATIONALE DE LA RECHERCHE TECHNIQUE (France)

(Contact for both: K. Govindaraju, CRPG, C.O. n° 1, 54500 Vandoeuvre-Nancy, France)

Mica Mg are from Govindaraju (1979). Those for BX-N, DR-N, DT-N, and UB-N are from de la Roche and Govindaraju (1973a), but an updated compilation is expected in 1980. The most recent information on FK-N and GS-N is that of de la Roche and Govindaraju (1976a); the same authors (1976b) reported on GL-O.

In the report on the two micas, Govindaraju (1979) reported totals of 99.22 and 98.91 per cent respectively for Mica Fe and Mica Mg, but addition of the appropriate oxides of the "trace elements" yields the more acceptable totals of 100.29 and 99.70, as shown in Table 3.

VS-N is a special case, being a synthetic sample. Unfortunately, it contains relatively high concentrations of all the additives, many of them at levels much higher than normally encountered in rock samples. As a result, interferences are possible in some analytical procedures. Further, even if an "additive-free" glass of similar composition was available for preparing dilutions, the additive elements would remain in constant proportion to one another, making it difficult to detect interference effects.

VS-N values listed in Table 5 are rounded versions of elemental equivalents of the oxide concentrations recommended by de la Roche and Govindaraju (1973b); those with question marks are similarly based on their "proposed values".

The latest ANRT samples to become available (and for which no quantitative data are as yet published) are anorthosite AN-G, basalt BE-N (a replacement for BR), and granite MA-N (which contains unusually high concentrations of Ag, Be, Cs, Li, Rb, Sn and W).

The first reference sample produced by CRPG was the experimental granite GR, which has been long since exhausted. Three other rocks, basalt BR and granites GA and GH, were produced subsequently and are generally considered among the best-established reference materials. They were followed by biotite Mica Fe and phlogopite Mica Mg. Later, the CRPG reference materials program was identified with ANRT and produced diorite DR-N, serpentine UB-N and two "non-rock" samples, kyanite DT-N and bauxite BX-N. Still later operations produced a synthetic glass VS-N, a potash feldspar FK-N and a granite GS-N, intended as a replacement for GR. A glauconite sample, GL-O, was prepared in limited quantity, mainly as a reference standard in geochronology, but because much analytical work was done on the sample, the originators assigned a number of values that may prove useful in general rock analysis.

In assigning values, the following general procedure has been used by this group. For each constituent of each sample, a mean and standard deviation of reported values were computed. Results differing from that mean by more than one standard deviation were rejected and the mean of the remainder was taken as "recommended value". However, some subjective factors were also considered in special cases, particularly with trace elements. In cases where there was sufficient uncertainty, the originators have given "proposed" rather than "recommended" values. Their "proposed values" are listed in this paper with question marks.

With samples other than BR, GA, and GH, assigned values for ferric iron appear to be based on the difference between assigned values for total and ferrous, rather than on the actual results reported for ferric. In those cases, the iron-oxide compatibility test no longer applies.

Values in Tables 3 and 4 for BR, GA, and GH are those of Roubault et al. (1970), those in Table 5 are from Govindaraju and de la Roche (1977). Values for Mica Fe and

IAEA – INTERNATIONAL ATOMIC ENERGY AGENCY

(Contact: IAEA, Analytical Quality Control Services, Kärntner Ring 11, P.O. Box 590, A-1011 Vienna, Austria)

A soil sample, designated SOIL-5, has been produced, and described by Dybczynski et al. (1979). Their compilation includes "laboratory averages" for each constituent, with from 2 to 60 laboratories reporting, depending on the constituent. The authors used several different statistical tests to identify and eliminate outliers, then utilized means of the remainders and their standard deviations to establish recommended values and confidence limits. Those described as "with a relatively high degree of confidence" are listed in the Tables in this paper without qualification; those "with a reasonable degree of confidence", with question marks. The originators non-certified "information values" are shown as such in Table 3 only.

As was the case with the CCRMP soil samples, the originators of SOIL-5 attempted to correlate results with "method used", but again the "methods" referred to are mere techniques of final measurement. They did indicate wh.ch of the reported results involved some 'chemical pre-treatment of the sample, but did not specify any details of the pre-treatments. It also appears that some of the collaborating analysts did not provide sufficient information on their methods.

UNS – ÚSTAV NEROSTNÝCH SUROVIN (Czechoslovakia)

(Contact: RN Dr. Václav Zýka, Director, Institute of Mineral Raw Materials, 28403 Kutna Hora, Czechoslovakia)

Two samples, a glass sand and a magnesite, have been available from this source for several years. The sand sample, hereafter referred to as SS (for the Czech designation Sklářský Písek, Střeleč), is included in this work

421

for the same reason as were the additional samples of NBS and BCS. The magnesite sample, whose composition is far removed from that of silicate rocks, is not included. This report, however, does include data on KK (for Kaolin, Karlovy Vary), for which recommended values have been made available to the author by the staff of the Institute at Kutna Hora. Similar information about SS is based on a report by Valcha (1972).

No information is available on individual results reported, but the recommended values are believed to have been derived by the method described by Dempir (1978).

ZGI – ZENTRALES GEOLOGISCHES INSTITUT (East Germany)

(Contact: Prof. K. Schmidt, ZGI, Invalidenstrasse 44, 104 Berlin, Deutsche Demokratische Republik)

This agency acts as co-ordinator of an Eastern European collaborative program of geological materials, involving the Czechoslovak institute mentioned above, among others. In general, collaborative analytical work done under that program involves a multiple-sampling scheme not unlike that of Sutarno and Faye (1975). They also use statistical tests to detect and reject outliers, then report means of the remainders, generally with confidence limits. In one case (Grassman, 1972), silica and alumina values are specifically referred to as "recommended", with no explanation. They do not publish compilations of individual results reported by contributing laboratories.

For basalt BM, granite GM, and slate TB, the tabulated values in this paper are those of Grassman (1972) for major and minor constituents, and of Schindler (1972) for trace elements. Question marks have been added to those values where the number of results reported appear to be too few to support firm values; in extreme cases, Schindler's values have not been listed here.

which a laboratory can be considered "select", to 30. The result has been a general slight lowering of most of the usable values and therefore a much improved summation.

Question marks were used according to the outline of the select laboratories method, as given above.

IGI – INSTITUT GEOKHIMII, IRKUTSK (U.S.S.R.)

(Contact: Prof. L.V. Tauson, Institute of Geochemistry, P.B. 701, Irkutsk 33, U.S.S.R.)

Three samples from this source were issued originally as trap 2001, gabbro 2003, and albitized granite 2005. No compilation of reported data has been published at this writing. However, Tauson et al. (1974) reported a set of "attested" values for three samples identified as trap ST-1A, gabbro SGD-1A, and albitized granite SG-1A. Comparison of those values with results obtained in the laboratories of the Geological Survey of Canada indicated that the two sets of samples were identical.

The values listed in the Tables in this report are those of Tauson et al. (1974). Those authors did not list individual results nor did they give details of the method used in deriving their tabulated values from the raw data.

Additional data on these samples were reported by Abbey and Govindaraju (1978).

In this paper, question marks have been added to those values listed by Tauson et al. (1974) in parentheses. It is noteworthy that a small but persistent positive bias appears in all of the summations. The fact that the bias is of similar magnitude and in the same direction for all three samples suggests that it is not due to random errors in the recommended values, but that some systematic error may exist in the method used for deriving those values from the raw data.

422

For feldspar sand FK, greisen GnA, serpentine SW and shale TS, the tabulated values are from certificates supplied with the samples. Question marks are used with values categorized by the originator as "non-certified".

The Eastern European program is expected to prepare, in coming years, a number of rock-like reference materials including gabbro, nepheline syenite, fire clay, monzonite, slate, skarn, and kieselguhr. Details are lacking, mainly because of difficulties in communication.

LEN – LENGOSUNIVERSITET (U.S.S.R.)

(Contact: Prof. A.A. Kukharenko, Department of Mineralogy, Leningrad State University, Leningrad V-164, U.S.S.R.)

The only sample from this source, a nepheline syenite designated here as NS-1, was originally identified as "Khibiny-Generalnaya" by Kukharenko et al. (1968). Unlike other Eastern European authors, they listed individual reported results, but no recommended values. In earlier papers in this series (Abbey, 1972, 1973, 1975a, 1977a), tabulated values were based on "adjusted means" – i.e. the means of values remaining for each constituent after rejecting the 20 per cent of the originally reported results that were farthest removed from the overall mean.

Because the above procedure led to somewhat unsatisfactory values – particularly a rather high summation – it was decided to apply the select laboratories approach for this paper, even though only one sample was involved. As a result of the limited amount of available data, it was necessary to lower the "specified limit" of rating, above

GSJ – GEOLOGICAL SURVEY OF JAPAN

(Contact: Dr. Atsushi Ando, Geochemical Research Section, Geological Survey of Japan, 135 Hisamoto-cho, Kawasaki-shi, Japan)

Analytical data for basalt JB-1 and granodiorite JG-1 were compiled and published by Ando et al. (1971, 1974), recommended values being given for only four elements in a later publication (Ando et al., 1975). "Estimated values' for a number of trace elements were also reported by Ando (personal communication, 1975). Some of the values in Tables 3, 4, and 5 are based on those recommended and estimated values; others were derived by the procedures outlined above.

Some difficulty was encountered in arriving at a usable value for silica in JB-1. One recent compilation (Abbey, 1975a) listed a value of 52.49 per cent (dry basis), but that value was based only on Ando's 1971 data. When his 1974 data were included, the same procedure yielded a value of 52.72 per cent. Changes for other components were less conspicuous. The higher silica value brought about a rather high summation, 100.31 per cent, although the availability of additional trace-element data may have been a factor in that case. However, it was felt that an increase from 52.49 to 52.72 per cent was too great. It was therefore decided to reject the "select mean" and to use the "adjusted mean" as a usable value, but to emphasize the uncertainty by adding a question mark. This was an example where failure to satisfy one of the "validity tests" mentioned above was considered sufficient grounds for departure from the established procedure.

The relatively high values for H_2O^- reported in Ando's compilations, averaging close to one per cent, may have been a source of discrepancy in the silica results reported by the collaborating analysts.

Two additional reference samples, andesite JA-1 and basalt JB-2, have been prepared recently. Kato et al. (1978) reported major and minor constituent results as determined by three analysts, but no other information has appeared, to my knowledge.

MRT – MINERAL RESOURCES, TANZANIA

(Contact: Commissioner, Mineral Resources Division, P.O. Box 903, Dodoma, Tanzania)

Tonalite sample T-1 was produced some years ago, and a compilation of analytical results published by Thomas and Kempe (1963). Those authors suggested that recommended values for each constituent be computed as the mean of those remaining after eliminating all results that differed from the overall mean by more than one standard deviation. That procedure was used for the values reported for this sample in earlier papers of this series (Abbey, 1972, 1973, 1975a, 1977a). Unfortunately, those values resulted in a somewhat high summation, and in a noticeable discrepancy between the two total-iron values. Because the select-laboratories method had been applied with some success to NS-1, the only other single-sample program, it was decided to do the same in this case.

Results are shown in the Tables. Table 3 reveals that the summation has shown no improvement over that in earlier editions, but the iron-oxide compatibility has been

out elsewhere (Abbey, 1977b). Although the certificates indicate that most of the listed values are mere "averages" or "magnitudes", the fact that they are referred to as "certified" can lead to erroneous conclusions on the part of some users.

In earlier work (Abbey, 1973, 1975a) usable values for these samples were calculated by the methods outlined above, but it was pointed out that some of the samples were of unusual composition, some of the collaborating analysts had apparently ignored that fact, and hence considerably more subjective judgment than usual was used in arriving at the tabulated values. Considerably less confidence was therefore placed in those values than in those listed for other samples.

Subsequent work in the Geological Survey of Canada laboratories cast further doubt on the earlier tabulated values for these samples. Through the kindness of T.W. Steele, of the National Institute for Metallurgy of South Africa, the author was provided with a computer printout of all available results on these samples to the end of 1975. The entire derivation procedure was repeated, using a more refined procedure than before. The results, listed in the Tables of the immediately preceding paper in this series (Abbey, 1977a) were in general closer to the NIM "certified" values than before, but were, as a rule, free from most of the objections to the latter values.

In apparent recognition of the questionable usefulness of the "certified values", the originators (Steele et al., 1978a, b; Steele and Hansen, 1979a, b) made a more detailed study, using essentially the same raw data as in the computer printout referred to above. They used a number of different significantly improved (all figures in per cent, dry basis):

	Fe$_2$O$_3$TR	Fe$_2$O$_3$TC
Earlier values	5.93?	6.03?
Current values	5.90	5.91

This improvement is noteworthy in view of the fact that the originally reported results for this sample were in closer agreement with one another than has been the case in most other rock programs. It may then be concluded that less satisfactory values may result from the use of "adjusted means", even where the raw data are comparatively coherent.

With reference to T-1, Bowden and Luena (1966) rightly warned against the dangers entailed in indiscriminate use of insufficiently well-established values. Little change may be expected with T-1, as it was reported by Flanagan (personal communication) that supplies of the sample are exhausted and no replacement is contemplated.

NIM – NATIONAL INSTITUTE FOR METALLURGY (South Africa)

(Contact: H.P. Beyers, South African Bureau of Standards, Private Bag 191, Pretoria, South Africa)

Russell et al. (1972) listed the then available results on the six rock samples, dunite NIM-D, granite NIM-G, lujavrite NIM-L, norite NIM-N, pyroxenite NIM-P, and syenite NIM-S. They calculated means, standard deviations, and adjusted means, but did not give recommended values. Later (Beyers, 1974), the originators issued a set of "certificates of analysis", listing values that are apparently means of values remaining after removal of those differing from the overall mean by more than three standard deviations. Several contradictions in those "certified" values have been pointed

tests for outliers, computed several different measures of central tendency (median, mean, dominant cluster mode, gamma central value, etc.), and selected a recommended value by subjective examination of those measures. In view of the thorough work done by Steele et al. (1978a, b) and Steele and Hansen (1979a, b), it was decided to favour their values (where there were differences) over those in Abbey (1977a).

More detailed study of the papers by Steele and Hansen (1979a, b) revealed some minor discrepancies: (a) Their "Others" did not include some trace elements whose oxide equivalents could be rounded to 0.01 per cent. (b) Some contributors of data had given results for only two of the three iron oxides (ferric, ferrous, total-as-ferric) without calculating the third. (c) Their total-iron-oxides compatibilities were not as good as in our own earlier work (Abbey, 1977a).

A comparison was therefore made between the various iron values recommended by Steele and Hansen (1979a, b) and those of Abbey (1977a) and a selection was made to provide the best possible combination of iron-oxide compatibility and summation. Table 6 shows a comparison between the values decided upon and those in the two sources. For all six samples, the values selected in this work provide better iron-oxide compatibility than do those of Steele and Hansen. For four of the six, this work gives summations that are as close to or closer to 100 per cent than are those of Steele and Hansen. For NIM-L, Steele and Hansen give a summation closer to 100 per cent, as they do for NIM-D, but in that case, it should be noted that with dunites and similar rocks high summations are the rule rather than the exception. The presence of small amounts of nickel in the metallic state, although it is reported as NiO, is generally regarded as the reason for high summations. The 100.17 per cent total in this work, therefore, may actually be superior to the 100.13 per cent of Steele and Hansen.

425

A further complication arose with the NIM samples. In Abbey (1977a), values were reported for certain trace elements for which Steele et al. (1978a, b) considered the available results inadequate to recommend values. Those trace element values therefore should have been omitted from this report. However, to do so would have meant that more information was available in an earlier paper. All values within that category therefore are shown in this paper with question marks, even though some of them appeared in the earlier compilation as unqualified usable values. Thus the quality of some listed values appears to have been "demoted".

Another special case occurred with gallium in NIM-L for which the median of the five available select values was 30 ppm; the mean was 35 ppm. The latter was listed as usable value in the preceding report. However, Steele et al. (1978a, b), after rejecting two extreme outliers, found a mean of 43 ppm, median 49 ppm, and dominant cluster mode of 54 ppm. They therefore recommended 54? ppm, a value supported by further results received after the computations were done. Thus, in this case, it was the select laboratories method that produced unsatisfactory results.

OTHER SOURCES

Ivanov (in press) reports on a Bulgarian granite, G-B. At this writing, no information is available.

Several of the agencies listed above prepare other reference materials that may be of use in general analysis of silicate rocks and minerals. Among the many other institutions that prepare inorganic reference materials, there may be some that could be useful in the analyses of rocks. Every effort has been made to render this compilation as comprehensive as possible, but some potentially useful samples have likely been overlooked.

Date's method can produce material which can be applied in some cases, but the usefulness of the product is somewhat limited.

USE OF THE TABLES

A number of citations of values from various compilations have strongly suggested an unfortunate tendency by some workers to accept any tabulated value without attempting to understand how it was derived, or its degree of reliability. It would appear that the time, effort, and money going into the establishment of reliable values for reference samples is not universally appreciated or even understood. Although lack of understanding by users may be blamed in some cases, even the originators and the compilers of data are not entirely innocent. For example, it does not help the situation when samples are offered as "standards" with little or no supporting analytical data, or with data from only the originating laboratories, or where the data are presented without sufficient emphasis on the degree of reliability involved.

For the above reasons, footnotes have been inserted on every page of Table 3. Readers are strongly urged not to use values from a Table before reading at least the notes concerning the issuing agencies of the samples of interest, or better still, the entire text of this paper. For example, it is important to understand the variation in the significance of the question mark in going from one issuing institution to another. In some cases, it would be advisable to study the compilations published by the originators of particular samples.

Another unfortunate tendency has become apparent in some papers where reference samples have been used in verifying new analytical methods. Some workers consider their results acceptable merely because they fall "within the range" of values listed in a compilation. In fact, such a situation merely indicates that the results in question are not as bad as the worst in the compilation. How bad that can be is clearly indicated in many of the original compilations.

The Tables in this paper are arranged in a manner intended to improve their usefulness to the reader. Thus anyone preparing a calibration curve would begin by scanning Tables 4 or 5, then examining Table 3 to learn about the overall composition of potentially useful samples, and finally Table 2 to find where further information may be found in the text. Similarly, anyone interested in a particular sample type would begin with Table 1, and so on.

ERRATA

Readers are requested to draw the author's attention to any errors they may observe in this paper. The preceding paper (Abbey, 1977a) had at least three errors: two different values for As in USGS-W-1, two for Be and two for Mo in NIM-L. One reader noticed the error in W-1; the others came to light only in the preparation of this paper.

ACKNOWLEDGMENTS

The author is indebted to all of the contact officers under the various originating institutions for the preparation and distribution of the samples listed in this paper. Special thanks are due to Dr. A.R. Date, Josef Dempir, K. Govindaraju, G. Jecko, J.J. Lynch, Dr. Rolf Schindler, T.W. Steele and Dr. Zdenek Valcha for supplying unpublished information and other valuable exchanges, as well as to Dr. J.A. Maxwell for critical reading of the manuscript.

At this writing, a collaborative effort is underway, involving the British Columbia Department of Energy, Mines and Petroleum Resources, the Canada Centre for Mineral and Energy Technology, and the Geological Survey of Canada, to produce reference samples of stream sediments, lake sediments, and possibly other materials which can be used in geochemical trace analysis. Preparation of reference samples of marine sediments is under study by the National Research Council of Canada.

A novel approach to the preparation of reference samples is that of Date (1978), who proposed a scheme for synthesizing such materials. It involves the preparation of two solutions, one an ethanolic solution of tetra-ethyl orthosilicate, the other a dilute acid solution of such salts of all other elements that can be converted to oxides on ignition. The two solutions are combined, and addition of a controlled amount of ammonium hydroxide results in a "flash hydrolysis", producing a gel in which the dissolved compounds are "frozen" in place. Careful evaporation, drying and ignition results in the production of a powdered mixture of silica and oxides of all other elements added.

The major advantage of such a scheme is the ability to vary composition at will. A disadvantage is that the physical characteristics of the product are very different from those of powdered rocks. Experiments carried out at the Geological Survey of Canada have also revealed that ignition must be carried out at a very high temperature (1200°C?) to minimize hygroscopic properties in the product. Unfortunately, such temperatures can result in the loss of some volatile constituents.

REFERENCES

Abbey, Sydney
1972: "Standard Samples" of silicate rocks and minerals — A review and compilation; Geological Survey of Canada, Paper 72-30, 13 p.

1973: Studies in "Standard Samples" of silicate rocks and minerals. Part 3: 1973 extension and revision of "usable" values; Geological Survey of Canada, Paper 73-36, 25 p.

1975a: Studies in "Standard Samples" of silicate rocks and minerals. Part 4: 1974 edition of "usable" values; Geological Survey of Canada, Paper 74-41, 23 p.

1975b: Critical Comment: 1972 values for geochemical reference samples; Geochimica et Cosmochimica Acta, v. 39, p. 535-537.

1977a: Studies in "Standard Samples" for use in the general analysis of silicate rocks and minerals. Part 5: 1977 edition of "usable" values; Geological Survey of Canada, Paper 77-34, 31 p. Also in X-ray Spectrometry (1978), v. 7, p. 99-121.

1977b: "Standard Samples" – How "standard" are they?; Geostandards Newsletter, v. 1, p. 39-45.

1978: U.S.G.S. II Revisited; Geostandards Newsletter, v. 2, p. 141-146.

— SY-2, SY-3 and MRG-1: Three rock samples certified as reference materials; Canada Centre for Mineral Energy Technology, Report 79-35. (in press)

Bowman, W.S., Faye, G.H., Sutarno, R., McKeague, J.A., and Kodama, H. (cont'd.)
1979b: New CCRMP reference soils SO-1 to SO-4; Geostandards Newsletter, v. 3, p. 109-113.

British Ceramic Research Association
1979: Announcement; Geostandards Newsletter, v. 3, p. 209.

Christie, O.H.J.
1975: Three trace-element geological materials certified as a result of a co-operative investigation; Talanta, v. 22, p. 1048-1050.

Christie, O.H.J. and Alfsen, K.H.
1977: Data transformation as a means to obtain reliable consensus values for reference materials; Geostandards Newsletter, v. 1, p. 47-49.

Date, A.R.
1978: Preparation of trace element reference materials by a co-precipitated Gel Technique; Analyst, v. 103, p. 84-92.

Dempir, J.
1978: Preparation of reference samples of mineral raw materials and their use in analytical practice; Geostandards Newsletter, v. 2, p. 117-120.

Dybczynski, R., Tugsavul, A., and Suschny, O.
1979: Soil-5, a new IAEA certified reference material for trace element determinations; Geostandards Newsletter, v. 3, p. 61-87.

Ellis, P.J., Copelowitz, I., and Steele, T.W.
1977: Estimation of the mode by the dominant cluster method; Geostandards Newsletter, v. 1, p. 123-130.

Abbey, Sydney and Govindaraju, K.
1978: Analytical Data on Three Rock Reference Samples from the Institute of Geochemistry, Irkutsk, U.S.S.R.; Geostandards Newsletter, v. 2, p. 15-22.

Abbey, Sydney, Meeds, R.A., and Bélanger, P.G.
1979: Reference samples of rocks – The search for "best values"; Geostandards Newsletter, v. 3, p. 121-133.

Ando, Atsushi; Kurasawa, Hajime; Ohmori, Teiko; and Takeda, Eizo
1971: 1971 compilation of data on rock standards JG-1 and JB-1 issued from the Geological Survey of Japan; Geochemical Journal, v. 5, p. 151-164.

1974: 1974 compilation of data on the GSJ geochemical reference samples JG-1 granodiorite and JB-1 basalt; Geochemical Journal, v. 8, p. 175-192.

Ando, Atsushi; Kurasawa, Hajime; and Uchinmi, Shigeru
1975: Evaluation of Rb, Sr, K and Na contents of the GSJ JG-1 granodiorite and JB-1 basalt; Geological Survey of Japan Bulletin, v. 26, p. 17 (335)-30(348).

Beyers, H.P.
1974: Announcement: "Igneous Rocks"; South African Bureau of Standards, Pretoria.

Bowden, P. and Luena, G.
1966: The use of T-1 as a geochemical standard; Geochimica et Cosmochimica Acta, v. 30, p. 361.

Bowman, W.S., Faye, G.H., Sutarno, R., McKeague, J.A., and Kodama, H.
1979a: Soil samples SO-1, SO-2, SO-3 and SO-4 – Certified reference materials; Canada Centre for Mineral Energy Technology, Report 79-3.

Flanagan, F.J.
1969: U.S. Geological Survey standards – II. First compilation of data for the new U.S.G.S. rocks; Geochimica et Cosmochimica Acta, v. 33, p. 81-120.

1973: 1972 values for international geochemical reference samples; Geochimica et Cosmochimica Acta, v. 37, p. 1189-1200.

1975: Author's reply to Critical Comment on "1972 values for geochemical reference samples"; Geochimica et Cosmochimica Acta, v. 39, p. 537-540.

1976a: G-1 et W-1: Requiescant in pace; U.S. Geological Survey Professional Paper 840, p. 189-192.

Flanagan, F.J. (ed.)
1976b: Description and analyses of eight new USGS rock standards; U.S. Geological Survey Professional Paper 840, p. 1-116, 119-126, 185-188.

Flanagan, F.J.
1976c: 1972 compilation of data on USGS standards; U.S. Geological Survey Professional Paper 840, p. 131-183.

Gladney, E.S., Perrin, D.R., Owens, J.W., and Knab, Darryl
1979: Elemental concentrations in the United States Geological Survey's Geochemical Exploration Reference Samples – A review; Analytical Chemistry, v. 51, p. 1557-1569.

Govindaraju, K.
1979: Report (1968-1978) on two mica reference samples: Biotite Mica-Fe and Phlogopite Mica-Mg; Geostandards Newsletter, v. 3, p. 3-24.

Govindaraju, K. and de la Roche, H.
1977: Rapport (1966-1976) sur les Elements en Traces Dans Trois Roches Standards Géochimiques Du CRPG: Basalte BR et Granites, GA et GH; Geostandards Newsletter, v. 1, p. 67-98.

Grassmann, Hans
1972: Die Standardgesteinsproben des ZGI. 6 Mitteilung: Neue Auswertung der Analysen auf Hauptkomponenten der Proben Granit GM, Basalt BM, Tonschiefer TB, Kalkstein KH und erste Auswertung der Proben Anhydrit AN und Schwarzschiefer TS; Zeitschrift fr angewandete Geologie, v. 18, p. 280-284.

Ingamells, C.O.
1978: Standard reference materials in geoexploration and extractive metallurgy research; Presented at Geoanalysis '78, Ottawa, Canada, May.

Ingamells, C.O. and Switzer, P.
1973: A proposed sampling constant for use in geochemical analysis; Talanta, v. 20, p. 547-568.

Ivanov, El. Al.
Rapport préliminaire sur un étalon géochimique de Bulgarie: Granite G-B; Geostandards Newsletter. (in press)

Jaffrezic, H.
1976: L'estimation de l'erreur, introduite dans le dosage des éléments à l'état de traces dans les roches, liée aux caractéristiques statistiques de leur repartition; Talanta, v. 23, p. 497-501.

Kato, Yuzo; Onuki, Hitoshi; and Aoki, Ken-ichiro
1978: Major element analyses of the Geochemical Standards JA-1 and JB-2; Journal of the Japanese Association of Mineralogists, Petrologists, and Economic Geologists, v. 73, p. 281-282. (In Japanese)

Mason, G.L. and Bowman, W.S.
1977: Blast furnace slag SL-1: Its preparation for use as a certified reference material; Canada Centre for Mineral Energy Technology, Report 77-57.

Meinke, W.W.
1965a: Provisional Certificate of Analysis, Standard reference material 70a, potash feldspar (National Bureau of Standards).
1965b: Provisional Certificate of Analysis, Standard reference material 99a, soda feldspar (National Bureau of Standards).

Myers, A.T., Havens, R.G., Connor, J.J., Conklin, N.M., and Rose, H.J., Jr.
1976: Glass Reference Standards for the Trace-Element Analysis of Geological Materials – Compilation of Interlaboratory Data; U.S. Geological Survey Professional Paper 1013.

Ondov, J.M., et al.
1975: Elemental concentrations in the National Bureau of Standards environmental coal and fly ash standard reference materials; Analytical Chemistry, v. 47, p. 1102-1109.

Poole, A.B.
1972: A third report on the Standard Rocks QMC I1, QMC I3, QMC M2 and QMC M3; Queen Mary College (Univ. London), Department of Geology.

Ridley, K.J.D., Turek, Andrew, and Riddle, Chris
1976: The variability of chemical analyses as a function of sample heterogeneity, and the implications to the analyses of rock standards; Geochimica et Cosmochimica Acta, v. 40, p. 1375-1379.

Kukharenko, A.A. et al.
1968:   Clarke values of the Khibinsky alkaline massif; Vsesoyuznoye Mineralogicheskoye Obshchestvo, Zapiski, v. 97 (pt. 2), p. 133-149. (In Russian)

La Roche, H. (de) and Govindaraju, K.
1973a: Rapport (1972) sur quatre standards géochimiques de l'Association Nationale de la Recherche Technique; diorite DR-N, serpentine UB-N, bauxite BX-N et disthène DT-N; Bulletin Société Française Céramique, n° 100, p. 49-75.

1973b: Etude coopérative sur un verre synthétique VS-N proposé comme étalon analytique pour le dosage des éléments en traces dans les silicates; Analusis (Paris), v. 2, p. 59-70.

1976a: Rapport préliminaire (1974) sur deux nouveaux standards géochimiques de l'A.N.R.T.: Granite GS-N et Feldspath FK-N; Analusis (Paris), v. 4, p. 347-372.

1976b: Rapport préliminaire sur l'étalon analytique glauconite GL-O de l'Association Nationale de la Recherche Technique; Analusis (Paris), v. 4, p. 385-397.

Maessen, F.J.M.J., Elgersma, J.W., and Boumans, P.W.J.M.
1976:   A systematic and rigorous approach for establishing the accuracy of analytical results and its application to a comparison of alternative d.c. arc procedures for trace analysis of geological materials; Spectrochimica Acta, v. 31B, p. 179-199.

Roubault, M., La Roche, H. (de), and Govindaraju, K.
1970:   Present status (1970) of the cooperative studies on the geochemical standards of the "Centre de Recherches Pétrographiques et Géochimiques"; Science Terre, v. 15, p. 351-393.

Russell, B.G., Goudvis, Rosemary G., Domel, Gisela, and Levin, J.
1972:   Preliminary report on the analysis of six NIMROC geochemical standard samples; National Institute for Metallurgy (Johannesburg), Report 1351.

Schindler, Rolf
1972:   Standardgesteinsproben des ZGI. 5 Mitteilung: Stand der Spurenelementanalyse der Gesteine GM, TB, BM und KH; Zeitschrift für angewandete Geologie, v. 18, p. 221-228.

Sine, N.M., Taylor, W.O, Webber, G.R., and Lewis, C.L.
1969:   Third report on analytical data for CAAS sulfide ore and syenite rock standards; Geochimica et Cosmochimica Acta, v. 33, p. 121-131.

Steele, T.W., Wilson, A., Goudvis, R., Ellis, P.J., and Radford, A.J.
1978a: Analyses of the NIMROC reference samples for minor and trace elements; National Institute for Metallurgy (Randburg, South Africa), Report 1945.

1978b: Trace element data (1966-1977) for the six "NIMROC" reference samples; Geostandards Newsletter, v. 2, p. 71-106.

Steele, T.W. and Hansen, R.G.
1979a: Analysis of the NIMROC reference samples for major elements; National Institute for Metallurgy (Randburg, South Africa), Report 2016.

431

Steele, T.W. and Hansen, R.G. (cont'd.)
1979b: Major element data (1966-1978) for the six "NIMROC" reference samples; Geostandards Newsletter, v. 3, p. 135-172.

Steger, H.F.
Certified and provisional reference materials available from the Canada Centre for Mineral and Energy Technology (1980); Canada Centre for Mineral and Energy Technology, Report 80-6E. (in press)

Sutarno, R. and Faye, G.H.
1975: A measurement for assessing certified reference ores and related materials; Talanta, v. 22, p. 676-681.

Tauson, L.V. et al.
1974: Geochemical standards of magmatic rocks; Yearbook of the Institute Geochemistry Irkutsk, v. 1974, p. 370-375. (In Russian)

Thomas, W.K.L. and Kempe, D.R.C.
1963: "Standard geochemical sample T-1"; (Government Printer, Dar-es-Salaam).

Valcha, Zdenek
1972: "Standard Reference Sample of Glass Sand"; Institute of Mineral Raw Materials, Kutna Hora, Czechoslovakia, 2 p.

Table 2

Alphabetical Listing of Samples

Sample No.	Type	Source	Country	Ref. Page
AGV-1	Andesite	USGS	U.S.A.	417
AN-G	Anorthosite	ANRT	France	420
ASK-1 ASK-2	Larvikite Schist	ASK	Scandinavia	419
BCR-1	Basalt	USGS	U.S.A	417
BCS-267 BCS-309 BCS-313 BCS-367 BCS-375 BCS-376	Silica Brick Sillimanite Pure Silica Blast Furnace Slag Soda Feldspar Potash Feldspar	BCS	U.K.	419
BE-N	Basalt	ANRT	France	420
BHVO-1 BIR-1	Basalt Basalt	USGS	U.S.A.	417
BM	Basalt	ZGI	East Germany	422
BR	Basalt	CRPG	France	420
BX-N	Bauxite	ANRT	France	420
DNC-1	Diabase	USGS	U.S.A.	417
DR-N DT-N	Diorite Kyanite	ANRT	France	420
DTS-1	Dunite	USGS	U.S.A.	417
ES-871-1	Iron Ore	BCS	U.K.	419
ES-878-1	Blast Furnace Slag	IRSID	France	420
FK	Feldspar Sand	ZGI	East Germany	422
FK-N	Potash Feldspar	ANRT	France	420
G-2	Granite	USGS	U.S.A.	417
GA GH	Granite Granite	CRPG	France	420
GL-O	Glauconite	ANRT	France	420
GM GnA	Granite Greisen	ZGI	East Germany	422
GS-N	Granite	ANRT	France	420
GSP-1	Granodiorite	USGS	U.S.A.	417
GXR-1 GXR-2 GXR-3 GXR-4 GXR-5 GXR-6	Jasperoid Soil "Deposit" "Mill-head" Soil Soil	USGS AEG	U.S.A.	417
I-1 I-3	Aplitic Granite Dolerite	QMC	U.K.	419
JA-1 JB-1 JB-2 JG-1	Andesite Basalt Basalt Granodiorite	GSJ	Japan	423
KK	Kaolinite	UNS	Czechoslovakia	421
M-2 M-3	Pelitic Schist Calc Silicate	QMC	U.K.	419
MAG-1	Marine Mud	USGS	U.S.A.	417
MA-N	Granite	ANRT	France	420
Mica Fe Mica Mg	Biotite Phlogopite	CRPG	France	420

Sample No.	Type	Source	Country	Ref. Page
MO8-1	Ferriferous Marl	IRSID	France	420
MRG-1	Gabbro	CCRMP	Canada	415
NBS-69a	Bauxite			
NBS-69b	Bauxite			
NBS-70a	Potash Feldspar			
NBS-81a	Glass Sand			
NBS-91	Opal Glass			
NBS-97a	Flint Clay	NBS	U.S.A.	418
NBS-98a	Plastic Clay			
NBS-99a	Soda Feldspar			
NBS-165a	Glass Sand			
NBS-697	Bauxite			
NBS-1633	Fly Ash			
NIM-D (SARM 6)	Dunite			
NIM-G (SARM 1)	Granite			
NIM-L (SARM 3)	Lujavrite	NIM	South Africa	425
NIM-N (SARM 4)	Norite			
NIM-P (SARM 5)	Pyroxenite			
NIM-S (SARM 2)	Syenite			
NS-1 (KG-1,etc.)	Syenite	LEN	U.S.S.R.	423
PCC-1	Peridotite			
QLO-1	Quartz Latite	USGS	U.S.A.	417
RGM-1	Rhyolite			
SG-1A (2005)	Albitized Granite			
SGD-1A (2003)	Gabbro	IGI	U.S.S.R.	422
SCo-1	Shale			
SDC-1	Mica Schist	USGS	U.S.A.	417
SGR-1	Shale			
SL-1	Blast Furnace Slag			
SO-1	Soil			
SO-2	Soil	CCRMP	Canada	415
SO-3	Soil			
SO-4	Soil			
SOIL-5	Soil	IAEA	International	421
SS	Glass Sand	UNS	Czechoslovakia	421
ST-1A (2001)	Trap	IGI	U.S.S.R.	422
STM-1	Syenite	USGS	U.S.A.	417
SW	Serpentine	ZGI	East Germany	422
SY-2	Syenite	CCRMP	Canada	415
SY-3	Syenite			
T-1	Tonalite	MRT	Tanzania	424
TB	Slate	ZGI	East Germany	422
TS	Shale			
UB-N	Serpentine	ANRT	France	420
VS-N	Synthetic Glass			
W-1	Diabase	USGS	U.S.A.	417
W-2	Diabase			

Table 3

Usable Values, "complete analysis" (per cent, dry basis)

C C R M P

	Gabbro MRG-1	Bl. Furn. Slag SL-1	Soil SO-1	Soil SO-2	Soil SO-3	Soil SO-4	Syenite SY-2	Syenite SY-3
SiO_2	39.32	35.73	55.02	53.46	33.93	68.5 ?	60.10	59.68
TiO_2	3.69	0.38?	0.88	1.43	0.32?	0.57	0.14	0.15
Al_2O_3	8.50	9.63	17.72	15.24	5.76	10.32	12.12	11.80
Fe_2O_3	8.26						2.28	2.44
FeO	8.63	0.92[4]					3.62	3.58
MnO	0.17	0.86?	0.11	0.09	0.07	0.08	0.32	0.32
MgO	13.49	12.27	3.83	0.90	8.47?	0.93	2.70	2.67
CaO	14.77	37.48	2.52	2.74	20.7 ?	1.55	7.98	8.26
Na_2O	0.71	0.39?	2.56?	2.35?	1.00	1.31?	4.34	4.15
K_2O	0.18	0.51?	3.23	2.95	1.40	2.08	4.48	4.20
H_2O^+	0.98						0.43	0.42
CO_2	1.00		4.4 ?[5]	11.5?[5]	25.3 ?[5]	10.4 ?[5]	0.46	0.38
P_2O_5	0.06	0.02?	0.14	0.69?	0.11?	0.21	0.43	0.54
F	0.025		0.07?	0.05?	0.03?	0.03?	0.51	0.66
S	0.06	1.26	0.01?	0.03?	0.02?	0.04?	0.011	0.05
Others[1]	0.33?	0.01?	0.27?	0.31?	0.10?	0.19?	0.43?	1.18?
Σ	100.18?	99.46?	99.33?	99.69?	99.37?	99.60?	100.35?	100.48?
O/F, S, Cl	0.04?	See text	0.03?	0.04?	0.02?	0.03?	0.22?	0.31?
Σ (corr.)	100.14?	?	99.30?	99.65?	99.35?	99.57?	100.13?	100.17?
Fe_2O_3 TR[2]	17.82	1.02	8.57	7.95	2.16	3.39	6.28	6.42
Fe_2O_3 TC[3]	17.85						6.27	6.42

[1] "Trace elements" (see Table 5), converted to oxides, where appropriate.

[2] Total iron, expressed as ferric oxide, derived from reported values for total iron.

[3] Total iron, expressed as ferric oxide, calculated from values derived for ferric and ferrous, based on reported values for ferric and ferrous.

[4] Total iron, expressed as ferrous oxide.

[5] Loss on ignition.

435

Table 3 (cont'd.)

Usable values, "complete analysis" (per cent, dry basis)

U S G S

	Andesite AGV-1	Basalt BCR-1	Basalt BHVO-1*	Dunite DTS-1	Granite G-2	Grano-diorite GSP-1	Marine Mud MAG-1*	Perid-otite PCC-1	Quartz Latite QLO-1*	Rhyolite RGM-1*	Shale SCo-1*
SiO_2	59.61	54.53	49.9 ?	40.61	69.22	67.32	50.9 ?	42.10	65.5 ?	73.4 ?	62.8 ?
TiO_2	1.06	2.26	2.7 ?	0.00?	0.48	0.66	0.73?	0.01	0.62?	0.27?	0.64?
Al_2O_3	17.19	13.72	13.7 ?	0.25	15.40	15.28	16.4 ?	0.73	16.2 ?	13.8 ?	13.6 ?
Fe_2O_3	4.56	3.48	2.7 ?	1.02?	1.07	1.70		2.54?	1.0 ?	0.5 ?	
FeO	2.03	8.96	8.5 ?	6.94?	1.44	2.32		5.17?	3.0 ?	1.2 ?	
MnO	0.10	0.18	0.17?	0.12	0.03	0.04	0.10?	0.12	0.09?	0.04?	0.05?
MgO	1.52	3.48	7.2 ?	49.80	0.75	0.97	3.0 ?	43.50	1.0 ?	0.28?	2.6 ?
CaO	4.94	6.97	11.4 ?	0.14	1.96	2.03	1.35?	0.55	3.2 ?	1.1 ?	2.6 ?
Na_2O	4.32	3.30	2.3 ?	0.01?	4.06	2.81	3.8 ?	0.01	4.2 ?	4.1 ?	0.9 ?
K_2O^+	2.92	1.70	0.53?	0.00	4.46	5.51	3.6 ?	0.00	3.6 ?	4.3 ?	2.7 ?
H_2O^+	0.78	0.67		0.42	0.50	0.58		4.70			
CO_2	0.02	0.02		0.07	0.08?	0.12		0.18			
P_2O_5	0.51	0.36		0.00?	0.13	0.28		0.01			
F	0.04	0.05	0.28?	0.00?	0.12	0.37		0.00?			
S	0.01?	0.04?		0.00?	0.01?	0.03?		0.01?			
Others[1]	0.32?	0.27?		0.93?	0.41?	0.42?		0.76?			
Σ	99.93?	99.99?		100.31?	100.12?	100.44?		100.39?			
O/F, S, Cl	0.03?	0.04?		0.00?	0.06?	0.18?		0.01?			
Σ (corr.)	99.90?	99.95?		100.31?	100.06?	100.26?		100.38?			
Fe_2O_3 TR[2]	6.78	13.41	12.0 ?	8.70	2.69	4.30	6.8 ?	8.28	4.3 ?	1.9 ?	5.1 ?
Fe_2O_3 TC[3]	6.82	13.44	12.1 ?	8.73?	2.67	4.28		8.29?	4.3 ?	1.8 ?	

[1] "Trace elements" (see Table 5), converted to oxides, where appropriate.

[2] Total iron, expressed as ferric oxide, derived from reported values for total iron.

[3] Total iron, expressed as ferric oxide, calculated from values derived for ferric and ferrous, based on reported values for ferric and ferrous.

* Incomplete.

	USGS				USGS–AEG					
	Mica Schist SDC-1[4]	Shale SGR-1[4,5]	Syenite STM-1[4]	Diabase W-1[4]	Jasperoid GXR-1[4]	Soil GXR-2[4]	"Deposit" GXR-3[4]	"Millhead" GXR-4[4]	Soil GXR-5[4]	Soil GXR-6[4]
SiO_2	66.0 ?	28.3 ?	59.5 ?	52.72	49.2	49.2	13.0	67.0	42.1	49.0
TiO_2	1.0 ?	0.26?	0.13?	1.07	0.11	0.47	0.17	0.43	0.35	0.83
Al_2O_3	16.0 ?	6.5 ?	18.5 ?	15.02	6.7	35.1		14.0	39.3	31.4
Fe_2O_3			2.9 ?	1.40						
FeO			2.1 ?	8.73						
MnO	0.12?	0.037?	0.22?	0.17	0.35	0.12		0.02	0.04	0.13
MgO	1.7 ?	4.5 ?	0.1 ?	6.63	1.22	1.45	2.9	2.7	2.0	1.02
CaO	1.4 ?		1.1 ?	10.98	0.07	1.15	1.05	1.26	1.05	0.14
Na_2O	2.0 ?	3.0 ?	9.0 ?	2.15	0.06	0.75	19.7	0.71	1.04	0.14
K_2O	3.3 ?	1.6 ?	4.3 ?	0.64		1.70	1.05	5.2	0.99	2.5
H_2O^+				0.53						
CO_2				0.06						
P_2O_5			0.16	0.14						
F				0.025						
S				0.012?						
Others[1]				0.21?						
Σ				100.48?						
O/F, S, Cl				0.02?						
Σ (corr.)				100.46?						
Fe_2O_3 TR[2]	6.9 ?	3.2 ?	5.2 ?	11.11	35.3	2.7	26.6	4.2	4.6	8.0
Fe_2O_3 TC[3]		3.2 ?	5.2 ?	11.10						

[1] "Trace elements" (see Table 5), converted to oxides, where appropriate.

[2] Total iron, expressed as ferric oxide, derived from reported values for total iron.

[3] Total iron, expressed as ferric oxide, calculated from values derived for ferric and ferrous, based on reported values for ferric and ferrous.

[4] Incomplete.

[5] Contains major amounts of petroleum.

Table 3 (cont'd.)

Usable values, "complete analysis" (per cent, dry basis)

N B S

	Bauxite 69a	Bauxite 69b	Potash Feldspar 70a	Glass Sand 81a[4]	Opal Glass 91	Flint Clay 97a	Plastic Clay 98a	Soda Feldspar 99a	Glass Sand 165a	Bauxite 697	Fly Ash 1633[4]
SiO₂	6.01	13.4	67.1	(99.+)[5]	67.53	43.67	48.94	65.2	(99.+)[5]	6.80	
TiO₂	2.78	2.0	0.01	0.12	0.019	1.90	1.61	0.007	0.011	2.6	
Al₂O₃	55.0	49.3	17.9	0.66	6.01	38.79	33.19	20.5	0.059	45.7	
Fe₂O₃											
FeO											
MnO	0.02	0.09			0.008	0.15				0.35	0.064
MgO	0.02				0.008	0.11	0.42	0.02		0.17	
CaO	0.29	0.12	0.11		10.48	0.037	0.31	2.14		0.60	
Na₂O	0.00	0.03	2.55		8.48	0.50	0.082	6.2		0.046	
K₂O	0.00	0.80	11.8		3.25	0.36	1.04	5.2		0.07	
L.O.I.	29.55	27.22	0.40		0.50 ?	13.32	12.44	0.02		22.2	2.07
P₂O₅	0.08	0.12			0.022		0.11			0.90	
F	0.04				5.72						
SO₃	0.27 ?	0.63			0.38 ?	0.46 ?	0.21 ?	0.49		0.15	
Others[1]		0.04 ?	0.08 ?							0.22 ?	
Σ	99.86 ?	100.85 ?	100.02 ?		102.49 ?	99.75 ?	99.69 ?	100.04 ?		99.81 ?	
O/F, S, Cl	?	?	?		2.41 ?	?	?	?			
Σ (corr.)	99.86 ?	100.85 ?	100.02 ?		100.08 ?	99.75 ?	99.69 ?	100.04 ?		99.81 ?	
Fe₂O₃ TR[2]	5.8	7.1	0.075	0.082	0.081	0.45	1.34	0.065	0.012	20.0	
Fe₂O₃ TC[3]			0.082								

[1] "Trace elements" (see Table 5), converted to oxides, where appropriate.

[2] Total iron, expressed as ferric oxide, derived from reported values for total iron.

[3] Total iron, expressed as ferric oxide, calculated from values derived for ferric and ferrous, based on reported values for ferric and ferrous.

[4] Incomplete.

[5] Bracketed figures are not usable values; they are given as a mere general indication of composition.

B C S

	Silica Brick 267	Sillimanite 309	"Pure" Silica 313	Bl. Furn. Slag 367	Soda Feldspar 375	Potash Feldspar 376	Iron Ore ES-681-1
SiO_2	95.9	34.1	99.6	34.4	67.1	67.1	17.8
TiO_2	0.17	1.92	0.022	0.75	0.38	0.01?	0.48
Al_2O_3	0.85	61.1	0.16	20.0	19.8	17.7	10.6
Fe_2O_3				1.00*			37.81 ??
FeO				1.16			8.70 ??
MnO	0.15	0.03?	0.00		0.05?		0.28
MgO	0.06	0.17	0.00	7.1	0.89		1.48
CaO	1.75	0.22	0.02	32.4	10.4	2.83	3.92
Na_2O	0.06	0.34	0.00	0.44	0.79	11.2	0.092
K_2O	0.14	0.46	0.04	1.17			0.59
L.O.I.		0.08?			0.39?	0.35?	10.24 ?[5]; 6.60 ?[5]
P_2O_5				0.14 ?			2.02
F						0.00?	0.19
S				0.94			0.103 ?
Others[1]							0.25 ?
Σ	99.87?	99.93?	99.84?	99.50? See text	99.92?	99.32?	See text
O/F, S, Cl							
Σ (corr.)	99.87?	99.93?	99.84?		99.92?	99.32?	47.48 Note[6]
Fe_2O_3 TR[2]	0.79	1.51	0.025	1.11	0.12	0.10	
Fe_2O_3 TC[3]							

Q M C

	Aplitic Granite 1-1	Dolerite 1-3	Pelitic Schist M-2	Calc-Silicate M-3
SiO_2	75.36 ?	49.75?	48.88?	55.59?
TiO_2	0.05	2.60	0.72	0.83
Al_2O_3	13.92 ?	13.07?	23.97?	17.62?
Fe_2O_3	0.33 ?	5.09?	2.31?	0.91?
FeO	0.20 ?	10.04?	6.30?	3.33?
MnO	0.03	0.22	0.26	0.28
MgO	0.11	4.18	2.45	1.21
CaO	0.80	8.20	1.75	12.01?
Na_2O	4.59	2.92	1.40	2.98
K_2O	4.28	1.43	7.90	0.71
H_2O^+	0.13 ?	1.71?	3.21?	0.78?
CO_2	?	?	?	2.98?
P_2O_5	0.02 ?	0.40?	0.50?	0.36?
F	0.005?	0.07?	0.10?	0.06?
S	0.11 ?			
Others[1]	0.11 ?	0.30?	0.34?	0.16?
Σ	99.94 ?	99.98?	100.09?	99.81?
O/F, S, Cl	0.03?	0.03?	0.04?	0.03?
Σ (corr.)	99.94 ?	99.95?	100.05?	99.78?
Fe_2O_3 TR[2]	0.54	16.22?	9.25	4.55
Fe_2O_3 TC[3]	0.55 ?	16.25?	9.31?	4.61?

[1] "Trace elements" (see Table 5), converted to oxides, where appropriate.

[2] Total iron, expressed as ferric oxide, derived from reported values for total iron.

[3] Total iron, expressed as ferric oxide, calculated from values derived for ferric and ferrous, based on reported values for ferric and ferrous.

[4] Total iron, expressed as ferrous oxide.

[5] Total carbon, expressed as carbon dioxide.

[6] Method used to derive Fe_2O_3 value invalidates comparison.

439

Table 3 (cont'd.)

Usable values, "complete analysis" (per cent, dry basis)

	ASK		IRSID		CRPG					
	Larvikite ASK-1[5]	Schist ASK-2[5]	Bl. Furn. Slag ES-878-1	Ferrf. Marl MO8-1	Basalt BR	Granite GA	Granite GH	Biotite MicaFe	Phlogopite MicaMg	
SiO_2	(59.5)*	(54.2)	33.65	60.39	38.39	69.96	75.85	34.55	38.42?	
TiO_2	(1.1)	(0.92)	0.619	0.714	2.61	0.38	0.08	2.51	1.64?	
Al_2O_3	(18.6)	(18.8)	16.15	9.94	10.25	14.51	12.51	19.58	15.25?	
Fe_2O_3				3.08	5.61	1.36	0.41	4.66	1.99?	
FeO				2.38	6.60	1.32	0.84	18.99	6.75?	
MnO	0.13	0.04	1.268	0.057	0.20	0.09	0.05	0.35	0.26?	
MgO	(1.1)	(2.0)	9.55	1.34	13.35	0.95	0.03	4.57	20.46?	
CaO	(3.2)	(0.75)	35.65	8.70	13.87	2.45	0.69	0.43	0.08?	
Na_2O	(6.5)	(0.8)	0.466	0.54?	3.07	3.55	3.85	0.30	0.12?	
K_2O^+	(4.2)	(5.3)	1.288	2.2?	1.41	4.03	4.76	8.79	10.03?	
H_2O^+				3.0?	2.31	0.87	0.46	2.92?	2.10?	
CO_2	(8.5)[7]			7.3?	0.86	0.11	0.14	0.19?	0.15?	
P_2O_5			0.034	0.014	1.05	0.12	0.01	0.45	0.01?	
F			0.149	0.014	0.10	0.05	0.35	1.59	2.86?	
S			0.812	0.455	0.039	?	?			
Others[1]			?	0.05?	0.56?	0.25?	0.13?	1.09?	0.80?	
Σ			100.50?	100.17?	100.28?	100.00?	100.16?	100.97?	100.92?	
O/F, S, Cl			1.03?	0.23?	0.07?	0.03?	0.15?	0.68?	1.22?	
Σ (corr.)			99.47?	99.94?	100.21?	99.97?	100.01?	100.29?	99.70?	
Fe_2O_3 TR[2]	(4.6)	(6.9)	0.861	5.72	12.90	2.77	1.35	25.76	9.49?	
Fe_2O_3 TC[3]			Note[6]	Note[6]	12.94	2.83	1.34	Note[6]	9.49? Note[6]	

[1] "Trace elements" (see Table 5), converted to oxides, where appropriate.

[2] Total iron, expressed as ferric oxide, derived from reported values for total iron.

[3] Total iron, expressed as ferric oxide, calculated from values derived for ferric and ferrous, based on reported values for ferric and ferrous.

[4] Bracketed figures are not usable values; they are given as a mere general indication of composition.

[5] Incomplete.

[6] Method used to derive Fe_2O_3 values invalidates comparison.

[7] Total carbon.

440

	ANRT								IAEA	UNS	
	Bauxite BX-N	Diorite DR-N	Kyanite DT-N	Potash Feldspar FK-N	Glauconite GL-O[4]	Granite GS-N	Serpentine UB-N	Synth. Glass VS-N[8]	Soil SOIL-5[8]	Kaolinite KK	Glass Sand SS
SiO_2	7.39	52.88	36.52	65.11	52.22	65.98	39.93	(55.50)[5]	(71)	47.06	99.35
TiO_2	2.41	1.10	1.40	0.02 ?	0.07 ?	0.68	0.2	(1.08)	(0.8)	0.166	0.036
Al_2O_3	54.53	17.56	59.21	18.64	7.75	14.71	2.97	(13.44)	15.47	36.77	0.249
Fe_2O_3	22.98	3.78	0.55	0.024?	17.61	1.93	5.46				
FeO	0.26	5.32	0.10	0.06 ?	2.25	1.66	2.69	0.09			
MnO	0.05	0.21	0.008	0.005?	0.008?	0.056	0.12		0.11?	0.015	0.001?
MgO	0.11	4.47	0.04	0.01 ?	4.58	2.31	35.4 ?	(4.51)	(2.5)	0.192	0.007
CaO	0.17	7.09	0.04	0.11	0.98	2.51	1.18	(4.53)	(3.1)	0.236	0.030
Na_2O	0.06	3.00	0.04	2.58	0.04 ?	3.78	0.10	(5.95)	2.59	0.032	0.043
K_2O	0.07	1.73	0.12	12.83	8.16	4.64	0.02	(8.12)	2.24	1.063	0.057
H_2O^+	11.69	2.20	0.95 ?	0.32 ?	5.72	1.07	11.23			12.75	0.169[9]
CO_2	0.45?	0.15	0.62 ?[6]	0.085?		0.18 ?	0.44			0.174	
P_2O_5	0.13	0.25	0.09	0.024?	0.38	0.28	0.03	(0.02)		0.090	
F		?			0.15 ?	?					
S		?								0.019	
Others[1]	?	0.20?	0.04 ?	0.14 ?	0.07 ?	0.33 ?	0.63?		(0.2)	0.22 ?	0.01 ?
Σ	100.48?	99.94?	99.73?	99.96 ?	99.99 ?	100.12 ?	100.39?			99.77 ?	99.99 ?
$O/F, S, Cl$	?	?			0.06 ?	?				0.01 ?	
Σ (corr.)	100.48?	99.94?	99.73?	99.96 ?	99.93 ?	100.12 ?	100.39?	(4.14)		99.76 ?	99.99 ?
Fe_2O_3 TR[2]	23.27	9.69	0.66	0.09	20.11	3.76	8.45		6.36?	0.982	0.038
Fe_2O_3 TC[3]	Note[7]		Note[7]		Note[7]	3.77	Note[7]			0.038	

[1] "Trace elements" (see Table 5), converted to oxides, where appropriate.

[2] Total iron, expressed as ferric oxide, derived from reported values for total iron.

[3] Total iron, expressed as ferric oxide, calculated from values derived for ferric and ferrous, based on reported values for ferric and ferrous.

[4] Dried for 15 h.

[5] Bracketed figures are not usable values; they are given as a mere indication of composition.

[6] Non-carbonate carbon.

[7] Method used to derive Fe_2O_3 values invalidates comparison.

[8] Incomplete.

[9] Loss on ignition.

441

Table 3 (cont'd.)

Usable values, "complete analysis" (per cent, dry basis)

	Z G I							L E N	I G I		
	Basalt BM	Feldspar Sand FK	Granite GM	Greisen GnA	Serpentine SW	Slate TB	Shale TS	Neph. Syenite NS-1	Albit. Granite SG-1A	Gabbro SGD-1A	Trap ST-1A
SiO_2	49.60	88.15 ?	73.55	71.45	39.05	60.30	62.71	53.22	73.36	46.39	49.12
TiO_2	1.14	0.059?	0.213	0.023	0.016	0.93	0.71	1.05	0.072	1.71	1.82
Al_2O_3	16.20	6.22 ?	13.50	14.7	0.66	20.55	15.94	21.31	13.84	14.88	14.23
Fe_2O_3	1.60		0.75	1.70	5.2 ?	0.91	6.72 ?	2.31	0.68	3.86	3.92
FeO	7.28		1.14	3.80	2.0 ?	5.43	0.66 ?	1.60	1.41	6.86	10.26
MnO	0.145	0.006?	0.166	0.083	0.052	0.04	0.006?	0.18	0.19	0.17	0.21
MgO	7.46	0.16 ?	0.377	0.033	38.5	1.94	1.77	0.64	0.054	7.09	5.74
CaO	6.46	0.12 ?	1.04	0.61	0.18	0.33	0.13	1.70	0.14	10.97 ?	10.24 ?
Na_2O	4.64	0.25 ?	3.76	0.08	0.013	1.31	0.09	9.85	5.46	2.82	2.49
K_2O	0.203	4.15 ?	4.74	2.63	0.014?	3.85	4.88	6.52	4.14	2.96	0.69
H_2O^+	3.62	0.41 ?	0.349	1.8 ?	13.6	3.82	4.03	0.66	0.21	0.83	0.97
CO_2	1.34	0.025?	0.278	0.04 ?	0.29 ?	0.13	0.04 ?	0.14?	0.073?	0.128?	0.099
P_2O_5	0.105	0.075?	0.063	0.016?	0.10	0.29	0.075?	0.28	0.013	1.01	0.21
F	0.026?		0.067?	3.36	0.007?	0.071?	0.118?	0.14	0.30	0.12	0.032?
S				0.016?					0.017	0.020	0.068
Others[1]	0.15 ?	0.02 ?	1.13?	0.68 ?	0.22 ?	1.89 ?	0.02 ?	0.54?	0.48?	0.63?	0.30 ?
Σ	99.97 ?	99.90 ?	100.00?	101.55?	100.32 ?	99.94 ?	100.04 ?	100.15?	100.44?	100.45?	100.36?
O/F, S, Cl	0.01 ?		0.03?	1.41 ?	0.00 ?	0.03 ?	0.05 ?	0.08?	0.13?	0.07?	0.06?
Σ (corr.)	99.96 ?	99.90 ?	99.97?	100.14 ?	100.32 ?	99.91 ?	99.99 ?	100.07?	100.31 ?	100.38 ?	
Fe_2O_3 TR[2]	9.68	0.26 ?	2.02	5.92	7.40	6.92	7.45	4.10	2.25	11.48	15.32
Fe_2O_3 TC[3]	9.69		2.02	Note[4]		6.94	7.45	4.09		Note[4]	

[1] "Trace elements" (see Table 5), converted to oxides, where appropriate.

[2] Total iron, expressed as ferric oxide, derived from reported values for total iron.

[3] Total iron, expressed as ferric oxide, calculated from values derived for ferric and ferrous, based on reported values for ferric and ferrous.

[4] Method used to derive Fe_2O_3 values invalidates comparison.

442

	GSJ		MRT	NIM					
	Basalt JB-1	Grano-diorite JG-1	Tonalite T-1	Dunite NIM-D	Granite NIM-G	Lujav-rite NIM-L	Norite NIM-N	Pyroxen-ite NIM-P	Syenite NIM-S
SiO_2	52.60?	72.36	62.70	38.96	75.70	52.40	52.64	51.10	63.63
TiO_2	1.34	0.27	0.58	0.02	0.09	0.48	0.20	0.20	0.04
Al_2O_3	14.62	14.20	16.69	0.3?	12.08	13.64	16.50	4.18	17.34
Fe_2O_3	2.36	0.37	2.71	0.71	0.6?	8.74[*]	0.8?	1.02[*]	1.07[*]
FeO	6.02	1.62	2.88	14.63	1.30	1.13	7.30[*]	10.59	0.30
MnO	0.15	0.06	0.10	0.22	0.067?	0.77	0.18	0.22	0.01
MgO	7.76	0.76	1.89	43.51	0.06?	0.28	7.50	25.33	0.46
CaO	9.35	2.17	5.08	0.28	0.78	3.22	11.50	2.66	0.68
Na_2O	2.79	3.39	4.39	0.04?	3.36	8.37	2.46	0.37	0.43
K_2O	1.42	3.96	1.24	0.01?	4.99	5.51	0.25	0.09	15.35
H_2O^+	1.01	0.54	1.52	0.30	0.10?	2.31	0.33	0.26	0.22
CO_2	0.18?	0.08?	0.07?	0.40?	0.01?	0.17	0.10?	0.08?	0.09
P_2O_5	0.26	0.09	0.14	0.01?	0.01?	0.06	0.03	0.02	0.12
F	0.04?	0.05?	0.05?	0.01?	0.42	0.44	0.03?	0.02?	0.01?
S	?	?	0.01?	0.02?	0.01?	0.065?	0.01?	0.02?	0.01?
Others[1]	0.28?	0.15?	0.19?	0.76?	0.19?	2.39?	0.12?	3.69?	0.34?
Σ	100.18?	100.07?	100.24?	100.19?	100.20?	99.98?	99.95?	99.85?	100.10?
O/F, S, Cl	0.02?	0.02?	0.03?	0.02?	0.19?	0.25?	0.02?	0.02?	0.01?
Σ (corr.)	100.16?	100.05?	100.21?	100.17?	100.01?	99.73?	99.93?	99.83?	100.09?
Fe_2O_3 TR[2]	9.01	2.16	5.90	16.96[*]	2.02[*]	9.96[*]	8.91[*]	12.76[*]	1.40[*]
Fe_2O_3 TC[3]	9.05	2.17	5.91	16.97	2.04?	10.00[*]	8.89[*]	12.79[*]	1.40[*]

[1] "Trace elements" (see Table 5), converted to oxides, where appropriate.

[2] Total iron, expressed as ferric oxide, derived from reported values for total iron.

[3] Total iron, expressed as ferric oxide, calculated from values derived for ferric and ferrous, based on reported values for ferric and ferrous.

[*] Not originators' values – see text.

Table 4

Usable values; arranged by major and minor components
(per cent, dry basis)

SiO₂

SiO₂		SiO₂ (cont'd.)		SiO₂ (cont'd.)	
99.6	BCS-313	55.59 ?	M-3	34.55	MicaFe
99.35	SS	55.02	SO-1	34.4	BCS-367
95.9	BCS-267	54.53	BCR-1	34.1	BCS-309
88.15 ?	FK	53.46	SO-2	33.93	SO-1
75.85	GH	53.22	NS-1	33.65	ES-878-1
75.70	NIM-G	52.88	DR-N	28.3 ?	SGR-1
75.36 ?	I-1	52.72	W-1	17.8	ES-681-1
73.55	GM	52.64	NIM-N	13.4	NBS-69b
73.4 ?	RGM-1	52.60 ?	JB-1	13.0	GXR-3
73.36	SG-1A	52.40	NIM-L	7.39	BX-N
72.36	JG-1	52.22	GL-O	6.80	NBS-697
71.45	GnA	51.10	NIM-P	6.01	NBS-69a
69.96	GA	50.9 ?	MAG-1		
69.22	G-2	49.9 ?	BHVO-1	**TiO₂**	
68.5 ?	SO-4	49.75 ?	I-3	3.69	MRG-1
67.53	NBS-91	49.60	BM	2.78	NBS-69a
67.32	GSP-1	49.2	GXR-1	2.7 ?	BHVO-1
67.1	BCS-375	49.2	GXR-2	2.61	BR
67.1	BCS-376	49.12	ST-1A	2.60	I-3
67.0	NBS-70a	49.0	GXR-6	2.6	NBS-697
67.0	GXR-4	48.94	NBS-98a	2.51	MicaFe
66.0 ?	SDC-1	48.88 ?	M-2	2.41	BX-N
65.98	GS-N	47.06	KK	2.26	BCR-1
65.5 ?	QLO-1	46.39	SGD-1A	2.0	NBS-69b
65.2	NBS-99a	43.67	NBS-97a	1.92	BCS-309
65.11	FK-N	42.10	PCC-1	1.90	NBS-97a
63.63	NIM-S	42.1	GXR-5	1.82	ST-1A
62.8 ?	SCo-1	40.61	UB-N	1.61	NBS-98a
62.71	TS	39.93	MRG-1	1.43	SO-2
62.70	T-1	39.32	SW	1.40	DT-N
60.39	MO8-1	39.05	NIM-D	1.34	JB-1
60.30	TB	38.96	MicaMg	1.14	BM
60.10	SY-2	38.42 ?	BR	1.10	DR-N
59.68	SY-3	38.39	DT-N	1.07	W-1
59.61	AGV-1	36.52	SL-1	1.06	AGV-1
59.5 ?	STM-1	35.73			

TiO₂ (cont'd.)		TiO₂ (cont'd.)		Al₂O₃	
1.05	NS-1	0.17	KK	61.1	BCS-309
1.0 ?	SDC-1	0.15	SY-3	59.21	DT-N
0.93	TB	0.14	SY-2	55.0	NBS-69a
0.88	SO-1	0.13 ?	STM-1	54.53	BX-N
0.83	GXR-6	0.12	NBS-81a	49.3	NBS-69b
0.83	M-3	0.12	UB-N	45.7	NBS-697
0.75	BCS-367	0.11	GXR-1	39.3	GXR-5
0.73 ?	MAG-1	0.09	NIM-G		
0.72	M-2	0.08	GH		
0.71	MO8-1	0.07	SG-1A		
0.71	TS	0.07 ?	GL-O		
0.68	GS-N	0.06 ?	FK		
0.66	GSP-1	0.05	I-1		
0.64 ?	SCo-1	0.04	NIM-S		
0.62	ES-878-1	0.04	SS		
0.62 ?	QLO-1	0.02	BCS-313		
0.58	T-1	0.02	GnA		
0.57	SO-4	0.02	NBS-91		
0.48	ES-681-1	0.02	NIM-D		
0.48	G-2	0.02	SW		
0.48	NIM-L	0.02 ?	FK-N		
0.47	GXR-2	0.01 ?	BCS-76		
0.43	GXR-4	0.01	NBS-70a		
0.38	BCS-375	0.01	NBS-99a		
0.38	GA	0.01	NBS-165a		
0.38 ?	SL-1	0.01	PCC-1		
0.35	GXR-5				
0.32 ?	SO-3				
0.27	JG-1				
0.27 ?	RGM-1				
0.26 ?	SGR-1				
0.21	GM				
0.20	NIM-N				
0.20	NIM-P				
0.17	BCS-267				
0.17	GXR-3				

Table 4 (cont'd.)

Usable values, arranged by major and minor components
(per cent, dry basis)

Al₂O₃ (cont'd.)		Fe₂O₃		FeO (cont'd.)		MnO (cont'd.)		MgO (cont'd.)	
38.79	NBS-97a	37.81	ES-681-1	2.88	T-1	0.07	SO-3	0.93	SO-4
36.77	KK	22.98	BX-N	2.69	UB-N	0.06	GS-N	0.90	SO-2
35.1	GXR-2	17.61	GL-O	2.38	MO8-1	0.06	JG-1	0.76	JG-1
33.19	NBS-98a	8.74	NIM-L	2.32	GSP-1	0.06	MO8-1	0.75	G-2
31.4	GXR-6	8.26	MRG-1	2.25	GL-O	0.06	NBS-1633	0.64	NS-1
23.97 ?	M-2	6.72 ?	TS	2.1 ?	STM-1	0.05	BX-N	0.46	NIM-S
21.31	NS-1	5.61	BR	2.03	AGV-1	0.05	GH	0.42	NBS-98a
20.55	TB	5.46	UB-N	2.0 ?	SW	0.05?	TB	0.38	GM
20.5	NBS-99a	5.20	SW	1.66	GS-N	0.04	SCo-1	0.35	GXR-1
20.0	BCS-367	5.09 ?	I-3	1.62	JG-1	0.04	ASK-2	0.28	NIM-L
19.8	BCS-375	4.66	MicaFe	1.60	NS-1	0.04	GM	0.28?	RGM-1
19.58	MicaFe	4.56	AGV-1	1.44	G-2	0.04	GSP-1	0.19	KK
18.64	FK-N	3.92	ST-1A	1.41	SG-1A	0.04	GXR-5	0.17	BCS-309
18.5 ?	STM-1	3.86	SGD-1A	1.32	GA	0.04?	TS	0.17	NBS-697
17.9	NBS-70a	3.78	DR-N	1.30	NIM-G	0.03	RGM-1	0.16?	FK
17.72	SO-1	3.48	BCR-1	1.2 ?	RGM-1	0.03	G-2	0.15	NBS-97a
17.7	BCS-376	3.09	MO8-1	1.14	GM	0.03	I-1	0.11	BX-N
17.62 ?	M-3	2.9 ?	STM-1	1.13	NIM-L	0.03	NIM-G	0.11	I-1
17.56	DR-N	2.71	T-1	0.84	GH	0.03?	BCS-309	0.1 ?	STM-1
17.34	NIM-S	2.7 ?	BHVO-1	0.66?	TS	0.03?	SGR-4	0.06	BCS-267
17.19	AGV-1	2.54 ?	PCC-1	0.30	NIM-S	0.02	GXR-4	0.06?	NIM-G
16.69	T-1	2.44	SY-3	0.26	BX-N	0.02	KK	0.05	SG-1A
16.50	NIM-N	2.36	JB-1	0.20?	I-1	0.02	NBS-69a	0.05?	BCS-375
16.4 ?	MAG-1	2.31	NS-1	0.10	DT-N	0.01	DT-N	0.04	DT-N
16.20	BM	2.31 ?	M-2	0.06	FK-N	0.01	NBS-91	0.03	GH
16.2 ?	QLO-1	2.28	SY-2			0.01	NIM-S	0.03	GnA
16.15	ES-878-1	1.99 ?	MicaMg			0.01?	FK	0.03?	BCS-376
16.0 ?	SDC-1	1.93	GS-N			0.01?	GL-O	0.02	NBS-69a
15.94	TS	1.70	GnA					0.02	NBS-99a
15.47	SOIL-5	1.70	GSP-1	MnO		MgO		0.01	NBS-91
15.40	G-2	1.60	BM	2.9	GXR-3	49.80	DTS-1	0.01	SS
15.28	GSP-1	1.40	W-1	1.27	ES-878-1	43.51	NIM-D	0.01?	FK-N
15.25 ?	MicaMg	1.36	GA	1.16	BCS-367	43.50	PCC-1		
15.24	SO-2	1.07	G-2	0.86	SL-1	38.5 ?	SW	CaO	
15.02	W-1	1.07	NIM-S	0.77	NIM-L	35.4 ?	UB-N	37.48	SL-1
14.88	SGD-1A	1.02 ?	DTS-1	0.35	MicaFe				
				0.35	NBS-697				
				0.32	SY-2				

446

Sample	Value
GS-N	14.71
GnA	14.7
JB-1	14.62
GA	14.51
ST-1A	14.23
JG-1	14.20
GXR-4	14.0
I-1	13.92 ?
SG-1A	13.84
RGM-1	13.8 ?
BCR-1	13.72
BHVO-1	13.7 ?
NIM-L	13.64
SCo-1	13.6 ?
GM	13.50
I-3	13.07 ?
GH	12.51
SY-2	12.12
NIM-G	12.08
SY-3	11.80
ES-681-1	10.60
SO-4	10.32
BR	10.25
MO8-1	9.94
SL-1	9.63
MRG-1	8.50
GL-O	7.75
GXR-J	6.7
SGR-1	6.5 ?
FK	6.22 ?
NBS-91	6.01
SO-3	5.76
NIM-P	4.18
UB-N	2.97
BCS-267	0.85
PCC-1	0.73
NBS-81a	0.66
SW	0.66
NIM-D	0.3 ?
DTS-1	0.25
SS	0.25
BCS-313	0.16
NBS-165a	0.06

Sample	Value
NIM-P	1.02 ?
QLO-1	1.0 ?
TB	0.91
NIM-N	0.91 ?
GM	0.8 ?
NIM-D	0.75
SG-1A	0.71
NIM-G	0.68
DT-N	0.6 ?
RGM-1	0.55 ?
GH	0.5 ?
JG-1	0.41
I-1	0.37
FK-N	0.33 ?
	0.02

FeO

Sample	Value
MicaFe	18.99
NIM-D	14.63
NIM-P	10.59
ST-1A	10.26
I-3	10.04?
BCR-1	8.96
W-1	8.73
ES-681-1	8.70?
MRG-1	8.63
BHVO-1	8.5 ?
NIM-N	7.30
BM	7.28
DTS-1	6.94?
SGD-1A	6.86
MicaMg	6.75?
BR	6.60
M-2	6.30?
JB-1	6.02
TB	5.43
DR-N	5.32
PCC-1	5.17
GnA	3.80
SY-2	3.62
SY-3	3.58
M-3	3.33?
QLO-1	3.0 ?

Sample	Value
SY-3	0.32
ES-681-1	0.28
M-3	0.28
M-2	0.26
MicaMg	0.26?
I-3	0.22
NIM-D	0.22
NIM-P	0.22
STM-1	0.22?
DR-N	0.21
ST-1A	0.21
BR	0.20
SG-1A	0.19
BCR-1	0.18
NIM-N	0.18
NS-1	0.18
GnA	0.17
MRG-1	0.17
SGD-1A	0.17
W-1	0.17?
BHVO-1	0.15
BCS-267	0.15
JB-1	0.15
BM	0.14
ASK-1	0.13
GXR-6	0.13
DTS-1	0.12
GXR-2	0.12
PCC-1	0.12
UB-N	0.12
SDC-1	0.12?
SO-1	0.11
SOIL-5	0.11?
AGV-1	0.10
T-1	0.10
MAG-1	0.10?
GA	0.09
NBS-69b	0.09
SO-2	0.09
VS-N	0.09
QLO-1	0.09?
GnA	0.08
SO-4	0.08
SW	0.08

Sample	Value
NIM-P	25.33
MicaMg	20.46?
MRG-1	13.49
BR	13.35
SL-1	12.27
ES-878-1	9.55
SO-3	8.47?
JB-1	7.76
NIM-N	7.50
BM	7.46
BHVO-1	7.2 ?
BCS-367	7.1
SGD-1A	7.09
W-1	6.63
GL-O	5.74
MicaFe	4.58
SGR-1	4.57
DR-N	4.5 ?
I-3	4.47
SO-1	4.18
BCR-1	3.83
MAG-1	3.48
SY-2	3.0 ?
GXR-4	2.70
SY-3	2.7
SCo-1	2.67
M-2	2.6 ?
GS-N	2.45
GXR-5	2.31
TB	2.0
T-1	1.94
TS	1.89
SDC-1	1.77
AGV-1	1.70
ES-681-1	1.52
GXR-2	1.48
MO8-1	1.45
M-3	1.34
GXR-3	1.21
GXR-6	1.05
QLO-1	1.02
GSP-1	1.0 ?
GA	0.97
	0.95

Sample	Value
ES-878-1	35.65
BCS-367	32.4
SO-3	20.7 ?
GXR-3	19.7
MRG-1	14.77
BR	13.87
M-3	12.01?
NIM-N	11.50
BHVO-1	11.4 ?
W-1	10.98
SGD-1A	10.97?
NBS-91	10.48
ST-1A	10.24?
JB-1	9.35
MO8-1	8.70
SY-3	8.26
I-3	8.20
SY-2	7.98
DR-N	7.09
BCR-1	6.97
BM	6.46
T-1	5.08
AGV-1	4.94
ES-681-1	3.92
NIM-L	3.22
QLO-1	3.2?
SO-2	2.74
NIM-P	2.66
SCo-1	2.6?
SO-1	2.52
GS-N	2.51
GA	2.45
JG-1	2.17
NBS-99a	2.14
GSP-1	2.03
G-2	1.96
BCS-267	1.75
M-2	1.75
NS-1	1.70
SO-4	1.55
SDC-1	1.4 ?
MAG-1	1.35?
GXR-4	1.26
GXR-1	1.22

Table 4 (cont'd.)

Usable values, arranged by major and minor components
(per cent, dry basis)

CaO (cont'd.)

Sample	Value
UB-N	1.18
GXR-2	1.15
RGM-1	1.1 ?
STM-1	1.1 ?
GXR-5	1.05
GM	1.04
GL-O	0.98
BCS-375	0.89
I-1	0.80
NIM-G	0.78
GH	0.69
NIM-S	0.68
GnA	0.61
NBS-697	0.60
PCC-1	0.55
MicaFe	0.43
TB	0.33
NBS-98a	0.31
NBS-69a	0.29
NIM-D	0.28
KK	0.24
BCS-309	0.22
SW	0.18
BX-N	0.17
DTS-1	0.14
GXR-6	0.14
SG-1A	0.14
TS	0.13
NBS-69b	0.12
FK	0.12?
FK-N	0.11
NBS-70a	0.11
NBS-97a	0.11
MicaMg	0.08?
DT-N	0.04
SS	0.03
BCS-313	0.02

Na2O (cont'd.)

Sample	Value
BHVO-1	2.3 ?
W-1	2.15
SDC-1	2.0 ?
M-2	1.40
TB	1.31
SO-4	1.31?
GXR-3	1.05
GXR-5	1.04
SO-3	1.00
SCo-1	0.9 ?
GXR-2	0.75
GXR-4	0.71
MRG-1	0.71
MO8-1	0.54?
ES-878-1	0.47
BCS-367	0.44
NIM-S	0.43
SL-1	0.39?
NIM-P	0.37
BCS-309	0.34
MicaFe	0.30
FK	0.25?
GXR-6	0.14
MicaMg	0.12?
UB-N	0.10
ES-681-1	0.09
TS	0.09
GnA	0.08
NBS-98a	0.08
GXR-1	0.07
BCS-267	0.06
BX-N	0.06
NBS-697	0.05
DT-N	0.04
NBS-97a	0.04
SS	0.04
GL-O	0.04?
NIM-D	0.04?

K2O (cont'd.)

Sample	Value
SGD-1A	2.96
SO-2	2.95
AGV-1	2.92
SCo-1	2.7 ?
GnA	2.63
GXR-6	2.5
SOIL-5	2.24
MO8-1	2.2 ?
SO-4	2.08
NBS-1633	2.07
DR-N	1.73
BCR-1	1.70
GXR-2	1.70
SGR-1	1.6 ?
I-3	1.43
JB-1	1.42
BR	1.41
SO-3	1.40
ES-878-1	1.29
T-1	1.24
BCS-367	1.17
KK	1.06
NBS-98a	1.04
GXR-5	0.99
NBS-69b	0.80
BCS-375	0.79
M-3	0.71
ST-1A	0.69
W-1	0.64
ES-681-1	0.59
BHVO-1	0.53?
SL-1	0.51
NBS-97a	0.50
BCS-309	0.46
NIM-N	0.25
BM	0.20
MRG-1	0.18
BCS-267	0.14

H2O+ (cont'd.)

Sample	Value
NS-1	0.66
GSP-1	0.58
JG-1	0.54
W-1	0.53
G-2	0.50
NIM-G	0.49
GH	0.46
SY-2	0.43
DTS-1	0.42
SY-3	0.42
FK	0.41?
GM	0.35
NIM-N	0.33
FK-N	0.32?
NIM-D	0.30
NIM-P	0.26
NIM-S	0.22
SG-1A	0.21
I-1	0.13?

CO2

Sample	Value
MO8-1	7.3 ?
M-3	2.98?
BM	1.34
MRG-1	1.00
BR	0.86
SY-2	0.46
BX-N	0.45
UB-N	0.44
NIM-D	0.40?
SY-3	0.38
SW	0.29?
GM	0.28
MicaFe	0.19?
PCC-1	0.18
GS-N	0.18?
JB-1	0.18?

P2O5 (cont'd.)

Sample	Value
BCR-1	0.36
NBS-97a	0.36
M-3	0.36?
TB	0.29
TS	0.29
GS-N	0.28
GSP-1	0.28
NS-1	0.28
BHVO-1	0.28?
JB-1	0.26
DR-N	0.25
SO-4	0.21
ST-1A	0.21
STM-1	0.16?
SO-1	0.14
T-1	0.21
W-1	0.14?
BCS-367	0.13
BX-N	0.13
G-2	0.13
GA	0.12
NBS-69b	0.12
NIM-S	0.12
NBS-98a	0.11
SO-3	0.11?
BM	0.10
TB	0.10
DT-N	0.09
JG-1	0.09
KK	0.09
NBS-69a	0.08
FK	0.08?
GM	0.06
MRG-1	0.06
NIM-L	0.06
UB-N	0.05
ES-878-1	0.03
NIM-N	0.03

Na₂O, continued

Value	Sample
0.03	KK
0.03	NBS-69b
0.01	PCC-1
0.01	SW
0.01?	DTS-1

K₂O

Value	Sample
15.35	NIM-S
12.83	FK-N
11.8	NBS-70a
11.2	BCS-376
10.03?	MicaMg
8.79	MicaFe
8.16	GL-O
7.90	M-2
6.52	NS-1
5.51	GSP-1
5.51	NIM-L
5.2	GXR-2
5.2	NBS-99a
4.99	NIM-G
4.88	TS
4.76	GH
4.74	GM
4.64	GS-N
4.48	SY-2
4.46	G-2
4.3 ?	RGM-1
4.3 ?	STM-1
4.28	I-1
4.20	SY-3
4.15?	FK
4.14	SG-1A
4.03	GA
3.96	JG-1
3.85	TB
3.6 ?	MAG-1
3.6 ?	QLO-1
3.3 ?	SDC-1
3.25	NBS-91
3.23	SO-1

H₂O⁺, continued

Value	Sample
0.12	DT-N
0.09	NIM-P
0.07	NBS-697
0.06	GXR-1
0.06	SS
0.04	BCS-313
0.02	UB-N
0.01?	NIM-D
0.01	SW

H₂O⁺

Value	Sample
13.6	SW
12.75	KK
11.69	BX-N
11.28	UB-N
10.24?	ES-681-1
5.72	GL-O
4.70	PCC-1
4.03	TS
3.82	TB
3.62	BM
3.21?	M-2
3.0 ?	MO8-1
2.92?	MicaFe
2.31	BR
2.31	NIM-L
2.20	DR-N
2.10?	MicaMg
1.8 ?	GnA
1.71?	I-3
1.52	T-1
1.07	GS-N
1.01	JB-1
0.98	MRG-1
0.97	SY-3
0.95?	ST-1A
0.87	DT-N
0.83	GA
0.78	SGD-1A
0.78?	AGV-1
0.67	BCR-1

Na₂O

Value	Sample
10.4	BCS-375
9.85	NS-1
9.0?	STM-1
8.48	NBS-91
8.37	NIM-L
6.2	NBS-99a
5.46	SG-1A
4.64	BM
4.59	I-1
4.39	T-1
4.34	SY-2
4.32	AGV-1
4.2 ?	QLO-1
4.15	SY-3
4.1 ?	RGM-1
4.06	G-2
3.85	GH
3.8 ?	MAG-1
3.78	GS-N
3.76	GM
3.55	GA
3.39	JG-1
3.36	NIM-G
3.30	BCR-1
3.07	BR
3.00	DR-N
3.0 ?	SGR-1
2.98	M-3
2.92	I-3
2.83	BCS-376
2.82	SGD-1A
2.81	GSP-1
2.79	JB-1
2.59	SOIL-5
2.58	FK-N
2.56?	SO-1
2.55	NBS-70a
2.49	NIM-N
2.46	ST-1A
2.35	SO-2

(continued)

Value	Sample
0.17	KK
0.17	NIM-L
0.15	DR-N
0.15?	MicaMg
0.14	GH
0.14?	NS-1
0.13	SGD-1A
0.13	TB
0.12	GSP-1
0.11	GA
0.10	ST-1A
0.10?	NIM-G
0.10?	NIM-N
0.09	NIM-S
0.08?	FK-N
0.08?	G-2
0.08?	JG-1
0.08?	NIM-P
0.07	DTS-1
0.07?	SG-1A
0.07?	T-1
0.06	W-1
0.04?	GnA
0.04?	TS
0.02	AGV-1
0.02	BCR-1
0.02?	FK

P₂O₅

Value	Sample
2.02	ES-681-1
1.05	BR
1.01	SGD-1A
0.90	NBS-697
0.69?	SO-2
0.54	SY-3
0.51	AGV-1
0.50?	M-2
0.45	MicaFe
0.43	SY-2
0.40?	I-3
0.38	GL-O

(continued)

Value	Sample
0.03	SW
0.02	NBS-91
0.02	NBS-99a
0.02	NIM-P
0.02?	FK-N
0.02?	GnA
0.02?	I-1
0.02?	SL-1
0.01	GH
0.01	MO8-1
0.01?	NIM-G
0.01	PCC-1
0.01	SG-1A
0.01?	MicaMg
0.01?	NIM-D

F

Value	Sample
5.72	NBS-91
3.36	GnA
2.86?	MicaMg
1.59	MicaFe
0.66	SY-3
0.51	SY-2
0.44	NIM-L
0.42	NIM-G
0.37	GSP-1
0.35	GH
0.30	SG-1A
0.19	ES-681-1
0.15?	ES-878-1
0.15?	GL-O
0.14	NS-1
0.12	G-2
0.12	SGD-1A
0.12?	TS
0.10	BR
0.10?	M-2
0.07?	GM
0.07?	I-3
0.07?	SO-1
0.07?	TB

449

Table 4 (cont'd.)

Usable values, arranged by major and minor components
(per cent, dry basis)

F (cont'd.)	
0.06?	M-3
0.05	BCR-1
0.05	GA
0.05?	JG-1
0.05?	SO-2
0.05?	T-1
0.04	AGV-1
0.04?	JB-1
0.03?	NIM-N
0.03?	BM
0.03?	SO-3
0.03?	SO-4
0.03?	ST-1A
0.02	MRG-1
0.02?	NIM-P
0.02	W-1
0.01	MO8-1
0.01?	NIM-S
0.01?	DTS-1
0.01?	SW

S	
1.26	SL-1
0.94	BCS-367
0.81	ES-878-1
0.46	MO8-1
0.10	ES-681-1
0.07	ST-1A

S (cont'd.)	
0.06	MRG-1
0.06?	NIM-L
0.05	SY-3
0.04	BR
0.04?	BCR-1
0.03?	GSP-1
0.02	GnA
0.02?	NIM-D
0.02?	NIM-P
0.02	SG-1A
0.02	SGD-1A
0.02?	KK
0.01?	NIM-G
0.01?	NIM-N
0.01	NIM-S
0.01?	SY-2
0.01?	AGV-1
0.01?	G-2
0.01?	NS-1
0.01?	PCC-1
0.01?	T-1
0.01?	W-1

SO₃	
0.63	NBS-69b
0.15	NBS-697
0.04	NBS-69a

Fe_2O_3 T	
47.48	ES-681-1
35.3	GXR-1
26.6	GXR-3
25.76	MicaFe
23.27	BX-N
20.11	GL-O
20.0	NBS-697
17.82	MRG-1
16.96	NIM-D
16.22?	I-3
15.32	ST-1A
13.41	BCR-1
12.90	BR
12.76	NIM-P
12.0?	BHVO-1
11.48	SGD-1A
11.11	W-1
9.96	NIM-L
9.69	DR-N
9.68	BM
9.49?	MicaMg
9.25	M-2
9.01	JB-1
8.91	NIM-N
8.70	DTS-1
8.57	SO-1
8.45	UB-N
8.28	PCC-1

Fe_2O_3 T (cont'd.)	
8.0	GXR-6
7.95	SO-2
7.45	TS
7.40	SW
7.1	NBS-69b
6.92	TB
6.9?	SDC-1
6.8?	MAG-1
6.78	AGV-1
6.42	SY-3
6.36	SO-5
6.28	SY-2
5.92	GnA
5.90	T-1
5.8	NBS-69a
5.72	MO8-1
5.2?	STM-1
5.1?	SCo-1
4.6	GXR-5
4.55	M-3
4.30	GSP-1
4.3?	QLO-1
4.2	GXR-4
4.10	NS-1
3.76	GS-N
3.39	SO-4
3.2?	SGR-1
2.77	GA
2.7	GXR-2

Fe_2O_3 T (cont'd.)	
2.69	G-2
2.25	SG-1A
2.16	JG-1
2.16	SO-3
2.02	GM
2.02	NIM-G
1.9?	RGM-1
1.51	BCS-309
1.40	NIM-S
1.35	GH
1.34	NBS-98a
1.11	BCS-367
1.02	SL-1
0.98	KK
0.86	ES-878-1
0.79	BCS-267
0.66	DT-N
0.54	I-1
0.45	NBS-97a
0.26	FK
0.12	BCS-375
0.10	BCS-376
0.08	NBS-70a
0.08	NBS-81a
0.08	NBS-91
0.06	NBS-99a
0.04	SS
0.02	BCS-313
0.01	NBS-165a

Table 5

Usable values, arranged by "trace elements"

Ag ppm	
0.4	ASK-2
0.14?	MRG-1

Ag ppb	
957?	AGV-1
83?	GSP-1
81	W-1
55?	JB-1
50	ASK-1
50?	JG-1
40?	G-2
35?	BCR-1
10?	DTS-1
10?	PCC-1

As ppm		pct As₂O₃T*
4000	GXR-3	0.53
1350	NBS-91	0.18
460	GXR-1	0.06
340	GXR-6	0.04
110?	ES-681-1	0.01
98	GXR-4	0.01
94	SOIL-5	0.01
61	NBS-1633	0.01
47?	GnA	0.01
31	GXR-2	0.01
30	MO8-1	
20	SY-3	
18	SY-2	
14?	BM	
12	GXR-5	
7.1?	SO-4	
4?	GM	
2.6?	SO-3	
1.9	W-1	

*Total arsenic, expressed as As₂O₃.

As ppm (cont'd.)		pct As₂O₃T
1.9?	SO-1	
1.2?	SO-2	
0.8?	AGV-1	
0.8?	BCR-1	
0.7	MRG-1	
0.25?	G-2	
0.09?	GSP-1	
0.05?	PCC-1	
0.03?	DTS-1	

Au ppb	
3.7	W-1
1?	G-2
1?	GSP-1
0.8	BCR-1
0.8?	DTS-1
0.7?	PCC-1
0.6?	AGV-1

B ppm		pct B₂O₃
300?	VS-N	0.10
180?	GXR-3	0.06
155	ASK-2	0.05
110	SY-3	0.04
85?	SY-2	0.03
85?	TS	0.03
44?	GXR-2	0.01
39?	SW	0.01
25?	GXR-5	0.01
22?	SO-3	0.01
20	GA	0.01
20?	GnA	0.01
20?	SO-1	0.01
16	SGD-1A	0.01
15	ST-1A	
15?	GXR-1	
15?	W-1	

B ppm (cont'd.)		pct B₂O₃
13?	GM	
13?	MRG-1	
13?	NIM-S	
12?	JB-1	
11?	GXR-6	
10	SG-1A	
10?	BR	
8?	NIM-G	
6?	AGV-1	
6?	JG-1	
6?	PCC-1	
4	BCR-1	
4?	GXR-4	
4?	NIM-D	
2?	G-2	

Ba ppm		pct BaO
4700	GXR-3	0.52
4000?	MicaMg	0.45
2700?	NBS-1633	0.30
2400	NIM-S	0.27
2300	NBS-99a	0.26
2000	GXR-2	0.22
1950?	TS	0.22
1900	G-2	0.21
1800	GXR-5	0.20
1550?	M-2	0.17
1400?	GS-N	0.16
1350	GXR-4	0.15
1300	GSP-1	0.15
1300	SGD-1A	0.15
1300?	QLO-1	0.15
1200	AGV-1	0.13
1200	NS-1	0.13
1150	ASK-1	0.13
1100	GXR-6	0.12
1050	BR	0.12
1000?	SO-2	0.11

Ba ppm (cont'd.)		pct BaO
900	VS-N	0.10
900?	SO-1	0.10
850	GA	0.09
780?	SO-4	0.09
760?	RGM-1	0.08
720	TB	0.08
690?	I-3	0.08
680	BCR-1	0.08
670	NBS-97a	0.07
670	T-1	0.07
630?	SDC-1	0.07
560	GXR-1	0.06
560	SOIL-5	0.06
560?	SCo-1	0.06
550?	STM-1	0.06
490	JB-1	0.05
480?	I-1	0.05
480?	MAG-1	0.05
460	JG-1	0.05
460	SY-2	0.05
450	NIM-L	0.05
430	SY-3	0.05
380	DR-N	0.05
330	GM	0.04
300?	SGR-1	0.03
290	ST-1A	0.03
280?	SO-3	0.03
260	BM	0.03
230	NBS-99a	0.03
210?	FK-N	0.02
180	NBS-70a	0.02
160	W-1	0.02
145	MicaFe	0.02
130?	BHVO-1	0.01
125?	M-3	0.01
120?	NIM-G	0.01
100	NIM-N	0.01
90	NBS-69a	0.01
80	NBS-697	0.01

Table 5 (cont'd.)

Usable values, arranged by "trace elements"

Ba ppm (cont'd.)

Sample	Ba ppm	pct BaO
MRG-1	50?	0.01
GnA	47?	0.01
NIM-P	46?	0.01
UB-N	40?	
GH	22	
SW	21	
SG-1A	19	
NIM-D	10?	
DTS-1	5?	
PCC-1	4?	

Be ppm

Sample	Be ppm	pct BeO
GXR-3	26?	0.01
SY-2	23	0.01
SY-3	22	0.01
NIM-L	20?	0.01
KK	12?	
NBS-1633	12?	
SG-1A	11	
STM-1	10?	
MicaFe	8?	
NIM-G	7?	
GH	6?	
NS-1	6?	
GnA	5?	
GM	4.4?	
ASK-1	4	
ASK-2	4	
TB	4?	
GA	3.6	
TS	3.5?	
G-2	2.4	
GXR-4	2.1	
SGD-1A	2	
AGV-1	2?	

C (non-carbonate)

Sample	ppm	pct
TS		1.39
DT-N	6200	0.62
KK	1400	0.14
ST-1A	360	0.04
SGD-1A	280	0.03
SY-2	270?	0.03
MRG-1	250?	0.02
SY-3	250?	0.02
SG-1A	240?	0.02
BCR-1	65?	0.01

C (total)

Sample	ppm	pct
SO-3		6.6?
SO-2		4.8?
SO-4		4.4?
ES-681-1		1.8
TS		1.4
MRG-1	3000?	0.30
SO-1	2500?	0.25
KK	1900	0.19
SY-2	1500?	0.15
SY-3	1300?	0.13
SGD-1A	630	0.06
ST-1A	630	0.06
BCR-1	440?	0.04
	120?	0.01

Cd ppm

Sample	Cd ppm	pct CdO
VS-N	900?	0.10
SS	3?	
NBS-1633	1.4	

Ce ppm (cont'd.)

Sample	Ce ppm
BM	23?
W-1	23?
ST-1A	22?
GXR-1	19?
GXR-3	16?
NIM-S	11?
NIM-N	10?

pct CeO2 / Cl ppm

Sample	pct CeO2	Cl ppm
MAG-1	3.0 ?	1200
NIM-L	0.12	800?
MicaMg	0.08	500?
MicaFe	0.05	500?
NS-1	0.05	500?
RGM-1	0.05	500?
STM-1	0.05	430?
ST-1A	0.04	400?
NIM-D	0.04	370?
BR	0.04	340
GSP-1	0.03	300?
GA	0.03	250?
QLO-1	0.03	220?
SGD-1A	0.02	200?
NIM-G	0.02	200?
NIM-P	0.02	200?
W-1	0.02	185
AGV-1	0.02	175?
JB-1	0.02	150?
MRG-1	0.02	140
NBS-91	0.02	140?
SY-3	0.01	130?
SY-2	0.01	100
ASK-1	0.01	100
G-2	0.01	100

(pct Cl)

Co ppm (cont'd.)

Sample	Co ppm	pct CoO
AGV-1	16	
GXR-4	16	
FK-N	16?	
SOIL-5	15	
SO-4	15?	
GXR-6	14	
DT-N	14?	
TB	13	
SO-2	13?	
T-1	13?	
SY-3	12	
SO-3	12?	
SY-2	11	
SGR-1	11?	
M-3	10?	
SCo-1	10?	
GXR-1	9	
GXR-2	9	
NS-1	8?	
GSP-1	7.8	
QLO-1	7?	
JG-1	6.4?	
ASK-1	6	
NIM-L	6?	
G-2	5	
GA	5	
NIM-G	4?	
GM	3.5	
NIM-S	3?	
RGM-1	3?	
GnA	2.1?	
GH	1.5?	
SG-1A	1.4	
SS	0.5?	

Bi ppm

Sample	Bi ppm
BCR-1	1.6?
GXR-2	1.6?
BM	1.3?
GXR-5	1.2?
GXR-1	1.1?
ST-1A	1
BR	1?
GSP-1	1?
GXR-6	1?
NIM-N	1?
NIM-S	1?
W-1	0.8?
MRG-1	0.6?

pct Bi₂O₃

Sample	Bi ppm	pct Bi₂O₃
VS-N	900?	0.10
GnA	220	0.02

Bi ppb

Sample	Bi ppb
AGV-1	50?
BCR-1	50?
W-1	50?
G-2	40?
GSP-1	40?
DTS-1	10?
PCC-1	10?

Br ppm

Sample	Br ppm
GXR-5	8?
SOIL-5	5?
GXR-2	3?
GXR-6	1.4?
PCC-1	0.6?
AGV-1	0.5?
GXR-4	0.5?
GXR-1	0.4?
W-1	0.4?
G-2	0.3?
BCR-1	0.2?
DTS-1	0.2?

Cd ppb

Sample	Cd ppb
SO-4	420?
SO-2	180?
W-1	150
SO-1	150?
SO-3	140?
DTS-1	120?
PCC-1	100?
AGV-1	90?
BCR-1	90?
GSP-1	60?
G-2	39?

Ce ppm / pct CeO₂

Sample	Ce ppm	pct CeO₂
SY-3	2200	0.27
VS-N	900?	0.11
MicaFe	370?	0.05
GSP-1	360	0.04
NIM-L	230?	0.03
SY-2	210?	0.03
NIM-G	200?	0.02
NS-1	185	0.02
G-2	160	0.02
SGD-1A	150	0.02
NBS-1633	145	0.02
GXR-4	115?	0.01
TB	115?	0.01
AGV-1	71	0.01
GA	70	0.01
JB-1	67?	0.01
SG-1A	67?	0.01
SOIL-5	60	0.01
GM	60?	0.01
BCR-1	53	0.01
GXR-2	50	0.01
GH	50?	0.01
JG-1	43?	0.01
GXR-5	40?	0.01
GXR-6	38?	
MRG-1	25?	

Co (top block)

Sample	value	pct CoO
NIM-N	100	
GH	100?	0.01
NIM-S	100?	0.01
PCC-1	80?	0.01
JG-1	59?	0.01
BCR-1	58?	0.01
ASK-2	14	
DTS-1	11?	

Co ppm / pct CoO

Sample	Co ppm	pct CoO
VS-N	700	0.09
NIM-D	210	0.03
DTS-1	135	0.02
NIM-P	110	0.01
PCC-1	110	0.01
UB-N	100	0.01
SW	86	0.01
MRG-1	80?	0.01
ES-681-1	65?	0.01
GS-N	58	0.01
NIM-N	50	0.01
BR	49?	0.01
I-3	48	0.01
GXR-3	47?	0.01
BHVO-1	47?	0.01
W-1	46	0.01
ST-1A	40	0.01
SGD-1A	40	
BX-N	40?	
NBS-1633	40?	
JB-1	39	
TS	38	
BCR-1	36	
DR-N	35	
BM	34	
SO-1	33?	
GXR-5	30	
M-2	30?	
ASK-2	27	
MicaFe	20	
MicaMg	20?	
SDC-1	20?	
MAG-1	18?	
GL-O	17?	

Cr ppm / pct Cr₂O₃

Sample	Cr ppm	pct Cr₂O₃
NIM-P	4200	3.50
DTS-1	2900	0.61
NIM-D	2800	0.42
PCC-1	2500	0.41
SW	2300	0.37
UB-N	700	0.34
VS-N	680?	0.10
NBS-697	450	0.10
MRG-1	400	0.07
JB-1	380	0.06
BR	340?	0.06
NBS-69a	320?	0.05
BHVO-1	280?	0.05
BX-N	270	0.04
TS	240?	0.04
DT-N	200?	0.04
NBS-97a	200?	0.03
NBS-98a	160	0.03
SO-1	140	0.02
ST-1A	140?	0.02
GL-O	130	0.02
NBS-1633	130?	0.02
MO8-1	125	0.02
BM	115?	0.02
W-1	105?	0.02
MAG-1	100	0.02
GXR-5	100?	0.01
MicaMg	96	0.01
GXR-6	90	0.01
ASK-2	90	0.01
MicaFe	80	0.01
TB	70?	0.01
NBS-69b	68?	0.01
SDC-1	65?	0.01
SCo-1	64	0.01
GXR-4	61	0.01
SO-4	60?	0.01
SL-1	56?	0.01
M-2	55?	0.01
GS-N	54?	0.01
M-3	53	0.01
JG-1		0.01

Table 5 (cont'd.)

Usable values, arranged by "trace elements"

Cr ppm (cont'd.)		pct Cr₂O₃
52	SGD-1A	0.01
45	DR-N	0.01
40	ASK-1	0.01
37	GXR-2	0.01
32?	SGR-1	
31	NBS-81a	
30?	NIM-N	
29?	SOIL-5	
26?	I-3	
20?	T-1	
19?	GXR-3	
16	SO-2	
15	BCR-1	
14	GnA	
12	GA	
12	GSP-1	
12	NIM-G	
12	NIM-S	
12	SG-1A	
12	SY-2	
10	AGV-1	
10	GM	
10	SY-3	
10?	GXR-1	
10?	NIM-L	
10?	NS-1	
9.5?	KK	
8	G-2	
7?	I-1	
6	GH	
4?	QLO-1	
3?	RGM-1	
3?	STM-1	

Cu ppm		pct CuO
6500?	GXR-4	0.81
1300?	GXR-1	0.16
800	VS-N	0.10
490?	TS	0.06
360?	GXR-5	0.05
220	ST-1A	0.03
165?	I-3	0.02
135	MRG-1	0.02
130	NBS-1633	0.02
130?	M-2	0.02
120	ASK-2	0.02
110	W-1	0.01
105?	GXR-6	0.01
77	SOIL-5	0.01
72	BR	0.01
68	SGD-1A	0.01
66?	SGR-1	0.01
61	SO-1	0.01
59	AGV-1	0.01
56	JB-1	0.01
52	DR-N	0.01
50	TB	0.01
48?	T-1	0.01
45	BM	0.01
35?	MO8-1	0.01
33	GSP-1	0.01
33?	MAG-1	0.01
31	SG-1A	0.01
30	UB-N	0.01
30?	QLO-1	
30?	SCo-1	
30?	SDC-1	
25?	M-3	

Dy ppm (cont'd.)		pct Dy₂O₃
4?	ST-1A	
3.5?	AGV-1	
3?	GXR-1	
3?	GXR-2	
3?	JG-1	
3?	MRG-1	
3?	NIM-L	
2.8?	GXR-6	
2.6?	GXR-4	
2.3	G-2	
2?	GXR-5	
2?	NS-1	
0.5?	NIM-S	

Dy ppb	
3?	DTS-1

Er ppm		pct Er₂O₃
50?	SY-3	0.01
12?	SY-2	
10?	NIM-G	
7?	SG-1A	
3.5?	BCR-1	
3?	GSP-1	
2.8?	SGD-1A	
2.4	W-1	
2.3?	JB-1	
2	NS-1	
2?	ST-1A	
1.3?	G-2	
1.2?	AGV-1	

Ga ppm (cont'd.)		pct Ga₂O₃
30?	GXR-6	
29	ASK-1	
28	SY-2	
27	NIM-G	
26	SY-3	
25	ASK-2	
25	DR-N	
25	TB	
24?	NS-1	
23	G-2	
23	GH	
23	GSP-1	
23?	M-2	
22	BCR-1	
22?	TS	
21?	MicaMg	
20	BR	
20?	T-1	
19	AGV-1	
19	SGD-1A	
18	SOIL-5	
18?	MRG-1	
17?	I-3	
17?	JB-1	
17?	M-3	
16	GA	
16	NIM-N	
16	ST-1A	
16	W-1	
15	BM	
15	GM	
15?	GXR-4	
15?	JG-1	
12?	GXR-1	

Cs

Cs ppm		pct Cs₂O	
900?	VS-N	0.10	VS-N
200	GXR-3	0.02	GXR-3
200?	MicaFe	0.02	MicaFe
57?	SOIL-5	0.01	SOIL-5
55?	MicaMg	0.01	MicaMg
46	GnA		
12	SG-1A		
12?	TS		
11	ASK-2		
10?	JG-1		
9?	NBS-1633		
7.6?	GM		
7?	FK-N		
6.8	TB		
6	GA		
6?	NIM-S		
5	GXR-2		
4.8	GXR-6		
4.0	SGD-1A		
4	GXR-1		
3.5	NIM-L		
3.3?	NS-1		
3	GXR-4		
2.5	GH		
2.5?	SY-3		
2.3?	SY-2		
2.2	GXR-5		
1.7?	BM		
1.5	ASK-1		
1.4	G-2		
1.3?	AGV-1		
1?	GSP-1		
1?	JB-1		
1?	NIM-G		
0.95?	BCR-1		
0.9	ST-1A		
0.9	W-1		
0.6?	MRG-1		
0.3?	SW		

Cs ppb	
25?	PCC-1
6?	DTS-1

Dy

Dy ppm		pct Dy₂O₃	
22	SO-4	0.01	SY-3
20?	BX-N		
20?	GS-N		
19	NIM-S		
18	GnA		
18	NIM-P		
17	SO-3		
16	BCR-1		
16	GA		
16	SY-3		
14	GH		
14	NIM-N		
13	GM		
13	NIM-L		
12	NIM-G		
11?	RGM-1		
10	G-2		
10	NIM-D		
10?	DT-N		
8.8	KK		
8	PCC-1		
8?	I-1		
8?	NS-1		
7	SS		
7	ASK-1		
7	SO-2		
5	SW		
5	DTS-1		
5?	SY-2		
4	GL-O		
4?	JG-1		
4?	MicaFe		
3?	MicaMg		
3?	FK-N		
3?	STM-1		

Dy ppm	
80?	SY-3
20?	SY-2
16?	NIM-G
7?	BCR-1
7?	GH
5.7?	GSP-1
5?	SG-1A
4	W-1
4?	JB-1
4?	SGD-1A
4?	SOIL-5

Eu

Eu ppm	
14?	SY-3
6	SGD-1A
3.7?	BR
3	ST-1A
2.5?	NBS-1633
2.4?	GSP-1
2.4?	SY-2
2.0?	BCR-1
2?	BHVO-1
1.6?	AGV-1
1.6?	GXR-4
1.5?	JB-1
1.4	G-2
1.4?	MRG-1
1.2?	SOIL-5
1.1	W-1
1.1?	BM
1?	NIM-L
0.9?	GXR-5
0.8?	GXR-2
0.8?	GXR-6
0.7?	GXR-1
0.7?	JG-1
0.6?	GM
0.5?	SG-1A
0.4?	GXR-3
0.4?	NIM-G
0.3?	NIM-S
0.2?	NIM-P

Eu ppb	
2?	PCC-1
0.9?	DTS-1

Ga

Ga ppm		pct Ga₂O₃	
400?	VS-N	0.05	VS-N
95?	MicaFe	0.01	MicaFe
60	GnA	0.01	GnA
54?	NIM-L	0.01	NIM-L
40	SG-1A	0.01	SG-1A
38?	STM-1	0.01	STM-1
32?	GXR-2		

Gd

Gd ppm		pct Gd₂O₃	
11	NIM-S	0.01	NIM-S
8?	NIM-P		
7?	UB-N		
1?	DTS-1		
0.7?	PCC-1		

Gd ppm	
55?	SY-3
15?	GSP-1
11?	NIM-G
6.6?	BCR-1
5.5?	AGV-1
5?	G-2
4?	W-1
0.7?	NIM-S

Ge

Ge ppm	
6.5?	GnA
3.3	SG-1A
2.5?	TB
1.6	ST-1A
1.6?	GM
1.5	BCR-1
1.5	SGD-1A
1.4?	W-1
1.2?	AGV-1
1?	G-2
1?	NS-1
0.9?	DTS-1
0.9?	GSP-1
0.9?	PCC-1

Hf

Hf ppm		pct HfC₂	
190?	NIM-L	0.02	NIM-L
17?	MicaFe		
14?	GSP-1		
12?	NIM-G		
10?	GXR-2		
9?	SY-3		
8?	G-2		
8?	GXR-4		
8?	NBS-1633		
8?	SY-2		
6.3?	SOIL-5		

Table 5 (cont'd.)

Usable values, arranged by "trace elements"

Hf ppm (cont'd.) | pct HfO2

Hf ppm (cont'd.)		pct HfO2
GXR-5	6?	
BCR-1	5	
AGV-1	5?	
GXR-6	5?	
GM	4.7?	
W-1	2.7?	
GXR-3	2.4?	
GXR-1	1.1?	

Hf ppb	
PCC-1	60?
DTS-1	10?

Hg ppm	
GXR-1	3.9?
GXR-2	3.2?

Hg ppb	
GXR-3	380?
W-1	220?
GXR-5	170?
NBS-1633	140?
GXR-4	130?
SO-2	82?
GXR-6	80?
G-2	44?
SO-4	33?
SO-1	22
SO-3	17
GSP-1	16?
AGV-1	15?
DTS-1	8?
BCR-1	7?
PCC-1	4?

La ppm		pct La2O3
SY-3	1350	0.16
VS-N	800?	0.09
NIM-L	240?	0.03
GSP-1	195	0.02
MicaFe	190?	0.02
STM-1	150?	0.01
M-2	120?	0.01
NIM-G	105?	0.01
NS-1	105?	0.01
G-2	92	0.01
SY-2	88	0.01
NBS-1633	82?	0.01
BR	80	0.01
SGD-1A	78	0.01
GXR-4	64	0.01
M-3	60?	0.01
SO-1	56?	0.01
TB	56?	
SO-2	48?	
SG-1A	41	
GA	38	
AGV-1	36	
JB-1	36?	
GnA	33?	
SO-4	33?	
I-3	30?	
SOIL-5	28	
BCR-1	27	
GH	25?	
GXR-2	25?	
JB-1	22?	
GXR-5	18	
GXR-6	14	
ST-1A	14	
MRG-1	10?	
W-1	9.8?	
BM	8.6?	
GXR-3	8.5?	

Li ppm (cont'd.)		pct Li2O
I-3	20?	
ASK-1	18?	
BCR-1	14	
SGD-1A	14	
ST-1A	14	
W-1	14?	
BR	13	
AGV-1	12	
NIM-G	12?	
JB-1	12?	
SO-2	11?	
NIM-N	9?	
SS	6?	
MRG-1	5?	
NIM-D	4?	
NIM-P	4?	
PCC-1	3?	
DTS-1	2?	
NIM-S	2?	

Lu ppm	
SY-3	8?
SY-2	3?
NIM-G	2?
BCR-1	0.5?
GM	0.45?
BM	0.4?
W-1	0.35?
SOIL-5	0.3?
AGV-1	0.3?
JB-1	0.3?
GSP-1	0.2?
JG-1	0.2?
MRG-1	0.2?
NIM-N	0.2?

N ppm		pct N
SO-4	4000?	0.40
SO-2	2200?	0.22
SO-1	400?	0.04
G-2	56?	0.01
W-1	52?	0.01
GSP-1	48?	
AGV-1	44?	
PCC-1	43?	
BCR-1	30?	
DTS-1	27?	

Nb ppm		pct Nb2O5
NIM-L	960	0.14
SG-1A	380	0.05
MicaFe	270?	0.04
NS-1	195?	0.03
SY-3	130	0.02
MicaMg	120?	0.02
BR	100?	0.01
GnA	90?	0.01
GH	85?	0.01
NIM-G	53	0.01
GSP-1	23?	0.01
SY-2	23?	
MRG-1	20?	
BCR-1	19?	
BHVO-1	19?	
GM	17?	
AGV-1	16?	
G-2	13?	
GA	10?	
W-1	9.5?	
SGD-1A	8	
ST-1A	8	
NIM-S	3.5?	
NIM-N	2?	
PCC-1	1?	

Ho ppm

	Ho ppm
SY-3	20?
NIM-G	3?
BCR-1	1.2?
SG-1A	1?
ST-1A	0.8
W-1	0.7?
AGV-1	0.6?
MRG-1	0.5?
SGD-1A	0.5?
G-2	0.4?

Ho ppb

	Ho ppb
DTS-1	3?

In ppb

	In ppb
BCR-1	95?
W-1	65
GSP-1	50?
AGV-1	40?
G-2	34?
DTS-1	2.5?

Ir ppb

	Ir ppb
PCC-1	6?
DTS-1	1?

Ir ppt

	Ir ppt
W-1	280?
GSP-1	12?
AGV-1	11?
BCR-1	4?
G-2	2?

La ppb

	La ppb
GXR-1	6?
NIM-S	4?
NIM-N	3?
NIM-P	2?
NIM-D	0.3?
PCC-1	150?
DTS-1	40?

Li ppm

	Li ppm	pct Li₂O
GnA	2200	0.47
MicaFe	1400?	0.30
NBS-97a	500?	0.11
VS-N	500?	0.11
SG-1A	390	0.08
NBS-98a	320?	0.07
KK	175?	0.04
MicaMg	120?	0.03
TB	115	0.02
JG-1	94?	0.02
SY-2	93	0.02
SY-3	92	0.02
GA	90	0.02
BM	70	0.02
GL-O	70?	0.02
GS-N	55?	0.01
GM	51	0.01
NIM-L	48?	0.01
RGM-1	46?	0.01
DR-N	45	0.01
GH	45?	0.01
TS	43?	0.01
SO-1	40?	0.01
G-2	35	0.01
ASK-2	30	0.01
GSP-1	30	0.01
UB-N	30?	0.01
STM-1	27?	0.01
NS-1	21?	0.01

Lu ppb

	Lu ppb
DTS-1	2?

Mo ppm

	Mo ppm	pct MoO₃
VS-N	700?	0.10
GXR-4	310	0.05
TS	130?	0.02
GnA	100?	0.02
ASK-2	60	0.01
GXR-5	30	
JB-1	20?	
NIM-D	4?	
AGV-1	3?	
BR	3?	
GH	3?	
NIM-C	3?	
NIM-L	3?	
NIM-N	3?	
SY-2	3?	
SY-3	2.5?	
JG-1	2?	
NS-1	2?	
SO-2	2?	
ST-1A	1.8	
GXR-6	1.7?	
SGD-1A	1.5	
BCR-1	1.5?	
GSP-1	1.5?	
SG-1A	1.3	
GM	1.1?	
DTS-1	1?	
SO-4	1?	
G-2	0.9?	
BM	0.6?	
W-1	0.57?	
PCC-1	0.5?	

Nd ppm

	Nd ppm	pct Nd₂O₃
SY-3	800?	0.09
GSP-1	190?	0.02
NS-1	71?	0.01
SY-2	71?	0.01
NIM-G	70?	0.01
SGD-1A	66?	0.01
BR	60?	0.01
G-2	58?	0.01
NIM-L	45?	
AGV-1	37?	
SOIL-5	30?	
BCR-1	26?	
GA	25?	
GH	25?	
JB-1	21?	
MRG-1	19?	
SG-1A	18	
W-1	15	
ST-1A	9?	
NIM-S	6?	
NIM-N	1.5?	

Ni ppm

	Ni ppm	pct NiO
PCC-1	2400	0.31
DTS-1	2300	0.29
SW	2200	0.28
NIM-D	2200	0.25
UB-N	2000	0.25
VS-N	2000	0.10
NIM-P	800	0.07
BR	560	0.07
BX-N	260	0.03
MRG-1	200?	0.03
TS	195	0.02
ES-681-1	185	0.02
ASK-2	160	0.02
BHVO-1	150	0.02
JB-1	140?	0.02
	135	

Table 5 (cont'd.)

Usable values, arranged by "trace elements"

Ni ppm (cont'd.)	
120	NIM-N
110	ASK-1
110?	MicaMg
98	NBS-1633
94	SO-1
90	ST-1A
90	W-1
76?	GXR-5
63?	BM
57	GXR-3
55?	MAG-1
52?	SGD-1A
50	MO8-1
50?	SDC-1
47?	GXR-1
42?	TB
40	GXR-4
38?	GL-O
36?	M-2
36?	MicaFe
35	GS-N
34?	SCo-1
30?	SGR-1
29?	SO-4
26	DR-N
22	GXR-6
22?	GXR-2
18?	I-3
17?	M-3
17?	SO-3
16	AGV-1
15	SO-2
12?	SG-1A
11	SY-3
10	BCR-1

pct NiO	
0.02	NIM-N
0.01	ASK-1
0.01	MicaMg
0.01	NBS-1633
0.01	SO-1
0.01	ST-1A
0.01	W-1
0.01	GXR-5
0.01	BM
0.01	GXR-3
0.01	MAG-1
0.01	SGD-1A
0.01	MO8-1
0.01	SDC-1
0.01	GXR-1
0.01	TB

Pb ppm (cont'd.)	
54	GSP-1
45	GH
43	NIM-L
40	NIM-G
40?	T-1
35?	MO8-1
33	AGV-1
30	G-2
30	GA
30	GM
28?	DT-N
26	JG-1
22?	GXR-5
20	SO-1
20	UB-N
20?	GnA
20?	M-3
19	SO-2
18	SGD-1A
17	FK
17?	M-2
16	SO-4
15?	GXR-3
14	BCR-1
13?	MicaFe
12	BM
12	JB-1
11	DTS-1
11	PCC-1
10	MRG-1
10?	I-3
9?	MicaMg
8	BR
7.8	W-1
7	TB

pct PbO	
0.01	GSP-1

Ra ppq	
1.8?	PCC-1
1.3?	DTS-1

pct Rb₂O	
0.24	MicaFe
0.22	GnA
0.14	MicaMg
0.12	SG-1A
0.10	VS-N
0.09	FK-N
0.06	NBS-70a
0.06	NIM-S
0.04	GH
0.03	NIM-G
0.03	M-2
0.03	GM
0.03	GSP-1
0.03	GL-O
0.02	SY-2
0.02	TS
0.02	NS-1
0.02	SY-3
0.02	NIM-L
0.02	GS-N
0.02	JG-1
0.02	TB
0.02	ASK-2
0.02	GA
0.02	GXR-4
0.02	G-2
0.02	RGM-1
0.02	KK
0.02	SO-1
0.02	SOIL-5

Rb ppm	
2200	MicaFe
2000	GnA
1300?	MicaMg
1100	SG-1A
900?	VS-N
850?	FK-N
550?	NBS-70a
530	NIM-S
390	GH
320	NIM-G
310?	M-2
250	GM
250	GSP-1
240	GL-O
220	SY-2
220?	TS
210	NS-1
208	SY-3
190	NIM-L
190?	GS-N
185	JG-1
180	TB
175	ASK-2
175	GA
175	GXR-4
170	G-2
165?	RGM-1
160?	KK
145?	SO-1
140	SOIL-5

Ru ppb	
9.5?	PCC-1
2.5?	DTS-1
1.0?	BCR-1

pct Sb₂O₅	
0.12	VS-N
0.02	GXR-1
0.01	GXR-2
0.01	GXR-3

Sb ppm	
900?	VS-N
125	GXR-1
48	GXR-2
40	GXR-3
14?	SOIL-5
7?	NBS-1633
4.4	GXR-4
4.3?	AGV-1
3.8	GXR-6
3.3?	TB
3.1?	GSP-1
2.1	GXR-5
2?	BM
1.4?	PCC-1
1.3?	SG-1A
1.0	ST-1A
1.0	W-1
1?	NIM-P
1?	SG-1A
0.6	BCR-1
0.6?	NIM-G
0.6?	NIM-S
0.5?	DTS-1
0.5?	GM
0.4	MRG-1
0.3	SY-3
0.3?	NIM-L
0.2	SY-2
0.06?	G-2

	Sc ppm	pct Sc₂O₃
VS-N	300?	0.05
I-3	50?	0.01
MRG-1	48?	0.01
ST-1A	43	0.01
NIM-N	38?	0.01
W-1	35?	0.01
BM	34	0.01
BCR-1	33	0.01
GXR-6	31?	
M-2	30?	
SGD-1A	27	
JB-1	27?	
NBS-1633	27?	
NIM-P	26?	
BR	22?	
TS	22?	
SDC-1	19?	
SO-1	19?	
GXR-3	18?	
MAG-1	18?	
SOIL-5	15?	
TB	13.5	
AGV-1	12.5	
SY-3	12?	
SCo-1	11?	
QLO-1	10?	
GnA	9?	
PCC-1	9?	
GXR-4	8?	
GXR-5	8?	
ASK-1	7	
GA	7?	
GXR-2	7?	
NIM-D	7?	
SY-2	7?	
GSP-1	6.6	
JG-1	6.5?	
GM	5.1	
SG-1A	5	
RGM-1	5?	
SGR-1	5?	
DTS-1	3.8	
NIM-S	3.6?	

	Re ppb	
FK	135?	0.01
I-1	130?	0.01
SDC-1	125?	0.01
SCo-1	120?	0.01
GXR-3	115?	0.01
NBS-1633	110?	0.01
STM-1	105?	0.01
GXR-2	100?	0.01
SGR-1	86?	0.01
ASK-1	85?	0.01
SO-2	81?	0.01
DR-N	75	0.01
SO-4	75?	0.01
SGD-1A	73	0.01
AGV-1	67	0.01
QLO-1	64?	0.01
BCR-1	47	
BR	47	
I-3	42?	
JB-1	41	
SO-3	40?	
GXR-5	32?	
T-1	29?	
GXR-1	25?	
M-3	21	
W-1	16	
ST-1A	12	
BM	9?	
BHVO-1	8	
MRG-1	5?	
NIM-N	5?	
NIM-P	0.3?	
PCC-1	0.05?	
DTS-1		

	Rh ppb
PCC-1	1.0?
DTS-1	0.9?
BCR-1	0.2?

	Pd ppb
NS-1	6
ST-1A	6?
NIM-N	6?
SS	5?
NIM-P	4?
W-1	25?
BCR-1	12?
PCC-1	5?
DTS-1	1?

	Pr ppm	pct Pr₆O₁₁
SY-3	120?	
GSP-1	50?	
NS-1	20?	
G-2	19?	
SGD-1A	7?	
AGV-1	7?	
BCR-1	7?	0.01
W-1	3.4?	0.01
SG-1A	3?	
ST-1A	2?	

	Pr ppb
DTS-1	6?

	Pt ppb
W-1	12?
PCC-1	10?
DTS-1	3?
BCR-1	2?
AGV-1	1?

	Ra ppt
G-2	0.71?
AGV-1	0.7?
GSP-1	0.66?
BCR-1	0.6?

	Os ppb
PCC-1	9?
DTS-1	1?
W-1	0.25?
BCR-1	0.1?

	Pb ppm	pct PbO
VS-N	900	0.10
NBS-91	750	0.08
GXR-1	670?	0.07
GXR-2	620?	0.07
FK-N	240?	0.03
SG-1A	230	0.02
BX-N	145?	0.02
SY-3	130	0.01
SOIL-5	130?	0.01
KK	120	0.01
GXR-6	110?	0.01
SY-2	80	0.01
I-1	74?	0.01
ES-681-1	72?	0.01
NBS-1633	70	0.01
DR-N	65	0.01
GS-N	60?	0.01

	Pb ppm
SY-2	10
GSP-1	10?
JG-1	9
NIM-G	8?
GM	8?
GA	7.5
NIM-S	7
NS-1	7?
QLO-1	7?
NIM-L	6?
RGM-1	6?
GnA	5.1?
STM-1	5?
G-2	4?
GH	3.5
FK-N	3?
SS	3?

Table 5 (cont'd.)

Usable values, arranged by "trace elements"

Sc ppm (cont'd)	pct Sc₂O₃
3.5	G-2
1.7?	GXR-1
1?	GH
0.3?	NIM-G
	NIM-L

Se ppm	
19	GXR-1
9.4	NBS-1633
6	GXR-4
1?	GXR-5
0.7	GXR-2
0.2	GXR-3
0.1?	BCR-1
0.1?	W-1

Sm ppm		pct Sm₂O₃
100?	SY-3	0.01
25	GSP-1	
17	SGD-1A	
16?	NIM-G	
15?	SY-2	
12?	BR	
12?	NBS-1633	
10?	GH	
10?	NS-1	
9?	TB	
7.2	G-2	
7	SG-1A	
6.5	BCR-1	
6	GXR-4	
6?	GM	
6?	NIM-L	
5.9	AGV-1	
5.4?	SOIL-5	
5?	GA	

Sn ppm (cont'd)	pct SnO₂
3.2	MRG-1
3.2	W-1
2.5	BCR-1
2?	JB-1
2?	NIM-D
1.7	BM
1.7?	DTS-1
1.6?	PCC-1
1.4?	G-2

Sr ppm		pct SrO
4600	NIM-L	0.53
2300	SGD-1A	0.27
1500	NBS-97a	0.18
1400?	NBS-1633	0.17
1300	BR	0.15
1150	NS-1	0.14
1150?	GXR-3	0.09
800?	VS-N	0.08
710?	STM-1	0.08
680	ASK-1	0.08
660	AGV-1	0.08
570?	GS-N	0.07
500?	M-3	0.06
480	G-2	0.06
440	JB-1	0.05
430?	SGR-1	0.05
400	DR-N	0.05
400?	BHVO-1	0.05
390?	T-1	0.05
350?	QLO-1	0.04
350?	SOIL-5	0.04
340	SO-2	0.04
330	BCR-1	0.04
330	NBS-98a	0.04
310	GA	0.04
306	SY-3	0.04
300?	SO-1	0.04

Ta ppm		pct Ta₂O₅
900?	VS-N	0.11
34?	MicaFe	
29?	GnA	
26	SG-1A	
11?	NS-1A	
1.4?	AGV-1	
1.2	ST-1A	
1.1	SGD-1A	
1?	GSP-1	
0.8?	BCR-1	
0.8?	G-2	
0.8?	GXR-2	
0.8?	GXR-4	
0.5	SOIL-5	
0.5?	W-1	
0.5?	GXR-5	
0.3?	GXR-6	
0.2?	GXR-3	
	GXR-1	

Tb ppm	
11?	SY-3
3?	NIM-G
2?	NBS-1633
2?	SY-2
1.4?	GSP-1
1.0	BCR-1
0.7?	AGV-1
0.7?	SOIL-5
0.65	W-1
0.5?	G-2
0.1?	NIM-S

Tb ppb	
3?	DTS-1
1?	PCC-1

Tl ppm	
4?	NBS-1633
1.6?	AGV-1
1.3?	GSP-1
1.2?	G-2
0.3?	BCR-1
0.11?	W-1

Tl ppb	
0.8?	DTS-1
0.5?	PCC-1

Tm ppm	
8?	SY-3
2?	NIM-G
2?	SY-2
0.6?	BCR-1
0.4?	AGV-1
0.3	W-1
0.3?	G-2
0.1?	MRG-1

Tm ppb	
1?	DTS-1

U ppm		pct U₃O₈
650	SY-3	0.08
290	SY-2	0.03
63	SG-1A	0.01
60?	MicaFe	0.01
35?	GXR-1	
22?	GnA	
22?	TS	
18?	GH	
15?	NIM-G	

5?	MRG-1	
4.8?	JB-1	
4.6?	JG-1	
4	ST-1A	
4?	BM	
3.6?	W-1	
3.3	GXR-2	
2.9	GXR-5	
2.4	GXR-6	
1.2?	NIM-S	
1?	GXR-3	
1?	NIM-N	

Sm ppb		
8?	PCC-1	
4?	DTS-1	

Sn ppm		pct SnO_2
1900	GnA	0.24
70?	MicaFe	0.01
33?	KK	
11	SG-1A	
10?	GH	
8?	BR	
7?	NIM-L	
6.4?	SW	
6?	SY-3	
5.7	TB	
5?	GSP-1	
4.9?	TS	
4.6	GM	
4	SY-2	
4?	GA	
4?	JG-1	
4?	NIM-G	
3.7	SGD-1A	
3.6	AGV-1	
3.5	ST-1A	
3.5?	NS-1	
280	GXR-1	0.03
275	SY-2	0.03
270	ST-1A	0.03
260	I-3	0.03
260	MRG-1	0.03
260	NIM-N	0.03
240	GSP-1	0.03
230	BM	0.03
220	GXR-4	0.03
220	SO-3	0.03
190	W-1	0.02
185	JG-1	0.02
185?	M-2	0.02
180?	SDC-1	0.02
175?	SCo-1	0.02
170	I-1	0.02
170	SO-4	0.02
160	GXR-2	0.02
155	TB	0.02
150?	MAG-1	0.02
135	GM	0.02
120	GXR-5	0.01
105?	RGM-1	0.01
100	ASK-2	0.01
93?	TS	0.01
76?	KK	0.01
62	NIM-S	0.01
42	GXR-6	
36?	DT-N	
35?	FK-N	
32	NIM-P	
25?	MicaMg	
20	SG-1A	
19	GL-O	
19?	GnA	
10	GH	
10	NIM-G	
10?	UB-N	
5	MicaFe	
3?	NIM-D	
0.4	PCC-1	
0.4?	DTS-1	

Th ppm		pct ThO_2
990	SY-3	0.11
380?	SY-2	0.04
150?	MicaFe	0.02
120	SG-1A	0.01
105	GSP-1	0.01
90?	GH	0.01
65	NIM-L	0.01
52	NIM-G	0.01
43?	BX-N	
35?	GM	
25	G-2	
24?	NBS-1633	
22	GXR-4	
19?	TB	
17	GA	
13.5	JG-1	
12?	BR	
11	SOIL-5	
9	SGD-1A	
9?	JB-1	
9?	NS-1	
8	GXR-2	
6.4	AGV-1	
6.1	BCR-1	
5.3	GXR-5	
5.2	GXR-6	
3?	BM	
3?	ST-1A	
2.9	GXR-3	
2.4	W-1	
2.3	GXR-1	
1?	MRG-1	
1?	NIM-N	
1?	NIM-P	
0.9?	NIM-S	
0.6?	NIM-D	

Th ppb		
10?	DTS-1	
10?	PCC-1	

14	NIM-L	
11.5	NBS-1633	
6.4?	GXR-4	
4	SGD-1A	
4?	GA	
4?	NS-1	
3.3	JG-1	
3.2?	SOIL-5	
3.1?	GXR-3	
3?	BR	
3?	GXR-2	
2.1	G-2	
2.1	GSP-1	
2.1?	GXR-5	
1.95	AGV-1	
1.8	JB-1	
1.7	BCR-1	
1.6?	GXR-6	
1.6?	SW	
1.0	ST-1A	
0.6?	NIM-S	
0.58?	W-1	
0.5?	MicaMg	
0.4?	BHVO-1	
0.4?	NIM-N	
0.4?	NIM-P	
0.3?	MRG-1	

U ppb		
5?	PCC-1	
4?	DTS-1	

V ppm		pct V_2O_5
930?	TS	0.17
770	ES-681-1	0.14
600?	VS-N	0.11
520	MRG-1	0.09
500?	I-3	0.09
420	BCR-1	0.07
400?	NBS-165a	0.07
320	ST-1A	0.06

Table 5 (cont'd.)

Usable values, arranged by "trace elements"

V ppm (cont'd.)		pct V₂O₅	Y ppm		pct Y₂O₃	Zn ppm		pct ZnO	Zr ppm		pct ZrO₂
310?	BX-N	0.06	800?	VS-N	0.10	1300	MicaFe	0.16	1350?	NIM-L	1.49
300?	BHVO-1	0.05	740	SY-3	0.09	800	VS-N	0.10	1200?	NBS-69a	0.18
260	W-1	0.05	180?	TS	0.02	740	GXR-1	0.09	800?	STM-1	0.16
240	BR	0.04	145	NIM-G	0.02	640?	NBS-91	0.08	790?	MicaFe	0.11
240	SGD-1A	0.04	130	SY-2	0.02	500	GXR-2	0.06	720	SO-2	0.11
230	NIM-P	0.04	70	GH	0.01	400	NIM-L	0.05	720	NS-1	0.10
220	ASK-2	0.04	69	SG-1A	0.01	370	SOIL-5	0.05	700?	SG-1A	0.10
220	DR-N	0.04	60?	M-2	0.01	300?	NBS-165a	0.04	500	VS-N	0.09
220	NIM-N	0.04	50?	I-3	0.01	290?	MicaMg	0.04	470?	GSP-1	0.07
210	JB-1	0.04	50?	STM-1	0.01	270	SG-1A	0.04	400	ASK-1	0.06
210	NBS-1633	0.04	40	BCR-1	0.01	250	SY-2	0.03	320	SY-3	0.05
180	BM	0.03	40?	SO-2	0.01	240	SY-3	0.03	310?	NBS-98a	0.04
180?	GXR-6	0.03	39?	TB		240?	STM-1	0.03	310?	SO-4	0.04
170?	NBS-69a	0.03	31?	JG-1		220	GXR-3	0.03	300	G-2	0.04
170?	NBS-69b	0.03	30	BR		210	NBS-1633	0.03	300	NIM-G	0.04
140	SO-1	0.02	30	SGD-1A		190	MRG-1	0.02	300?	SDC-1	0.04
140?	MAG-1	0.02	30?	ST-1A		185?	T-1	0.02	290?	M-3	0.04
135	MicaFe	0.02	29	GSP-1		165	ASK-2	0.02	280	SY-2	0.04
130?	MO8-1	0.02	26?	BM		150	BR	0.02	280?	TS	0.04
130?	SGR-1	0.02	26?	GM		150	DR-N	0.02	250	BR	0.03
125	AGV-1	0.02	26?	JB-1		150	ST-1A	0.02	250	NBS-81a	0.03
115?	SCo-1	0.02	25	W-1		145	SO-1	0.02	240	SGD-1A	0.03
105	TB	0.02	25?	MicaFe		140?	MAG-1	0.02	240?	GS-N	0.03
105?	SDC-1	0.02	25?	NIM-L		130?	M-2	0.02	230	AGV-1	0.03
99?	T-1	0.02	24?	SO-1		125	BCR-1	0.02	210?	RGM-1	0.03
92?	GXR-4	0.02	23?	SO-4		125	SO-2	0.02	200?	GXR-2	0.03
90	SO-4	0.02	23?	SOIL-5		120	GXR-6	0.01	200?	GXR-4	0.03
90?	MicaMg	0.02	22?	GnA		120	SGD-1A	0.01	190?	I-3	0.03
88?	GXR-1	0.02	21	GA		110?	SCo-1	0.01	185	BCR-1	0.02
81	NIM-L	0.01	19	AGV-1		105	ASK-1	0.01	175	TB	0.02
80?	M-2	0.01	17?	NS-1		105	BM	0.01	175?	QLO-1	0.02
75	UB-N	0.01	17?	SO-3		105	GSP-1	0.01	170	ASK-2	0.02
75?	M-3	0.01	16?	MRG-1		105?	SDC-1	0.01	165?	SCo-1	0.02
64	SO-2	0.01	13?	FK-N		100	NIM-P	0.01	160?	NIM-P	0.02
62?	GS-N	0.01	11	G-2		100?	BHVO-1	0.01	160?	BHVO-1	0.02
61?	QLO-1	0.01	6?	NIM-N		100?	I-3	0.01	160?	T-1	0.02
60?	GXR-5	0.01	4?	NIM-P		95	TB	0.01	155?	JB-1	0.02
57?	GXR-2	0.01	3?	NIM-S		94	SO-4	0.01	150	GA	0.02
54	GSP-1	0.01	0.05?	DTS-1		90	NIM-D	0.01			
						89?	TS				

Table 1

Det. limit (pct)	Sample	Value
0.02	GH	150
0.02	SO-3	150?
0.02	GM	145
0.02	GXR-5	140?
0.02	ST-1A	130
0.02	MAG-1	130?
0.02	DR-N	125?
0.01	JG-1	110?
0.01	BM	105
0.01	MRG-1	105
0.01	W-1	105
0.01	GXR-6	105?
0.01	M-2	100?
0.01	SO-1	81?
0.01	NBS-91	70
0.01	FK	70?
0.01	GnA	70?
0.01	GXR-1	66
0.01	SGR-1	62?
0.01	I-1	60?
0.01	NBS-165a	40?
	NIM-S	33?
	NIM-P	30?
	NBS-69b	25?
	NIM-N	23?
	MicaMg	20?
	DTS-1	12?
	NIM-D	10?
	PCC-1	7?

Table 2

Det. limit (pct)	Sample	Value
0.01	AGV-1	86
0.01	W-1	86
0.01	GH	85
0.01	UB-N	85?
0.01	G-2	84
0.01	JB-1	84
0.01	SGR-1	82?
0.01	GA	80
0.01	GnA	76
0.01	SS	74
0.01	NS-1	70?
0.01	NIM-N	68
0.01	GXR-4	64?
0.01	QLO-1	60?
0.01	BX-N	59?
0.01	SW	58
0.01	SO-3	52
0.01	NIM-G	50
0.01	GXR-5	50?
0.01	MO8-1	50?
	KK	49
	GS-N	48?
	DTS-1	46
	M-3	44?
	GL-O	43?
	PCC-1	41
	GM	40
	JG-1	40
	RGM-1	32?
	DT-N	29?
	FK-N	24?
	I-1	16?
	FK	12?
	NIM-S	10?

Table 3

Sample	Yb ppm	pct Yb$_2$O$_3$
VS-N	900?	0.10
SY-3	65	0.01
SY-2	17	
NIM-G	14	
GH	8?	
SG-1A	6	
ST-1A	4	
TB	4?	
BM	3.5?	
NIM-L	3.5?	
BCR-1	3.4	
GM	3?	
SGD-1A	2.9	
GXR-6	2.77?	
SOIL-5	2.2	
GXR-2	2.2?	
W-1	2.1	
JB-1	2.1?	
BR	2.0?	
GA	2.0?	
GXR-5	2.0?	
AGV-1	1.9	
GSP-1	1.9	
GXR-1	1.8?	
GXR-4	1.8?	
JG-1	1.5?	
MRG-1	1?	
G-2	0.86	
GXR-3	0.87?	
NIM-N	0.6?	
NIM-P	0.6?	
NIM-S	-0.1?	
PCC-1	0.02	
DTS-1	0.01?	

Table 4

Det. limit (pct)	Sample	Value
0.01	SY-2	52
0.01	NS-1	51
0.01	SY-3	51
0.01	ASK-1	49
0.01	SO-3	44?
0.01	NIM-D	40
0.01	GXR-3	39?
0.01	GA	38
	G-2	36
	PCC-1	29
	JG-1	24
	SL-1	20?
	SW	19?
	RGM-1	15?
	DTS-1	11
	GM	11
	NIM-S	10
	I-1	7?
	SG-1A	5
	GH	5?
	NIM-G	2?

Table 5

Sample	W ppm	pct WO$_3$
GnA	520?	0.07
GXR-4	28	
TB	3?	
SG-1A	2.3	
GM	2?	
GXR-2	1.8	
GXR-3	1.1	
BM	1?	
GXR-6	0.9	
AGV-1	0.55?	
W-1	0.5?	
BCR-1	0.4?	
G-2	0.1?	
GSP-1	0.1?	
PCC-1	0.06	
DTS-1	0.04?	

Table 6

Comparison of recommended iron oxide values on NIM samples

Sample	Per cent	Abbey (1977a)	Steele and Hansen (1979a,b)	This work
NIM-D	Fe_2O_3	0.90	0.71	0.71
	FeO	14.46	14.63	14.63
	Fe_2O_3TR*	16.96	17.00	16.96
	Fe_2O_3TC	16.97	16.97	16.97
	Σ (corr.)	100.22?	100.13?	100.17?
NIM-G	Fe_2O_3	0.58	0.6?	0.6?
	FeO	1.30	1.30	1.30
	Fe_2O_3TR	2.02	2.00	2.02
	Fe_2O_3TC	2.02	2.04	2.04
	Σ (corr.)	99.96?	99.99?	100.01?
NIM-L	Fe_2O_3	8.74	8.78	8.74
	FeO	1.08	1.13	1.13
	Fe_2O_3TR	9.96	9.91	9.96
	Fe_2O_3TC	9.94	10.04	10.00
	Σ (corr.)	99.72?	99.85	99.73?
NIM-N	Fe_2O_3	0.76	0.8?	0.8?
	FeO	7.30	7.47	7.30
	Fe_2O_3TR	8.91	8.97	8.91
	Fe_2O_3TC	8.87	9.10	8.89
	Σ (corr.)	99.80?	100.13?	99.93?
NIM-P	Fe_2O_3	1.02?	0.87	1.02?
	FeO	10.59?	10.59	10.59
	Fe_2O_3TR	12.76	12.70	12.76
	Fe_2O_3TC	12.79?	12.64	12.79?
	Σ (corr.)	99.77?	99.66?	99.83?
NIM-S	Fe_2O_3	1.07	1.11	1.07
	FeO	0.30	0.30	0.30
	Fe_2O_3TR	1.40	1.40	1.40
	Fe_2O_3TC	1.40	1.44	1.40
	Σ (corr)	100.17?	100.13	100.09?

*See Table 3, note 2 for explanation of these terms.

APPENDIX 2

COMPARISON TABLE OF UNITED STATES, TYLER, CANADIAN, BRITISH, FRENCH, AND GERMAN STANDARD SIEVE SERIES

Reproduced by the courtesy of C. E. Tyler, W. S. Tyler Incorporated, Mentor, OH.

U.S.A. (1)		TYLER (2)	CANADIAN (3)		BRITISH (4)		FRENCH (5)		GERMAN (6)
*Standard	Alternate	Mesh Designation	Standard	Alternate	Nominal Aperture	Nominal Mesh No.	Opg. M.M.	No.	Opg.
125 mm	5″								
106 mm	4.24″								
100 mm	4″								
90 mm	3½″								
75 mm	3″								
63 mm	2½″								
53 mm	2.12″								
50 mm	2″								
45 mm	1¾″								
37.5 mm	1½″								
31.5 mm	1¼″								
26.5 mm	1.06″	1.05″	26.9 mm	1.06″					
25.0 mm	1″								25.0 mm
22.4 mm	⅞″	.883″	22.6 mm	⅞″					
19.0 mm	¾″	.742″	19.0 mm	¾″					20.0 mm
									18.0 mm
16.0 mm	⅝″	.624″	16.0 mm	⅞″					16.0 mm
13.2 mm	.530″	.525″	13.5 mm	.530″					
12.5 mm	½″								12.5 mm
11.2 mm	7/16″	.441″	11.2 mm	7/16″					
									10.0 mm
9.5 mm	⅜″	.371″	9.51 mm	⅜″					
8.0 mm	5/16″	2½	8.00 mm	5/16″					8.0 mm
6.7 mm	.265″	3	6.73 mm	.265″					
6.3 mm	¼″								6.3 mm
5.6 mm	No. 3½	3½	5.66 mm	No. 3½					
4.75 mm							5.000	38	5.0 mm
	4	4	4.76 mm	4					
4.00 mm	5	5	4.00 mm	5			4.000	37	4.0 mm
3.35 mm	6	6	3.36 mm	6	3.35 mm	5			
							3.150	36	3.15 mm
2.80 mm	7	7	2.83 mm	7	2.80 mm	6			
2.36 mm	8	8	2.38 mm	8	2.40 mm	7	2.500	35	2.5 mm
2.00 mm	10	9	2.00 mm	10	2.00 mm	8	2.000	34	2.0 mm
1.70 mm	12	10	1.68 mm	12	1.68 mm	10	1.600	33	1.6 mm
1.40 mm	14	12	1.41 mm	14	1.40 mm	12			
1.18 mm							1.250	32	1.25 mm
1.00 mm	16	14	1.19 mm	16	1.20 mm	14			
	18	16	1.00 mm	18	1.00 mm	16	1.000	31	1.0 mm
850 µm	20	20	841 µm	20	850 µm	18			
							.800	30	800 µm
710 µm	25	24	707 µm	25	710 µm	22			
							.630	29	630 µm
600 µm	30	28	595 µm	30	600 µm	25			
500 µm	35	32	500 µm	35	500 µm	30	.500	28	500 µm
425 µm	40	35	420 µm	40	420 µm	36			
							.400	27	400 µm
355 µm	45	42	354 µm	45	355 µm	44			
							.315	26	315 µm
300 µm	50	48	297 µm	50	300 µm	52			
250 µm	60	60	250 µm	60	250 µm	60	.250	25	250 µm
212 µm	70	65	210 µm	70	210 µm	72			
							.200	24	200 µm
180 µm	80	80	177 µm	80	180 µm	85			
							.160	23	160 µm
150 µm	100	100	149 µm	100	150 µm	100			
125 µm	120	115	125 µm	120	125 µm	120	.125	22	125 µm
106 µm	140	150	105 µm	140	105 µm	150			
							.100	21	100 µm
90 µm	170	170	88 µm	170	90 µm	170			90 µm
							.080	20	80 µm
75 µm	200	200	74 µm	200	75 µm	200			
									71 µm
63 µm	230	250	63 µm	230	63 µm	240	.063	19	63 µm
									56 µm
53 µm	270	270	53 µm	270	53 µm	300			
							.050	18	50 µm
45 µm	325	325	44 µm	325	45 µm	350			45 µm
							.040	17	40 µm
38 µm	400	400	37 µm	400					

(1) U.S.A. Sieve Series - ASTM Specification E-11-70
(2) Tyler Standard Screen Scale Sieve Series.
(3) Canadian Standard Sieve Series 8-GP-1b.
(4) British Standards Institution, London BS-410-62.
(5) French Standard Specifications, AFNOR X-11-501.
(6) German Standard Specification DIN 4188.

**These sieves correspond to those recommended by ISO (International Standards Organization) as an International Standard and this designation should be used when reporting sieve analysis intended for international publication.*

APPENDIX 3

SOURCE MATERIALS FOR STANDARD STOCK SOLUTIONS

Reprinted from Reference 30 of Chapter 7, p. 332, by courtesy of Marcel Dekker, Inc.

467

Element	Procedure
Aluminum	Dissolve 1.000 g Al wire in minimum amount of 2 M HCl; dilute to volume.
Antimony	Dissolve 1.000 g Sb in (1) 10 ml HNO_3 plus 5 ml HCl, and dilute to volume when dissolution is complete; or (2) 18 ml HBr plus 2 ml liquid Br_2, when dissolution is complete add 10 ml $HClO_4$, heat in a well-ventilated hood while swirling until white fumes appear and continue for several minutes to expel all HBr, then cool and dilute to volume.
Arsenic	Dissolve 1.3203 g of As_2O_3 in 3 ml 8 M HCl and dilute to volume; or treat the oxide with 2 g NaOH and 20 ml water, after dissolution dilute to 200 ml, neutralize with HCl (pH meter), and dilute to volume.
Barium	(1) Dissolve 1.7787 g $BaCl_2 \cdot 2H_2O$ (fresh crystals) in water and dilute to volume. (2) Dissolve 1.516 g $BaCl_2$ (dried at $250°C$ for 2 hr) in water and dilute to volume. (3) Treat 1.4367 g $BaCO_3$ with 300 ml water, slowly add 10 ml of HCl and, after the CO_2 is released by swirling, dilute to volume.
Beryllium	(1) Dissolve 19.655 g $BeSO_4 \cdot 4H_2O$ in water, add 5 ml HCl (or HNO_3), and dilute to volume. (2) Dissolve 1.000 g Be in 25 ml 2 M HCl, then dilute to volume.
Bismuth	Dissolve 1.000 g Bi in 8 ml of 10 M HNO_3, boil gently to expel brown fumes, and dilute to volume.
Boron	Dissolve 5.720 g fresh crystals of H_3BO_3 and dilute to volume.
Bromine	Dissolve 1.489 g KBr (or 1.288 g NaBr) in water and dilute to volume.
Cadmium	(1) Dissolve 1.000 g Cd in 10 ml of 2 M HCl; dilute to volume. (2) Dissolve 2.282 g $3CdSO_4 \cdot 8H_2O$ in water; dilute to volume.
Calcium	Place 2.4973 g $CaCO_3$ in volumetric flask with 300 ml water, carefully add 10 ml HCl, after CO_2 is released by swirling, dilute to volume.
Cerium	(1) Dissolve 4.515 g $(NH_4)_4Ce(SO_4)_4 \cdot 2H_2O$ in 500 ml water to which 30 ml H_2SO_4 had been added, cool, and dilute to volume. Advisable to standardize against As_2O_3. (2) Dissolve 3.913 g $(NH_4)_2Ce(NO_3)_6$ in 10 ml H_2SO_4, stir 2 min, cautiously introduce 15 ml water and again stir 2 min. Repeat addition of water and stirring until all the salt has dissolved, then dilute to volume.
Cesium	Dissolve 1.267 g CsCl and dilute to volume. Standardize: Pipette 25 ml of final solution to Pt dish, add 1 drop H_2SO_4, evaporate to dryness, and heat to constant weight at $>800°C$. Cs (in $\mu g/ml$) =(40)(0.734)(wt of residue)
Chlorine	Dissolve 1.648 g NaCl and dilute to volume.
Chromium	(1) Dissolve 2.829 g $K_2Cr_2O_7$ in water and dilute to volume. (2) Dissolve 1.000 g Cr in 10 ml HCl, and dilute to volume.
Cobalt	Dissolve 1.000 g Co in 10 ml of 2 M HCl, and dilute to volume.
Copper	(1) Dissolve 3.929 g fresh crystals of $CuSO_4 \cdot 5H_2O$, and dilute to volume. (2) Dissolve 1.000 g Cu in 10 ml HCl plus 5 ml water to which HNO_3 (or 30% H_2O_2) is added dropwise until dissolution is complete. Boil to expel oxides of nitrogen and chlorine, then dilute to volume.
Dysprosium	Dissolve 1.1477 g Dy_2O_3 in 50 ml of 2 M HCl; dilute to volume.

[a] 1000 $\mu g/ml$ as the element in a final volume of 1 liter unless stated otherwise.

Element	Procedure
Erbium	Dissolve 1.1436 g Er_2O_3 in 50 ml of 2 M HCl; dilute to volume.
Europium	Dissolve 1.1579 g Eu_2O_3 in 50 ml of 2 M HCl; dilute to volume.
Fluorine	Dissolve 2.210 g NaF in water and dilute to volume.
Gadolinium	Dissolve 1.152 g Gd_2O_3 in 50 ml of 2 M HCl; dilute to volume.
Gallium	Dissolve 1.000 g Ga in 50 ml of 2 M HCl; dilute to volume.
Germanium	Dissolve 1.4408 g GeO_2 with 50 g oxalic acid in 100 ml of water; dilute to volume.
Gold	Dissolve 1.000 g Au in 10 ml of hot HNO_3 by dropwise addition of HCl, boil to expel oxides of nitrogen and chlorine, and dilute to volume. Store in amber container away from light.
Hafnium	Transfer 1.000 g Hf to Pt dish, add 10 ml of 9 M H_2SO_4, and then slowly add HF dropwise until dissolution is complete. Dilute to volume with 10% H_2SO_4.
Holmium	Dissolve 1.1455 g Ho_2O_3 in 50 ml of 2 M HCl; dilute to volume.
Indium	Dissolve 1.000 g In in 50 ml of 2 M HCl; dilute to volume.
Iodine	Dissolve 1.308 g KI in water and dilute to volume.
Iridium	(1) Dissolve 2.465 g Na_3IrCl_6 in water and dilute to volume. (2) Transfer 1.000 g Ir sponge to a glass tube, add 20 ml of HCl and 1 ml of $HClO_4$. Seal the tube and place in an oven at $300°C$ for 24 hr. Cool, break open the tube, transfer the solution to a volumetric flask, and dilute to volume. Consult (1) and observe all safety precautions in opening the glass tube.
Iron	Dissolve 1.000 g Fe wire in 20 ml of 5 M HCl; dilute to volume.
Lanthanum	Dissolve 1.1717 g La_2O_3 (dried at $110°C$) in 50 ml of 5 M HCl, and dilute to volume.
Lead	(1) Dissolve 1.5985 g $Pb(NO_3)_2$ in water plus 10 ml HNO_3, and dilute to volume. (2) Dissolve 1.000 g Pb in 10 ml HNO_3, and dilute to volume.
Lithium	Dissolve a slurry of 5.3228 g Li_2CO_3 in 300 ml of water by addition of 15 ml HCl, after release of CO_2 by swirling, dilute to volume.
Lutetium	Dissolve 1.6079 g $LuCl_3$ in water and dilute to volume.
Magnesium	Dissolve 1.000 g Mg in 50 ml of 1 M HCl and dilute to volume.
Manganese	(1) Dissolve 1.000 g Mn in 10 ml HCl plus 1 ml HNO_3, and dilute to volume. (2) Dissolve 3.0764 g $MnSO_4 \cdot H_2O$ (dried at $105°C$ for 4 hr) in water and dilute to volume. (3) Dissolve 1.5824 g MnO_2 in 10 ml HCl in a good hood, evaporate to gentle dryness, dissolve residue in water and dilute to volume.
Mercury	Dissolve 1.000 g Hg in 10 ml of 5 M HNO_3 and dilute to volume.
Molybdenum	(1) Dissolve 2.0425 g $(NH_4)_2MoO_4$ in water and dilute to volume. (2) Dissolve 1.5003 g MoO_3 in 100 ml of 2 M ammonia, and dilute to volume.
Neodymium	Dissolve 1.7373 g $NdCl_3$ in 100 ml 1 M HCl and dilute to volume.
Nickel	Dissolve 1.000 g Ni in 10 ml hot HNO_3, cool, and dilute to volume.
Niobium	Transfer 1.000 g Nb (or 1.4305 g Nb_2O_5) to Pt dish, add 20 ml HF, and heat gently to complete dissolution. Cool, add 40 ml H_2SO_4, and evaporate to fumes of SO_3. Cool and dilute to volume with 8 M H_2SO_4.
Osmium	Dissolve 1.3360 g OsO_4 in water and dilute to 100 ml. Prepare only as needed as solution loses strength on standing unless Os is reduced by SO_2 and water is replaced by 100 ml 0.1 M HCl.
Palladium	Dissolve 1.000 g Pd in 10 ml of HNO_3 by dropwise addition of HCl to hot solution; dilute to volume.
Phosphorus	Dissolve 4.260 g $(NH_4)_2HPO_4$ in water and dilute to volume.

APPENDIX 3 (Continued).

Element	Procedure
Platinum	Dissolve 1.000 g Pt in 40 ml of hot aqua regia, evaporate to incipient dryness, add 10 ml HCl and again evaporate to moist residue. Add 10 ml HCl and dilute to volume.
Potassium	Dissolve 1.9067 g KCl (or 2.8415 g KNO_3) in water and dilute to volume.
Praseodymium	Dissolve 1.1703 g Pr_2O_3 in 50 ml of 2 M HCl; dilute to volume.
Rhenium	Dissolve 1.000 g Re in 10 ml of 8 M HNO_3 in an ice bath until initial reaction subsides, then dilute to volume.
Rhodium	Dissolve 1.000 g Rh by the sealed-tube method described under iridium (1).
Rubidium	Dissolve 1.4148 g RbCl in water. Standardize as described under cesium. Rb (in $\mu g/ml$) = (40)(0.320)(wt of residue).
Ruthenium	Dissolve 1.317 g RuO_2 in 15 ml of HCl; dilute to volume.
Samarium	Dissolve 1.1596 g Sm_2O_3 in 50 ml of 2 M HCl; dilute to volume.
Scandium	Dissolve 1.5338 g Sc_2O_3 in 50 ml of 2 M HCl; dilute to volume.
Selenium	Dissolve 1.4050 g SeO_2 in water and dilute to volume or dissolve 1.000 g Se in 5 ml of HNO_3, then dilute to volume.
Silicon	Fuse 2.1393 g SiO_2 with 4.60 g Na_2CO_3, maintaining melt for 15 min in Pt crucible. Cool, dissolve in warm water, and dilute to volume. Solution contains also 2000 $\mu g/ml$ sodium.
Silver	(1) Dissolve 1.5748 g $AgNO_3$ in water and dilute to volume. (2) Dissolve 1.000 g Ag in 10 ml of HNO_3; dilute to volume. Store in amber glass container away from light.
Sodium	Dissolve 2.5421 g NaCl in water and dilute to volume.
Strontium	Dissolve a slurry of 1.6849 g $SrCO_3$ in 300 ml of water by careful addition of 10 ml of HCl, after release of CO_2 by swirling, dilute to volume.
Sulfur	Dissolve 4.122 g $(NH_4)_2SO_4$ in water and dilute to volume.
Tantalum	Transfer 1.000 g Ta (or 1.2210 g Ta_2O_5) to Pt dish, add 20 ml of HF, and heat gently to complete the dissolution. Cool, add 40 ml of H_2SO_4 and evaporate to heavy fumes of SO_3. Cool and dilute to volume with 50% H_2SO_4.
Tellurium	(1) Dissolve 1.2508 g TeO_2 in 10 ml of HCl; dilute to volume. (2) Dissolve 1.000 g Te in 10 ml of warm HCl with dropwise addition of HNO_3, then dilute to volume.
Terbium	Dissolve 1.6692 g of $TbCl_3$ in water, add 1 ml of HCl, and dilute to volume.
Thallium	Dissolve 1.3034 g $TlNO_3$ in water and dilute to volume.
Thorium	Dissolve 2.3794 g $Th(NO_3)_4 \cdot 4H_2O$ in water, add 5 ml HNO_3, and dilute to volume.
Thulium	Dissolve 1.142 g Tm_2O_3 in 50 ml of 2 M HCl; dilute to volume.
Tin	Dissolve 1.000 g Sn in 15 ml of warm HCl; dilute to volume.
Titanium	Dissolve 1.000 g Ti in 10 ml of H_2SO_4 with dropwise addition of HNO_3; dilute to volume with 5% H_2SO_4.
Tungsten	Dissolve 1.7941 g of $Na_2WO_4 \cdot 2H_2O$ in water and dilute to volume.
Uranium	Dissolve 2.1095 g $UO_2(NO_3)_2 \cdot 6H_2O$ (or 1.7734 g uranyl acetate dihydrate) in water and dilute to volume.
Vanadium	Dissolve 2.2963 g NH_4VO_3 in 100 ml of water plus 10 ml of HNO_3; dilute to volume.

Element	Procedure
Ytterbium	Dissolve 1.6147 g $YbCl_3$ in water and dilute to volume.
Yttrium	Dissolve 1.2692 g Y_2O_3 in 50 ml of 2 M HCl and dilute to volume.
Zinc	Dissolve 1.000 g Zn in 10 ml of HCl; dilute to volume.
Zirconium	Dissolve 3.533 g $ZrOCl_2 \cdot 8H_2O$ in 50 ml 2 M HCl, and dilute to volume. Solution should be standardized.

INDEX

473

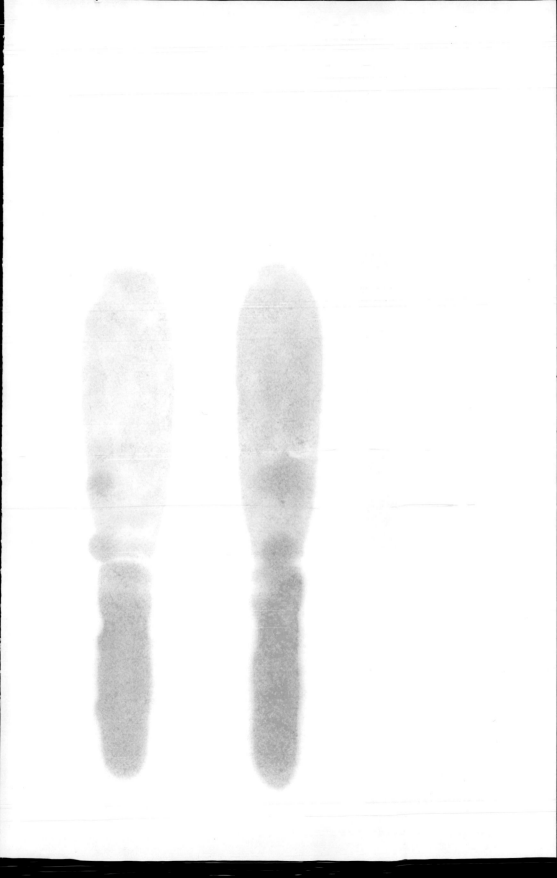